Lecture Notes in Computer Science  3

Edited by G. Goos and J. Hartmanis

Advisory Board: W. Brauer   D. Gri

David L. Dill (Ed.)

Computer Aided Verification

6th International Conference, CAV '94
Stanford, California, USA, June 21-23, 1994
Proceedings

Springer-Verlag

Berlin Heidelberg New York
London Paris Tokyo
Hong Kong Barcelona
Budapest

Series Editors

Gerhard Goos
Universität Karlsruhe
Postfach 69 80
Vincenz-Priessnitz-Straße 1
D-76131 Karlsruhe, Germany

Juris Hartmanis
Cornell University
Department of Computer Science
4130 Upson Hall
Ithaca, NY 14853, USA

Volume Editor

David L. Dill
Department of Computer Science, Stanford University
Stanford, CA 94305, USA

CR Subject Classification (1991): F.3, D.2.4, D.2.2, B.7

ISBN 3-540-58179-0 Springer-Verlag Berlin Heidelberg New York
ISBN 0-387-58179-0 Springer-Verlag New York Berlin Heidelberg

CIP data applied for

© Springer-Verlag Berlin Heidelberg 1994
Printed in Germany

Typesetting: Camera-ready by author
SPIN: 10127511   45/3140-543210 - Printed on acid-free paper

Preface

This volume contains the proceedings of the Sixth Conference on Computer-Aided Verification (CAV'94), held at Stanford University on June 21–23.

The purpose of the CAV conference is to bring together researchers and practitioners who are interested in the theory and practice of formal verification. Traditionally, CAV has placed a strong emphasis on finite-state concurrent systems; but as the art advances, the line between "finite-state" and unbounded or infinite state systems is becoming increasingly fuzzy. There has always been an emphasis at CAV on methods that can be applied in practice to important problems.

Of the 121 papers submitted this year, 37 were accepted for publication in the proceedings. There were papers on a wide variety of topics. This year, there were many papers on verification of real-time systems, introducing new results of both practical and theoretical importance. There were also several papers on hybrid systems which broke new ground. There were papers proposing new symbolic representations for large state spaces. There were many papers that advanced the capabilities of verification for particular application areas (for example, verification of pipelined processors and mobile processes), and several studies of verification methods applied to real examples from hardware and protocols. There were also various papers on improving the efficiency or generality of verification techniques, and papers advancing the theory of synthesis, probabilistic checking, and branching-time temporal logic.

CAV was sponsored by six generous (and forward-looking) companies who provided essential financial support: AT&T, IBM, Intel, Motorola, Redwood Design Automation, and Sun Microsystems.

The papers to be presented were chosen by the conference Program Committee. The Steering Committee consisting of conference founders Edmund M. Clarke (Carnegie Mellon University), Robert Kurshan (AT&T Bell Laboratories), Amir Pnueli (Weizmann Institute), and Joseph Sifakis (VERIMAG) helped with reviewing and provided advice in the organization of the conference. Also on the program committee were R. Alur (AT&T Bell Laboratories), R. Brayton (U. of California, Berkeley), E. Brinksma (U. of Twente), R. Bryant (Carnegie Mellon U.), R. Cleaveland (N. Carolina St. U.), C. Courcoubetis (U. of Crete), R. de Simone (INRIA), A. Emerson (U. of Texas, Austin), M. Fujita (Fujitsu Laboratories), S. German, O. Grumberg (Technion), N. Halbwachs (IMAG), G. Holzmann (AT&T Bell Laboratories), K. Larsen (Aalborg U.), K. McMillan (AT&T Bell Laboratories), L. Paulson (Cambridge U.), N. Shankar (SRI International), F. Somenzi (U. of Colorado, Boulder), B. Steffen (Techical U. of Aachen), P. Varaiya (U. of California, Berkeley), P. Wolper, (U. de Liege), and T. Yoneda (Tokyo Inst. of Tech.).

David Dill of Stanford University is the General and Program Chair of the conference. The conference was held on the Stanford campus under the sponsorship of the Department of Computer Science. Nancy Millikin, Stanford's Conference Coordinator, and Cecilia Sanchez of Events Plus assisted in planning of the local arrangements. Jerry Burch helped with the assembly of the final camera-ready copy.

We would also like to thank the following additional reviewers, who helped with the evaluation of one or more papers: A. Alkar, H. Reif Andersen, P. Attie, A. Aziz, F. Balarin, I. Beer, S. Ben-David, O. Bernholtz, R. Bharadwaj, G. Bhat, B. Boigelot, A. Bouajjani, A. Bouali, F. Boussinot, J.R. Burch, O. Burkart, K. C. Chen, A. Carruth, S.T. Cheng, F. Corno, P. Curzon, D. Cyrluk, A. Deshpande, A. Dicky, H. Eertink, J.-C. Fernandez, G. Gamage', H. Garavel, A. Geser, P. Godefroid, D. Goldschlag, A. Gollu, S. Graf, B. Graham, G. Hachtel, L. Heerink, T. Henzinger, H. Higuchi, H. Hiraishi, R. Hojati, I. Honma, A.J. Hu, H. Hungar, H. Huttel, N. C-W. Ip, M. Kaltenbach, S. Kalvala, P. Kars, J.-P. Katoen, D. Kirkpatrick, S. Krishnan, Y. Kukimoto, C. Laneve, T. Langerak, G. Luettgen, E. Macii, E. Madelaine, T. Margaria, Y. Matsunaga, J. McManis, L. Mounier, K. Nakane, K. Namjoshi, M. Nesi, X. Nicollin, A. Nymeyer, S. Owre, A. Pardo, S. Park, D. Peled, D. Pirottin, A. Puri, R. Ranjan, Y.S. Ramakrishna, H. Schlingloff, T. Shiple, J. Sifakis, V. Singhal, A. Skou, M. Lara de Souza, M.K. Srivas, P. Stephan, A. Takahara, S. Tasiran, R. Trefler, L. Tumlin, C. Weise, W. Wong, E. Wu, and M. Yoeli.

Stanford, April 1994   David L. Dill

Table of Contents

A Determinizable Class of Timed Automata

Rajeev Alur[1]   Limor Fix[2]*   Thomas A. Henzinger[2]**

[1] AT&T Bell Laboratories, Murray Hill, NJ
[2] Department of Computer Science, Cornell University, Ithaca, NY

Abstract. We introduce the class of *event-recording timed automata* (ERA). An event-recording automaton contains, for every event a, a clock that records the time of the last occurrence of a. The class ERA is, on one hand, expressive enough to model (finite) timed transition systems and, on the other hand, determinizable and closed under all boolean operations. As a result, the language inclusion problem is decidable for event-recording automata. We present a translation from timed transition systems to event-recording automata, which leads to an algorithm for checking if two timed transition systems have the same set of timed behaviors.

We also consider *event-predicting timed automata* (EPA), which contain clocks that predict the time of the next occurrence of an event. The class of *event-clock automata* (ECA), which contain both event-recording and event-predicting clocks, is a suitable specification language for real-time properties. We provide an algorithm for checking if a timed automaton meets a specification that is given as an event-clock automaton.

1 Introduction

Finite automata are instrumental for the modeling and analysis of many phenomena within computer science. In particular, automata theory plays an important role in the verification of concurrent finite-state systems [10, 16]. In the trace model for concurrent computation, a system is identified with its behaviors. Assuming that a behavior is represented as a sequence of states or events, the possible behaviors of a system can be viewed as a formal language, and the system can be modeled as an automaton that generates the language (a complex system is modeled as the product of automata that represent the component systems). Since the admissible behaviors of the system also constitute a formal language, the requirements specification can be given by another automaton (the adequacy of automata as a specification formalism is justified by the fact that competing formalisms such as linear temporal logic are no more expressive). The verification problem of checking that a system meets its specification,

* Supported in part by the Office of Naval Research under contract N00014-91-J-1219, the National Science Foundation under grant CCR-8701103, and by DARPA/NSF under grant CCR-9014363.
** Supported in part by the National Science Foundation under grant CCR-9200794 and by the United States Air Force Office of Scientific Research under contract F49620-93-1-0056.

then, reduces to testing language inclusion between two automata. The decision procedure for language inclusion typically involves the complementation of the specification automaton, which in turn relies upon determinization [9, 15].

To capture the behavior of a real-time system, the model of computation needs to be augmented with a notion of time. For this purpose, *timed automata* [3] provide a simple, and yet powerful, way of annotating state-transition graphs with timing constraints, using finitely many real-valued variables called *clocks*. A timed automaton, then, accepts *timed words*—strings in which each symbol is paired with a real-valued time-stamp. While the theory of timed automata allows the automatic verification of certain real-time requirements of finite-state systems [1, 3, 4, 8], and the solution of certain delay problems [2, 6], the general verification problem (i.e., language inclusion) is undecidable for timed automata [3]. This is because, unlike in the untimed case, the nondeterministic variety of timed automata is strictly more expressive than the deterministic variety. The notion of nondeterminism allowed by timed automata, therefore, seems too permissive, and we hesitate to accept timed automata as the canonical model for finite-state real-time computation [5].

In this paper, we obtain a determinizable class of timed automata by restricting the use of clocks. The clocks of an *event-clock automaton* have a fixed, predefined association with the symbols of the input alphabet (the alphabet symbols typically represent events). The *event-recording clock* of the input symbol a is a history variable whose value always equals the time of the last occurrence of a relative to the current time; the *event-predicting clock* of a is a prophecy variable whose value always equals the time of the next occurrence of a relative to the current time (if no such occurrence exists, then the clock value is undefined). Thus, unlike a timed automaton, an event-clock automaton does not control the reassignments of its clocks and, at each input symbol, all clock values of the automaton are determined solely by the input word. This property allows the determinization of event-clock automata, which, in turn, leads to a complementation procedure. Indeed, the class ECA of event-clock automata is closed under all boolean operations (timed automata are not closed under complement), and the language inclusion problem is decidable for event-clock automata.

While event-predicting clocks are useful for the specification of timing requirements, automata that contain only event-recording clocks (*event-recording automata*) are a suitable abstract model for real-time systems. We confirm this claim by proving that event-recording automata are as powerful as another popular model for real-time computation, *timed transition systems* [7]. A timed transition system associates with each transition a lower bound and an upper bound on the time that the transition may be enabled without being taken (many related real-time formalisms also use lower and upper time bounds to express timing constraints [13, 14]). A run of a timed transition system, then, is again a timed word—a sequence of time-stamped state changes. We construct, for a given timed transition system T with a finite set of states, an event-recording automaton that accepts precisely the runs of T. This result leads to an algorithm for checking the equivalence of two finite timed transition systems.

2 Event-clock Automata

Timed words and timed languages

We study formal languages of timed words.[3] A *timed word* $\bar{w}$ over an alphabet Σ is a finite sequence $(a_0, t_0)(a_1, t_1) \ldots (a_n, t_n)$ of symbols $a_i \in \Sigma$ that are paired with nonnegative real numbers $t_i \in \mathbb{R}^+$ such that the sequence $\bar{t} = t_1 t_2 \ldots t_n$ of time-stamps is nondecreasing (i.e., $t_i \leq t_{i+1}$ for all $0 \leq i < n$). Sometimes we denote the timed word $\bar{w}$ by the pair $(\bar{a}, \bar{t})$. A *timed language* over the alphabet Σ is a set of timed words over Σ. The boolean operations of union, intersection, and complement of timed languages are defined as usual. Given a timed language $\mathcal{L}$ over the alphabet Σ, the projection *Untime*$(\mathcal{L})$ is obtained by discarding the time-stamps: *Untime*$(\mathcal{L}) \subseteq \Sigma^*$ consists of all strings $\bar{a}$ for which there exists a sequence $\bar{t}$ of time-stamps such that $(\bar{a}, \bar{t}) \in \mathcal{L}$.

Automata with clocks

Timed automata are finite-state machines that are constrained with timing requirements so that they accept (or generate) timed words (and thus define timed languages); they were proposed in [3] as an abstract model for finite-state real-time systems. A timed automaton operates with finite control—a finite set of locations and a finite set of real-valued variables called *clocks*. Each edge between locations specifies a set of clocks to be reset (i.e., restarted). The value of a clock always records the amount of time that has elapsed since the last time the clock was reset: if the clock z is reset while reading the i-th symbol of a timed input word $(\bar{a}, \bar{t})$, then the value of z while reading the j-th symbol, for $j > i$, is $t_j - t_i$ (assuming that the clock z is not reset at any position between i and j). The edges of the automaton put certain arithmetic constraints on the clock values; that is, the automaton control may proceed along an edge only when the values of the clocks satisfy the corresponding constraints.

Each clock of a timed automaton, therefore, is a real-valued variable that records the time difference between the current input symbol and another input symbol, namely, the input symbol on which the clock was last reset. This association between clocks and input symbols is determined dynamically by the behavior of the automaton. An event-clock automaton, by contrast, employs clocks that have a tight, predefined, association with certain symbols of the input word. Suppose that we model a real-time system so that the alphabet symbols represent events of the system. In most cases, it will suffice to know, for each event, the time that has elapsed since the last occurrence of the event. For example, to model a delay of 1 to 2 seconds between the input and output events of a device, it suffices to use a clock z that records the time that has elapsed since the last input event, and require the constraint $1 < z < 2$ when the output event occurs. This observation leads us to the definition of clocks that have a fixed association with input symbols and cannot be reset arbitrarily.

---

[3] For the clarity of exposition, we limit ourselves to finite words. Our results can be extended to the framework of ω-languages.

Event-recording and event-predicting clocks

Let Σ be a finite alphabet. For every symbol $a \in \Sigma$, we write x_a to denote the *event-recording clock* of a. Given a timed word $\bar{w} = (a_0, t_0)(a_1, t_1) \ldots (a_n, t_n)$, the value of the clock x_a at the j-th position of $\bar{w}$ is $t_j - t_i$, where i is the largest position preceding j such that a_i equals a. If no occurrence of a precedes the j-th position of $\bar{w}$, then the value of the clock x_a is "undefined," denoted by $\bot$. We write $\mathrm{R}_\bot^+ = \mathrm{R}^+ \cup \{\bot\}$ for the set of nonnegative real numbers together with the special value $\bot$. Formally, we define for all $0 \le j \le n$,

$$val(\bar{w}, j)(x_a) = \begin{cases} t_j - t_i & \text{if there exists } i \text{ such that } 0 \le i < j \text{ and } a_i = a \\ & \text{and for all } k \text{ with } i < k < j, \ a_k \ne a, \\ \bot & \text{if } a_k \ne a \text{ for all } k \text{ with } 0 \le k < j. \end{cases}$$

That is, the event-recording clock x_a behaves exactly like an automaton clock that is reset every time the automaton encounters the input symbol a. The value of x_a, therefore, is determined by the input word, not by the automaton. Auxiliary variables that record the times of last occurrences of events have been used extensively in real-time reasoning, for example, in the context of model-checking for timed Petri nets [17], and in assertional proof methods [11, 14].

Event-recording clocks provide timing information about events in the past. The dual notion of event-predicting clocks provides timing information about future events. For every symbol $a \in \Sigma$, we write y_a to denote the *event-predicting clock* of a. At each position of the timed word $\bar{w}$, the value of the clock y_a indicates the time of the next occurrence of a relative to the time of the current input symbol; the special value $\bot$ indicates the absence of a future occurrence of a. Formally, we define for all $0 \le j \le n$,

$$val(\bar{w}, j)(y_a) = \begin{cases} t_i - t_j & \text{if there exists } i \text{ such that } j < i \le n \text{ and } a_i = a \\ & \text{and for all } k \text{ with } j < k < i, \ a_k \ne a, \\ \bot & \text{if } a_k \ne a \text{ for all } k \text{ with } j < k \le n. \end{cases}$$

The event-predicting clock y_a can be viewed as an automaton clock that is reset, every time the automaton encounters the input symbol a, to a nondeterministic negative starting value, and checked for 0 at the subsequent occurrence of a.

We write C_Σ for the set $\{x_a, y_a \mid a \in \Sigma\}$ of event-recording and event-predicting clocks. For each position j of a timed word $\bar{w}$, the *clock-valuation function* $val(\bar{w}, j)$, then, is a mapping from C_Σ to $\mathrm{R}_\bot^+$. The clock constraints compare clock values to rational constants or to the special value $\bot$. Let $\mathrm{Q}_\bot$ denote the set of nonnegative rational numbers together with $\bot$. Formally, a *clock constraint* over the set C of clocks is a boolean combination of atomic formulas of the form $z \le c$ and $z \ge c$, where $z \in C$ and $c \in \mathrm{Q}_\bot$. The clock constraints over C are interpreted with respect to clock-valuation functions from C to $\mathrm{R}_\bot^+$: the atom $\bot = \bot$ evaluates to true, and all other comparisons that involve $\bot$ (e.g., $\bot \ge 3$) evaluate to false. For a clock-valuation function γ and a clock constraint ϕ, we write $\gamma \models \phi$ to denote that according to γ the constraint ϕ evaluates to true.

Syntax and semantics of event-clock automata

An event-clock automaton is a (nondeterministic) finite-state machine whose edges are annotated both with input symbols and with clock constraints over event-recording and event-predicting clocks. Formally, a *event-clock automaton* A consists of a finite input alphabet Σ, a finite set L of locations, a set $L_0 \subseteq L$ of start locations, a set $L_f \subseteq L$ of accepting locations, and a finite set E of edges. Each edge is a quadruple (ℓ, ℓ', a, ϕ) with a source location $\ell \in L$, a target location $\ell' \in L$, an input symbol $a \in \Sigma$, and a clock constraint ϕ over the clocks C_Σ.

Now let us consider the behavior of an event-clock automaton over the timed input word $\bar{w} = (a_0, t_0)(a_1, t_1) \ldots (a_n, t_n)$. Starting in one of the start locations and scanning the first input pair (a_0, t_0), the automaton scans the input word from left to right, consuming, at each step, an input symbol together with its time-stamp. In location ℓ scanning the i-th input pair (a_i, t_i), the automaton may proceed to location ℓ' and the $i + 1$-st input pair iff there is an edge (ℓ, ℓ', a, ϕ) such that a equals the current input symbol a_i and $val(\bar{w}, i)$ satisfies the clock constraint ϕ. Formally, a *computation* of the event-clock automaton A over the timed input word $\bar{w}$ is a finite sequence

$$\ell_0 \xrightarrow{e_0} \ell_1 \xrightarrow{e_1} \ell_2 \xrightarrow{e_2} \cdots \xrightarrow{e_{n-1}} \ell_n \xrightarrow{e_n} \ell_{n+1}$$

of locations $\ell_i \in L$ and edges $e_i = (\ell_i, \ell_{i+1}, a_i, \phi_i) \in E$ such that $\ell_0 \in L_0$ and for all $0 \le i \le n$, $val(\bar{w}, i) \models \phi_i$; the computation is *accepting* if $\ell_{n+1} \in L_f$. The timed language $\mathcal{L}(A)$ defined by the event-clock automaton A, then, consists of all timed words $\bar{w}$ such that A has an accepting computation over $\bar{w}$. We write ECA for the class of timed languages that are definable by event-clock automata.

The event-clock automaton A is an *event-recording automaton* if all clock constraints of A contain only event-recording clocks; A is an *event-predicting automaton* if the clock constraints of A contain only event-predicting clocks. The class of timed languages that can be defined by these two restricted types of event-clock automata are denoted ERA and EPA, respectively.

Examples of event-clock automata

The event-clock automaton A_1 of Figure 2 uses two event-recording clocks, x_a and x_b. The location ℓ_0 is the start location of A_1, and also the sole accepting location. The clock constraint $x_a < 1$ that is associated with the edge from ℓ_2 to ℓ_3 ensures that c occurs within 1 time unit of the preceding a. A similar mechanism of checking the value of x_b while reading d ensures that the time difference between b and the subsequent d is always greater than 2. Thus, the timed language $\mathcal{L}(A_1)$ defined by A_1 consists of all timed words of the form $((abcd)^m, \bar{t})$ such that $m \ge 0$ and for all $0 \le j < m$, $t_{4j+2} < t_{4j} + 1$ and $t_{4j+3} > t_{4j+1} + 2$. Note that the timed language $\mathcal{L}(A_1)$ can also be defined using event-predicting clocks: require $y_c < 1$ while reading a, and $y_d > 2$ while reading b.

The duality of the two types of clocks is further illustrated by the automata of Figure 2. The event-recording automaton A_2 accepts all timed words of the

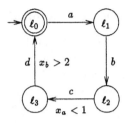

Fig. 1. Event-recording automaton A_1

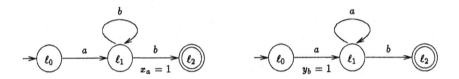

Fig. 2. Event-recording automaton A_2 and event-predicting automaton A_3

form $(ab^*b, \bar{t})$ such that the time difference between the two extreme symbols is 1, which is enforced by the event-recording clock x_a. It is easy to check that there is no event-predicting automaton that defines the timed language $\mathcal{L}(A_2)$. The event-predicting automaton A_3, on the other hand, accepts all timed words of the form $(aa^*b, \bar{t})$ such that the time difference between the two extreme symbols is 1; for this purpose, the event-predicting clock y_b is used to predict the time of the first b. There is no event-recording automaton that defines $\mathcal{L}(A_3)$.

3 Deterministic Event-clock Automata

A finite-state machine (with a single start location) is deterministic iff all input symbols that label edges with the same source location are pairwise distinct. We consider for event-clock automata the notion of determinism that was proposed for timed automata in [3]. The event-clock automaton $A = (\Sigma, L, L_0, L_f, E)$ is *deterministic* if A has a single start location (i.e., $|L_0| = 1$) and any two edges with the same source location and the same input symbol have mutually exclusive clock constraints; that is, if $(\ell, \ell', a, \phi_1) \in E$ and $(\ell, \ell'', a, \phi_2) \in E$ then for all clock-valuation functions γ, if $\gamma \models \phi_1$ then $\gamma \not\models \phi_2$. The determinism condition ensures that at each step during a computation, the choice of the next edge is uniquely determined by the current location of the automaton, the input word, and the current position of the automaton along the input word. It is easy to check that every deterministic event-clock automaton has at most one computation over any given timed input word.

Of our examples from the previous section, the event-clock automata A_1 and A_3 are deterministic. While the automaton A_2 is nondeterministic, it can be determinized without changing its language, by adding the clock constraint $x_a < 1$ to the self-loop at location ℓ_1.

In the theory of finite-state machines, it is well-known that every nondeterministic automaton can be determinized; that is, the deterministic and nondeterminstic varieties of finite-state machines define the same class of languages (the regular languages). In the case of timed automata, however, the nondeterministic variety is strictly more expressive than its deterministic counterpart [3]. We now show that the event-clock automata form a determinizable subclass of the timed automata.

Theorem 1 (Determinization). *For every event-clock (event-recording; event-predicting) automaton A, there is a deterministic event-clock (event-recording; event-predicting) automaton that defines $\mathcal{L}(A)$.*

Proof. Let A be the given event-clock automaton with the location set L. The locations of the determinized automaton $Det(A)$ are the nonempty subsets of L. Consider a location $L' \subseteq L$ of $Det(A)$, and an input symbol $a \in \Sigma$. Let $E' \subseteq E$ be the set of all a-labeled edges of A whose source locations are in L'. Then, for every nonempty subset $E'' \subseteq E'$, there is an edge from L' to L'' labeled with the input symbol a and the clock constraint ϕ iff L'' contains precisely the target locations of the edges in E'', and ϕ is the conjunction of all clock constraints of E''-edges and all negated clock constraints of $(E' - E'')$-edges. It is easy to check that the clock constraints on different a-labeled edges starting from L' are mutually exclusive. ∎

Notice that the determinization of an event-clock automaton causes an exponential blow-up in the number of locations, but changes neither the number of clocks nor the constants that occur in clock constraints.

The key for the determinization of event-clock automata is the property that at each step during a computation, all clock values are determined solely by the input word. We therefore obtain derminizable superclasses of event-clock automata if we add more clocks that do not violate this property. For example, for each input symbol a and each natural number i, we could employ a clock z_a^i that records the time since the i-th occurrence of a, and a clock x_a^i that records the time since the i-th-to-last occurrence of a (i.e., $x_a = x_a^1$). Or, more ambitiously, we may want to use for each linear temporal formula φ a *formula-recording clock* x_φ that measures the time since the last position of the input word at which φ was true, and a *formula-predicting clock* y_φ that measures the time until the next position at which φ will be true.

4 Properties of Event-Clock Automata

Event-clock automata as labeled transition systems

We now consider an alternative semantics for event-clock automata, using labeled transition systems. Let $A = (\Sigma, L, L_0, L_f, E)$ be an event-clock automaton. A *state* of A is a pair (ℓ, γ) that consists of a location $\ell \in L$ and a clock-valuation function γ from C_Σ to $R_\perp^+$, which determines the values of all clocks. The state (ℓ, γ) is *initial* if $\ell \in L_0$ and $\gamma(x_a) = \perp$ for all input symbols $a \in \Sigma$; (ℓ, γ) is *final* if $\ell \in L_f$ and $\gamma(y_a) = \perp$ for all $a \in \Sigma$. We write S_A for the (infinite) set of states of the event-clock automaton A and define a labeled transition relation over S_A to capture the behavior of A over timed words.

For two states $s, s' \in S_A$, an input symbol $a \in \Sigma$, and a real-valued time delay $\delta \in R^+$, let $s \xrightarrow{a} s'$ if the automaton A may proceed from the state s to the state s' by reading the input symbol a, and let $s \xrightarrow{\delta} s'$ if A may proceed from s to s' by letting time δ pass. Formally, $(\ell, \gamma) \xrightarrow{a} (\ell', \gamma')$ if there is a clock-valuation function γ'' and an edge $(\ell, \ell', a, \phi) \in E$ such that $\gamma = \gamma''[y_a := 0]$ (i.e., γ agrees with γ'' on all clocks except y_a, which in γ evaluates to 0), $\gamma' = \gamma''[x_a := 0]$, and $\gamma \models \phi$; and $(\ell, \gamma) \xrightarrow{\delta} (\ell', \gamma')$ if $\ell = \ell'$ and for all input symbols $b \in \Sigma$,

1. if $\gamma(x_b) = \perp$ then $\gamma'(x_b) = \perp$, else $\gamma'(x_b) = \gamma(x_b) + \delta$;
2. if $\gamma'(y_b) = \perp$ then $\gamma(y_b) = \perp$, else $\gamma(y_b) = \gamma'(y_b) + \delta$.

We inductively extend the labeled transition relation to timed words: $s \xrightarrow{(a_0, t_0)} s'$ if there is a state $s'' \in S_A$ such that $s \xrightarrow{t_0} s''$ and $s'' \xrightarrow{a_0} s'$; if $\bar{w} = (a_0, t_0) \ldots (a_n, t_n)$ and $\bar{w}' = \bar{w}(a_{n+1}, t_{n+1})$, then $s \xrightarrow{\bar{w}'} s'$ if there is a state s'' such that $s \xrightarrow{\bar{w}} s''$ and $s'' \xrightarrow{(a_{n+1}, t_{n+1} - t_n)} s'$. The following lemma shows the correctness of the labeled-transition-system semantics for event-clock automata.

Lemma 2. *The event-clock automaton A accepts the timed word $\bar{w}$ iff $s \xrightarrow{\bar{w}} s'$ for some initial state s and some final state s' of A.*

The region construction

The analysis of timed automata builds on the so-called region construction, which transforms a timed automaton into an untimed finite-state machine [1, 3]. Here we apply the region construction to event-clock automata. We consider again the given event-clock automaton A and begin with defining the region-equivalence relation $\cong_A$ as a finite partition of the infinite state space S_A.

We assume that all clock constraints of A contain only integer constants (otherwise, all constants need to be multiplied by the least common multiple of the denominators of all rational numbers that appear in the clock constraints of A). Let c be the largest integer constant that appears in a clock constraint of A. Informally, two clock-valuation functions γ and γ' from C_Σ to $R_\perp^+$ are *region-equivalent*, written $\gamma \cong_A \gamma'$, if γ and γ' agree on which clocks have the undefined value $\perp$, agree on the integral parts of all defined clock values that are at most c, and agree on the ordering of the fractional parts of all defined clock values (the fractional part of the event-recording clock x_a according to γ is $\gamma(x_a) - \lfloor \gamma(x_a) \rfloor$; the fractional part of the event-predicting clock y_a is $\lceil \gamma(y_a) \rceil - \gamma(y_a)$). Two states $(\ell, \gamma), (\ell', \gamma') \in S_A$ are *region-equivalent* if $\ell = \ell'$ and $\gamma \cong_A \gamma'$. A formal definition of the region-equivalence relation $\cong_A$ is given in [3].

A *region* of the event-clock automaton A is an $\cong_A$-equivalence class of states in S_A. The number of A-regions is finite—linear in the number of locations, exponential in the number of clocks (that is, exponential in the size of the input alphabet), and exponential in the size of the clock constraints of A. The region equivalence is instrumental for analyzing event-clock automata, because $\cong_A$ is a bisimulation.

Lemma 3. *For all states $s_1, s_1', s_2 \in S_A$ of an event-clock automaton A, all input symbols a of A, and all real-valued time delays $\delta \in \mathbb{R}^+$, if $s_1 \cong_A s_1'$ and $s_1 \xrightarrow{(a,\delta)} s_2$, then there is a state $s_2' \in S_A$ and a time delay $\delta' \in \mathbb{R}^+$ such that $s_2 \cong_A s_2'$ and $s_1' \xrightarrow{(a,\delta')} s_2'$.*

Now we are ready to define the *region automaton* $Reg(A)$ of A, an untimed finite-state machine over the input alphabet Σ. The locations of $Reg(A)$ are the regions of A. A region is starting if it contains an initial state of A, and accepting if it contains a final state of A. There is an edge from the region ρ to the region ρ' labeled with the input symbol a if there are two states $s \in \rho$ and $s' \in \rho'$, and a time delay $\delta \in \mathbb{R}^+$, such that $s \xrightarrow{(a,\delta)} s'$. From Lemmas 2 and 3 it follows that the region automaton $Reg(A)$ defines the untimed language $Untime(\mathcal{L}(A))$.

Theorem 4 (Untiming). *For every event-clock automaton A, the untimed language $Untime(\mathcal{L}(A))$ is regular.*

Closure properties and decision problems

While the class of timed automata is not closed under complement, and the language inclusion (verification) problem for timed automata is undecidable, the subclass of event-clock automata is well-behaved.

Theorem 5 (Closure properties). *Each of the classes ECA, ERA, and EPA of timed languages are closed under union, intersection, and complement.*

Proof. Closure under union is trivial, because event-clock automata admit multiple start locations. Closure under intersection is also straightforward, because the standard automata-theoretic product construction $A_1 \times A_2$ for two given event-clock (event-recording; event-predicting) automata A_1 and A_2 yields an event-clock (event-recording; event-predicting) automaton. Closure under complement relies on the determinization construction: given an event-clock (event-recording; event-predicting) automaton A, the event-clock (event-recording; event-predicting) automaton $\neg Det(A)$ that results from complementing the acceptance condition of $Det(A)$ (interchange the accepting and the nonaccepting states of $Det(A)$) defines the complement of the timed language $\mathcal{L}(A)$. ∎

Unlike (nondeterministic) timed automata, however, event-clock automata are not closed under hiding and renaming of input symbols. This is because the timed language $\mathcal{L}$ that contains all timed words $\bar{w} = (\bar{a}, \bar{t})$ over a unary alphabet in which no two symbols occur with time difference 1 (i.e., $t_j - t_i \neq 1$ for all positions i and j of $\bar{w}$) cannot be defined by a timed automaton [3]. With complementation and renaming (or hiding), on the other hand, $\mathcal{L}$ is easily definable from a language in ERA ∩ EPA.

The determinization, closure properties, and region construction can be used to solve decision problems for event-clock automata. To check if the timed language of an event-clock automaton A is empty, we construct the region automaton $Reg(A)$ and check if the untimed language of $Reg(A)$ is empty. To check if the language of the event-clock automaton A_1 is included in the language of the event-clock automaton A_2, we determinize A_2, complement $Det(A_2)$, take the product with A_1, and check if the language of the resulting event-clock automaton $A_1 \times \neg Det(A_2)$ is empty by constructing the corresponding region automaton.

Theorem 6 (Language inclusion). *The problem of checking if $\mathcal{L}(A_1) \subseteq \mathcal{L}(A_2)$ for two event-clock automata A_1 and A_2 is decidable in PSPACE.*

On the other hand, the problem of checking if the language of a given event-recording (or event-predicting) automaton is empty can be shown to be PSPACE-hard (similar to the hardness proof for emptiness of timed automata [3]).

Relationship between classes of timed automata

We briefly review the definition of a timed automaton [3]. A (nondeterministic) *timed automaton A* consists of a finite input alphabet Σ, a finite set L of locations, a set $L_0 \subseteq L$ of start locations, a set $L_f \subseteq L$ of accepting locations, a finite set C of clocks, and a finite set E of edges. Each edge e is labeled with an input symbol, a clock constraint over C, and a *reset condition* $C_e \subseteq C$ that specifies the clocks that are reset to 0 when the edge e is traversed. Every timed automaton A, then, defines a timed language $\mathcal{L}(A)$, and we write NTA for the class of timed languages that are definable by timed automata. The class NTA is closed under union and intersection, but not under complement.

The definition of determinism for timed automata is the same as for event-clock automata. We write DTA for the class of timed languages that are definable by deterministic timed automata. Since DTA is closed under all boolean operations, DTA is strictly contained in NTA.

Theorem 7 (Relationship between classes).

(1) ERA $\not\subseteq$ EPA	(2) EPA $\not\subseteq$ ERA	(3) ERA $\cup$ EPA $\subset$ ECA
(4) ECA $\subset$ NTA	(5) ERA $\subset$ DTA	(6) EPA $\not\subseteq$ DTA
(7) DTA $\not\subseteq$ ECA		

Proof. For (1), the language of the event-recording automaton A_2 of Figure 2 is not definable by an event-predicting automaton. For (2), the language of the event-predicting automaton A_3 of Figure 2 cannot be defined by an event-recording automaton. Similarly, for (3) it is possible to combine A_2 and A_3 into an event-clock automaton whose language is neither in ERA nor in EPA.

Every event-clock automaton can be tranlated into a timed automaton. While the translation preserves determinism for event-recording automata, event-predicting clocks introduce nondeterminism. The inclusions (4) and (5) follow. Inclusion (4) is strict, because ECA is closed under complement while NTA is not. Inclusion (5) is strict because of (7). For (6), the timed language $\{(a^n b, t_0 \ldots t_n) \mid \exists 0 \leq i < n.t_n - t_i = 1\}$ is in EPA but not in DTA. For (7), the timed language $\{(aaa, t_0 t_1 t_2) \mid t_2 - t_0 = 1\}$ is in DTA but not in ECA. ∎

In [5], we defined another subclass of NTA that is closed under all boolean operations, namely, the class 2DTA of timed languages that are definable by deterministic twoway automata that can read the input word a bounded number of times. While ECA is easily seen to be contained in 2DTA, and while there are obvious similarities between event-predicting clocks and the twoway reading of the timed input word, the exact relationship between event-clock automata and deterministic twoway automata remains to be studied. However, because they admit nondeterminism, event-clock automata are perhaps more suited for specification than deterministic twoway automata.

5 Timed Transition Systems as Event-clock Automata

Timed transition systems

A *transition system* T consists of a set S of states, a set $S_0 \subseteq S$ of initial states, and a finite set $\mathcal{T}$ of transitions. Each transition $\tau \in \mathcal{T}$ is a binary relation over S. For each state $s \in S$, the set $\tau(s)$ gives the possible τ-successors of s; that is, $\tau(s) = \{s' \mid (s, s') \in \tau\}$. The transition system T is *finite* if the set S of states is finite. A *run* $\bar{s}$ of the transition system T is a finite sequence $s_0 \rightarrow s_1 \rightarrow \cdots \rightarrow s_n$ of states such that $s_0 \in S_0$ and for all $0 \leq i < n$, there exists a transition $\tau_i \in \mathcal{T}$ such that $s_{i+1} \in \tau_i(s_i)$. The transition τ is *enabled* at the i-th step of the run $\bar{s}$ if $\tau(s_i)$ is nonempty, and τ is *taken* at the i-th step if $s_i \in \tau(s_{i-1})$ (i.e., multiple transitions may be taken at the same step). A variety of programming systems, such as message-passing systems and shared-memory systems, can be given a transition-system semantics [12].

The model of transition systems is extended to timed transition systems so that it is possible to express real-time constraints on the transitions [7]. A *timed transition system* T consists of a transition system $(S, S_0, \mathcal{T})$ and two functions l and u from $\mathcal{T}$ to R^+ that associate with each transition $\tau \in \mathcal{T}$ a *lower bound* $l(\tau)$ and an upper bound $u(\tau)$. Informally, the transition τ must be enabled continuously for at least $l(\tau)$ time units before it can be taken, and τ must not be enabled continuously for more than $u(\tau)$ time units without being taken. Formally, we associate a real-valued time-stamp with each state change along a run. A *timed run* $\bar{r}$ of the timed transition system T is a finite sequence

$$\xrightarrow{t_0} s_0 \xrightarrow{t_1} s_1 \xrightarrow{t_2} \cdots \xrightarrow{t_n} s_n$$

of states $s_i \in S$ and nondecreasing time-stamps $t_i \in \mathrm{R}^+$ such that $\bar{s}$ is a run of the underlying transition system and

1. *Upper Bound*: if τ is enabled at all steps k for $i \leq k < j$, and not taken at all steps k for $i < k < j$, then $t_j - t_i \leq u(\tau)$;
2. *Lower Bound*: if τ is taken at the j-th step then there is some step $i < j$ such that $t_j - t_i \geq l(\tau)$ and τ is enabled at all steps k for $i \leq k < j$, and not taken at all steps k for $i < k < j$.

In other words, t_0 is the initial time, and the transition system proceeds from the state s_i to the state s_{i+1} at time t_{i+1}. The semantics of the timed transition system T is the set of timed runs of T. Two timed transition systems are *equivalent* if they have the same timed runs.

From timed transition systems to event-recording automata

We now show that the set of timed runs of a finite timed transition system can be defined by an event-recording automaton. For this purpose, we need to switch from the state-based semantics of transition systems to an event-based semantics. With the given timed run $\bar{r}$, we associate the timed word

$$\bar{w}(\bar{r}) = (\langle \perp, s_0 \rangle, t_0) \, (\langle s_0, s_1 \rangle, t_1) \, (\langle s_1, s_2 \rangle, t_2) \, \cdots \, (\langle s_{n-1}, s_n \rangle, t_n),$$

where $\perp$ is a special symbol not in S (as usual, $S_\perp = S \cup \{\perp\}$). Notice that the timed run $\bar{r}$ and the corresponding timed word $\bar{w}(\bar{r})$ contain the same information: each event (i.e., state change) of $\bar{r}$ is modeled by a pair of states—a source state and a target state. Every finite timed transition system $T = (S, \mathcal{T}, S_0, l, u)$, then, defines a timed language $\mathcal{L}(T)$ over the alphabet $S_\perp \times S$, namely, the set of timed words $\bar{w}(\bar{r})$ that correspond to timed runs $\bar{r}$ of T. It is easy to check that two timed transition systems are equivalent iff they define the same timed language.

Theorem 8 (Timed transition systems). *For every finite timed transition system T, there is an event-recording timed automaton A_T that defines the timed language $\mathcal{L}(T)$.*

Proof. Consider the given finite timed transition system T. Each location of the corresponding event-clock automaton A_T records a state $s \in S$ and, for each transition $\tau \in \mathcal{T}$, a pair of states $\langle \alpha(\tau), \beta(\tau) \rangle \in S_\perp \times S$ such that if τ is enabled in s, then τ has been enabled continuously without being taken since the last state change from $\alpha(\tau)$ to $\beta(\tau)$. In addition, we use a special location ℓ_0 as the sole start location of A_T. Every location is an accepting location.

For every initial state $s_0 \in S_0$, there is an edge from ℓ_0 to $(s_0, \langle \alpha, \beta \rangle)$ labeled with the input symbol $\langle \perp, s_0 \rangle$ and the trivial clock constraint *true*, where $\alpha(\tau) = \perp$ and $\beta(\tau) = s_0$ for all transitions $\tau \in \mathcal{T}$. In addition, there is an edge from $(s, \langle \alpha, \beta \rangle)$ to $(s', \langle \alpha', \beta' \rangle)$ labeled with the input symbol $\langle s, s' \rangle$ and the clock constraint ϕ iff there is a transition $\tau \in \mathcal{T}$ such that $(s, s') \in \tau$, and for all transitions $\tau \in \mathcal{T}$,

1. if τ is enabled in s and $s' \notin \tau(s)$, then $\langle \alpha'(\tau), \beta'(\tau) \rangle = \langle \alpha(\tau), \beta(\tau) \rangle$, else $\langle \alpha'(\tau), \beta'(\tau) \rangle = \langle s, s' \rangle$;
2. if τ is enabled in s, then ϕ contains the conjunct $x_{\langle \alpha(\tau), \beta(\tau) \rangle} \leq u(\tau)$;
3. if $s' \in \tau(s)$, then ϕ contains the conjunct $x_{\langle \alpha(\tau), \beta(\tau) \rangle} \geq l(\tau)$.

Notice that the size of the event-recording automaton A_T is exponential in the size of the timed transition system T. ∎

To check if two timed transition systems T_1 and T_2 are equivalent, we construct the corresponding event-recording automata A_{T_1} and A_{T_2} and check if they define the same timed language.

Corollary 9. *The problem of checking if two finite timed transition systems are equivalent is decidable in EXPSPACE.*

References

1. R. Alur, C. Courcoubetis, and D. Dill. Model-checking in dense real-time. *Information and Computation*, 104:2–34, 1993.

2. R. Alur, C. Courcoubetis, and T. Henzinger. Computing accumulated delays in real-time systems. In *Proceedings of the Fifth Conference on Computer-Aided Verification*, Lecture Notes in Computer Science 697, pages 181–193. Springer-Verlag, 1993.

3. R. Alur and D. Dill. Automata for modeling real-time systems. In *Proceedings of the 17th International Colloquium on Automata, Languages, and Programming*, Lecture Notes in Computer Science 443, pages 322–335. Springer-Verlag, 1990.

4. R. Alur, T. Feder, and T. Henzinger. The benefits of relaxing punctuality. In *Proceedings of the Tenth ACM Symposium on Principles of Distributed Computing*, pages 139–152, 1991.

5. R. Alur and T. Henzinger. Back to the future: Towards a theory of timed regular languages. In *Proceedings of the 33rd IEEE Symposium on Foundations of Computer Science*, pages 177–186, 1992.

6. C. Courcoubetis and M. Yannakakis. Minimum and maximum delay problems in real-time systems. In *Proceedings of the Third Workshop on Computer-Aided Verification*, Lecture Notes in Computer Science 575, pages 399–409, 1991.

7. T. Henzinger, Z. Manna, and A. Pnueli. Temporal proof methodologies for real-time systems. In *Proceedings of the 18th ACM Symposium on Principles of Programming Languages*, pages 353–366, 1991.

8. T. Henzinger, X. Nicollin, J. Sifakis, and S. Yovine. Symbolic model-checking for real-time systems. In *Proceedings of the Seventh IEEE Symposium on Logic in Computer Science*, pages 394–406, 1992.

9. J. Hopcroft and J. Ullman. *Introduction to Automata Theory, Languages, and Computation*. Addison-Wesley, 1979.

10. R. Kurshan. Reducibility in analysis of coordination. In *Lecture Notes in Computer Science*, volume 103, pages 19–39. Springer-Verlag, 1987.

11. N. Lynch and H. Attiya. Using mappings to prove timing properties. *Distributed Computing*, 6:121–139, 1992.

12. Z. Manna and A. Pnueli. *The Temporal Logic of Reactive and Concurrent Systems*. Springer-Verlag, 1991.

13. M. Merritt, F. Modugno, and M. Tuttle. Time-constrained automata. In *Proceedings of the Workshop on Theories of Concurrency*, Lecture Notes in Computer Science 527, pages 408–423. Springer-Verlag, 1991.

14. F. Schneider, B. Bloom, and K. Marzullo. Putting time into proof outlines. In *Real-Time: Theory in Practice*, Lecture Notes in Computer Science 600, pages 618–639. Springer-Verlag, 1991.

15. A. Sistla, M. Vardi, and P. Wolper. The complementation problem for Büchi automata with applications to temporal logic. *Theoretical Computer Science*, 49:217–237, 1987.

16. P. Wolper, M. Vardi, and A. Sistla. Reasoning about infinite computation paths. In *Proceedings of the 24th IEEE Symposium on Foundations of Computer Science*, pages 185–194, 1983.

17. T. Yoneda, A. Shibayam, B. Schlingloff, and E. Clarke. Efficient verification of parallel real-time systems. In *Proceedings of the Fifth Conference on Computer-Aided Verification*, Lecture Notes in Computer Science 697, pages 321–332. Springer-Verlag, 1993.

Real-Time System Verification using P/T Nets [*]

Roberto Gorrieri[†]   Glauco Siliprandi[‡]

[†]Dipartimento di Matematica, Università di Bologna
Piazza di Porta S. Donato 5, I-40127 Bologna, Italy
[‡]Dipartimento di Matematica, Università di Siena
Via del Capitano 15, I-53100 Siena, Italy
e-mail:{gorrieri,silipran}@cs.unibo.it

Abstract

Timed Nets are proposed to model the behavior of real-time systems. Net transitions are annotated by timing constraints, using finitely many real-valued *clocks*. A timed net accepts *timed words*, i.e. infinite sequences in which a time of occurrence is associated with each symbol. We study expressiveness, closure properties and decision problems of such nets, where the acceptance condition is based on actions. The main result of the paper is an algorithm for deciding the inclusion problem for timed languages.

1   Introduction

There is a great variety of automata-based approaches to the specification and verification of real-time systems, depending mainly on the assumptions about the nature of time. The use of *dense-time* domains (events may happen arbitrarily close to each other) is an important feature whenever one is interested in modeling heterogeneous systems, i.e., systems composed of digital and analogical (hence, continuous) devices.

Among the models based on dense-time, particular interest has been stirred up by *Timed Automata*, proposed by Alur and Dill [1, 2]. This model is essentially the timed version of finite-state ω-automata, hence recognizing *timed words* – infinite sequences in which a real-valued time is associated with each symbol. A timed word is recognized by one of such automata if, for instance, the set of those states passed through infinitely often is equal to a given set of accepting states. The behavior of a real-time system is modeled by a timed language $\mathcal{L}_{imp}$; since also the requirements the system must satisfy can be expressed as a timed language $\mathcal{L}_{spec}$, the problem of verifying that a system satisfies a certain property essentially reduces to the inclusion problem of the implementation timed language $\mathcal{L}_{imp}$ into the specification timed language $\mathcal{L}_{spec}$. Due to the decidability of the inclusion problem, Timed Automata have been profitably used for the automatic formal verification of real-time finite-state systems [2].

[*]This research has been partially supported by CNR grant N.92.00069.CT12.115.25585 and by MURST.

Despite of their elegant characterization and strong properties, Timed Automata are not expressive enough to model systems with an infinite number of states. Here, we extend the approach of Alur and Dill to a more general class of automata, namely Place/Transition Petri Nets, where the number of tokens in each place can increase unboundedly. As the global state of a system is the collection of the tokens in the places, the number of global states of the system a P/T net represents is infinite.

Our approach follows an action based acceptance condition: a timed word is accepted iff the set of those transitions fired infinitely often belongs to a given family of sets of transitions. The choice of such an acceptance condition is due to the fact that the set of transitions of a (Timed) P/T Net is finite. This permits to prove some relevant decidability results: the problem of language emptiness and the problem of inclusion of a timed P/T net language in a (deterministic) timed regular language. These are at the base of automatic verification of real-time properties, expressed as Timed Automata, of real-time systems, expressed as Timed P/T Nets.

An example in Section 7 shows a simple real-time system which is modeled by a Timed P/T Net and which cannot be represented through a Timed Automaton, hence proving that the class of systems we can model is strictly larger. The final section introduces a simple example of application of our theoretical results.

2 ω-languages and ω-automata.

In this section we give some basic definitions about ω-words and their recognizing automata. For more details see [4, 5].

Let Σ be a finite alphabet. Σ^* is the set of all finite sequences over Σ. An infinite sequence over Σ, also called ω-word, is a map $\sigma : I\!\!N^+ \to \Sigma$. Since σ_i denotes the ith symbol of σ, we also write $\sigma = \sigma_1 \sigma_2 \dots$. The set of all ω-words over Σ is denoted by Σ^ω. If A is another alphabet, $h : \Sigma \to A$ and $\sigma \in \Sigma^\omega$, we use $h(\sigma)$ to denote the ω-word $h(\sigma_1)h(\sigma_2)\dots \in A^\omega$. $L \subseteq \Sigma^\omega$ will be called a ω-language. With $\exists^\infty n.P(n)$ we denote a property P which holds infinitely often: $\forall m \in I\!\!N.\exists n > m.P(n)$ holds. We define $In(\sigma) \stackrel{\text{def}}{=} \{a \in \Sigma | \exists_i^\infty.\sigma_i = a\}$.

In the literature various types of finite state ω-automata have been studied: among these, we recall Büchi automata. A Transition Table (TT for short) is a tuple $\mathcal{A} = (\Sigma, S, S_0, E)$, where Σ is an alphabet, S is a finite set of states, $S_0 \subseteq S$ is a set of start states, and $E \subseteq S \times S \times \Sigma$ is a set of transitions. A *run* $r = (s_0, s_1, \sigma_1)(s_1, s_2, \sigma_2)\dots \in E^\omega$ on an ω-word σ is such that $s_0 \in S_0$. The sequence of *reached states* in the run r is the ω-word $s_1 s_2 s_3 \dots \in S^\omega$, denoted by $St(r)$. A Büchi Automata (BA for short) is a tuple $\mathcal{A} = (\Sigma, S, S_0, E, F)$ such that (Σ, S, S_0, E) is a TT and $F \subseteq S$. A run r is *accepted* iff $In(St(r)) \cap F \neq \emptyset$. The accepted ω-language is $\mathcal{L}(\mathcal{A}) \stackrel{\text{def}}{=} \{\sigma \in \Sigma^\omega | $ there is a run on σ accepted by $\mathcal{A}\}$. The class of languages accepted by BAs, called *regular ω-languages*, is denoted by $\mathcal{B}\mathcal{A}$.

Definition 2.1 (Place/Transition Petri Nets) $N = (\Sigma, P, T, F, W, h, m_0)$ is a PTP, where Σ is an input alphabet, P is a finite set of *places*, T is a finite set of *transitions* (disjoint from P), $F \subseteq (P \times T) \cup (T \times P)$ is the *flow relation*, $W : F \rightarrow I\!N^+$ is the *weight function*, $h : T \rightarrow \Sigma$ is the *labeling function*, $m_0 \in I\!N^P$ is the *initial marking* (the functions from P to $I\!N$, are called *markings* for N; such a set will be usually denoted by M and ranged over by m).

The *pre-set* of a transition t is ${}^\bullet t \overset{\text{def}}{=} \{p|\,(p,t) \in F\}$, and its *post-set* is $t^\bullet \overset{\text{def}}{=} \{p|\,(t,p) \in F\}$. N *has concession* in m iff $m(p) \geq W(p,t)$ for all $p \in {}^\bullet t$. If t has concession in m then t *fires* m to m', where $m'(p) = m(p) - W(p,t)$ if $p \in {}^\bullet t \setminus t^\bullet$, $m'(p) = m(p) + W(t,p)$ if $p \in t^\bullet \setminus {}^\bullet t$, $m'(p) = m(p) - W(p,t) + W(t,p)$ if $p \in {}^\bullet t \cap t^\bullet$ and $m'(p) = m(p)$ if $p \in P \setminus ({}^\bullet t \cup t^\bullet)$. This relation on $M \times T \times M$ is the *firing relation*, denoted by $m\,(t)m'$.

We say that $r \in T^\omega$ is a *run* of N over $\sigma \in \Sigma^\omega$ iff $h(r) = \sigma$ and there are $m_1, m_2, \ldots \in M$ such that $\forall i \in I\!N.m_i\,(r_{i+1})m_{i+1}$. $m_0, m_1, \ldots$ is the *sequence of markings associated to* r. □

In [6] PTPs are provided with acceptance conditions based on transitions fired infinitely often in a run. Among the various notions proposed there, we consider the one for which the class of accepted languages is the largest.

Definition 2.2 (Action Based PTP) $N = (\Sigma, P, T, F, W, h, m_0, \mathcal{F})$ is an ABP, where $(\Sigma, P, T, F, W, h, m_0)$ is a PTP and $\mathcal{F} \subseteq \wp^+(T)$. A run r is accepted iff $In(r) \in \mathcal{F}$. $\mathcal{ABP}$ denotes the class of ω-languages accepted by ABPs. □

Theorem 2.3 *[6] The emptiness problem is decidable for $\mathcal{ABP}$.*

3 Timed Languages

In this section we present the timed languages, as introduced in [2].

Definition 3.1 (Time Sequence) A *time sequence* $\tau = \tau_1\tau_2\ldots$ is an infinite sequence of positive reals $\tau_i \in I\!R^{>0}$, satisfying

Monotonicity: $\tau_i < \tau_{i+1}$ for all $i \geq 1$.

Progress: for every $t \in I\!R^{\geq 0}$ there is some $i \geq 1$ such that $\tau_i > t$.

For any time sequence $\tau = \tau_1\tau_2\ldots$ we assume $\tau_0 \overset{\text{def}}{=} 0$. □

The progress condition is introduced to avoid the Zeno paradox: it will never be the case that infinitely many events happen in a bounded time interval.

Definition 3.2 (Timed Words and Languages) A *timed word* over Σ is a pair (σ, τ) where $\sigma \in \Sigma^\omega$ and τ is a time sequence. Σ^t, ranged over by ψ, denotes the set of all timed words over Σ. A *timed language* is a set $L \subseteq \Sigma^t$. □

The *Untime* operation discards the time values associated with the symbols, i.e., it considers the projection of a timed word (σ, τ) on the first component.

Definition 3.3 (Untimed Language) Given a timed language L, we define $Untime[L] \overset{\text{def}}{=} \{\sigma|\,\exists\tau.(\sigma, \tau) \in L\}$, also denoted by $\pi_1(L)$. □

4   Timed Finite State Automata

In [1, 2] the definition of finite state ω-automata is augmented, so that they accept timed languages. In this section we recall some results presented there and a new result of ours.

Transition tables are extended to *timed* TTs so that they can read timed words. When executing a transition, the choice of the next state depends also on the time of the input symbol w.r.t. the times of the previously read symbols. For this purpose a finite set of *clocks* is associated with each TT. A clock can be set to zero simultaneously with the execution of a transition, while its value is equal to the time elapsed since the last time of reset. A *clock constraint* is associated with each transition and a transition may be taken only if the current values of the clocks satisfy its constraint.

Definition 4.1 (Clocks) For a set C of clock variables, the set $\Phi(C)$ of *clock constraints* is defined inductively by

$$\delta := c \leq q \mid q \leq c \mid \neg\delta \mid \delta \wedge \delta$$

where c is a *clock* in C and q is a nonnegative rational constant. A *clock interpretation* (CI for short) for C is a function $\nu : C \to I\!\!R^{\geq 0}$. We say that ν satisfies δ iff δ evaluates to true. ν_0 denotes the CI such that $\forall c \in C.\nu_0(c) \stackrel{\text{def}}{=} 0$. If $t \in I\!\!R^{\geq 0}$, then $\nu + t$ is the CI mapping every clock c to $\nu(c) + t$, and $t \cdot \nu$ the one assigning $t \cdot \nu(c)$ to each clock c. For $Y \subseteq C$, $\nu[t/Y]$ denotes the CI assigning t to each $c \in Y$ and agreeing with ν over the other clocks. □

A Timed Transition Table (TTT for short) is a tuple $\mathcal{A} = (\Sigma, S, S_0, C, E)$, where C is a finite set of clocks, and $E \subseteq S \times S \times \Sigma \times 2^C \times \Phi(C)$ is the set of transitions. (s, s', a, y, δ) represents a transition from s to s' on input symbol a. The set $y \subseteq C$ gives the clocks to be reset when executing this transition, and δ is a clock constraint to be satisfied. A *run* of $\mathcal{A}$ on (σ, τ) is an infinite sequence of transitions $r = (s_0, s_1, \sigma_1, y_1, \delta_1)(s_1, s_2, \sigma_2, y_2, \delta_2)\ldots$ such that (i) $s_0 \in S_0$ and (ii) there are $\nu_1, \nu_2, \ldots$ CIs such that δ_{i+1} is satisfied by $\nu_i + (\tau_{i+1} - \tau_i)$ and $\nu_{i+1} \stackrel{\text{def}}{=} (\nu_i + (\tau_{i+1} - \tau_i))[0/y_{i+1}]$. We say that $\nu_0, \nu_1, \ldots$ is the *sequence of CIs* associated to r, and $St(r) \stackrel{\text{def}}{=} s_1 s_2 s_3 \ldots$ is the sequence of *reached states*.

Definition 4.2 (Timed Büchi Automata)
A TBA is a tuple $\mathcal{A} = (\Sigma, S, S_0, C, E, F)$, where (Σ, S, S_0, C, E) is a TTT, and $F \subseteq S$. A run r is accepted iff $In(St(r)) \cap F \neq \emptyset$. The accepted timed language is $\mathcal{L}(\mathcal{A}) \stackrel{\text{def}}{=} \{\psi \in \Sigma^t \mid \mathcal{A} \text{ has an accepting run over } \psi\}$. The class of timed languages accepted by TBAs, called *timed regular languages*, is denoted by $T\mathcal{BA}$. □

Definition 4.3 (Timed Muller Automata)
A TMA is a tuple $\mathcal{A} = (\Sigma, S, S_0, C, E, \mathcal{F})$, where (Σ, S, S_0, C, E) is a TTT, and $\mathcal{F} \subseteq \wp^+(S)$. A run r is accepted iff $In(St(r)) \in \mathcal{F}$. The set of timed languages accepted by TMAs is denoted by $T\mathcal{MA}$. □

Theorem 4.4 *[2]* (i) $TBA = TMA$
(ii) TBA *is closed under finite union and intersection.*

In order to define a class of timed languages closed under all boolean operations, deterministic automata are investigated. A TMA $(\Sigma, S, S_0, C, E, \mathcal{F})$ is called *deterministic* (DTMA) if and only if $|S_0| = 1$ and for all $s \in S$, for all $a \in \Sigma$, for every pair of transitions of the form $(s, -, a, -, \delta_1)$ and $(s, -, a, -, \delta_2)$, we have that $\delta_1 \wedge \delta_2$ is unsatisfiable. The class of timed languages accepted by DTMA is denoted by $DTMA$.

Theorem 4.5 *[2]* (i) $DTMA \subset TMA$
(ii) $DTMA$ *is closed under finite union, intersection and complementation.*

We now study new acceptance conditions based on transitions rather than states. We define the *action based timed Büchi automata* and the *action based timed Muller automata*, and we prove that these have the same expressive power of TBAs and TMAs.

Definition 4.6 (Action Based TBA) $\mathcal{A} = (\Sigma, S, S_0, C, E, F)$ is an ABTBA, where (Σ, S, S_0, C, E) is a TTT, and $F \subseteq E$ is the set of *accepting transitions*. A run r is accepted iff $In(r) \cap F \neq \emptyset$. The set of accepted timed languages is denoted by $ABTBA$. $\qquad\square$

Definition 4.7 (Action Based TMA) $\mathcal{A} = (\Sigma, S, S_0, C, E, \mathcal{F})$ is an ABTMA, where (Σ, S, S_0, C, E) is a TTT, and $\mathcal{F} \subseteq \wp^+(E)$ specifies an *acceptance family*. A run r is accepted iff $In(r) \in \mathcal{F}$. The set of timed languages accepted by ABTMAs is denoted by $ABTMA$. $\qquad\square$

Theorem 4.8 *[3]* $ABTBA = TBA = TMA = ABTMA$.

5 Timed Place/Transition Nets

In this section we augment the definition of PTPs following the same idea illustrated in the previous section, so that they can recognize timed words.

Definition 5.1 (Timed P/T Net) $N = (\Sigma, P, T, F, W, h, m_0, C, c^c, c^r)$ is a TP, where $N' = (\Sigma, P, T, F, W, h, m_0)$ is a PTP, C is a finite set of clocks, $c^c : T \rightarrow \Phi(C)$ gives the clock constraint associated to each transition, and $c^r : T \rightarrow 2^C$ gives the set of clocks to be reset when a transition is executed.
A *snapshot* for N is a couple (m, ν) such that m is a marking for N' and ν is a CI for C. The set of snapshots is denoted by Γ. A transition t of N *has concession* in (m, ν) for N iff t has concession in m for N' and ν satisfies $c^c(t)$. If t has concession in (m, ν), then t *fires* (m, ν) to (m', ν') where $m \langle t \rangle m'$ and $\nu' = \nu[0/c^r(t)]$. This relation on $\Gamma \times T \times \Gamma$ is the *timed firing relation* denoted by $(m, \nu) \langle t \rangle (m', \nu')$. A *run* of N over (σ, τ) is an infinite sequence $r = t_1 t_2 \ldots \in T^\omega$ such that $h(r) = \sigma$ and there are $(m_1, \nu_1), (m_2, \nu_2), \ldots$ snapshots for N such that $\forall i \in \mathbb{N}.(m_i, \nu_i + (\tau_{i+1} - \tau_i)) \, (t_{i+1})(m_{i+1}, \nu_{i+1})$. $(m_0, \nu_0), (m_1, \nu_1), \ldots$ is the *sequence of snapshots associated to r.* $\qquad\square$

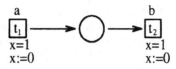

Figure 1: An example of timed P/T net.

For example, consider the TP in Figure 1, where a and b are the labels for t_1 and t_2. In every run, in each moment of the execution, the number m of fired t_1 transitions is not less than the number n of executed t_2 transitions; indeed the place keeps exactly $m - n$ tokens. The timing constraints impose that a transition must be fired after each time unit, thus the first transition in the run fires at time 1, the second at time 2 and so on.

We provide TPs with an action based acceptance condition.

Definition 5.2 (Action Based TP) $N = (\Sigma, P, T, F, W, h, m_0, C, c^c, c^r, \mathcal{F})$ is an ATP, where $(\Sigma, P, T, F, W, h, m_0, C, c^c, c^r)$ is a TP and $\mathcal{F} \subseteq \wp^+(T)$ is an *acceptance family*. A run r is accepted iff $In(r) \in \mathcal{F}$. The accepted timed language is $\mathcal{L}(N) \stackrel{\text{def}}{=} \{\psi \in \Sigma^t | N$ has an accepting run over $\psi\}$. The class of accepted timed languages is denoted by $\mathcal{ATP}$. $\qquad \Box$

Consider again the net in Figure 1, with acceptance family $\mathcal{F} \stackrel{\text{def}}{=} \{\{t_1\}, \{t_1, t_2\}\}$: it can be proved that every run is accepted and the recognized timed language is $\{(\sigma, \tau) \in \{a, b\}^t | \forall i \in \mathbb{N}.\#\{n \le i | \sigma_n = a\} \ge \#\{n \le i | \sigma_n = b\} \wedge \tau = 1\,2\,3\ldots\}$.

6 Emptiness

In this section we develop an algorithm that, given an ATP accepting a timed language L, constructs an ABP recognizing $Untime[L]$ (in [2] an analogous procedure was defined for TBAs). This important result allows us to state the decidability of the emptiness problem for timed languages in $\mathcal{ATP}$, and defines a limit to the expressive power of ATPs, in the sense that a timed language may be accepted by an ATP only if its $Untime$ is recognized by an ABP. We use this fact to prove the nonclosure of $\mathcal{ATP}$ under complementation.

First of all we want to prove that it suffices to consider those nets in which only integer constants are used in clock constraints.

Definition 6.1 (Multiplied Clock Constraints)
Given a TP N and $t \in \mathbb{R}^{>0}$, N_t is the TP obtained by replacing each constant q, in each δ appearing in N, by $t \cdot q$. $\qquad \Box$

Lemma 6.2 *Consider a TP N, a timed word (σ, τ), and a positive rational t. Then r is a run of N over (σ, τ) iff r is a run of N_t over $(\sigma, t \cdot \tau)$.*

If we choose t to be the least common multiple of all nonzero constants appearing in the clock constraints of N, then the clock constraints in N_t use only nonnegative integers; furthermore, if N is an ATP then $Untime[\mathcal{L}(N)] = Untime[\mathcal{L}(N_t)]$;

as a consequence, $\mathcal{L}(N) = \emptyset$ iff $\mathcal{L}(N_t) = \emptyset$. Thus, in the remainder of the section we assume that clock constraints use nonnegative integers only.

The problem of checking the emptiness of the timed language recognized by an ATP N can be more easily coped with if we show that this is equivalent to checking the emptiness of the language recognized by an ABP N'. This could be done, in principle, by adding places for clock interpretations: a snapshot $\langle m, \nu \rangle$ in N is simulated in N' by putting a token in the place corresponding to ν. Even if the CIs for N are infinite, we can define an equivalence relation $\cong$ on CIs, with a finite number of classes, and then add a new place for each of such classes. In [2] $\cong$ has been defined in the case of TBAs.

Definition 6.3 (Clock Regions) Let q_c denote the largest integer q such that $c \leq q$ or $q \leq c$ is a subformula of some δ appearing in the net N; for any nonnegative real t, $fract(t)$ denotes the fractional part of t, and $\lfloor t \rfloor$ denotes the integer part of t; that is $t = \lfloor t \rfloor + fract(t)$.
$\nu \cong \nu'$ iff the following three conditions hold:

1. $\forall c \in C. \lfloor \nu(c) \rfloor = \lfloor \nu'(c) \rfloor \vee (\nu(c) > q_c \wedge \nu'(c) > q_c)$
2. $\forall c, d \in C. (\nu(c) \leq q_c \wedge \nu(d) \leq q_d) \Rightarrow$
$$(fract(\nu(c)) \leq fract(\nu(d)) \Leftrightarrow fract(\nu'(c)) \leq fract(\nu'(d)))$$
3. $\forall c \in C. \nu(c) \leq q_c \Rightarrow (fract(\nu(c)) = 0 \Leftrightarrow fract(\nu'(c)) = 0)$ □

It can be proved that $\cong$ is an equivalence relation. An equivalence class is called a *clock region* (CR for short). $[\nu]$ denotes the CR which ν belongs to. The set of CRs is ranged over by α.

Proposition 6.4 *Let $c \in C$, $\nu \cong \nu'$ and $t \in T$. Then*

1. *ν satisfies $c^c(t)$ iff ν' satisfies $c^c(t)$*
2. *$\nu[0/c^r(t)] \cong \nu'[0/c^r(t)]$*
3. *$\nu(c) = 0$ iff $\nu'(c) = 0$ and $\nu(c) > q_c$ iff $\nu'(c) > q_c$.*

Thus in the remainder of the section we shall say that $[\nu]$ satisfies $c^c(t)$ if ν satisfies $c^c(t)$, and we shall use $[v][0/c^r(t)]$ instead of $[\nu[0/c^r(t)]]$.
In [3] an algorithm producing a representative CI for each clock region is provided: $\overline{\alpha}$ denotes the representative of class α. As a consequence

Proposition 6.5

1. *The set of clock regions is finite.*
2. *If $\nu \cong \nu'$ and $t > 0$, then there exists $t' > 0$ such that $\nu + t \cong \nu' + t'$.*

Then a new place is introduced for each clock region. When a transition fires, it puts a token in a clock region place. This means that the CI reached just after the firing of the transition belongs to the clock region corresponding to that place. Moreover, the previous proposition states that, if $\nu \cong \nu'$, ν and ν' visit the same clock regions as time progresses.

Definition 6.6 (Time-successor) A CR α' is a *time-successor* of a CR α iff for each $\nu \in \alpha$ there exists $t > 0$ such that $\nu + t \in \alpha'$. □

In [3] we provide an algorithm building the representative of any time successor of a given CR. Now we define the ABP recognizing $Untime[\mathcal{L}(N)]$.

Definition 6.7 (Region Net) Given a TP $N = (\Sigma, P, T, F, W, h, m_0, C, c^c, c^r)$ we define the PTP $\mathcal{R}(N) \stackrel{\text{def}}{=} (\Sigma, P', T', F', W', h', m_0')$ where

- $P' \stackrel{\text{def}}{=} P \cup \{\alpha \,|\, \alpha \text{ is a CR for } N\}$

- $T' \stackrel{\text{def}}{=} \{(t, \alpha, \alpha') \,|\, t \in T \text{ and there is an } \alpha'' \text{ time-successor of } \alpha, \text{ such that } \alpha''$ satisfies $c^c(t)$ and $\alpha' = \alpha''[0/c^r(t)]\}$

- $F' \stackrel{\text{def}}{=} \{(p, (t, \alpha, \alpha')) \,|\, (p, t) \in F\} \cup \{((t, \alpha, \alpha'), p) \,|\, (t, p) \in F\} \cup$
  $\{(\alpha, (t, \alpha, \alpha')) \,|\, (t, \alpha, \alpha') \in T'\} \cup \{((t, \alpha, \alpha'), \alpha') \,|\, (t, \alpha, \alpha') \in T'\}$

- $W'(p, (t, \alpha, \alpha')) \stackrel{\text{def}}{=} \begin{cases} W(p, t) & \text{if } (p, t) \in F \\ 1 & \text{otherwise} \end{cases}$

  $W'((t, \alpha, \alpha'), p) \stackrel{\text{def}}{=} \begin{cases} W(t, p) & \text{if } (t, p) \in F \\ 1 & \text{otherwise} \end{cases}$

- $h'(t, \alpha, \alpha') \stackrel{\text{def}}{=} h(t)$

- $m_0'(p) \stackrel{\text{def}}{=} \begin{cases} m_0(p) & \text{if } p \in P \\ 1 & \text{if } p = [\nu_0] \\ 0 & \text{otherwise} \end{cases}$ $\qquad \square$

In the net $\mathcal{R}(N)$, any reachable marking has one token in only one of the clock region places, while all the other CR places are empty.

Definition 6.8 ($\mathcal{R}(N)$ Markings) Let m be a marking of N. $\langle m, \alpha \rangle$ denotes the marking in $\mathcal{R}(N)$ such that $\langle m, \alpha \rangle(p) \stackrel{\text{def}}{=} m(p)$ if $p \in P$, $\langle m, \alpha \rangle(\alpha) \stackrel{\text{def}}{=} 1$, and $\langle m, \alpha \rangle(p) \stackrel{\text{def}}{=} 0$ otherwise. $\qquad \square$

Now we want to establish a correspondence between each run in N and some run in $\mathcal{R}(N)$.

Definition 6.9 (Run Projection)
Let $r = t_1 t_2 \ldots$ be a run of N on (σ, τ) and $(m_0, \nu_0), (m_1, \nu_1), \ldots$ be the sequence of snapshots associated to r. We define the *run projection* of r to be $[r] \stackrel{\text{def}}{=} (t_1, [\nu_0], [\nu_1])(t_2, [\nu_1], [\nu_2]) \ldots .$ $\qquad \square$

Lemma 6.10 $[r]$ *is a run of the region net* $\mathcal{R}(N)$.

Because of the progress condition, every clock in a run is either reset infinitely often, or from a certain time onwards it increases unboundedly.

Definition 6.11 (Progressive Run) A run of $\mathcal{R}(N)$, with associated sequence of markings $\langle m_0, [\nu_0] \rangle, \langle m_1, [\nu_1] \rangle, \ldots$, is *progressive* iff $\forall c \in C, \exists^\infty i.(\nu_i(c) = 0 \lor \nu_i(c) > q_c)$. $\qquad \square$

Lemma 6.12 *If r is a run of N, then $[r]$ is a progressive run of $\mathcal{R}(N)$.*

Conversely, we now claim that, given a progressive run r' of $\mathcal{R}(N)$, a run r of N can be defined such that r' is its projection.

Lemma 6.13 *If* $r' = (t_1, \alpha_0, \alpha_1)(t_2, \alpha_1, \alpha_2) \ldots$ *is a progressive run of* $\mathcal{R}(N)$ *over* σ*, then there exist a time sequence* τ *and a run* r *of* N *over* (σ, τ) *such that* $r' = [r]$.

Finally we define the acceptance family. A set of accepting transitions has to ensure that a run in $\mathcal{R}(N)$ may be accepted only if it is progressive.

Definition 6.14 (Progressive Acceptor) A set $R \subseteq T'$ is a *progressive acceptor* if and only if $\forall c \in C. \exists (t, \alpha', \alpha) \in R. \overline{\alpha}(c) = 0 \vee \overline{\alpha}(c) > q_c$. $\qquad \square$

Theorem 6.15 *Given an ATP* N*, there exists an ABP* N' *such that* $\mathcal{L}(N') = Untime[\mathcal{L}(N)]$.

N' is $\mathcal{R}(N)$ provided with the acceptance family $\mathcal{F}' \stackrel{\text{def}}{=} \{R' \subseteq T' | \pi_1(R') \in \mathcal{F}$ and R' is a progressive acceptor $\}$. As a consequence, the following holds

Theorem 6.16 *The emptiness problem for ATPs is decidable.*

Proof: $L = \emptyset$ iff $Untime[L] = \emptyset$; the thesis follows by Th.6.15 and 2.3. $\qquad\blacksquare$

7 Expressiveness and Closure Properties

In this section we compare ATPs with timed finite state automata and show some closure properties w.r.t. boolean operations.

Theorem 7.1 *[3]* $TBA \subset ATP$.

The basic idea underlying the proof is that every ABTMA can be simulated by an ATP: a place in the net corresponds to a state in the ABTMA. On the contrary, there exist nets recognizing non regular timed languages. A counterexample is illustrated by the net N in Figure 1, with acceptance family $\mathcal{F} \stackrel{\text{def}}{=} \{\{t_1\}, \{t_1, t_2\}\}$. It can be proved that $Untime[\mathcal{L}(N)] = \{\sigma \in \{a, b\}^\omega | \forall k \in I\!N. \#\{i \leq k | \sigma_i = a\} \geq \#\{i \leq k | \sigma_i = b\}\}$ is not a regular ω-language. In [2] Alur and Dill proved that this fact implies that $\mathcal{L}(N)$ is not a timed regular language.

Theorem 7.2 *[3]* ATP *is closed under finite union and intersection.*

$\mathcal{ATP}$ is not closed under complementation; consider the ATP N in Figure 1 *with no clock constraints*, and acceptance family $\mathcal{F} \stackrel{\text{def}}{=} \{\{t_1\}, \{t_1, t_2\}\}$. It can be proved that $\mathcal{L}(N) \stackrel{\text{def}}{=} \{(\sigma, \tau) \in \{a, b\}^t | \forall k \in I\!N. \#\{i \leq k | \sigma_i = a\} \geq \#\{i \leq k | \sigma_i = b\}\}$. Thus $\overline{\mathcal{L}(N)} \stackrel{\text{def}}{=} \{(\sigma, \tau) \in \{a, b\}^t | \exists k \in I\!N. \#\{i \leq k | \sigma_i = a\} < \#\{i \leq k | \sigma_i = b\}\}$. Suppose there exists an ATP N' such that $\mathcal{L}(N') = \overline{\mathcal{L}(N)}$. Then, for Theorem 6.15 there exists an ABP N'' such that $\mathcal{L}(N'') = \{\sigma \in \{a, b\}^\omega | \exists k \in I\!N. \#\{i \leq k | \sigma_i = a\} < \#\{i \leq k | \sigma_i = b\}\}$, but it can be proved that such an ABP does not exist.

8 Verification

Here we discuss how to use the theory of timed P/T nets to prove correctness of some infinite-state real-time systems. A trace will be a timed word over sets of events: if two events a and b happen simultaneously, the trace will have the set $\{a, b\}$.

Definition 8.1 (Timed Traces and Timed Processes)
$\psi \in \wp^+(A)^t$ is a *timed trace* over alphabet A. A *timed process* is a pair (A, L) where A is a finite set of observable events, and L is a timed language over $\wp^+(A)$. The set of timed processes is denoted by $\mathcal{TP}$. The class of processes *modeled* by ATPs is $\mathcal{ATPP} \stackrel{\text{def}}{=} \{(A, L) \in \mathcal{TP} \mid L \in \mathcal{ATP}\}$. □

Remark 8.2 *A property Π can be represented as the set of traces satisfying it. Hence verifying that a process $P = (A, L)$ satisfies property Π is equivalent to check that $L \subseteq \Pi$, i.e. that all the execution traces satisfy the property Π.*

Various operations can be defined on processes; these are useful for describing complex systems as composed of simpler ones. We will consider only parallel composition, which prescribes the joint behavior of a set of processes running concurrently.

The parallel composition operator can be conveniently defined using the *projection* operation. The projection of $(\sigma, \tau) \in \wp^+(A)^t$ onto $B \subseteq A$ is formed by intersecting each event set in σ with B and deleting all the empty sets from the sequence. Notice that the projection operation may result in a finite sequence, thus we define the set of timed traces *projectable* onto B.

Definition 8.3 (Projectable Timed Traces)
$A \lceil B \stackrel{\text{def}}{=} \{(\sigma, \tau) \in \wp^+(A)^t \mid \exists^\infty i \, . \, \sigma_i \cap B \neq \emptyset\}$. If $B_1, \ldots, B_n \subseteq A$, we define $A \lceil_i B_i \stackrel{\text{def}}{=} \{\psi \in \wp^+(A)^t \mid \wedge_i \psi \in A \lceil B_i\}$. □

Definition 8.4 (Projection) Assume $(\sigma, \tau) \in A \lceil B$. We define $\xi_1 \stackrel{\text{def}}{=} \min\{i \in I\!N^+ \mid \sigma_i \cap B \neq \emptyset\}$, $\xi_{k+1} \stackrel{\text{def}}{=} \min\{i > \xi_k \mid \sigma_i \cap B \neq \emptyset\}$, $\sigma' \stackrel{\text{def}}{=} \sigma_{\xi_1} \sigma_{\xi_2} \ldots$, $\tau' \stackrel{\text{def}}{=} \tau_{\xi_1} \tau_{\xi_2} \ldots$, and $(\sigma, \tau) \lceil B \stackrel{\text{def}}{=} (\sigma', \tau')$. □

Definition 8.5 (Parallel Composition) Assume $P_i = (A_i, L_i)$ is a timed process for $i = 1, 2, \ldots, n$. Their parallel composition is the timed process $\|_i P_i \stackrel{\text{def}}{=} (\cup_i A_i, \|_i L_i)$ where $\|_i L_i \stackrel{\text{def}}{=} \{\psi \in (\cup_i A_i) \lceil_j A_j \mid \wedge_j \psi \lceil A_j \in L_j\}$. □

We want to prove that $\mathcal{ATPP}$ is closed under parallel composition. Assume $P_i = (A_i, L_i) \in \mathcal{ATPP}$ for $i = 1, 2, \ldots, n$, and N_i is an ATP such that $\mathcal{L}(N_i) = L_i$. $\mathcal{G}(L_k)$ denotes the timed language $\{\psi \in (\cup_i A_i) \lceil_j A_j \mid \psi \lceil A_k \in L_k\}$. The first step will be, for each $k \in \{1, \ldots, n\}$, to define an ATP $\mathcal{G}(N_k)$ such that $\mathcal{L}(\mathcal{G}(N_k)) = \mathcal{G}(L_k)$.

Definition 8.6 (Generalization Construction)
Given the net $N_k = (\wp^+(A_k), P_k, T_k, F_k, W_k, h_k, m_k, c_k^c, c_k^r, \mathcal{F}_k)$, we define below the net $\mathcal{G}(N_k) \stackrel{\text{def}}{=} (\wp^+(\cup_i A_i), P_k, T_k', F_k', W_k', h_k', m_k, c_k'^c, c_k'^r, \mathcal{F}_k')$ as follows.

- $T_k' \stackrel{\text{def}}{=} \overline{A}_k \cup \{(t, A) \mid A = \emptyset \vee A \in \overline{A}_k\}$ where $\overline{A}_k \stackrel{\text{def}}{=} \wp^+(\cup_i A_i \setminus A_k)$; [1]

- $F_k' \stackrel{\text{def}}{=} \{(p, (t, A)) \mid (p, t) \in F_k\} \cup \{((t, A), p) \mid (t, p) \in F_k\}$;

- $W_k'(p, (t, A)) \stackrel{\text{def}}{=} W_k(p, t)$ and $W_k'((t, A), p) \stackrel{\text{def}}{=} W_k(t, p)$;

- $h_k'(A) \stackrel{\text{def}}{=} A$ and $h_k'(t, A) \stackrel{\text{def}}{=} h_k(t) \cup A$.

- $c_k'^c(A) \stackrel{\text{def}}{=} true$ and $c_k'^c(t, A) \stackrel{\text{def}}{=} c_k^c(t)$;

- $c_k'^r(A) \stackrel{\text{def}}{=} \emptyset$ and $c_k'^r(t, A) \stackrel{\text{def}}{=} c_k^r(t)$;

- Given $T \subseteq T_k'$, we say that T is *complete* iff $\forall j. \exists t \in T. A_j \cap h_k'(t) \neq \emptyset$. When $R \in \mathcal{F}_k$, we say that T *emulates* R iff $\pi_1(T \setminus \overline{A}_k) = R$. Then $\mathcal{F}_k' \stackrel{\text{def}}{=} \{T \subseteq T_k' \mid T \text{ is complete and there exists } R \in \mathcal{F}_k \text{ such that } T \text{ emulates } R\}$. $\square$

It can be proved that $\mathcal{L}(\mathcal{G}(N_k)) = \mathcal{G}(L_k)$; since $\mathcal{ATP}$ is closed under intersection, the following holds

Theorem 8.7 *$\mathcal{ATPP}$ is closed under parallel composition.*

Typically, a system implementation is described as a composition of several components; if each component is a timed process modeled by an ATP and the specification is given as a property modeled by a DTMA, then the correctness of the implementation can be checked, due to the following

Theorem 8.8 *Given $P_i = (A_i, \mathcal{L}(N_i)) \in \mathcal{ATPP}$, modeled by ATPs N_i (for $i = 1, \ldots, n$), and a specification as a DTMA $\mathcal{A}$, the inclusion of $\|_i \mathcal{L}(N_i)$ in $\mathcal{L}(\mathcal{A})$ can be decided effectively.*

For deciding if $\|_i \mathcal{L}(N_i) \subseteq \mathcal{L}(\mathcal{A})$, in fact, it suffices to check if $\|_i \mathcal{L}(N_i) \cap \overline{\mathcal{L}(\mathcal{A})} = \emptyset$, and this can be done due to Theorems 8.7, 4.5, 7.2, and 6.16.

9 A Verification Example

As an example of automatic verification using ATPs, we consider a simple process to produce wooden black horses. In Figure 2 we show an ATP modeling a workman that carves a raw block of wood producing a wooden horse. We impose that both transitions must be fired infinitely often: thus the process ensures that if a new block is provided, then a new horse is refined within 2 minutes. Notice that the workman does not accept a new block until it has completely carved the previous one. This process could have been modeled by a TMA, too.

---

[1]If $A \in \overline{A}_k$, then events in A can happen at any time, thus we introduce a new transition for each of such an A; furthermore, events in A can happen contemporaneously to events in A_k hence, for each transition $t \in T_k$, we introduce a new transition (t, A).

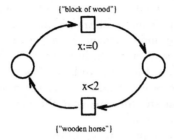

Figure 2: CARVER.

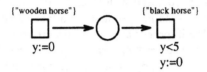

Figure 3: PAINTER.

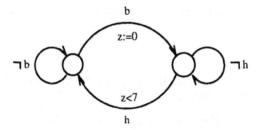

Figure 4: Strong Efficiency Property.

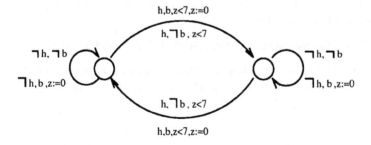

Figure 5: Weak Efficiency Property.

In Figure 3 we show an ATP modeling a workman that obtains a wooden horse and paints it. Again, we impose that both transitions must be repeated infinitely often. The painter can accept more than one horse (we could say that he can put them on a shelf, and this operation makes him busy). However, if no new horse arrives for at least 5 minutes, then the workman has the time to concentrate on painting, and a black horse is produced. Since the process 'remembers' the number of carved horses that are still unpainted, this process cannot be modeled by a timed finite state automaton.

The complete process (Horse Manufactory) for the production of black horses is HM $\stackrel{\text{def}}{=}$ CARVER $\parallel$ PAINTER. The factory administrator could be tempted to require that if a new block of wood is provided then, in no more than 7 minutes, a new black horse is produced. This property is modeled by the DTMA in Figure 4 (b denotes "block of wood", and h denotes "black horse"): a b-labeled edge stands for any event set containing b, and a ¬b-labeled edge means any event set not containing b. We impose that both states are reached infinitely often in an accepted run. Using Theorem 8.8, it can be proved that HM does not satisfy this property. On the other hand, the administrator could be pleased with the HM ensuring that if a new block is provided and for at least 7 minutes no new block is supplied, allowing all workmen to concentrate on the refining job, then at least a black horse is completed. This property is modeled by the DTMA in Figure 5, in which a run is accepted only if it reaches both states infinitely often. Using Theorem 8.8, it can be proved that HM satisfies this weaker property.

References

[1] R. Alur, D. Dill. "Automata for Modeling Real-Time Systems". In *Proc. ICALP 90*, LNCS 443, 322-335. Springer, 1990.

[2] R. Alur, D. Dill. "The Theory of Timed Automata". In *Proc. of the REX workshop "Real-Time: Theory in Practice"*, LNCS 600, 45-73. Springer, 1992.

[3] R. Gorrieri, G. Siliprandi. "A Theory of P/T Nets for Timed Languages". In preparation.

[4] W. Thomas. "Automata on Infinite Objects". Handbook of Theoretical Computer Science. Elsevier, 133-191, 1990.

[5] R. Valk. "Infinite Behavior of Petri Nets". TCS 25, 311-341, 1983.

[6] R. Valk, M. Jantzen. "The Residue of Vector Sets with Applications to Decidability Problems in Petri Nets". Acta Informatica 21, 643-647, 1985.

Criteria for the Simple Path Property in Timed Automata[*]

William K.C. Lam[1] and Robert K. Brayton[2]

[1] Hewlett-Packard Co., Palo Alto, California
[2] Department of EECS, University of California, Berkeley, California

Abstract. Timed automata have been studied in the past and have been found to have a complexity dependent on the relative scale of the time constants involved in the timing constraints imposed, even if the timing constraints are restricted to the form $x < k$ where x is a clock variable and k is a constant. We have previously shown that this complexity dependence on the time constants can be eliminated if the timed automaton has the simple path property (state A is reachable from state B if and only if it is reachable along a path with no cycles), and gave a set of conditions on the placement of clock queries and resets which imply this simple path property. These automata were called alternating RQ timed automata. We gave a technique for using this property to iteratively constrain an untimed automaton to rule out simple paths which cannot meet their timing constraints. The simple path property means that only simple paths need be constrained. In this paper, we give conditions for a timed automaton with arbitrary constraint equations to have the simple path property. As far as we know all practical examples in the literature meet these criteria. For example, this includes all automata with constraints of the form for each state s, a trace must remain in s for a time t where $t_{s_{min}} < t < t_{s_{max}}$. We are currently working on an efficient implementation for timed automata where arbitrary linear inequalities among the clock values are allowed. Using linear programming, we iteratively detect simple paths which are not traversable and construct untimed automata which disallow these paths. The present paper serves to extend this approach to a wide class of applications. In addition, we define extended RQ timed automata which include all the examples in the literature and are easily tested for this property.

1 Introduction

To model real-time behavior of finite state automata, such as that each transition takes at least 1 second and at most 2 seconds, timed automata, proposed by Allur and Dill [6, 3], add resetable clocks and conditions on the values of the clocks to the transitions of ordinary finite state automata. If a reset statement of a clock appears on a transition, then on completion of the transition the value of the clock becomes zero. Once a clock is reset, the value of the clock increments as a real clock until it is reset again. A timed automaton can have several clocks which may increment at different rates.[3] If a condition on the values of clocks, called timing constraint or query, appears on a transition, then the transition is enabled only if the condition is satisfied. Hence, to transit from the

---

[*] Project supported by SRC under contract 93-DC-008

[3] In this paper, we assume all clocks have the same rate.

present state to a next state, one must spend enough time in the present state so that the values of clocks satisfy any query on the transition. Thus, a sequence of states in a timed automaton can only be traversed in a such timely manner that all the queries along the traversal are satisfied.

The introduction of resetable clocks into finite state automata complicates the traversal problem: no longer can a state be decided reachable from another state by examining only whether there is a path connecting the two states — timing constraints or queries on a path must also be determined satisfiable to conclude that the path is traversable. Thus, a path is traversable only if there exists a schedule of stay-times on the states on the path such that the transition from the present state to the next state on the path is always enabled, i.e. the query (if any) is satisfied, upon the completion of staying in the present state for the amount of time according to the schedule. Hence, reachability analysis can not be performed on simple paths alone (simple paths are the paths on which each state is visited at most once.). In the following example, state S_4 is reachable from S_1 through non-simple paths only. Thus, in deciding whether a state is reachable one may need to examine all paths to the state, which may be infinite. In addition, the following example also shows that the traversal problem can depend on timing constraints.

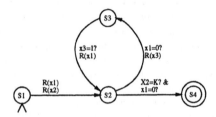

Fig. 1. Time Constant Dependent Traversal

Example 1. In this timed automaton, if K is a positive integer, then the accepting state S_4 can be reached. And the only way to get from the initial state S_1 to the final state S_4 is to go around the loop K times, i.e. only by traversing a non-simple path. During each visit of the loop, the automaton stays at S_3 for 1 unit of time to get clock X_2 to increment by 1. Thus, to satisfy the timing constraints on the transition between S_2 and S_4, e.g. $X_2 = K$, the loop needs to be traversed K times. If K is not an integer, S_4 is not reachable, demonstrating that traversability is intimately related to the time constants.

Traversal in timed automata is a key part of timing verification. With both a design and a specification expressed with timed automata, D and S, respectively, showing that the design meets the specification amounts to showing that the language of the design automaton is contained in that of the specification automaton, i.e. $\mathcal{L}(D) \subseteq \mathcal{L}(S)$; equivalently, $\mathcal{L}(D \otimes S^c) = \phi$, where S^c denotes the complement of S, and $\otimes$ is the product operation. The language of a timed automaton is empty if and only if there is no input sequence accepted; that is, there is no traversable accepting path.

2   Previous Work

To verify timed automata, one can make several restrictions to reduce verification complexity. One is the allowable contents of timing constraints, and another the allowable placement of timing constraints. Allur and Dill [6, 3] restricted the contents of timing constraints to $x \leq k$ or $x \geq k$ or $x - y \leq k$, where x and y are clock variables and k is a constant. However there is no restriction on the placement of these constraints. It was shown that any timed automaton satisfying this restriction can be converted to an ordinary untimed automaton, called a region automaton, by augmenting the state space to account for the effects of timing constraints. With this conversion, a generic (untimed) verifier can verify the original timed automaton. However, the number of extra states in the conversion is proportional to the relative magnitudes of the time constants k's; thus verification complexity depends on the time constants. The works in [1, 11] proposed algorithms to minimize the number of states in the region automata.

Balarin and Sangiovanni-Vincentelli [4] considered the class of timed automata whose only timing constraints are on the amount of time an automaton can stay in a state. This amounts to restriction on both the contents and placement of timing constraints. With this simplification, the class of timed automata can be verified iteratively by adding timing constraints to eliminate traces that failed specifications in the preceding iteration. Each iteration uses a generic verifier. If all failure traces are eventually eliminated, the timing verification is successful. If there are still failure traces with all the timing constraints added or if a trace is produced that satisfies its timing constraints, the verification fails. A limitation of this class of timed automata is its expressiveness.

The approaches in [6, 3, 2] restrict the content of timing constraints, while the approach in [4] restricts both the contents and placement of timing constraints. A third method proposes alternating RQ timed automata, which allow *arbitrary* timing constraints but restrict their placement, [9]. It was shown that the traversal problem for alternating RQ timed automata can be reduced to simple paths. That is, a state is reachable from another state if and only if it is reachable through a *simple* path. Further, alternating RQ automata can be verified using generic verifiers with the introduction of constraining automata which delete untraversable simple paths. Moreover, verification complexity for this class is independent of the contents of timing constraints. A limitation is their expressiveness for general timing conditions. Although many practical situations can be represented with alternating RQ timed automata, the scope of this expressiveness is not known previously.

3   Timing Verification Paradigm Using Generic Automata

Here we discuss a paradigm of verifying timed automata using generic verifiers, first proposed by Kurshan [2] and used by others [4]. First, we examine the effects of timing constraints for *general* timed automata with arbitrary timing constraints and placement. Let P be the set of all paths, P_t the set of traversable paths, and P_c (c stands for constrained.) the set of untraversable paths; thus, $P = P_t + P_c$. Without timing constraints, $P_c = 0$. Hence, the contribution of the timing constraints is to partition the paths into P_t and P_c. Let M be a timed automaton, M_0 derived from M by removing all timing constraints, and M_c a generic automaton which accepts only the paths in P_c. Then, a path is traversable in M if and only if it is also a path in M_0 and is not accepted by M_c. Further, let M and M_0 have the same acceptance conditions. Then, the language of M is the language of M_0 minus that of M_c. Symbolically,

$$L(M) = L(M_0 \otimes M_c^c).$$

That is, timed automaton M is modelled by the product of M_0 and the complement of M_c. We call M_0 the untimed automaton of M, and M_c the constraining automaton of M, both of which are ordinary automata with no timing constraints. If M_c can be constructed, general timed automaton M can be verified using generic verifiers.

Using constraining automata to model timing constraints has the following advantages. First, timing verification can be performed using existing generic verifiers. Second, since not all timing constraints may effect verification, it may not be necessary to explicitly take into account all timing constraints. With the constraining automaton paradigm, timing constraints can be incorporated iteratively. At each iteration, new timing constraints can be added by creating a new constraining automaton, treating the present $M_0 \otimes M_c^c$ as a new M_0. So at the ith iteration, we check:

$$L(M_0(i) \otimes M_c^c(i)) = \phi.$$

If it is empty, it is verified successfully. Otherwise, we add more timing constraints. For the next stage $i + 1$, let $M_0(i + 1) = M_0(i) \otimes M_c^c(i)$, $M_c^c(i + 1)$ accept the paths made untraversable by the new timing constraints, and repeat the language emptiness checking. If no timing constraints are left and the language is not empty, it fails the verification. Adding new timing constraints can also be done by augmenting the acceptance condition of M_c to accept the paths made untraversable by the new timing constraints. An advantage of this iterative approach is that a design may be verified without using all timing constraints, translating to faster run time and smaller memory usage; of course, all timing constraints may have to be used to prove a design fails. Also, if the verification succeeds, M_c contains a minimal set of timing constraints. Then the design can be optimized by eliminating redundant timing constraints not in M_c or by relaxing over-constraining conditions in M_c. If the verification fails, M_c can be readily augmented to accommodate new timing constraints or a new M_c is created.

4 Complexity Reduction of M_c

The crucial step in the above paradigm is the construction of M_c. For general timed automata, enumerating all untraversable paths is difficult. To make this tractable, we restrict the placement of timing constraints for the following reason. We observe that for an arbitrary distribution of timing constraints, it is hard for the designer to determine the exact conditions being imposed; thus, placing timing constraints arbitrarily may result in over-constraining or incorrect specification. Hence, for many practical situations, regular patterns of timing constraints are observed.

Alternating RQ timed automata, defined in [9], are a class of timed automata that allow arbitrary timing constraints but restrict the patterns of resets and queries to be alternating. It is proved that alternating RQ timed automata have the so-called simple path property that, independent of timing constraints:

1. a state is reachable from another if and only if it is reachable via a simple path.
2. a loop is traversable infinitely often if and only if it is traversable once.

We illustrate the importance of the first property. Consider two automata, one S with the simple path property, and the other N without. When the generic verifier returns a simple error trace $a \rightarrow b \rightarrow c$ whose timing constraints can not be satisfied, for N we can only throw out the path $a \rightarrow b \rightarrow c$ and no other paths; however, for S we can throw out an infinite number of paths, i.e. any path whose projection to a simple path is $a \rightarrow b \rightarrow c$, e.g. $a \rightarrow b \rightarrow \ldots \rightarrow b \rightarrow c$, where ... represents any sequence of states. Hence, with this simple path property, M_c **can compactly represent the untraversable paths**, because to decide reachability of a state in the original timed automaton M, we can simply check the traversability of all the simple paths to the state; and if there is such a simple path not accepted by M_c, then the state is reachable; otherwise, i.e. if all simple paths to the state are accepted by M_c, then the state is not reachable. Because this simple path property is independent of timing constraints, reachability analysis, and thus verification complexity is independent of timing constraints.

The restriction on the placement of timing constraints means that alternating RQ timed automata can not express all timing conditions, and their exact modeling scope is not known. In this paper, we generalize alternating RQ timed automata and show that this generalized class is the largest class of timed automata that a) allow arbitrary timing constraints and b)traversability can be decided based on simple paths only, *independent of timing constraints.*

5   Reachability Analysis with Arbitrary Timing Constraints

Denote a reset of clock variables $x_1, \ldots, x_n$ by $R(x_1, \ldots, x_n)$, and a query on clock variables $x_1, \ldots, x_n$ by $Q(x_1, \ldots, x_n)$. We want to find a condition such that, if satisfied, reachability of a state can be decided efficiently. By arbitrary timing constraints, we mean that the function of $Q(x_1, \ldots, x_n)$ can be any function involving clock variables $x_1, \ldots, x_n$.

Definition 1. 1. Given a set of distinct states $S = \{s_1, \ldots, s_n\}$ and an order on s_i's, e.g. $s_1, \ldots, s_n$, a path **traverses** S if the path traverses the states in S according to the order. A **simple path through** S is a path such that every state appears at most once in the sub-path between two consecutive s_i's, e.g. s_i and s_{i+1}.
2. If path π traverses through a set of states S, let $Sim(\pi)$ denote the set of all simple paths through S derived from path π by deleting cycles.
3. A **RQ sequence** of a path is the sequence of resets and queries on the path.
4. A **symbol** in an RQ sequence is either a reset or a query.
5. The **interarrival** time between two symbols is the amount of time between when the transition of the first symbol is activated and when the transition of the second symbol is activated.

6. The **support** of a query $Q(x_1, \ldots, x_n)$ is the set of clock variables in $Q(x_1, \ldots, x_n)$, i.e. $x_1, \ldots, x_n$.

7. Given a path, a reset $R(x_i)$ is **effective** if after $R(x_i)$ there is a $Q(x_1, \ldots, x_n)$ whose support contains x_i and there is no other $R(x_i)$ between them. That is, $R(x_i)$ is the closest reset on x_i preceding $Q(x_1, \ldots, x_n)$.

8. The **effective RQ sequence** of a path is derived from the RQ sequence of the path by deleting all ineffective resets.

9. Consider two symbols c_1 and c_2 in an RQ sequence. Let $c_1 \prec c_2$ denote that c_1 appears before the transition of c_2, $c_1 = c_2$, that c_1 and c_2 appear on the same transition, and $c_1 \preceq c_2$, that c_1 appears before or on the same transition as c_2, Let Γ_1 and Γ_2 be two RQ sequences. Γ_1 **dominates** Γ_2 if the effective RQ sequences of Γ_1 and Γ_2 have the same set of symbols and $c_1 \preceq c_2$ in the effective RQ sequence of Γ_2 implies $c_1 \preceq c_2$ in that of Γ_1, i.e. $c_1 \preceq c_2$ in the effective Γ_1 may correspond to $c_1 = c_2$ in the effective Γ_1.

10. Let $\pi_l = v_a, \ldots, v_k, v_l, \ldots, v_l, v_m, \ldots, v_z$ contain cycle $L = v_l, \ldots, v_l$, and $\pi = v_a, \ldots, v_k, v_l, v_m, \ldots, v_z$ be derived from π_l by deleting L. π_l is π with **cycle expansion (in path)**. Path π with **cycle reset expansion** to π_l is the path π_l except that all queries in the cycle L are removed. Queries not in L remain.

11. The RQ order of path π is **preserved under cycle reset expansion** if the effective RQ sequence of π dominates the effective RQ sequence of π with cycle reset expansion.

Example 2. Consider the automaton in Figure 1. The seven symbols $\{c_1, \ldots, c_7\}$ are:

$$\{R(x_1), R(x_2), R(x_3), x_1 = 0?, R(x_1), x_3 = 1?, x_2 = K? \wedge x_1 = 0?\};$$

Symbols c_1 and c_5 are both equal to $R(x_1)$. The RQ sequence of the path S_1, S_2, S_3, S_2, S_4 is

$$R(x_1) \wedge R(x_2), R(x_3) \wedge x_1 = 0?, R(x_1) \wedge x_3 = 1?, x_2 = K? \wedge x_1 = 0?,$$

in symbols,

$$c_1 \wedge c_2, c_3 \wedge c_4, c_5 \wedge c_6, c_7.$$

Reset $R(x_1)$ of c_5 is effective, because it is followed by c_7, a query involving x_1, and there is no other $R(x_1)$ between them. Reset $R(x_1)$ of c_1 is also effective, because it is followed by a query involving x_1, namely c_4 and there is no other $R(x_1)$ c_1 and c_4.

Let t_i be the interarrival time between the ith and the $i + 1$th symbols. For example, t_2 is the time between $R(x_2)$ and $R(x_3)$, i.e. the time between the transition $S_1 \rightarrow S_2$ and the transition $S_2 \rightarrow S_3$. Because c_3 and c_4 are on the same transition, $t_3 = 0$. The inequalities of all queries involve interarrival times only. For example, the query of $c_7 = (x_2 = K? \wedge x_1 = 0?)$, in above the RQ sequence, produces the following constraints:

$$t_2 + t_3 + t_4 + t_5 + t_6 = K$$
$$t_5 + t_6 = 0$$
$$t_3 = 0$$
$$t_5 = 0$$

Consider the path $\pi = S_1, S_2, S_4$ and a reset expansion of the cycle S_2, S_3, S_2. The RQ sequence of path π is:

$$(c_1, c_2, c_7) = (R(x_1), R(x_2), x_2 = K? \wedge x_1 = 0?),$$

which is effective, i.e. all resets on π are effective. A reset expansion of the cycle S_2, S_3, S_2 adds the reset in the cycle, namely c_3 and c_5, to the RQ sequence of path π, resulting the RQ sequence of path π with the cycle reset expansion:

$$(c_1, c_2, c_3, c_5, c_7) = (R(x_1) \wedge R(x_2), R(x_3), R(x_1), x_2 = K? \wedge x_1 = 0?).$$

Now, the reset $R(x_1)$ of c_1 is ineffective, because there is another $R(x_1)$, namely c_5 between it and its succeeding query c_7. Reset $R(x_3)$ of c_3 is also ineffective, because there is no succeeding query involving x_3. Eliminating these two ineffective resets, the effective RQ sequence of path π with the cycle reset expansion is:

$$(c_2, c_5, c_7) = (R(x_2), R(x_1), x_2 = K? \wedge x_1 = 0?).$$

This effective RQ sequence is not dominated by that of π, because $R(x_2) \preceq R(x_1)$ in this sequence but $R(x_1) \preceq R(x_2)$ in that of path π (note that c_1 and c_5 are the same). Thus, the RQ order of path π is not preserved under cycle reset expansion.

Theorem 2. *If the RQ orders of all simple paths from state s_1 to state s_2 are preserved under single cycle reset expansion, then s_2 is reachable from s_1 if and only if it is reachable through a simple path.*

In verification, sometimes traversability of a set, or a subset, of states needs to be decided, e.g. cycle sets in L-automata. The orders of the states in the set to be traversed may or may not be specified. A path is said to traverse through the states in a given set of states if the path traverses the states in the set in a specified order (if any). We extend the simple path property between two states to a set of states. By a simply path through a set of given states, we mean that the sub-path between any two consecutive states is a simply path. Note that a simple path through a set of states may traverse a state (not in the set) more than once, i.e. a simple path through a specified set of states may contain a loop. For L-automata, a cycle set is traversable if there is a traversable cycle of a subset of states consisting of only the states in the cycle set.

Theorem 3. *Given a set of states S, if the RQ orders of all simple paths through S are preserved under single cycle reset expansion, then the set of states in S are traversable if and only if they are traversable through a simple path.*

6   Scopes and Classes of Simple Path Timed Automata

The following simple path property allows efficient reachability analysis and hence verification. Here, we find classes of timed automata with arbitrary timing constraints that have the simple path property. By arbitrary timing constraints, we mean queries of the form $Q(x_1, \ldots, x_n)$ where $Q(x_1, \ldots, x_n)$ can be any function of the clock variables in the support of $Q(x_1, \ldots, x_n)$ only, and the support of $Q(x_1, \ldots, x_n)$, namely $x_1, \ldots, x_n$, can be a subset of all clock variables in an automaton. For example, if $Q(x_1, x_2)$ is a query in an automaton having clock variables x_1, x_2, x_3, then $Q(x_1, x_2)$ can be any function involving only x_1 or x_2 or both, but not x_3. For this case, we give a sufficient condition for a timed automaton to have the simple path property. However, if there are no restrictions on both the support and form of the queries, i.e. a query can be any function of any subset of the clock variables in an automaton, we give a necessary and sufficient condition for a timed automaton to have the simple path property. We call the following property the **simple path property** for timed automata:

1. A set of states is traversable if and only if there is a traversable simple path through the entire set.
2. A cycle of states is traversable infinitely often if and only if it is traversable once through a simple path.

The first statement of the simple property becomes "a state is reachable from another state if and only if it is reachable via a simple path" when the set of states consists of two states and the order is that one state is before the other.

Definition 4. 1. A cycle is **separable** if, for each clock, there is a state (entry state) in the cycle such that in traversing the cycle, starting at that state, each query occurs after all its effective resets.

If a cycle is not separable, then, every time the cycle is entered, a query in the cycle is encountered before some of its resets in the cycle are encountered; thus, in the first time the cycle is entered, that query involves interarrival times outside of the cycle. The query can have or have no resets in the cycle. If it does, second and further traversals of the cycle involve constraints on the interarrival times inside the cycle. Hence, the first traversal constrains the interarrival times outside the cycle, while further traversals constrain the interarrival times inside the cycle; this partitioning of constraints could be done more systematically by using different clocks and queries for those outside and inside the cycle. If the query does not have a reset inside the cycle, then the number of times the cycle can be traversed is timing constraint dependent, because if the query is of the form $x < k$ then after a finite number of traversals the query will not be satisfied (it is reasonable to assume that every cycle in a real system takes a finite amount of time to traverse). Therefore, to have a condition on infinite cycle traversal with arbitrary timing constraints, it is mild to assume that every cycle has an entry state. All timed automata in literature meet this assumption, except the one in Figure 1.

Theorem 5. *Assume that every cycle is separable in a timed automaton with arbitrary timing constraints. The timed automaton has the simple path property if the RQ sequence of each simple path is preserved under single cycle reset expansion.*

Note that only those cycles of states that need to be traversed infinitely often, e.g. those in a cycle set of a L-automaton, need to be separable. The condition that the RQ order of each simple path is preserved under single cycle reset expansion is not a necessary condition to have the simple path property. However, if we augment the supports of all queries to be all clock variables in a timed automaton, i.e. a query can be an arbitrary function of any subset of all clock variables, then the above is necessary and sufficient.

Example 3. The timed automaton in Figure 1 does not satisfy the RQ order preserving condition under single cycle reset expansion, because by expanding the loop, resets $R(x_1)$ and $R(x_3)$ are added to the RQ sequence of the simple path. But the newly added reset $R(x_3)$ is not effective because there is no query on the simple path involving x_3. The newly added $R(x_1)$ becomes effective and eliminates the original $R(x_1)$ from the effective RQ sequence. The order of $R(x_1)$ and $R(x_2)$ have changed after the cycle expansion, from $R(x_1) = R(x_2)$ to $R(x_2) \preceq R(x_1)$, violating the RQ order preserving condition; hence, the timed automaton is not guaranteed to have the simple path property and, in this case, does not have the simple path property.

Theorem 6. *Assume that the queries in a timed automaton can be an arbitrary function of **any** subset of clock variables and that each cycle is separable. The timed automaton has the simple path property if and only if the RQ order of each simple path is preserved under single cycle reset expansion.*

The above theorems show that, assuming separable cycles, the class of timed automata satisfying the RQ order preserving condition is a large class allowing arbitrary timing constraints and having the simple path property, and is the *largest* such class when timing constraints can be functions of any clock variable.

7 Example Class: Alternating RQ Timed Automata

A class of timed automata satisfying the RQ order preserving condition for simple paths is the alternating RQ timed automata [9], which satisfy the following two properties:

1. For each clock x_i, there is only one pair of $R(x_i)$ and $Q(\ldots, x_i, \ldots)$, i.e. a distinct clock is used for each query.
2. For each path π starting from an initial state, the RQ sequence of each clock on π alternates.

In [9], it is proved that alternating RQ timed automata have the simple path property. We would like to know whether this property can also be deduced from the RQ order preserving condition. An alternating RQ automaton has only one pair of reset and query for each clock variable. The alternating condition forces the pair to be in a cycle together if either one is in the cycle. Therefore, in an expansion of a simple path to include a cycle, there can be no query, on the simple path, that involves a reset variable from the cycle; thus, the resets from the cycle are not effective and do not alter the RQ order of the RQ sequence on the simple path. Hence the RQ sequence on the simple path is preserved under single cycle reset expansion. Therefore, alternating RQ timed automata have the simple path property and are a sub-class of simple path automata.

The timed automaton in [3] describing the language

$$\{((ab)^\omega, \tau) : \exists i, j \geq i : \tau_{2j+2} \leq \tau_{2j+1} + 2\}$$

is not an alternating RQ timed automaton (Figure 2) (and cannot be converted into one), but satisfies the RQ order preserving condition and thus is a simple path timed automaton. For example, the RQ order of the effective RQ sequence of the simple path S_0, S_2, S_3 under single cycle reset expansion, e.g. cycle S_2, S_1, S_2, remains the same, i.e. $R(x)$, $x <= 2?$. Therefore, the RQ order preserving condition is more general than the RQ alternating condition.

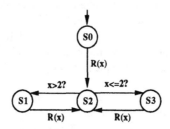

Fig. 2. A Non-alternating RQ Timed Automaton

However, checking whether a timed automaton satisfies the RQ order preserving condition is more complicated than checking the condition for alternating RQ timed automata which can be done with a simple graphical criterion [9]. Hence, it may be desirable to generalize alternating RQ timed automata. An extension is to relax the requirement that there is only one query for each clock variable.

DEFINITION (Extended Alternating RQ Timed Automata)

1. *For each clock x_i, there is only one $R(x_i)$.*
2. *Each path π from an initial state starts with a reset $R(x_i)$ and every reset $R(x_i)$ is followed by a query involving x_i.*
3. *For each cycle that needs to be traversed infinitely often, a query on x_i in the cycle implies a reset $R(x_i)$ also in the cycle.*

The second condition says that every path's first symbol is a reset so that all the clock variables involved in a query on the path are reset before being queried. Of course, if x_i is never queried on a path, $R(x_i)$ may be missing on the path. The third condition is required only for the cycles whose infinite traversability is a part of the specifications to be verified, e.g. cycles with fairness constraints. That is, cycles other than those in the specifications may not have to meet this requirement. This condition ensures that infinite traversal of the cycle is timing constraint independent. Without this condition, a cycle can easily depend on timing constraints. For example, if a cycle has a query of the form $x_i < k$ and not a reset $R(x_i)$, then the cycle can be traversed only a finite number of times because each traversal takes a finite amount of time for a real system; then after a finite number of traversals, the query $x_i < k$ will never be satisfied.

Theorem 7. *The extended alternating RQ timed automata have the simple path property.*

With this extension, the timed automaton in Figure 2 can be converted into an extended alternating RQ timed automaton, as shown in Figure 3.

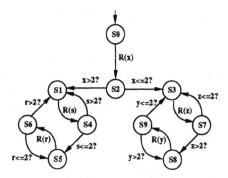

Fig. 3. Extended Alternating RQ Timed Automaton

In addition, this extension allows the use of the following common branching condition, which sends the automaton to the appropriate state depending on the value of the clock variable x. It is assumed that paths are reset before entering the *switch* statement and once going through one of the transition T_i, $i = 1, \ldots, n$ the paths can only re-enter the statement through the *switch*.

```
switch (x):{
    case x < k₁:              activate transition T₁.
    case k₁ < x < k₂:         activate transition T₂.
    ⋮
    case kₙ₋₁ < x < kₙ:        activate transition Tₙ.
}
```

This branching condition is not allowed in the unextended alternating RQ timed automata, because there is more than one query on clock variable x.

Checking whether a timed automaton is an extended alternating RQ timed automaton is similar to that for (unextended) alternating RQ timed automata, [9]. The following theorem expresses the alternating condition in terms of *cut* requirements on the reset and the query edges.

Definition 8. In a graph, a set of edges $\{e_i\}$ is a **cut** for an ordered vertex pair (v_1, v_2) if either v_2 is not reachable from v_1 or the removal of all the edges in $\{e_i\}$ from the graph makes v_2 unreachable from v_1. Denote the cut by $\{e_i\}|(v_1, v_2)$.

Thus, if $e_i|(v_1, v_2)$, then all paths from v_1 to v_2 must pass through e_i.

Theorem 9. *Assume that a given timed automaton with initial states $\{s_j\}$ satisfies the extended alternating RQ condition 1 and 3 (which can be easily checked). It is an extended alternating RQ timed automaton, if and only if for each clock x and each initial state s_j,*

$$e_r|(s_j, e_{q_i}^1), \quad \{e_{q_i}\}|(e_r^2, e_r^1).$$

where $e_r = (e_r^1, e_r^2)$ is the edge where $R(x)$ resides, and $e_{q_i} = (e_{q_i}^1, e_{q_i}^2)$, where $Q_i(x)$ resides.

The first condition of Theorem 9, a cut requirement on reset edges, requires that every path from an initial state encounters a reset of a clock before its query. The second condition, a cut requirement on query edges, says that a rest on x must be followed by at least a query on x before the reset is encountered again. All these conditions involve reachability analysis in (untimed) graphs; thus, they can be easily implemented using techniques such as depth first search.

8 Application in Non-simple Traversability

Here, we examine how the RQ order preserving condition can be used in deciding traversability in general timed automata which may not have the simple path property. First, even if a timed automaton does not have the simple path property, the RQ order preserving condition can detect the simple paths that violate these condition and only these simple paths need to be examined in more details, possibly including non-simple paths. Second, a general timed automaton can be iteratively converted to a simple path timed automaton by adding simple paths which are cycle expansions in path of the simple paths that violate the RQ order preserving condition. Then verification can be done using the above paradigm.

Example 4. The timed automaton in Figure 1 does not satisfy the RQ order preserving condition; we try to convert it to one satisfying the condition by adding simple paths which are cycle expansions of the loop. Since the simple path from s_1 to s_4 is not traversable and does not satisfy the RQ order preserving condition, meaning going through the loop may help, we expand the cycle once to obtain the timed automaton in Figure 4. The new simple path again does not satisfy the RQ order preserving condition and is not traversable. This procedure may be repeated. For this particular timed automaton, each newly introduced simple path does not satisfy the RQ order preserving condition, because each expansion changes the order $R(x_1) = (x_3 = 1?)$ to $(x_3 = 1?) \preceq R(x_1)$; thus, no finite number of cycle expansions will convert this automaton to a simple path automaton. This is expected, because the number of times the loop needs to be traversed depends on the time constant K, i.e. it is timing constraint dependent. If K is an integer, then the automaton with the loop expanded K times has a traversable path from S_1 to S_4.

9 Simple Path Property of Composite Timed Automata

As discussed, preservation of RQ orders under cycle reset expansion of timed automata implies the simple path property. Here we examine the simple path property of product timed automata, namely, whether preservation of RQ orders under cycle reset expansion in component timed automata implies the simple path property in the product timed automata. Let $M = M_1 \times \ldots \times M_n$ be a product timed automaton of components $M_1, \ldots, M_n$.

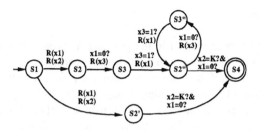

Fig. 4. Non-simple Path Conversion

Definition 10.   1. Let π be a path in M. The **projected path** of π on M_i, denoted by $\pi|_{M_i}$ is the sequence of states in M_i visited by π.

Theorem 11. *Assume M is a product timed automaton with arbitrary timing constraints and every cycle of states in M is separable. If π_s is a simple path in the product timed automaton M and the RQ order of $\pi_s|_{M_i}$ is preserved under single cycle reset expansion in M_i $i = 1, \ldots, n$, then M has the simple path property.*

Note that a simple path π_s in M does not imply that $\pi_s|_{M_i}$ in M_i is a simple path. Thus, non-simple paths in M_i may need to be checked for RQ order preservation under single cycle reset expansion. However, if every simple path in M consists of simple paths in the component automata, i.e. π_s in M implies that $\pi_s|_{M_i}$ is a simple path in M_i, then RQ order preservation of the simple paths in the component automata implies RQ order preservation of the simple paths in the product automaton.

Corollary 12. *Assume that every simple path in M consists of simple paths in all of the component automata and every cycle of states in M is separable. If the RQ order of every simple path in M_i is preserved under single cycle reset expansion, then M has the simple path property with arbitrary timing constraints.*

10   Construction of Constraining Automaton, M_c

As discussed in Section 3, verification of timed automata can be done with generic verifiers if constraining automata M_c can be constructed. Constraining automata accept the paths that can not be traversed, because they are blocked by timing constraints. The complexity of constraining automata is drastically reduced if the associated timed automata have the simple path property because only the untraversable simple paths need to be represented. The construction consists of two steps. First, untraversable simple paths are identified by determining satisfiability of the queries on simple paths. If the timing constraints are linear, linear programming can be used for this purpose. Once all the untraversable simple paths (from initial states) are determined, constraining automata are built to recognize these paths. Heuristics may be used to further reduce the complexity of constraining automata by identifying untraversable partial paths. In [9], a more detailed explanation and algorithms are given for constructing constraining automata and their use in verifying alternating RQ timed L-automata.

11 Conclusion

Timed automata have been studied in the past and have been found to have a complexity dependent on the relative values of the time constants involved in the timing constraints imposed, even if the timing constraints are restricted to the form $x < k$. We have previously shown that this complexity dependence on the time constants can be eliminated if the timed automaton has the simple path property and proposed alternating RQ timed automata which have this property. In this paper, we gave conditions for a timed automaton with arbitrary constraint equations to have the simple path property. As far as we know all practical examples in the literature meet these criteria. We are currently working on an efficient implementation for timed automata where arbitrary linear inequalities among the clock values are allowed. Using linear programming, we iteratively detect simple paths which are not traversable and construct untimed automata which disallow these paths. The present paper serves to extend this approach to a wide class of applications. In addition, we defined extended RQ timed automata which include all the examples in the literature and are easily tested for this property.

References

1. R. Alur, C. Courcoubetis, N. Halbwachs, D. Dill, and H. Wong-Toi. Minimization of timed transition systems. *International Conference on Computer-Aided Verification*, 1992.
2. R. Alur, A. Itai, R. Kurshan, and M. Yannakakis. Timing verification by successive approximation. *International Conference on Computer-Aided Verification*, 1992.
3. Rajeev Alur and David Dill. Automata for modeling real-time systems. *1990 ACM International Workshop on Timing Issues In the Specification and Synthesis of Digital Systems*, 1990.
4. Felice Balarin and Alberto Sangiovanni-Vincentelli. A verification strategy for timing constrainted systems. *International Conference on Computer-Aided Verification*, 1992.
5. E. Clarke, O. Grumberg, and R. Kurshan. A synthesis of two approaches for verifying finite state concurrent systems. *Workshop on Automatic Verification Methods for Finite State Systems*, 1989.
6. David Dill. Timing assumptions and verification of finite-state concurrent systems. *Workshop on Automatic Verification Methods for Finite State Systems*, 1989.
7. R.Kurshan E.M.Clarke, I.A.Draghicescu. A unified approach for showing language containment and equivalence between various types of ω-automata. *Tech. report, CMU,*, 1989.
8. R. Hojati, H. Touati, R. Kurshan, and R. Brayton. Efficient ω-regular language containment. *International Conference on Computer-Aided Verification*, 1992.
9. W. Lam and R. Brayton. Alternating rq timed automata. *International Conference on Computer-Aided Verification*, 1993.
10. W. Lam and R. Brayton. Criteria for the simple path property in timed automata. *UC Berkeley ERL memorandum: UCB/ERL*, 1994.
11. Mihalis Yannakakis and David Lee. An efficient algorithm for minimizing real-time transition systems. *International Conference on Computer-Aided Verification*, 1993.

Hierarchical representations of discrete functions, with application to model checking

K. L. McMillan

AT&T Bell Laboratories
Murray Hill, NJ

Abstract. BDD trees provide a hierarchically structured canonical representation for boolean functions, based on ordered binary decision diagrams (OBDD's). We describe algorithms for function application and boolean quantification on BDD trees, allowing them to be used in applications such as symbolic model checking. Experimentally, we find that BDD trees can be greatly more efficient than ordinary OBDD's in verifying tree structured systems using symbolic model checking. In one case, sublinear growth is observed in the size of the transition relation representation. Analytically, we find that for a class of circuits of fixed tree width, BDD trees are asymptotically efficient.

1 Introduction

Binary decision diagrams have been widely used as a representation for boolean formulas in the verification of digital systems [Bry86]. They can be applied to the comparison of switching functions, or to the verification of temporal properties using a technique called symbolic model checking [BCM+90]. The sharing of substructure within BDD's allows them to efficiently capture and exploit certain kinds of regularity within the truth table of a boolean function. Heuristically, they provide a compact representation for most switching functions occurring in digital systems, given an appropriate total ordering on the boolean variables.

OBDD's are asymptotically efficient for any class of circuits of fixed *path width*. The path width is the optimum wiring channel width for any linear arrangement of the circuit elements. For such a class of circuits, the size of the optimum OBDD representing the circuit's function is linearly bounded in the size of the circuit. Tree width is the analog of path width for tree structured arrangements. For fixed tree width, the optimum OBDD size is polynomial in the circuit size, but of degree that is exponential in the tree width [McM92].

This paper proposes a canonical representation that is *linear* in the tree width and like OBDD's, has quadratic algorithms for applying logical operations. It is based on the notion of a hierarchical decomposition of a discrete function. The representation uses a generalized form of binary decision diagrams as part of its structure. For this reason, we will refer to representations of this type as *BDD trees*. In some sense, BDD trees are a natural generalization of OBDD's from linear structure to tree structure. In fact, one can draw an analogy from BDD trees to deterministic finite tree automata in much the same way one can from OBDD's to ordinary DFA's.

To use the BDD tree representation, one must provide a hierarchical decomposition of the boolean variables rather than the linear order required for ordered binary decision diagrams. This decomposition is easily obtained in the case of hierarchically structured circuits, since it simply follows from the modular structure of the circuit. Two examples are presented here of the use of BDD trees in verifying tree structured systems by symbolic model checking. In one case, the ability of BDD trees to capture redundancy in the system hierarchy results in logarithmic growth in the representation of the transition relation as the system is scaled up. Significant speedups are obtained over the OBDD method, even for small system sizes. In the other case, BDD trees are significantly better for representing the transition relation, but do not improve the performance of checking temporal logic formulas.

2 Informal description of BDD trees

A BDD tree is essentially a hierarchical decomposition of a discrete function. As an example of a hierarchical decomposition, suppose that we wish to represent a function of some set of boolean variables. We can divide the variables into three parts – those variables belonging to the *root* of the hierarchy, those belonging to the *left child* of the hierarchy, and those belonging to the *right child*. The truth assignments to the left child variables can be partitioned into equivalence classes. Two such partial truth assignments are in the same equivalence class if they are equivalent *cofactors* of our function – that is, if fixing the values of the left child variables according to the two truth assignments yields the same function of the remaining variables. These equivalence classes of truth assignments to the left child variables can be numbered in some canonical way. We can then define a discrete function for the left child of the hierarchy that takes a truth assignment and returns the number corresponding to its equivalence class. This left child function tells us exactly the information we need to know about the left child variables in order to evaluate the function. We can define a similar function for the right child variables. The root function then takes the result of the right and left child functions, and an assignment to the root variables, and returns the value of the original function. This decomposition of the function is depicted in figure 1.

Assuming we have fixed a canonical numbering of the equivalence classes, this hierarchical representation is canonical. That is, for each function we have a unique representation. The left and right child functions may themselves be represented hierarchically by dividing their variables into left, right and root functions recursively, until the child functions become constants (*i.e.*, functions of zero variables). Such a decomposition leaves us only with the question of how to represent the root function at each point in the hierarchy. For this, we will use *ordered binary decision diagrams* (OBDD's), in order to take advantage of the very simple and efficient algorithms available for manipulating them [Bry86]. To represent the root function, we will use a matrix of OBDD's, with one row for each left child equivalence class and one column for each right child equivalence class. Thus, the left child variables determine the row of the matrix, the right child variables determine the column, and the OBDD in the given row and

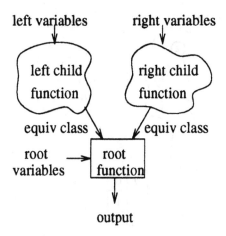

Fig. 1. Hierarchical decomposition of a function.

column represents the resulting function of the remaining variables. Since the child functions are numeric valued rather than boolean, and are also represented recursively using OBDD trees, we will generalize OBDD's slightly, using natural numbers as the OBDD leaves rather than booleans. In general, a *boolean function* in this paper will refer to a function that takes a truth assignment to some boolean variables and returns a natural number, rather than a truth value.

An example of the BDD tree representation is shown in figure 2. This BDD tree represents the boolean function $ab + cd$. The left child contains variables a and b, the right child variable c and the root variable d. Note that the truth assignments $\{00, 11, 10\}$ to ab fall into the same equivalence class, which is numbered 0. The remaining equivalence class, containing only 11 is numbered 1. This reflects the canonical numbering of the equivalence classes. We fix a total order on the boolean variables, in this case (a, b, c, d). This defines a lexical order on truth assignments. That is, two truth assignments are ordered based on the first variable in which they differ. This is the order in which the truth assignments would appear in a dictionary if represented as boolean vectors. The equivalence classes are numbered starting from zero, ordered by the least truth assignment they contain. Hence, the set $\{00, 01, 10\}$ comes before $\{11\}$.

3   Formal definition of BDD trees

We now give a formal definition of the BDD tree representation, defining a function $\langle\rangle$ that maps each boolean function onto its BDD tree representation. We prove that this function is invertible. Hence two functions are equal exactly when their BDD tree representations are equal. Each definition will be illustrated using the example function, $f = ab + cd$.

Definition 1. Let $V = \{v_1, \ldots, v_n\}$ be an ordered set (of variables).

For our example function, let $V = \{a, b, c, d\}$.

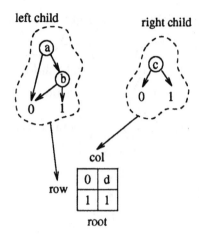

Fig. 2. BDD tree example (the functions in the matrix are also represented by OBDD's, though this representation is not shown).

Definition 2. Let t be a function $V \to \{0,1\}^*$.

This function defines the decomposition. The function t maps every variable to its "tree address", which is a boolean vector. We can think of the boolean vector as a vertex in a binary tree. If ν is a vertex of the tree, then $\nu 0$ is the left child of ν and $\nu 1$ is the right child of ν. The empty string ϵ is the root. In our example, $t(a) = t(b) = 0$, $t(c) = 1$ and $t(d) = \epsilon$. This is because a and b are assigned to the left child, c to the right child, and d to the root. If the left child in turn had a left child, its tree address would be 00.

Definition 3. For any $\nu \in \{0,1\}^*$ let $S(\nu)$ be the set

$$\{v \in V \mid \nu \text{ is a prefix of } t(v)\}.$$

If ν is a binary tree vertex, $S(\nu)$ is the set of variables assigned to any descendent of ν. In our example, $S(0) = \{a,b\}$, while $S(\epsilon) = \{a,b,c,d\}$. Now we define the lexical order on truth assignments that gives us the canonical numbering of equivalence classes:

Definition 4. For any functions $\sigma, \sigma' : V \to \{0,1\}$, $\sigma < \sigma'$ exactly when there is some $1 \le i \le n$ such that $\sigma(v_i) = 0$ and $\sigma'(v_i) = 1$ and for all $1 \le j < i$, $\sigma(v_j) = \sigma'(v_j)$.

This is just the ordinary lexical order on truth assignments expressed as boolean vectors. So, for example, $0010 < 0011$. Now we define what it means to *cofactor* a function by a partial truth assignment:

Definition 5. For any function $f : (V \to \{0,1\}) \to \mathbf{N}$, any $V' \subseteq V$ and any $\sigma : V' \to \{0,1\}$, let $f\sigma$ be the function $(V \setminus V' \to \{0,1\}) \to \mathbf{N}$ such that $(f\sigma)(\rho) = f(\sigma \cup \rho)$.

In the example, suppose σ is a partial truth assignment that assigns 01 to ab. Then $f\sigma = 01 + cd = cd$.

Definition 6. For any function $f : (V \to \{0,1\}) \to \mathbf{N}$, any vertex $\nu \in \{0,1\}^*$ and any truth assignment σ to V, let $f_\nu(\sigma) = |\{f\sigma' \mid \sigma' \le \sigma\}| - 1$.

The function f_ν is the child function at vertex ν of the tree. It yields the number of the equivalence class of a given truth assignment σ to $S(\nu)$. This is one less than the number of functions that can be obtained by cofactoring f by truth assignments to $S(\nu)$ that are lexically less than or equal to σ. In our example, $S(0) = \{a, b\}$. The equivalence classes of assignments to ab are, in lexical order, $\{00, 01, 10\}, \{11\}$. The first equivalence class corresponds to the cofactor cd, while the second corresponds to 1 (true). Thus, for example $f_0(01) = 0$ and $f_0(11) = 1$.

Definition 7. For any function $f : (V \to \{0,1\}) \to \mathbf{N}$ and any $\nu \in \{0,1\}^*$, let $|f|_\nu = |\{f\sigma \mid \sigma : S(\nu) \to \{0,1\}\}|$.

The notation $|f|_\nu$ stands for the number of equivalence classes of f at vertex ν. In the example, $|f|_0 = 2$.

Definition 8. For any function $f : (V \to \{0,1\}) \to \mathbf{N}$ and any $\nu \in \{0,1\}^*$ and any $0 \le j < |f|_\nu$, let f_ν^j be the lexically least σ, such that $f_\nu(\sigma) = j$.

Thus, f_ν^j is a representative of the jth equivalence class of assignments at vertex ν. That fact that is it the least in its class is immaterial. In the example, $f_0^0 = 00$ and $f_0^1 = 11$.

The function R_ν, defined below, yields the BDD tree representation of a function f at binary tree vertex ν. This representation is a triple, consisting of the root matrix, the left and the right child functions.

Definition 9. For any function $f : (V \to \{0,1\}) \to \mathbf{N}$ and any $\nu \in \{0,1\}^*$, let

- $R_\nu(f) = x$ if f is a constant function yielding x, for $x \in \mathbf{N}$,
- otherwise $R_\nu(f)$ is a triple (M, c_0, c_1) where M is a $|f|_{\nu 0} \times |f|_{\nu 1}$ matrix,
  - $M(j, k) = f f_{\nu 0}^j f_{\nu 1}^k$
  - $c_0 = R_{\nu 0}(f_{\nu 0})$
  - $c_1 = R_{\nu 1}(f_{\nu 1})$

In the example, $R_{00}(f_{00}) = 0$. This is because $S(00)$ is the empty set, hence there is only one equivalence class of truth assignments to $S(00)$, hence f_{00} is a constant function returning 0. Likewise $R_{01}(f_{01}) = 0$. Thus $|f|_{00} = |f|_{01} = 1$ (that is, both children of 0 have only one equivalence class). This means that $R_0(f_0)$ is a triple (M, c_0, c_1), where M is a one-by-one matrix and $c_0 = c_1 = 0$. The only element of this matrix is just the function f_0, since f_{00}^0 and f_{01}^0 are both empty assignments.

Definition 10. For any function $f : (V \to \{0,1\}) \to \mathbf{N}$, let $\langle f \rangle = R_\epsilon(f)$.

Now consider $R_\epsilon(f)$ in our example. Since $|f|_0 = |f|_1 = 2$, M is a 2-by-2 matrix. The 01 element of this matrix, for example, is f cofactored by f_0^0 and f_1^1 (that is, by representatives of equivalence class number 0 of the left child and equivalence class number 1 of the right child). This gives us 00 for ab and 1 for c, which yields $00 + 1d = d$. See the figure 2 for the rest of the representation.

We will refer to $\langle f \rangle$ as the *abstract representation* of f. The *concrete representation* will use OBDD's to stand for the boolean functions occurring in the matrices. For most purposes, there will be no need to distinguish between the two.

Theorem 11. $\langle \rangle$ *is one-to-one.*

4 BDD tree algorithms

In this section, we set out algorithms that allow us to build to the concrete BDD tree representations for the boolean functions of circuits. The boolean functions can be generated from literals using a suitable set of basis functions (*i.e.*, gates). A literal is a boolean function that yields true for σ if a given variable v is true in σ, and false if v is false in σ. Thus, there is a literal for each variable, a literal being a boolean function, whereas a variable is just an arbitrary element from a finite collection. As to basis functions, the "nand" function by itself is a complete basis. "And", "or" and "not" are also complete. To generate the representation of any circuit, we require algorithms to generate the representation of literals, and to apply the basis functions to the representations of their arguments.

We begin with the literals. The representation of a literal consists of one OBDD node at a suitable place in the hierarchy. For example, figure 3 shows a BDD tree representing the literal v, where the tree address of v is $t(v) = 01$. Literals are constructed by the following function:

function BDDT_literal(v,t)
    if $t = \epsilon$ **return** $([\text{BDD\_literal}(v)], 0, 0)$
    else if $\text{head}(t) = 0$
        return BDDT_find($\begin{bmatrix} 0 \\ 1 \end{bmatrix}$, BDDT_literal(v, tail(t)), 0)
    else return BDDT_find($[\,0\ 1\,]$, 0, BDDT_literal(v, tail(t)))
end

Note that all the leaves of the BDD tree are the boolean function 0. In the above procedure, the function BDD_literal returns an OBDD consisting of one decision node labeled with v. The function BDDT_find looks up a triple, consisting of a matrix and two BDD tree pointers, in a hash table. Finding a match, it returns the previously created BDD tree structure, otherwise it creates a new one and returns it. This guarantees that *two BDD trees that are equal always have the same memory address*. Thus equality comparisons between BDD trees can be made in constant time. Significant space can also be saved by sharing of subtrees, as we will see later. If the matrix M is simply $[n]$, where n is a constant, then BDDT_find returns n.

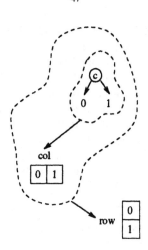

Fig. 3. BDD tree representing the literal v.

The function "apply" is used to combine two BDD trees using a given basis function. For example, given the BDD trees for a and b, "apply" can be called to build the BDD tree for $a \wedge b$ or $a \vee b$. We will consider the case of binary operations only. Algorithms for unary and ternary operations can easily be derived. The binary "apply" function takes any binary function o on the natural numbers, and the representation of two boolean functions $\langle f \rangle$ and $\langle g \rangle$, and returns $\langle o(f, g) \rangle$ (where $o(f, g)(\sigma) = o(f(\sigma), g(\sigma))$). A major component of "apply" is a function BDDT_pairs that takes two BDD trees representing functions f and g, and returns the list of pairs (j, k), such that for some truth assignment σ, $f(\sigma) = j$ and $g(\sigma) = k$. The list is ordered lexically by the least truth assignment producing each pair. Note: in the following f_l is used to denote the left child component of f, and f_r the right child, while f_M is used to denote the matrix (*i.e.*, root function).

```
function BDDT_pairs(f,g)
    if f and g are constants return ((f, g))
    if f is a constant, f = ([f], 0, 0)
    if g is a constant, g = ([g], 0, 0)
    if (f, g, r) stored in hash table return r
    let p_l = BDDT_pairs(f_l, g_l)
    let p_r = BDDT_pairs(f_r, g_r)
    create empty hash table H, let i = 0
    for j = 1 … length(p_l), let (r_f, r_g) = p_l(j) in
        for k in 1 … length(p_r), let (c_f, c_g) = p_r(k) in
            BDD_pairs(f_M(r_f, c_f), g_M(r_f, c_g), H, i)
    let r be the list (p_0, …, p_{i-1}) where (j, p_j) in H
    store (f, g, r) in hash table
    return r
end function
```

Note that we first check a hash table to see if BDDT_pairs has previously been computed for the same inputs. This means that BDDT_pairs will execute $O(|f| + |g|)$ times during the execution of "apply", no matter how many times it is called. Note also that we traverse the list of left child pairs in the *outer* loop. This means that we are assuming the left child variables are more significant in the lexical order than the right child variables, which in turn are more significant than the root variables. The procedure BDD_pairs (figure 4) traverses a pair of binary decision diagrams, storing each new pair of leaves reached in the hash table H, and incrementing the counter i with each new pair. The traversal order guarantees that the pairs are stored in lexical order.

```
procedure BDD_pairs(f,g,var H, var i)
    procedure recurse(f,g)
        if (f, g) stored in J return
        else if f and g are both leaves then
            if (?, (f, g)) not in H, store (i, (f, g)) in H, i = i + 1
            return
        else
        let f_0,f_1,g_0,g_1 be the cofactors of f and g
            w.r.t. their least common variable.
        recurse(f_0,g_0), recurse(f_1, g_1)
        end if
    end procedure
    let J be an empty hash table
    recurse(f,g)
end procedure
```

Fig. 4. The function BDD_pairs.

The BDDT_apply function (figure 5) uses BDDT_pairs to compute a list of output pairs for the left and right child functions. It then computes a "product matrix" P, that yields the value of the function $o(f, g)$ for each combination of a left child pair and a right child pair. Having the product matrix makes it possible to group the left and right child pairs into equivalence classes, and number the equivalence classes canonically. Two left child pairs are in the same equivalence class if they index equal rows in the product matrix, and similarly, two right child pairs are equivalent if they index equal columns of the product matrix. BDDT_apply then generates a map ϕ_l (ϕ_r) that takes each left (right) child pair and returns the number of its equivalence class. An inverse map ρ_l (ρ_r) is also generated that maps each equivalence class to some left (right) child pair in that class. BDDT_apply is then called recursively with ϕ_l (ϕ_r) to produce the left (right) child function. The root matrix can then be culled from the product matrix, using the inverse function ρ_l (ρ_r) to select a row (column) of the product matrix for each row (column) of the root matrix.

```
function BDDT_apply(o,f,g)
    if f and g are constants return o(f,g)
    if f is a constant, f = ([f],0,0)
    if g is a constant, g = ([g],0,0)
    let pₗ = BDDT_pairs(fₗ,gₗ)
    let pᵣ = BDDT_pairs(fᵣ,gᵣ)
    create matrix P of dimension length(pₗ) × length(pᵣ)
    for j = 1...length(pₗ), let (rf,rg) = pₗ(j) in
        for k in 1...length(pᵣ), let (cf,cg) = pᵣ(k) in
            P(j,k) = BDD_apply(o,fₘ(rf,cf),gₘ(rf,cg))
    let H be an empty hash table, nₗ = 0
    for j in 1...length(pₗ)
        let r be row j of P, p = pₗ(j)
        if (r,k) is stored in H (for some k), then let φₗ(p) = k
        else let φₗ(p) = nₗ, ρₗ(nₗ) = j, nₗ = nₗ + 1
    {compute φᵣ and ρᵣ in similar manner, using columns instead of rows}
    let rₗ = BDDT_apply(φₗ,fₗ,gₗ)
    let rᵣ = BDDT_apply(φᵣ,fᵣ,gᵣ)
    create a matrix M of dimension nₗ × nᵣ
    for j in 0...nₗ - 1
        for k in 0...nᵣ - 1
            M(j,k) = P(ρₗ(j),ρₗ(k))
    return BDDT_find(M,rₗ,rᵣ)
end function
```

Fig. 5. The function BDDT_apply. Note, BDD_apply is "apply" for OBDD's, as described in [Bry86].

Theorem 12. *If o is a binary operator in the natural numbers, and f and g are boolean functions on V, then*

$$BDDT\_apply(o, \langle f \rangle, \langle g \rangle) = \langle o(f,g) \rangle$$

Theorem 13. *BDDT_apply(o, $\langle f \rangle$, $\langle g \rangle$) runs in time $O(|\langle f \rangle| \times |\langle g \rangle|)$.*

5 Tree width and BDD trees

BDD trees efficiently represent a class of boolean circuits that are hierarchically structured in a certain quantifiable sense. The concept of the tree width of a graph was introduced by Robertson and Seymour in their work on graph minors [RS86]. For circuits, we can define an analogous concept. For any natural number k, the class of circuits of tree width k can be represented by BDD trees using linear space.

A *tree decomposition* of a circuit associates the gates of a circuit with vertices of a tree. Intuitively, the width of a tree decomposition is the maximum number of wires that would run through any vertex if the circuit were wired through the tree. The tree width of a circuit is the least width of any tree decomposition of that circuit.

To make this definition precise, we first need a formal definition of a circuit.

Definition 14. Given a collection of boolean *operators* and *variables*, a *circuit* is either

1. v, where v is a variable, or
2. $(f, x_1, \ldots, x_n)$ where f is an operator of arity n and $x_1, \ldots, x_n$ are circuits. The circuits $x_1, \ldots, x_n$ are said to be the *children* of $(f, x_1, \ldots, x_n)$.

The set of subcircuits of a circuit g, denoted $S(g)$ is either

1. $\{g\}$, if g is a variable, or
2. $\{g\} \cup_{i=1}^{n} S(x_i)$ where $x_1, \ldots, x_n$ are the children of g.

Definition 15. The *function* $\bar{g}$ associated with a circuit g is defined recursively, such that

- $\bar{v}(\sigma) = \sigma(v)$ if v is a variable and
- $\bar{g}(\sigma) = f(\bar{x}_1(\sigma), \ldots, \bar{x}_n(\sigma))$ if $g = (f, x_1, \ldots, x_n)$.

Definition 16. A tree decomposition of a circuit g is an undirected tree T with vertices $V(T)$ and edges $E(T)$ where each vertex $i \in V(T)$ is associated with $X_i \subseteq S(g)$ such that

1. for all circuits $(f, x_1, \ldots, x_n)$ in $S(g)$, for some $i \in V(T)$,

$$\{(f, x_1, \ldots, x_n), x_1, \ldots, x_n\} \subseteq X_i$$

and
2. for all $i, j, k \in V(T)$, if j lies on the path in T from i to k, then $X_i \cap X_k \subseteq X_j$.

Remark: In other words, each vertex of the tree is labeled with a subset of the gates. The set of vertices labeled with a given gate is connected, and every gate is somewhere in the same set with all of its inputs.

Definition 17. The *width* of a tree decomposition is the max of $|X_i|$ for $i \in V(T)$. The tree width of a formula g is the least w such that g has a tree decomposition of width w.

Theorem 18. *Let C_k be the class of circuits of tree width k. The optimum BDD tree representation of $\bar{g}$, where $g \in C_k$, has size $O(|S(g)|)$.*

Space prohibits presenting a proof of this theorem. The intuition behind it is simple, however – if a given subtree of the decomposition communicates with its environment through at most k wires, then the number of equivalence classes of assignments to its variables is at most the number of boolean functions of k inputs and k outputs, which is doubly exponential in k. This bounds the size of every node of the BDD tree by a constant, making the total size linear in the circuit size. One must also show, of course, that a tree decomposition can be made to have binary branching and size proportional to $|S(g)|$ without changing its width.

6   Experimental results

We now consider the application of the above results to the verification of tree structured circuits and systems, using the symbolic model checking technique. This technique uses operations on boolean formulas to construct a symbolic representation of the transition relation of a model, and to compute fixed point series that characterize formulas in temporal logic that are given as specifications [BCM+90, McM92].

The first example we will use is a tree structured speed-independent arbiter circuit studied by Dill [Dil88] and based on work of Seitz [Sei80]. Dill likened the operation of this circuit to an elimination tournament. The arbiter cells are organized in a hierarchy, each cell arbitrating between two cells at the next lower level of the hierarchy (see figure 6). Requests and acknowledgements follow a four phase signaling protocol. The implementation of a single cell as a speed independent circuit (due to Seitz) is also depicted in the figure.

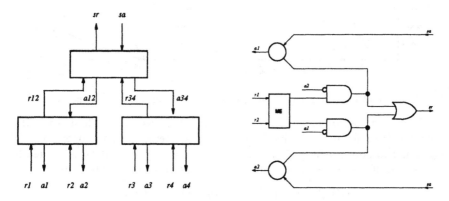

Fig. 6. A tree arbiter with four users and three cells; a speed-independent tree arbiter cell.

The first requirement of the symbolic model checking method is to represent the transition relation as a boolean formula. There is an obvious tree decomposition of this formula, where we simply group all of the gates and inputs of a single arbiter cell in one vertex of the tree. The width of this decomposition is the number of inputs of the cell plus twice the number of gates in the cell, since we have two boolean variables for each gate, representing the old and new logic values of its output[1]. Since this number is a constant independent of the number of cells, we expect a linear upper bound on the size of the transition relation represented as a BDD tree, using a suitable tree ordering of the variables. In fact,

---

[1] Note that the transition relation is represented by a boolean formula, but this does not correspond to any combinational logic function computed by the circuit. In this case, the transition relation is a conjunction of formulas, each constraining the output behavior of one gate. See, for example, [McM92, BCM+90] on how to represent the transition relation of asynchronous circuits.

something more interesting happens – we find the total number of OBDD nodes used in the BDD tree representation is bounded by a constant. This is because all cells of the arbiter are identical, hence all of the OBDD's (except for those representing the users and server) are identical and share the same space. This is accomplished by using the same OBDD variables in the boolean representations at each point in the tree. For a related reason, the total space used by the matrices M is logarithmic in the number of cells, since all BDD trees at a given level of the hierarchy are identical, hence share the same space. As a result, the total space used to represent the transition relation is logarithmic in the number of arbiter cells. Figure 7 plots the space used to represent the transition relation usings OBDD's and using BDD trees, as a function of the number of users of the arbiter.

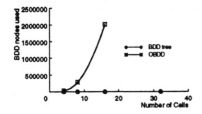 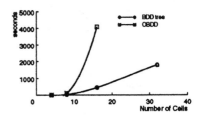

Fig. 7. For model checking tree arbiter, transition relation size and verification time.

The second requirement of the symbolic model checking method is to compute the fixed point series that characterize the temporal logic operators. Of particular importance in this process is computing a representation of the reachable states of the model. The basic step in this computation is to compute

$$q_i(V') = q_{i-1}(V') \vee \exists V. \, (q_{i-1}(V) \wedge R(V, V'))$$

where q_i represents the ith approximation to the reachable states (q_0 is the set of initial states), and R represents the transition relation (note – the boolean quantification requires an addition BDD tree algorithm which is not presented here due to space considerations). This step is repeated until a fixed point is reached.

Figure 7 shows a plot of the time required for this computation using the OBDD representation and the BDD tree representation, as a function of the number of users of the arbiter. There are several factors that go into this time. First, the number of iterations required to reach a fixed point increases as roughly $n \log n$ where n is the number of users. The most distant state is one where every user has requested the resource, received acknowledgement, removed its request, and is about to receive a negation of its acknowledgement. For the case of 32 users, this required 433 steps. A second factor is the size of the transition relation, and a third is the size of the fixed point approximations.

As a second example, we consider a distributed cache consistency protocol, studied in [MS91] and [McM92]. This system also has a natural tree decomposition, as depicted in figure 8. For the transition relation, we find a result similar to that of the tree arbiter – the number of BDD nodes used remains constant as the number of cluster buses increases. The time to construct the transition relation using BDD trees increases roughly linearly. Thus, as the size increases, BDD trees soon outperform OBDD's by an order of magnitude (see figure 9). However, in the reachable states computation, we find the size of the fixed point approximations increasing exponentially for both methods. This appears to be due to "artificial correlations" between clusters in the fixed point approximations caused by interleaving on the global bus. This could possibly be corrected using a "modified search order" [BCL91], but clearly BDD trees by themselves are not sufficient to allow us to model check large examples of the cache protocol with the basic symbolic method. On the other hand, BDD trees would be very effective in proving invariants of the protocol, which does not require fixed point calculations.

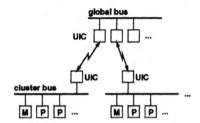

Fig. 8. Gigamax memory architecture.

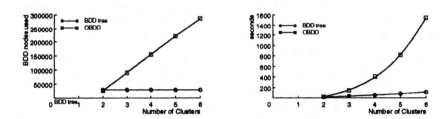

Fig. 9. For Gigamax verification, transition relation size, and transition relation construction time.

7 Conclusion

BDD trees provide a canonical boolean representation, which has a linear upper space bound for circuits of fixed tree width. This is a generalization of a similar result for OBDD's which applies only to path width. At the present time, tree decompositions of a given width appear to be obtainable algorithmically only for very small widths [Bod93], but certain systems have a natural hierarchical structure that lends itself to easy tree decomposition. In fact, the variable orderings for the above experiments were obtained by adding a simple primitive called "group" to the description language that creates a subtree in the tree decomposition. Finding good locations for these "group" commands was a trivial matter.

Since hierarchical structure occurs in many systems (the author is aware of at least three other hierarchically structured cache protocols), there is some prospect that BDD trees will be useful in practice, in spite of their relatively complicated structure in comparison to OBDD's. As a practical matter, one would expect the BDD tree algorithms to be somewhat less efficient in their implementation because of their greater complexity. This inefficiency is mainly offset, however by the use of OBDD's inside the representation. Profiling shows that most of the time is spent in the OBDD calculations – thus, optimizing the BDD tree code is not a major concern.

For general circuits, it is possible that practical algorithms can developed for tree decomposition for widths at least as large as can be handled by BDD trees – at least there are no theoretical results precluding this. Any progress in this area would make BDD trees more widely applicable.

Acknowledgements I would like to acknowledge discussions with Randy Bryant and Jerry Burch, who suggested a structure similar to BDD trees in 1992. I would also like to thank Bob Kurshan and David Long for critical reading.

References

[BCL91]   Jerry R. Burch, Edmund M. Clarke, and David E. Long. Symbolic model checking with partitioned transition relations. In A. Halaas and P. B. Denyer, editors, *Proceedings of the IFIP International Conference on Very Large Scale Integration*, Edinburgh, Scotland, August 1991.

[BCM+90] J. R. Burch, E. M. Clarke, K. L. McMillan, D. L. Dill, and J. Hwang. Symbolic model checking: 10^{20} states and beyond. In *Proceedings of the Fifth Annual Symposium on Logic in Computer Science*, June 1990.

[Bod93]   H. L. Bodlaender. A linear time agorithm for finding tree-decompositions of small treewidth. In *ACM STOC '93 (25th)*, CA, USA, May 1993.

[Bry86]   R. E. Bryant. Graph-based algorithms for boolean function manipulation. *IEEE Transactions on Computers*, C-35(8), 1986.

[Dil88]   D. Dill. Trace theory for automatic hierarchical verification of speed-independent circuits. Technical Report 88-119, Carnegie Mellon University, Computer Science Dept, 1988.

[McM92]   K. L. McMillan. Symbolic model checking: an approach to the state explosion problem. Technical Report 92-131, Carnegie Mellon University, Computer Science Dept, 1992.

[MS91]   K. L. McMillan and J. Schwalbe. Formal verification of the Encore Gigamax cache consistency protocol. In *International Symposium on Shared Memory Multiprocessors*, 1991.

[RS86]   N. Robertson and P. D. Seymour. Graph minors. II. algorithmic aspects of tree-width. *J. Algorithms*, 7:309–322, 1986.

[Sei80]   C. L. Seitz. Ideas about arbiters. *Lambda*, 10(14), 1980.

Symbolic Verification with Periodic Sets*

Bernard Boigelot** and Pierre Wolper

Université de Liège, Institut Montefiore, B28, 4000 Liège Sart Tilman, Belgium.
Email : {pw,boigelot}@montefiore.ulg.ac.be

Abstract. Symbolic approaches attack the state explosion problem by introducing implicit representations that allow the simultaneous manipulation of large sets of states. The most commonly used representation in this context is the Binary Decision Diagram (BDD). This paper takes the point of view that other structures than BDD's can be useful for representing sets of values, and that combining implicit and explicit representations can be fruitful. It introduces a representation of complex periodic sets of integer values, shows how this representation can be manipulated, and describes its application to the state-space exploration of protocols. Preliminary experimental results indicate that the method can dramatically reduce the resources required for state-space exploration.

1 Introduction

Verification by state-space exploration is an old technique [Wes78] whose many advantages have long been outweighed by its main drawback: the state explosion problem. However, in recent years, this central problem has been attacked from several directions with sufficient success to give, if not the hope of total victory, at least the possibility of containment in some contexts. Two main tactics have been used against the state explosion problem. The first is to limit the search to a reduced state-space that is still sufficient for verifying the property of interest. Among these, one can cite on-the-fly approaches [Hol85, VW86, JJ89, FM91, CVWY92], partial-order methods [Val90, God90, HGP92, McM92, GW93, Pel93, WG93], and abstraction techniques [GL93]. The second tactic avoids handling each state individually by using symbolic representations. This makes it possible to manipulate very large sets of states simultaneously [BCM+90, CMB90].

The main representation of sets of states that has been used in symbolic verification is the Binary Decision Diagram (BDD) [Bry92]. While this representation is simple and general, and can be extremely effective [BCM+90], it is not a panacea. Indeed, for fundamental reasons, not all sets of states can be

* This work was supported by the Esprit BRA action REACT and by the Belgian Incentive Program "Information Technology" - Computer Science of the future, initiated by the Belgian State - Prime Minister's Office - Science Policy Office. The scientific responsibility is assumed by its authors.
** "Aspirant" (Research Assistant) for the National Fund for Scientific Research (Belgium).

represented by small BDD's. Whether BDD's will be effective or not depends on the nature and structure of the sets of states that have to be represented. Symptomatically of this, there are more success stories about the use of BDD's for hardware verification than for other applications such as protocol verification.

Another family of verification approaches that can broadly be classified as symbolic are those that have been developed for the verification of real-time properties [ACD93]. Indeed, in these approaches, sets of time values are represented with the help of polyhedra rather than explicitly. However, as opposed to what is done with BDD's the rest of the state information is represented explicitly rather than symbolically.

The work we present here is based on a similar intuition: it can be fruitful to represent states partly explicitly and partly symbolically. The question then is: which symbolic representation can be used effectively in such a combined approach?

This paper attempts to give an answer to this question in a specific context: protocols using integer variables. Indeed, when analyzing protocols, it often turns out that the state space explodes due to the presence of integer variables used for instance as counters. Even though this cause of state explosion is often not inherent to the protocol design, it is resistant to existing techniques. However, a close look at the sets of values taken by these variables reveals that they often are periodic or projected from a periodic set. Based on this motivation, we introduce a representation of periodic subsets of $\mathbf{Z}^n$ (which we name *periodic vector sets*) as a tool to be used in protocol verification.

The representation we propose is derived fairly naturally from the main source of periodicity in sets of reachable integer values: the iteration of linear transformations. It allows the representation of finite and infinite sets and is based on a few simple concepts: linear combinations, linear constraints, and projection. We show that it is closed under the application of iterated linear transformations. Moreover, we establish that it can also represent the composition of transformations, as well as their iterations, though not always their nested iterations.

Next, we turn to the use of our new representation in the context of the exploration of the state-space of protocols. There are several approaches to using a representation of periodic vector sets in this context. We present one that is based on the selective precomputation of the fixpoint of some program transitions. These fixpoints are then added as generalized transitions to the program, and a traditional search of this modified program is then performed with the help of our representation of periodic sets. This can dramatically reduce the resources needed for the state-space search compared to a simple enumerative exploration. Moreover, state reachability questions can still be fully answered, which makes it possible to use the method for the verification of a large class of properties [CVWY92].

Finally, we present some experimental results that were obtained with a preliminary implementation, compare our method to existing work, and discuss its benefits and limits.

2 Defining Periodic Vector Sets

Consider the simple program represented in Figure 1, where x is an integer variable. Although its number of *control* states is limited to four, it is easily seen that

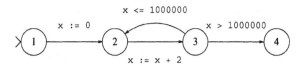

Fig. 1. A simple program.

its state-space contains more than one million reachable states. However, this does not make it impractical to deal with this set of states: it is straightforwardly represented as follows

$$S = \{(①, \perp), (②, 2k), (③, 2k + 2), (④, 1000002) \mid k \in \mathbf{N} \wedge 0 \leq k \leq 500000\}$$

Moreover, this "symbolic" description of the set of reachable states can be used to validate simple properties of the state-space, for instance the reachability of $(③, 456)$.

Intuitively, the reason for which the set of reachable states of the program of Figure 1 can be easily represented is that it has a *periodic* structure. Such periodic sets of reachable states are due to the repeated execution of instructions forming a *cycle* in the control graph. For instance, the set $\{2k \mid k \in \mathbf{N} \wedge 0 \leq k \leq 500000\}$ of reachable values at control state $②$ may be seen as the result of multiple executions of the cycle $②$–$③$–$②$, starting with the initial reachable value 0.

Our goal is to provide a general symbolic representation system for sets of values such as those appearing in the example above. This system has to be able to represent single values as well as periodic sets resulting from cyclic executions, and it should allow elementary operations to be easily applied to the representations. The requirements on the representation and on the operations it supports are linked to the operations that are allowed in the programs to be analyzed. Let us thus briefly define the programming formalism within which we will work.

We consider extended finite-state machines with unbounded integer variables on which the only allowed operations are constant assignment ($x := k$), adding a constant to a variable ($x := x + k$), and testing linear equalities ($x \leq 2y + z - 1$). In other words, each transition between control states can be labeled with a normalized instruction of the form $Tx \leq u \rightarrow x := Ax + b$, where each component of the vector x is a variable, $Tx \leq u$ is the set of linear inequalities serving as precondition to the transition and $x := Ax + b$ is the linear transformation associated with the transition. A is a diagonal matrix whose non-zero elements are equal to 1 and b is an integer vector.

Such normalized instructions have nice properties. First, it is always possible to express the *composition*[3] of two instructions as a normalized instruction. Hence, the body of any cycle in the control graph is always equivalent to a single instruction whose representation may easily be computed by successive compositions. Second, given that the matrix A is idempotent ($A^2 = A$), executing such an instruction repeatedly always leads to a periodic set[4]. Indeed, for a vector $\mathbf{x}_0$ of initial values, the sequence of values obtained by the repeated execution of a normalized instruction is

$$A\mathbf{x}_0 + \mathbf{b}, \; A\mathbf{x}_0 + A\mathbf{b} + \mathbf{b}, \; A\mathbf{x}_0 + 2A\mathbf{b} + \mathbf{b}, \; \ldots, A\mathbf{x}_0 + kA\mathbf{b} + \mathbf{b}, \ldots, \quad (1)$$

which is periodic with period $A\mathbf{b}$ (the *periodicity vector* of the set). The constant k appearing in an element $A\mathbf{x}_0 + kA\mathbf{b} + \mathbf{b}$ of the set is called the *repetition count* of that element. Notice that only a subset of (1) is generally reached. Indeed, the iteration process only proceeds up to repetition counts for which $A\mathbf{x}_0 + kA\mathbf{b} + \mathbf{b}$ satisfies the precondition $T\mathbf{x} \leq \mathbf{u}$. Notice that this last condition can be written as a set of linear inequalities to be satisfied by k: $TA\mathbf{x}_0 + kTA\mathbf{b} + T\mathbf{b} \leq \mathbf{u}$. Generalizing this to multiple periodicity vectors and repetition counts, we obtain our definition of a periodic vector set.

Definition 1. A *periodic vector set* is a set of vectors $\mathbf{x} \in \mathbf{Z}^n$ such that

$$\exists \mathbf{k} \in \mathbf{Z}^m : \mathbf{x} = C\mathbf{k} + \mathbf{d} \wedge P\mathbf{k} \leq \mathbf{q}.$$

In this definition, m periodicity vectors are grouped in the matrix C, and $\mathbf{k}$ gathers the corresponding repetition counts. Similarly, the linear equalities bounding the repetition are gathered into a single linear system $P\mathbf{k} \leq \mathbf{q}$. To represent a periodic vector set, one needs to represent the values of m, C and $\mathbf{d}$ as well as the linear system $P\mathbf{k} \leq \mathbf{q}$. For the latter, it is often useful for efficiency reasons to use representations other than the direct syntactic one.

The set of solutions of a linear system of the form $P\mathbf{k} \leq \mathbf{q}$ is a *closed convex polyhedron* (or polyhedron, for short) which can be represented as follows [CH78, Hal93]. It is the set of points $\mathbf{v}$ satisfying

$$\mathbf{v} = \sum_{i=1}^{\sigma}(\lambda_i \mathbf{s}_i) + \sum_{j=1}^{\rho}(\mu_j \mathbf{r}_j) + \sum_{k=1}^{\delta}(\nu_k \mathbf{d}_k)$$

with the constraints

$$\begin{cases} 0 \leq \lambda_i & i = 1, 2, \ldots, \sigma \\ \sum_{i=1}^{\sigma} \lambda_i = 1 \\ 0 \leq \mu_j & j = 1, 2, \ldots, \rho \end{cases}$$

---

[3] The *composition* of two instructions is the instruction equivalent to their sequential execution.

[4] Actually, it is possible to extend the set of operations allowed in instructions as long as this leads to an idempotent matrix A. Going further in this direction, one can allow any linear assignment in instructions, and then apply the method we describe only to transitions for which A is idempotent.

where $V = \{s_1, s_2, \ldots, s_\sigma\}$ is the set of *vertices*, $R = \{r_1, r_2, \ldots, r_\rho\}$ is the set of *rays*, and $D = \{d_1, d_2, \ldots, d_\delta\}$ is the set of *lines* of the polyhedron. Thus a polyhedron is entirely characterized by the sets V, R, and D. In practice, we maintain both the direct representation of the system of linear inequalities as well as the sets V, R, and D characterizing the corresponding polyhedron. This allows us to choose the most convenient representation for each operation that has to be performed.

3 Operations on Periodic Vector Sets

The use of periodic vector sets in verification involves applying a number of operations to these sets. In this section, we describe the required operations as well as the corresponding algorithms.

3.1 Execution of a Single Instruction

Given the representation of a periodic vector set S and a normalized instruction I, our goal is to compute the representation of the set of values that is obtained by applying I to each element x of S, i.e. the set $S' = \{I(x) : x \in S\}$. Recall that S is the set of all the points x satisfying

$$\exists k \in \mathbf{Z}^m : x = Ck + d \wedge Pk \le q.$$

Consequently, the image of S obtained by applying the instruction $Tx \le u \rightarrow x := Ax + b$ is the set of all x' such that

$$\exists k \in \mathbf{Z}^m : x' = ACk + (Ad + b) \wedge T(Ck + d) \le u \wedge Pk \le q.$$

Each element x' of S' thus satisfies

$$\exists k \in \mathbf{Z}^m : x' = C'k + d' \wedge P'k \le q',$$

with $C' = AC$, $d' = Ad + b$, and where the linear system $P'k \le q'$ is the conjunction of the systems $TCk \le u - Td$ and $Pk \le q$. There are known effective algorithms for computing the intersection of the corresponding polyhedra described by their vertices, rays and lines [CH78, Hal93, LeV92].

3.2 Repeated Execution of an Instruction

The purpose of this operation is to compute the set S' of all values resulting from executing one or more times a normalized instruction I on a given periodic vector set S. In other words, we have $S' = I^+(S)$, where $I^+(X)$ denotes the infinite union $I(X) \cup I^2(X) \cup I^3(X) \cup \cdots$. The set S' can be viewed as the least fixpoint containing $I(S)$ of $f(X) = X \cup I(X)$.

If $Ax + b$ is the linear transformation of I and A is idempotent, the elements of S' are of the form $x' = Ax + kAb + b$ for $x \in S$ (cf Equation 1 in Section 2). Now, for a given $x \in S$, such an element is only in S' for values of k (the repetition count) such that the enabling condition $T\mathbf{x} \leq \mathbf{u}$ of the instruction is satisfied before each execution of the instruction, in other words if

$$T\mathbf{x} \leq \mathbf{u} \wedge \forall i \in \{0, 1, \ldots, k-1\} : T(A\mathbf{x} + iA\mathbf{b} + \mathbf{b}) \leq \mathbf{u}.$$

It can easily be seen that, given that we are dealing with convex sets of constraints, this expression can be rewritten as

$$(k = 0 \wedge T\mathbf{x} \leq \mathbf{u})$$
$$\vee (k \geq 0 \wedge T\mathbf{x} \leq \mathbf{u} \wedge T(A\mathbf{x} + \mathbf{b}) \leq \mathbf{u} \wedge T(A\mathbf{x} + (k-1)A\mathbf{b} + \mathbf{b}) \leq \mathbf{u})$$

Since this expression is disjunctive, it cannot in general be represented by a closed convex polyhedron. To avoid this problem, we notice that the term $(k = 0 \wedge T\mathbf{x} \leq \mathbf{u})$ is needed only to ensure that all values resulting from a single execution of the instruction I are represented. So, dropping this term amounts at most to dropping part of $I(S)$, i.e. of computing a set S' such that $II^+(S) \subset S' \subset I^+(S)$. This is not a real problem since the whole of $I^+(S)$ can always be obtained by computing separately $I(S)$ and our approximation S'.

Recalling that the periodic vector set S is the set of all $\mathbf{x}$ satisfying $\exists \mathbf{k} \in \mathbf{Z}^m : \mathbf{x} = C\mathbf{k} + \mathbf{d} \wedge P\mathbf{k} \leq \mathbf{q}$, the set S' we are computing can be expressed as the set of all $\mathbf{x}'$ such that

$$\exists \mathbf{k} \in \mathbf{Z}^m, k \in \mathbf{Z} : \mathbf{x}' = AC\mathbf{k} + kA\mathbf{b} + A\mathbf{d} + \mathbf{b} \wedge P\mathbf{k} \leq \mathbf{q} \wedge k \geq 0$$
$$\wedge T(C\mathbf{k} + \mathbf{d}) \leq \mathbf{u}$$
$$\wedge T(AC\mathbf{k} + A\mathbf{d} + \mathbf{b}) \leq \mathbf{u}$$
$$\wedge T(AC\mathbf{k} + (k-1)A\mathbf{b} + A\mathbf{d} + \mathbf{b}) \leq \mathbf{u}. \qquad (2)$$

Thus, each element $\mathbf{x}'$ of S' satisfies $\exists \mathbf{k}' \in \mathbf{Z}^{m+1} : \mathbf{x}' = C'\mathbf{k}' + \mathbf{d}' \wedge P'\mathbf{k}' \leq \mathbf{q}'$, where we have $\mathbf{k}' = \begin{bmatrix} \mathbf{k} \\ k \end{bmatrix}$, $C' = [AC; A\mathbf{b}]$, $\mathbf{d}' = A\mathbf{d} + \mathbf{b}$, and where the system $P'\mathbf{k}' \leq \mathbf{q}'$ represents the conjunction of the constraints present in (2). Hence, we have a direct method of computing the representation of S' given S and I. Notice that one of the effects of applying an iterative transformation to a periodic vector set is the creation of a new column in its C matrix. In practice, it is often possible to limit the size of C by first checking whether the new column is identical to an existing one, and simplifying the linear system accordingly.

3.3 Repeated Execution of Nested Instructions

Programs often contain nested loops. If we want to apply the construction described in the previous section in this context, we need to be able to compute a normalized instruction that is equivalent to the body of a cycle, itself containing a cycle. In this section, we show that this can be done provisionally. One condition that needs to be satisfied is that the precondition and the exit condition of a cycle are mutually exclusive (not at all uncommon in practice). For instance, let us consider the program depicted in Figure 2 where the cycle ④–③–④ has the precondition $y < x$ and the exit condition $y \geq x$. Any repeated execution of the cycle, starting in ④ and followed by the transition ending in ①, is equivalent

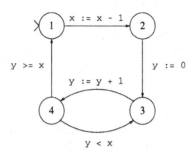

Fig. 2. A program with nested loops.

to the simpler instruction y := x. We present a systematic way of computing such an equivalent instruction.

Formally, given a cycle instruction I and an exit condition φ, we are looking for a normalized instruction I' such that for each periodic vector set S we have $I'(S) = \varphi(I^+(S))$, where $\varphi(X)$ denotes the subset of X containing the values satisfying φ. Using the results of the previous section[5] $I'(S)$ is the set of all $\mathbf{x}'$ such that

$$\exists k \in \mathbf{Z}, \mathbf{x} \in S : \mathbf{x}' = A\mathbf{x} + kA\mathbf{b} + \mathbf{b} \wedge T' \begin{bmatrix} \mathbf{x} \\ k \end{bmatrix} \leq \mathbf{u}',$$

where T' and $\mathbf{u}'$ are a linear system equivalent to the conjunction of $k \geq 0$, $T\mathbf{x} \leq \mathbf{u}$, $T(A\mathbf{x} + \mathbf{b}) \leq \mathbf{u}$, $T(A\mathbf{x} + (k-1)A\mathbf{b} + \mathbf{b}) \leq \mathbf{u}$ and $\varphi(A\mathbf{x} + kA\mathbf{b} + \mathbf{b})$.

Assume now that the linear system defined by T' and $\mathbf{u}'$ contains at least one equation $\alpha_1 \cdot \mathbf{x} + \alpha_k k = \beta$ for which $\alpha_k \neq 0$. This means that the number of iterations is determined by the initial values, i.e. that the loop is deterministic. In this situation, k can be expressed as a linear function $k(\mathbf{x}) = \frac{1}{\alpha_k}(\beta - \alpha_1 \cdot \mathbf{x})$, and if all the coefficients in $k(\mathbf{x})$ are integers, the values $\mathbf{x}' \in I'(S)$ can be expressed as

$$\exists \mathbf{x} \in S : \mathbf{x}' = A\mathbf{x} + (k(\mathbf{x})A\mathbf{b} + \mathbf{b}) \wedge T' \begin{bmatrix} \mathbf{x} \\ k(\mathbf{x}) \end{bmatrix} \leq \mathbf{u}',$$

which can be turned into the canonical form of a normalized instruction by expanding $k(\mathbf{x})$. In practice, the equation used to compute $k(\mathbf{x})$ is chosen at random among the suitable ones. These are obtained straightforwardly from the representation of the system [CH78, Hal93, LeV92].

---

[5] The same approximation applies, namely that we compute a set S' such that $II^+(S) \subset S' \subset I^+(S)$.

3.4 Inclusion Between Periodic Vector Sets

Testing inclusion between periodic vector sets is essential for detecting termination of a state-space exploration. Unfortunately, it is a very difficult problem. We take the pragmatic approach of using an approximate test which never gives false positives (claims that there is inclusion when there is not) but which can give false negatives (not detecting inclusion when it actually occurs). This is not too bothersome given the way we use periodic vector sets (see next section): false negatives when testing inclusion can lengthen the exploration of the state-space, but will never lead to incorrect results.

Consider a periodic vector set S, i.e. the set of all $\mathbf{x}$ such that $\exists \mathbf{k} \in \mathbf{Z}^m$: $\mathbf{x} = C\mathbf{k} + \mathbf{d} \wedge P\mathbf{k} \leq \mathbf{q}$. We want to test whether all elements of this set are included in a periodic vector set $S' = \{\mathbf{x} \mid \exists \mathbf{k} \in \mathbf{Z}^m : \mathbf{x} = C'\mathbf{k} + \mathbf{d}' \wedge P'\mathbf{k} \leq \mathbf{q}'\}$. If we manage to find an integer matrix M and an integer vector $\mathbf{n}$ such that $\mathbf{d} - \mathbf{d}' = C'\mathbf{n}$ and $C = C'M$, we have for all $\mathbf{x}$ in $S : \exists \mathbf{k} \in \mathbf{Z}^m : \mathbf{x} = C'(M\mathbf{k} + \mathbf{n}) + \mathbf{d}' \wedge P\mathbf{k} \leq \mathbf{q}$. Hence, if all the solutions of $P\mathbf{k} \leq \mathbf{q}$ are solutions of $P'(M\mathbf{k} + \mathbf{n}) \leq \mathbf{q}'$, we have $\exists \mathbf{k}' \in \mathbf{Z}^{m'} : \mathbf{x} = C'\mathbf{k}' + \mathbf{d}' \wedge P'\mathbf{k}' \leq \mathbf{q}'$ and $\mathbf{x}$ belongs to S'.

The existence of suitable M and $\mathbf{n}$ is related to the possibility of expressing each periodicity vector of S, as well as the difference of initial values $\mathbf{d} - \mathbf{d}'$, as linear combinations of the periodicity vectors of S'. This allows us to associate a vector of repetition counts of S' to each point of S. The inclusion test can be carried out by first finding M and $\mathbf{n}$ and thereafter performing an inclusion test between linear systems of inequalities. The latter problem is equivalent to an inclusion test between polyhedra and can be easily solved in the dual representation of polyhedra [CH78, Hal93]. The former is equivalent to solving a general linear diophantine equation where each component of M and $\mathbf{n}$ is considered as a variable. This problem can be simplified by looking only for solutions where $\mathbf{n}$ and each column of M have no more than one non-zero component. In that case, each column of C is constrained to be an integer multiple of a column of C' and the computation of M is reduced to determining the coefficients of the mapping. This heuristic has shown itself to be powerful enough in most cases.

3.5 Testing for Emptiness

A periodic vector set is called *empty* if it does not contain any point. Testing for emptiness is useful, for instance, in order to determine if there exists a reachable state belonging to a reachable group of states. A periodic vector set is empty if and only if its linear system of inequalities $P\mathbf{k} \leq \mathbf{q}$ does not admit any integer solution (in other words, the polyhedron determined by this linear system does not contain any integer point). The test of emptiness can thus be reduced to an *integer programming problem* [Mur76, Gre71].

4 Verification with Periodic Vector Sets

The simplest way to explore the state-space of a program is to use a search to enumerate all reachable states one-by-one. Of course, this approach fails when the state-space is too large. In what follows we describe how periodic vector sets can be used to significantly extend its applicability.

All enumerative state-space searches are based on a common principle, they spread the reachability information along the transitions of the system to be analyzed. The exploration process starts with the initial state of the system, and tries at every step to enlarge its current set of reachable states by propagating these states through transitions. The procedure terminates when a stable set is reached.

The idea behind the use of periodic vector sets is to process sets of states rather than individual states which makes it possible to reach a stable set faster. The problem is that, starting with a single state, and applying only deterministic transitions to individual states, one never generates sets of states that can be manipulated simultaneously. A solution is the use of *meta-transitions* able to generate large sets of reachable states from a single reachable state. This is exactly what we have done while computing the effect of the repeated application of a transition. So, we have the tools for using the following approach.

One first analyses the cycles and nested cycles of the control graph of the program. For each cycle of interest, a corresponding meta-transition is added to the program. Now, the state-space search algorithm is rewritten in a such a way that it works with sets of states (represented as a finite union of periodic vector sets) rather than with individual states. One still starts with a single initial state, but each time a meta-transition is encountered a periodic vector set is produced. It is thus a good heuristic to give priority to meta-transitions. The exploration terminates when the representation of the set of reachable states stabilizes. This happens when every new deductible periodic vector set is included in the current list of reachable sets. Notice that all approximations we have made in the implementation of operations on periodic vector sets are conservative. They will reduce the benefit of using periodic vector sets, but will not lead to states being missed or to states being incorrectly considered as reachable.

In general, finding all cycles in a graph is an NP-hard problem. However, the cycle analysis needed to introduce meta-transitions does not need to be exhaustive and there are a number of techniques that can be used to make it reasonably efficient. Many programming languages include explicit instructions for loops, and thus allow cycles to be detected during the compilation of the program into a transition system. Another idea is to make the most of the exploration algorithm; for instance, if a depth-first search of the control state-space is used, one may detect duplicate control states on the search stack and compute the periodicity vector of the corresponding cycle.

5 Experimental Results

An experimental system that allows the manipulation of periodic vector sets has been implemented. This system has been used to do state-space searches for programs represented as extended finite-state machines. The method that was used to introduce meta-transitions was to consider only the cycles found by detecting duplicate control states on the search stack during a depth-first search.

The first results are quite encouraging. As an example, we applied our method to the analysis of the lift program described in [Val89]. This program is composed of two parallel processes intended to model the interaction between the motor of a lift and one of its control panels. Each process is represented by an extended finite-state machine, and communicates with its counterpart via global integer variables. The interleaving semantics of parallelism is assumed, hence there exists an easily derivable extended finite-state machine equivalent to the parallel composition of the two processes. The program, expressed in a Promela-like language[6] [Hol91], is given in Figure 3. The global variable c is intended to

```
int c = 1, g = 1, a = 0, N = 10;

process motor {
  do
     :: atomic { a == 1 -> a = 0;  c = c + 1 }
     :: atomic { a == 2 -> a = 0;  c = c - 1 }
  od
}

process control {
  do
     :: atomic { c < g   -> a = 1 } ; a == 0 ->
     :: atomic { c > g   -> a = 2 } ; a == 0 ->
     :: atomic { c == g ->
        do
          :: g < N -> g = g + 1
          :: g > 1 -> g = g - 1
          :: break
        od }
  od
}
```

Fig. 3. The lift example program.

store at any time the *current* floor. The motor program works by waiting for an order from the control part, expressed as a non-zero value of the *aim* variable a, then by updating c accordingly. The control program repeatedly compares the values of c and g, the latter expressing the *goal* floor, then sends appropriate commands to the motor, and finally chooses a new goal floor at random when

---

[6] The purpose of **atomic** statements is to define sequences of states that cannot be interleaved.

the current one is reached. The valid floor numbers are the integers between 1 and the value of the constant N.

It can be shown that the number of reachable states of this program grows quadratically with N, and exceeds 20000 for N = 50. Using our implementation, we were able to construct a representation of these reachable states in only 43 analysis steps and by consuming less than one second of CPU time. A noteworthy observation was the independence between the analysis time and the value of N.

6 Conclusions and Comparison with Other Work

We have suggested periodic vector sets as a representation of the sets of data-values a program can generate. This representation was developed pragmatically with the goal of improving the efficiency of state-space exploration. It can represent sets with a complex periodic structure, whether they are finite or infinite. It is far from perfect. We have been able to show how a number of operations can be applied to periodic vector sets, but we sometimes have had to make conservative approximations. Moreover, we cannot claim good general worst-case complexity bounds on the operations. The positive side is that, it works! We have been able to analyze systems for which enumerative methods are hopeless.

One major advantage of our representation, and of the way we suggest using it, is that it is very flexible and perfectly compatible with a number of other techniques. First, there is complete flexibility as to which part of the state descriptors is represented as a periodic vector set. Second, since verification is viewed as a form of extended enumerative state-space search, the techniques that have been developed for the latter approach can still be used in conjunction with periodic vector sets. This is for instance the case for partial-order methods [Val90, God90, HGP92, McM92, GW93, Pel93, WG93]. The approach we have presented thus allows the combination of symbolic and partial-order approaches which could be extremely fruitful. In summary, periodic vector sets are not *the* solution to the state-space explosion problem, but rather a useful technique that can give excellent results for particular types of large sets of states and, crucially, can be combined with other techniques.

Linear constraints have been repeatedly suggested as a useful tool in program analysis and verification [CH78, Kri93, Lub84] and, recently for the verification of real-time properties of systems [Hal93, YL93]. The work presented here is definitely in this tradition. The main innovation in our work is the introduction of periodicity in the representation which technically amounts to working with the projection of linear transformations of convex polyhedra bounded sets of integers. Representation systems for periodic sets of integers have already been proposed for instance in [Mer90], but only in the framework of single-dimensional spaces and in the particular case of infinite periodic sets. These systems are thus unable to deal with polyhedral boundaries involving more than one variable. A type of representation much closer to our work has also been considered in the context of temporal databases [KSW90]. However, the representation introduced there is more restrictive than ours and would be unsufficient for our verification goal. The idea of analyzing cycles also appears in [Kri93], [Lub84] and [Val89]. In [Kri93] transitions do not have preconditions which substantially simplifies the problem. Moreover, no systematic representation of periodic sets is introduced. In [Lub84] and [Val89], cycles are treated with an inductive argument rather than with a powerful representation. The method presented there thus does not have the advantage of being a direct extension of a systematic state-space search.

References

[ACD93]   R. Alur, C. Courcoubetis, and D. Dill. Model-checking in dense real-time. *Information and Computation*, 104(1):2–34, May 1993.

[BCM+90]   J.R. Burch, E.M. Clarke, K.L. McMillan, D.L. Dill, and L.J. Hwang. Symbolic model checking: 10^{20} states and beyond. In *Proceedings of the 5th Symposium on Logic in Computer Science*, pages 428–439, Philadelphia, June 1990.

[Bry92]   Randal E. Bryant. Symbolic boolean manipulation with ordered binary-decision diagrams. *ACM Computing Surveys*, 24(3):293–318, 1992.

[CH78]   P. Cousot and N. Halbwachs. Automatic discovery of linear restraints among variables of a program. In *Proc. 5th ACM Symposium on Principles of Programming Langages*, 1978.

[CMB90]   O. Coudert, J. C. Madre, and C. Berthet. Verifying temporal properties of sequential machines without building their state diagram. In *Proc. 2nd Workshop on Computer Aided Verification*, volume 531 of *Lecture Notes in Computer Science*, pages 23–32, Rutgers, June 1990. Springer-Verlag.

[CVWY92]   C. Courcoubetis, M.Y. Vardi, P. Wolper, and M. Yannakakis. Memory efficient algorithms for the verification of temporal properties. *Formal Methods in System Design*, 1:275–288, 1992.

[FM91]   J.C. Fernandez and L. Mounier. On the fly verification of behavioural equivalences and preorders. In *Proc. 3rd Workshop on Computer Aided Verification*, volume 575 of *Lecture Notes in Computer Science*, pages 181–191, Aalborg, July 1991.

[GL93]   S. Graf and C. Loiseaux. A tool for symbolic program verification and abstraction. In *Computer Aided Verification, Proc. 5th Int. Workshop*, volume 697, pages 71–84, Elounda, Crete, June 1993. Lecture Notes in Computer Science, Springer-Verlag.

[God90]   P. Godefroid. Using partial orders to improve automatic verification methods. In *Proc. 2nd Workshop on Computer Aided Verification*, volume 531 of *Lecture Notes in Computer Science*, pages 176–185, Rutgers, June 1990. Springer-Verlag.

[Gre71]   H. Greenberg. *Integer Programming*. Academic Press, New York, 1971.

[GW93]   P. Godefroid and P. Wolper. Using partial orders for the efficient verification of deadlock freedom and safety properties. *Formal Methods in System Design*, 2(2):149–164, April 1993.

[Hal93]   N. Halbwachs. Delay analysis in synchronous programs. In *Proc. 5th Workshop on Computer Aided Verification*, volume 697, Elounda, Crete, June 1993. Lecture Notes in Computer Science, Springer-Verlag.

[HGP92]   G. J. Holzmann, P. Godefroid, and D. Pirottin. Coverage preserving reduction strategies for reachability analysis. In *Proc. 12th International Sym-*

posium on Protocol Specification, Testing, and Verification, Lake Buena Vista, Florida, June 1992. North-Holland.

[Hol85] G. J. Holzmann. Tracing protocols. *AT&T Technical Journal*, 64(12):2413–2434, 1985.

[Hol91] G. Holzmann. *Design and Validation of Computer Protocols*. Prentice-Hall International Editions, 1991.

[JJ89] C. Jard and T. Jeron. On-line model-checking for finite linear temporal logic specifications. In *Automatic Verification Methods for Finite State Systems, Proc. Int. Workshop, Grenoble*, volume 407, pages 189–196, Grenoble, June 1989. Lecture Notes in Computer Science, Springer-Verlag.

[Kri93] A.S. Krishnakumar. Reachability and recurrence in extended finite state machines: Modular vector addition systems. In *Proc. 5th Workshop on Computer Aided Verification*, volume 697, pages 110–122, Elounda, Crete, June 1993. Lecture Notes in Computer Science, Springer-Verlag.

[KSW90] F. Kabanza, J.-M. Stévenne, and P. Wolper. Handling infinite temporal data. In *Proc. of the 9th ACM Symposium on Principles of Database Systems*, pages 392–403, Nashville Tennessee, 1990.

[LeV92] H. LeVerge. A note on Chernikova's algorithm. Research Report 1662, INRIA, Rennes, April 1992.

[Lub84] B. Lubachevsky. An approach to automating the verification of compact parallel coordination programs. I. *Acta Informatica*, 21:125–169, 1984.

[McM92] K. McMillan. Using unfolding to avoid the state explosion problem in the verification of asynchronous circuits. In *Proc. 4th Workshop on Computer Aided Verification*, Montreal, June 1992.

[Mer90] N. Mercouroff. Analyse sémantique de communications entre processus de programmes parallèles. Rapport de Recherche LIX/RR/90/09, Ecole Polytechnique, Palaiseau, France, September 1990.

[Mur76] K. Murty. *Linear and Combinatorial Programming*. Wiley, New York, 1976.

[Pel93] D. Peled. All from one, one for all: on model checking using representatives. In *Proc. 5th Conference on Computer Aided Verification*, Elounda, June 1993. Lecture Notes in Computer Science, Springer-Verlag.

[Val89] A. Valmari. State space generation with induction. In *Proc. Scandinavian Conference on Artificial Intelligence - 89*, pages 99–115, Tampere, Finland, June 1989.

[Val90] A. Valmari. A stubborn attack on state explosion. In *Proc. 2nd Workshop on Computer Aided Verification*, volume 531 of *Lecture Notes in Computer Science*, pages 156–165, Rutgers, June 1990. Springer-Verlag.

[VW86] M.Y. Vardi and P. Wolper. An automata-theoretic approach to automatic program verification. In *Proceedings of the First Symposium on Logic in Computer Science*, pages 322–331, Cambridge, June 1986.

[Wes78] C.H. West. Generalized technique for communication protocol validation. *IBM J. of Res. and Devel.*, 22:393–404, 1978.

[WG93] P. Wolper and P. Godefroid. Partial-order methods for temporal verification. In *Proc. CONCUR '93*, volume 715 of *Lecture Notes in Computer Science*, pages 233–246, Hildesheim, August 1993. Springer-Verlag.

[YL93] M. Yannakakis and D. Lee. An efficient algorithm for minimizing real-time transition systems. In *Proc. 5th Workshop on Computer Aided Verification*, volume 697, pages 210–224, Elounda, Crete, June 1993. Lecture Notes in Computer Science, Springer-Verlag.

Automatic Verification of Pipelined Microprocessor Control

Jerry R. Burch and David L. Dill

Computer Science Department
Stanford University

Abstract. We describe a technique for verifying the control logic of pipelined microprocessors. It handles more complicated designs, and requires less human intervention, than existing methods. The technique automatically compares a pipelined implementation to an architectural description. The CPU time needed for verification is independent of the data path width, the register file size, and the number of ALU operations. Debugging information is automatically produced for incorrect processor designs. Much of the power of the method results from an efficient validity checker for a logic of uninterpreted functions with equality. Empirical results include the verification of a pipelined implementation of a subset of the DLX architecture.

1   Introduction

The design of high-performance processors is a very expensive and competitive enterprise. The speed with which a design can be completed is a crucial factor in determining its success in the marketplace. Concern about design errors is a major factor in design time. For example, each month of additional design time of the MIPS 4000 processor was estimated to cost $3-$8 million, and 27% of the design time was spent in "verification and test" [13].

We believe that formal verification methods could eventually have a significant economic impact on microprocessor designs by providing faster methods for catching design errors, resulting in fewer design iterations and reduced simulation time. For maximum economic impact, a verification methodology should:

- be able to handle modern processor designs,
- be applicable to the aspects of the design that are most susceptible to errors,
- be relatively fast and require little labor, and
- provide information to help pinpoint design errors.

The best-known examples of formally verified processors have been extremely simple processor designs, which were generally unpipelined [7, 8, 15, 16]. The verification methods used rely on theorem-provers that require a great deal of very skilled human guidance (the practical unit of for measuring labor in these studies seems to be the person-month). Furthermore, the processor implementations that were verified were so simple that they were able to avoid central problems such as control complexity. There are more recent verification techniques [1, 17] that are much more automatic, but they have not been demonstrated on pipelined processors.

The verification of modern processors poses a special problem. The natural specification of a processor is the programmer-level functional model, called the *instruction set architecture*. Such a specification is essentially an operational description of a processor that executes each instruction separately, one cycle per instruction. The implementation, one the other hand, need not execute each instruction separately; several instruction might be executing simultaneously because of pipelining, etc. Formal verification requires proving that the specification and implementation are in a proper relationship, but that relationship is not necessarily easy to define.

Recently, there have been successful efforts to verify pipelined processors using human-guided theorem-provers [11, 19, 20, 22]. However, in all of these cases, either the processor was extremely simple or a large amount of labor was required.

Although the examples we have attacked are still much simpler than current high-performance commercial processors, they are significantly beyond the capabilities of automatic verification methods reported previously. The method is targeted towards errors in the microprocessor *control,* which, according to many designers, is where most of the bugs usually exist (datapaths are usually not considered difficult, except for floating point operations). Labor is minimized, since the procedure is automatic except for the development of the descriptions of the specification and implementation. When the implementation of the processor is incorrect, the method can produce a specific example showing how the specification is violated.

Since we wish to focus on the processor control, we assume that the combinational logic in the data path is correct. Under this assumption (which can be formally checked using existing techniques), the differences between the specification and implementation behaviors are entirely in the timing of operations and the transfer of values. For example, when the specification stores the sum of two registers in a destination register, the implementation may place the result in a pipe register, and not write the result to its destination until after another instruction has begun executing.

The logic we have chosen is the quantifier-free logic of uninterpreted functions and predicates with equality and propositional connectives. Uninterpreted functions are used to represent combinational ALUs, for example, without detailing their functionality. Propositional connectives and equality are used in describing control in the specification and the implementation, and in comparing them.

The validity problem for this logic is decidable. In practice, the complexity is dominated by handling Boolean connectives, just as with representations for propositional logic such as BDDs [2]. However, the additional expressiveness of our logic allows verification problems to be described at a higher level of abstraction than with propositional logic. As a result, there is a substantial reduction in the CPU time needed for verification.

Corella has also observed that uninterpreted functions and constants can be used to abstract away from the details of datapaths, in order to focus on control issues [9, 10]. He has a program for analyzing logical expressions which he has used for verifying a non-pipelined processor and a prefetch circuit. Although the details are not presented, his analysis procedure appears to be much less efficient than ours, and he does not address the problems of specifying pipelined processors.

Our method can be split into two phases. The first phase compiles operational descriptions of the specification and implementation, then constructs a logical formula that is valid if and only if the implementation is correct with respect to the specification. The second phase is a decision procedure that checks whether the formula is valid. The next two sections describe these two phases. We then give experimental results and concluding remarks.

2 Correctness Criteria

The verification process begins with the user providing behavioral descriptions of an implementation and a specification. For processor verification, the specification describes how the programmer-visible parts of the processor state are updated when one instruction is executed every cycle. The implementation description should be at the highest level of abstraction that still exposes relevant design issues, such as pipelining.

Each description is automatically compiled into a *transition function*, which takes a state as its first argument, the current inputs as its second argument, and returns the next state. The transition function is encoded as a vector of symbolic expressions with one entry for each state variable. Any HDL could be used for the descriptions, given an appropriate compiler. Our prototype verifier used a simple HDL based on a small subset of Common LISP. The compiler translates behavioral descriptions into transition functions through a kind of symbolic simulation.

We write F_{Spec} and F_{Impl} to denote the transition function of the specification and the implementation, respectively. We require that the implementation and the specification have corresponding input wires. The processors we have verified have no explicit output wires since the memory was modeled as part of the processor and we did not model I/O.

Almost all processors have an input setting that causes instructions already in the pipeline to continue execution while no new instructions are initiated. This is typically referred to as *stalling* the processor. If I_{Stall} is an input combination that causes the processor to stall, then the function $F_{\text{Impl}}(\cdot, I_{\text{Stall}})$ represents the effect of stalling for one cycle. All instructions currently in the pipeline can be completed by stalling for a sufficent number of cycles. This operation is called *flushing* the pipeline, and it is an important part of our verification method.

Intuitively, the verifier should prove that if the implementation and specification start in any matching pair of states, then the result of executing any instruction will lead to a matching pair of states. The primary difficulty with matching the implementation and specification is the presence of partially executed instructions in the pipeline. Various parts of the implementation state are updated at different stages of the execution of an instruction, so it is not necessarily possible to find a point where the implementation state and the specification state can be compared easily. The verifier solves this problem by simulating the effect of completing every instruction in the pipeline before doing the comparison. The natural way to complete every instruction is to flush the pipeline.

All of this is made more precise in figure 1. The implementation can be in an arbitrary state Q_{Impl} (labeled "Old Impl State" in the figure). To complete the partially executed instructions in Q_{Impl}, the pipeline is flushed, producing "Flushed Old Impl State". Then, all but the programmer-visible parts of the implementation state are stripped off (we define the function *proj* for this purpose) to produce Q_{Spec}, the "Old Spec State". Because of the way Q_{Spec} is constructed from Q_{Impl}, we say that Q_{Spec} *matches* Q_{Impl}.

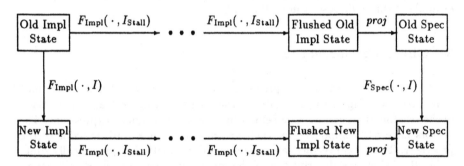

Fig. 1. Commutative diagram for showing our correctness criteria.

Let I be an arbitrary input combination to the pipeline (recall that the specification and the implementation are required to have corresponding input wires). Let $Q'_{\text{Impl}} = F_{\text{Impl}}(Q_{\text{Impl}}, I)$, the "New Impl State", and let $Q'_{\text{Spec}} = F_{\text{Spec}}(Q_{\text{Spec}}, I)$, the "New Spec State". We consider the implementation to satisfy the specification if and only if Q'_{Spec} matches Q'_{Impl}. To check this, flush and project Q'_{Impl}, then see if the result is equal to Q'_{Spec}, as shown at the bottom of figure 1.

It is often convenient to use a slightly different (but equivalent) statement of our correctness criteria. In figure 1, there are two different paths from "Old Impl State" to "New Spec State". The path that involves $F_{\text{Impl}}(\,\cdot\,, I)$ is called the *implementation side* of the diagram; the path that involves $F_{\text{Spec}}(\,\cdot\,, I)$ is called the *specification side*. For each path, there is a corresponding function that is the composition of the functions labeling the arrows on the path. We say that the implementation satisfies the specification if and only if the function corresponding to the implementation side of the diagram is equal to the function corresponding to the specification side of the diagram. More succinctly, the diagram must *commute*.

The reader may notice that figure 1 has the same form as commutative diagrams used with *abstraction functions*. In our case, the abstraction function represents the effect of flushing an implementation state and then applying the *proj* function. Typical verification methods require that there exist some abstraction function that makes the diagram commute. In contrast, we require that our specific abstraction function makes the diagram commute.

In some cases, it may be necessary during verification to restrict the set of "Old Impl States" considered in figure 1. In this case, an *invariant* could be provided (the invariant must be closed under the implementation transition function). All of the examples in this paper were proved correct without having to use an invariant.

Notice that the same input I is applied to both the implementation and the specification in figure 1. This is only appropriate in the simple case where the implementation requires exactly one cycle per instruction (once the pipeline is filled). If more than one cycle is sometimes required, then on the extra cycles it is necessary to apply I_{Stall} to the inputs of the specification rather than I. An example of this is discussed in section 4.2.

3 Checking Correctness

As described above, to verify a processor we must check whether the two functions corresponding to the two sides of the diagram in figure 1 are equal. Each of the two functions can be represented by a vector of symbolic expressions. The vectors have one component for each programmer-visible state variable of the processor. These expressions can be computed efficiently by symbolically simulating the behavioral descriptions of the implementation and the specification. The implementation is symbolically simulated several times to model the effect of flushing the pipeline.

Let $\langle s_1, \ldots, s_n \rangle$ and $\langle t_1, \ldots, t_n \rangle$ be vectors of expressions. To verify that the functions they represent are equal, we must check whether each formula $s_k = t_k$ is valid, for $1 \leq k \leq n$. Before describing our algorithm for this, we define the logic we use to encode the formulas.

3.1 Uninterpreted Functions with Equality

Many quantifier-free logics that include uninterpreted functions and equality have been studied. Unlike most of those logics [18, 21], ours does not include addition or any arithmetical relations. For our application of verifying microprocessor control, there does not appear to be any need to have arithmetic built into the logic (although the ability to declare certain uninterpreted functions to be associative and/or commutative would be useful).

We begin by describing a subset of the logic we use. This subset has the following abstract syntax (where *ite* denotes the if-then-else operator):

$$\langle formula\rangle ::= ite(\langle formula\rangle, \langle formula\rangle, \langle formula\rangle)$$
$$| \; (\langle term\rangle = \langle term\rangle)$$
$$| \; \langle predicate\ symbol\rangle(\langle term\rangle, \dots, \langle term\rangle)$$
$$| \; \langle propositional\ variable\rangle \; | \; true \; | \; false$$

$$\langle term\rangle ::= ite(\langle formula\rangle, \langle term\rangle, \langle term\rangle)$$
$$| \; \langle function\ symbol\rangle(\langle term\rangle, \dots, \langle term\rangle)$$
$$| \; \langle term\ variable\rangle.$$

Notice that the *ite* operator can be used to construct both formulas and terms. We included *ite* as a primitive because it simplifies our case-splitting heuristics and because it allows for efficient construction of transition functions without introducing auxiliary variables.

There is no explicit quantification in the logic. Also, we do not require specific interpretations for function symbols and predicate symbols. A formula is valid if and only if it is true for all interpretations of variables, function symbols and predicate symbols.

Although the *ite* operator, together with the constants *true* and *false*, is adequate for constructing all Boolean operations, we also include logical negation and disjunction as primitives in our decision procedure. This simplifies the rewrite rules used to reduce our formulas, especially rules involving associativity and commutativity of disjunction.

Verifying a processor usually requires reasoning about *stores* such as a register file or main memory. We model stores as having an unbounded address space. If a processor design satisfies our correctness criteria in this case, then it is correct for any finite register file or memory. If certain conventions are followed, the above logic is adequate for reasoning about stores. However, we found it more efficient to add two primitives, *read* and *write*, for manipulating stores. These primitives are essentially the same as the *select* and *store* operators used by Nelson and Oppen [18]. If *regfile* is a variable representing the initial state of a register file, then

$$write(regfile, addr, data)$$

represents the store that results from writing the value *data* into address *addr* of *regfile*. The value at address *addr* in the original state of the register file is denoted by

$$read(regfile, addr).$$

Any expression that denotes a store, whether it is constructed using variables, *write*'s or *ite*'s, can be used as the first argument of a *read* or a *write* operation.

3.2 Validity Checking Algorithm

Pseudo-code for a simplified version of our decision procedure for checking validity, along with a description of its basic operation, is given in figure 2. This procedure is still preliminary and may be improved further, so we will just sketch the main ideas behind it.

Our decision procedure differs in several respects from earlier work [18, 21]. Arithmetic is not a source of complexity for our algorithm, since it is not included in our logic. In our applications, the potentially complex Boolean structure of the formulas we check is the primary bottleneck. Thus, we have concentrated on handling Boolean structure efficiently in practice.

Another difference is that we are careful to represent formulas as directed acyclic graphs (DAGs) with no distinct isomorphic subgraphs. For this to reduce the time complexity of the validity checker, it is necessary to memoize (cache) intermediate results. As shown in figure 2, the caching scheme is more complicated than in standard BDD algorithms [2] because formulas must be cached relative to a set of assumptions.

The final major difference between our algorithm and previous work is that we do not require formulas to be rewritten into a Boolean combination of atomic formulas. For example, formulas of the form $e_1 = e_2$, where e_1 and e_2 may contain an arbitrary number of ite operators, are checked directly without first being rewritten.

Detlefs and Nelson [12] have recently developed a new decision procedure based on a conjunctive normal form representation that appears to be efficient in practice. We have not yet been able to do a thorough comparison, however.

As $check$ does recursive case analysis on the formula p, it accumulates a set of assumptions A that is used as an argument to deeper recursive calls (see figure 2). This set of assumptions must not become inconsistent. To avoid such inconsistency, we require that if p_0 is the first result of $simplify(p, A_0)$, then neither $choose\_splitting\_formula(p_0)$ nor its negation is logically implied by A_0. We call this the $consistency\ requirement$ on $simplify$ and $choose\_splitting\_formula$. Maintaining the consistency requirement is made easier by restricting the procedure $choose\_splitting\_formula$ to return only atomic formulas (formulas containing no ite, or, not or $write$ operations).

As written in figure 2, our algorithm is indistinguishable from a propositional tautology checker. Dealing with uninterpreted functions and equality is not done in the top level algorithm. Instead, it is done in the $simplify$ routine. For example, given the assumptions $e_1 = e_2$ and $e_2 = e_3$, it must be possible to simplify the formula $e_1 \neq e_3$ to $false$; otherwise, the consistency requirement would not be maintained. As a result, $simplify$ and associated routines require a large fraction of the code in our verifier.

In spite of the consistency requirement, there is significant latitude in how aggressively formulas are simplified. It may seem best to do as much simplification as possible, but our experiments indicate otherwise. We see two reasons for this. If $simplify$ does the minimal amount of simplification necessary to meet the consistency requirement, then it may use less CPU time than a more aggres-

```
check(p: formula, A: set of formula): set of formula;
    var
            s, p₀, p₁: formula;
            A₀, A₁, U₀, U₁, U: set of formula;
            G: set of set of formula;
    begin
            if p ≡ true then return ∅;
            if p ≡ false then
                    print "not valid";
                    terminate unsuccessfully;
            G := cache(p);
            if ∃U ∈ G such that U ⊆ A then return U;       /* cache hit */
            /* cache miss */
            s := choose_splitting_formula(p);       /* prepare to case split */
            /* do s false case */
            A₀ := A ∪ {¬s};
            (p₀, U₀) := simplify(p, A₀);
            U₀ := U₀∪check(p₀, A₀);       /* assumptions used for s false case */
            /* do s true case */
            A₁ := A ∪ {s};
            (p₁, U₁) := simplify(p, A₁);
            U₁ := U₁∪check(p₁, A₁);       /* assumptions used for s true case */
            U := (U₀ − {¬s}) ∪ (U₁ − {s});       /* assumptions used */
            cache(p) := G ∪ {U};       /* add cache entry */
            return U;
    end;
```

Fig. 2. The procedure *check* terminates successfully if the formula p is logically implied by the set of assumptions A; otherwise, it terminates unsuccessfully. Not all of the assumptions in A need be relevant in implying p; when *check* terminates successfully it returns those assumptions that were actually used. The set of assumptions used need not be one of the minimal subsets of A that implies p. Checking whether p is valid is done by letting A be the emptyset. Initially, the global lookup table *cache* returns the emptyset for every formula p. Later, $cache(p)$ returns the set containing those assumption sets that have been sufficient to imply p in previous calls of *check*. The procedure *choose_splitting_formula* heuristically chooses a formula to be used for case splitting. The call $simplify(p, A_0)$ returns as its first result a formula formed by simplifying p under the assumptions A_0. The second result is the set of formulas in A_0 that were actually used when simplifying p.

sive simplification routine. Thus, even if slightly more case splitting is needed (resulting in more calls to *simplify*), the total CPU time used may be reduced.

The second reason is more subtle. Suppose we are checking the validity of a formula p that has a lot of shared structure when represented as a DAG. Our hope is that by caching intermediate results, the CPU time typically needed for validity checking grows with the size of the DAG of p, rather than with the size of its tree representation. This can be important in practice; for the DLX example (section 4.2) it is not unusual for the tree representation of a formula to be two orders of magnitude larger than the DAG representation. The more aggressive *simplify* is, the more the shared structure of p is lost during recursive case analysis, which appears to result in worse cache performance. We are continuing to experiment with different kinds of simplification strategies in our prototype implementation.

Unlike the algorithm in figure 2, our validity checker produces debugging information for invalid formulas. This consists of a satisfiable set of (possibly negated) atomic formulas that implies the negation of the original formula. When verifying a microprocessor, the debugging information can be used to construct a simulation vector that demonstrates the bug.

There is another important difference between our current implementation and the algorithm in figure 2. Let (p_0, U_0) be the result of *simplify*(p, A_0). Contrary to the description in figure 2, in our implementation U_0 is not required to be a subset of A_0. All that is required is that all of the formulas in U_0 are logically implied by A_0, and that the equivalence of p and p_0 is logically implied by U_0. As a result, something more sophisticated than subtracting out the sets $\{s\}$ and $\{\neg s\}$ must be done to compute a U that is weak enough to be logically implied by A (see figure 2). A second complication is that finding a cache hit requires checking sufficient conditions for logical implication between sets of formulas, rather than just set containment. However, dealing with these complications seems to be justified since the cache hit ratio is increased by having *simplify* return a U_0 that is weaker than it could be if it had to be a subset of A_0. We are still experimenting with ideas for balancing these issues more efficiently.

4   Experimental Results

In this section, we describe empirical results for applying our verification method to a pipelined ALU [5] and a subset of the DLX processor [14].

4.1   Pipelined ALU

The 3-stage pipelined ALU we considered (figure 3) has been used as a benchmark for BDD-based verification methods [3, 4, 5, 6]. A natural way to compare the performance of these methods is to see how the CPU time needed for verification grows as the pipeline is increased in size by (for example) increasing its datapath width w or its register file size r. For Burch, Clarke and Long [4] the CPU time grew roughly quadratically in w and cubically in r. Clarke, Grumberg and Long [6], using a simple abstraction provided by the user, demonstrated linear growth in both w and r. Sublinear growth in r and subquadratic growth in w was achieved by Bryant, Beatty and Seger [3].

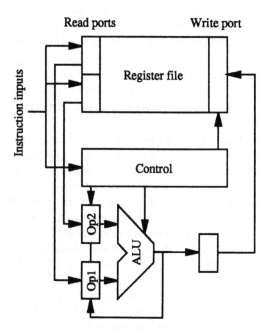

Fig. 3. 3-stage pipelined ALU. If the *stall* input is true, then no instruction is loaded. Otherwise, the *src1* and *src2* inputs provide the address of the arguments in the register file, the *op* input specifies the ALU operation to be performed on the arguments, and the *dest* input specifies were the result is to be written.

In our verification method, the width of the data path and the number of registers and ALU operations can be abstracted away. As a result, one verification run can check the control logic of pipelines with any combination of values for these parameters. A total of 370 milliseconds of CPU time (running compiled Lucid Common LISP on a DECstation 5000/240) is required to do a complete verification run, including loading and compiling behavioral descriptions, automatically constructing the abstraction function and related expressions, and checking the validity of the appropriate formula. The validity checking itself, the primary bottleneck on larger examples, only required 50 milliseconds for the pipelined ALU.

4.2 DLX Processor

Hennessy and Patterson [14] designed the DLX architecture to teach the basic concepts used in the MIPS 2000 and other RISC processors of that generation.

The subset of the DLX that we verified had six types of instructions: store word, load word, unconditional jump, conditional branch (branch when the source register is equal to zero), 3-register ALU instructions, and ALU immediate instructions. As with the pipelined ALU described earlier, the specifics of the ALU operations are abstracted away in both the specification and the implementation. Thus, our verification covers any set of ALU operations, assuming that the combinational ALU in the processor has been separately verified.

Our DLX implementation has a standard 5-stage pipeline. The DLX architecture has no branch delay slot; our implementation uses the "assume branch not taken" strategy. No pipelining is exposed in the DLX architecture or in our specification of it. Thus, it is the responsibility of the implementation to provide forwarding of data and a load interlock.

The interlock and the lack of a branch delay slot mean that the pipeline executes slightly less than one instruction per cycle, on average. This complicates "synchronizing" the implementation and the specification during verification, since the specification executes exactly one instruction per cycle. We address the problem in a manner similar to that used by Saxe et al. [20]. The user must provide a predicate on the implementation states that indicates whether the instruction to be loaded on the current cycle will actually be executed by the pipeline. While this predicate can be quite complicated, it is easy to express in our context, using internal signals generated by the implementation. In particular, our pipeline will not execute the current instruction if and only if one or more of the following conditions holds: the stall input is asserted, the signal indicating a taken branch is asserted, or the signal indicating that the pipeline has been stalled by the load interlock is asserted.

When internal signals are used in this way, it is possible for bugs in the pipeline to lead to a false positive verification result. In particular, the pipeline may appear correct even if it can get into a state where it refuses to ever execute another instruction (a kind of livelock). To avoid the possibility of a false positive, we automatically check a progress condition that insures that livelock cannot occur. The CPU time needed for this check is included in the total given below.

Our specification has four state variables: the program counter, the register file, the data memory and the instruction memory. If the data memory and the instruction memory are combined into one store in the specification and the implementation, then the verifier will detect that the pipeline does not satisfy the specification for certain types of self-modifying code (this has been confirmed experimentally). Separating the two stores is one way to avoid this inappropriate negative result.

For each state variable of the specification, the verifier constructs an appropriate formula and checks its validity. Since neither the specification nor the implementation write to the instruction memory, checking the validity of the corresponding formula is trivial. Checking the formulas for the program counter, the data memory and the register file requires 15.5 seconds, 34 seconds and 9.5 seconds of CPU time, respectively. The total CPU time required for the full verification (including loading and compiling the behavioral descriptions, etc.) is less than 66 seconds.

In another test, we introduced a bug in the forwarding logic of the pipeline. The verifier required about 8 seconds to generate 3 counter-examples, one each for the three formulas that had to be checked. These counter-examples provided sufficient conditions on a initial implementation state where the effects of the bug would be apparent. This information can be analyzed by hand, or used to construct a start state for a simulator run that would expose the bug.

5 Concluding Remarks

The need for improved debugging tools is now obvious to everyone involved in producing a new processor implementation. It is equally obvious that the problem is worsening rapidly: driven by changes in semiconductor technology, architectures are moving steadily from the simple RISC machines of the 1980s towards very complex machines which aggressively exploit concurrency for greater performance.

Although we have demonstrated that the techniques presented here can verify more complex processors with much less effort than previous work, examples such as our DLX implementation are still not nearly as complex as commercial microprocessor designs. We have also not yet dealt with memory systems and interrupts, which are rich source of bugs in practice.

It will be very challenging to increase the capacity of verification tools as quickly as designers are increasing the scale of the problem. Clearly, the computational efficiency of logical decision procedures (in practice, not in the worst case) will be a major bottleneck. If decision procedures cannot be extended rapidly enough, it may still be possible to use some of the same techniques for partial verification or in a mixed simulation/verification tool.

References

1. D. L. Beatty. *A Methodology for Formal Hardware Verification, with Application to Microprocessors.* PhD thesis, School of Computer Science, Carnegie Mellon University, Aug. 1993.
2. K. S. Brace, R. L. Rudell, and R. E. Bryant. Efficient implementation of a BDD package. In *27th ACM/IEEE Design Automation Conference*, 1990.
3. R. E. Bryant, D. L. Beatty, and C.-J. H. Seger. Formal hardware verification by symbolic ternary trajectory evaluation. In *28th ACM/IEEE Design Automation Conference*, 1991.
4. J. R. Burch, E. M. Clarke, and D. E. Long. Representing circuits more efficiently in symbolic model checking. In *28th ACM/IEEE Design Automation Conference*, 1991.
5. J. R. Burch, E. M. Clarke, K. L. McMillan, and D. L. Dill. Sequential circuit verification using symbolic model checking. In *27th ACM/IEEE Design Automation Conference*, 1990.

6. E. M. Clarke, O. Grumberg, and D. E. Long. Model checking and abstraction. In *Nineteenth Annual ACM Symposium on Principles on Programming Languages*, 1992.

7. A. J. Cohn. A proof of correctness of the Viper microprocessors: The first level. In G. Birtwistle and P. A. Subrahmanyam, editors, *VLSI Specification, Verification and Synthesis*, pages 27–72. Kluwer, 1988.

8. A. J. Cohn. Correctness properties of the Viper block model: The second level. In G. Birtwistle, editor, *Proceedings of the 1988 Design Verification Conference*. Springer-Verlag, 1989. Also published as University of Cambridge Computer Laboratory Technical Report No. 134.

9. F. Corella. Automated high-level verification against clocked algorithmic specifications. Technical Report RC 18506, IBM Research Division, Nov. 1992.

10. F. Corella. Automatic high-level verification against clocked algorithmic specifications. In *Proceedings of the IFIP WG10.2 Conference on Computer Hardware Description Languages and their Applications*, Ottawa, Canada, Apr. 1993. Elsevier Science Publishers B.V.

11. D. Cyrluk. Microprocessor verification in PVS: A methodology and simple example. Technical Report SRI-CSL-93-12, SRI Computer Science Laboratory, Dec. 1993.

12. D. Detlefs and G. Nelson. Personal communication, 1994.

13. J. L. Hennessy. Designing a computer as a microprocessor: Experience and lessons from the MIPS 4000. A lecture at the Symposium on Integrated Systems, Seattle, Washington, March 14, 1993.

14. J. L. Hennessy and D. A. Patterson. *Computer Architecture: A Quantitative Approach*. Morgan Kaufmann, 1990.

15. W. A. Hunt, Jr. FM8501: A verified microprocessor. Technical Report 47, University of Texas at Austin, Institute for Computing Science, Dec. 1985.

16. J. Joyce, G. Birtwistle, and M. Gordon. Proving a computer correct in higher order logic. Technical Report 100, Computer Lab., University of Cambridge, 1986.

17. M. Langevin and E. Cerny. Verification of processor-like circuits. In P. Prinetto and P. Camurati, editors, *Advanced Research Workshop on Correct Hardware Design Methodologies*, June 1991.

18. G. Nelson and D. C. Oppen. Simplification by cooperating decision procedures. *ACM Trans. Prog. Lang. Syst.*, 1(2):245–257, Oct. 1979.

19. A. W. Roscoe. Occam in the specification and verification of microprocessors. *Philosophical Transactions of the Royal Society of London, Series A: Physical Sciences and Engineering*, 339(1652):137–151, Apr. 15, 1992.

20. J. B. Saxe, S. J. Garland, J. V. Guttag, and J. J. Horning. Using transformations and verification in circuit design. Technical Report 78, DEC Systems Research Center, Sept. 1991.

21. R. E. Shostak. A practical decision procedure for arithmetic with function symbols. *J. ACM*, 26(2):351–360, Apr. 1979.

22. M. Srivas and M. Bickford. Formal verification of a pipelined microprocessor. *IEEE Software*, 7(5):52–64, Sept. 1990.

Using Abstractions for the Verification of Linear Hybrid Systems [*]

A. Olivero, J. Sifakis and S. Yovine

VERIMAG [**]
Miniparc-Zirst rue Lavoisier
38330 Montbonnot St. Martin, France

1 Introduction

Hybrid systems are dynamical systems consisting of interacting discrete and continuous components [NSY91, MMP91]. They are used to model the combined behavior of embedded real-time systems and their physical environments. Recently, there have been attempts to develop verification methods for hybrid systems by working in two complementary directions:

- The first direction concerns the identification of subclasses of hybrid systems for which there exist decidability results and effective verification methods for various classes of properties. The main decidability results concern very restricted classes of hybrid systems like timed automata [AD90, ACD90, Alu91, NSY91] and integration graphs [KPSY93].
- The second direction concerns the elaboration of a general verification methodology for classes of systems for which tractable semi-decision methods are applicable. One interesting class is linear hybrid systems [ACHH93, NOSY93], i.e., systems where conditions on discrete transitions and evolution laws of continuous variables are linear constraints.

This paper deals with the verification of linear hybrid systems by bringing results in the two mentioned directions:

- It identifies a class of linear hybrid systems which are shown to be bisimilar to timed automata for which the satisfaction of TCTL properties can be decided by model-checking.
- It studies transformations of general linear hybrid systems inducing abstractions which are timed automata. The properties expressed in $\forall$TCTL, a fragment of TCTL, are preserved on these systems.

The paper is organized as follows:

Section 2 presents the model of Linear Hybrid Systems (LHS) where continuous variables evolve according to linear functions with slopes varying within an interval. We define two sub-classes of LHS: Constant Slope Hybrid Systems (CSHS) [KPSY93], where the slopes of the evolution functions are constant and K-Timed Graphs (K-TG) that are CSHS where for all variables the slope is equal to a constant K.

[*] Supported by the ESPRIT project N°6021 REACT-P
[**] VERIMAG is a joint laboratory of CNRS, INPG, UJF and VERILOG S.A.

In Sect. 3 we review the syntax and the semantics of the logic TCTL used to describe system properties.

In Sect. 4 we give a simple transformation that maps the sub-class of CSHS with non-zero slopes into equivalent timed graphs in the sense that the underlying models are bisimilar. This implies in particular, that the initial and the transformed system satisfy the same TCTL formulas modulo a linear time transformation. These results are used to show that satisfaction of TCTL is decidable for transformable CSHS.

In Sect. 5 we show that the results of Sect. 4 cannot be extended to general LHS. We propose an extension of the transformation of Sect. 4 which applied to a LHS yields a timed graph which is an abstraction of it. Then we prove that the validity of $\forall$TCTL formulas is preserved for timed graphs that are abstractions of LHS.

2 Linear Hybrid Systems

2.1 Syntax

Let X be a set of *variables* ranging over the set $\mathbb{R}$ of real numbers. A *valuation* X_0 of X assigns to each variable $x \in X$ a real number $x_0 \in \mathbb{R}$. A *linear predicate* over X is a boolean combination of linear inequalities with integer coefficients over X.

A *linear hybrid system* (LHS) H is a structure $\langle S, X, E, A, B, \phi \rangle$ where:

- S is a finite set of *locations*. In a graphical representation of the system they are drawn as the nodes of the graph.
- X is a finite set of real-valued *variables*. These variables change *continuously* at locations and *discretely* via transitions.
- E is a finite set of *edges*. Each edge $e \in E$ is a tuple $\langle s, a, \psi, v, s' \rangle$, where $s \in S$ is the *source*, $s' \in S$ is the *target*, a is the *label*, ψ is a linear predicate called the *guard*, and v is an *assignment* represented as a set $\{x := v_x \mid x \in Y \subseteq X\}$ where v_x is an interval with end-points in $\mathbb{Z} \cup \{-\infty, +\infty\}$. In the graphical representation, each $e \in E$ is drawn as an edge from the source location to the target location with the label (a, ψ, v).
- A and B associate to each location $s \in S$ the functions A_s and B_s from X to $\mathbb{Z}$ such that $A_s \leq B_s$. The rate of change of the variable x at the location s belongs to the closed interval $[A_s(x), B_s(x)]$.
- ϕ associates with each location $s \in S$ a linear predicate ϕ_s over X called the *invariant* at s.

A *Constant Slope Hybrid System* (CSHS) [KPSY93] is a linear hybrid system such that $A = B$. That is, any variable x changes continuously at a constant rate $A_s(x)$ at s. We denote a CSHS as a tuple $H = \langle S, X, E, A, \phi \rangle$ rather than $H = \langle S, X, E, A, A, \phi \rangle$.

A *K-Timed Graph* (*K*-TG) is a constant slope hybrid system where all variables are nonnegative and change *uniformly* at the same rate $K > 0$, that is $A_s(x) = K$. Guards are boolean combinations of inequalities of the form

$l \prec x \prec u$ where $\prec \in \{<, \leq\}$ and $l \in \mathbb{N}$ and $u \in \mathbb{N} \cup \{+\infty\}$. We denote a K-TG as a tuple $H = \langle S, X, E, \mathbf{K}, \phi \rangle$.

A 1-*Timed Graph* differs from a timed graph [ACD90, Alu91, NSY91] in the sense that assignments are intervals rather than resets to zero. 1-Timed Graphs are an extension of timed graphs for which all the decidability and model-checking results concerning timed graphs can also be applied. For this reason 1-timed graphs are hereafter simply called timed graphs.

Notation: If A and B are tuples of scalars and c is a scalar, we write:
- $A(i)$ to represent the i-th component of A.
- $\frac{c}{A}$ to represent the tuple with i-th component $\frac{c}{A(i)}$.
- $A \prec B$ if $A(i) \prec B(i)$ for any component and $\prec \in \{<, \leq\}$.

2.2 Semantics

Let $H = \langle S, X, E, A, B, \phi \rangle$ be a linear hybrid system. A *state* of H is a pair (s, X_0) such that $\phi_s(X_0)$. We denote by Q the set of states of H. For $q = (s, X_0) \in Q$, $q(x)$ is the value x_0 of x at the valuation X_0.

H changes its state either by a *discrete* and *instantaneous* move through an edge, or by a *continuous* transformation of its variables while time elapses. During the continuous transformations at location s we require that the associated invariant ϕ_s continuously holds. The semantics of H is a labeled transition system $\langle Q, \rightarrow \rangle$ where the transition relation $\rightarrow$ between states is defined by the following rules:

- An edge $\langle s_0, a, \psi, v, s_1 \rangle \in E$ is *enabled* at a state (s_0, X_0) if X_0 satisfies the guard ψ. Whenever an edge is enabled at a state it may be taken, and the resulting state is a pair (s_1, X_1) where X_1 is such that for every variable $x \in X$, $x_1 \in v_x$ if $x := v_x \in v$, and $x_1 = x_0$ otherwise. We denote by $v(X_0)$ the set of such valuations.

$$\frac{\langle s_0, a, \psi, v, s_1 \rangle \in E, \psi(X_0), X_1 \in v(X_0)}{(s_0, X_0) \xrightarrow{a} (s_1, X_1)}$$

- Time can progress by t at (s, X_0), if there is a valuation X_1 reached from X_0 for a rate $\Lambda \in [A_s, B_s]$ and all the intermediate states satisfy the invariant associated to s.

$$\frac{X_1 = X_0 + \Lambda t, \Lambda \in [A_s, B_s], \forall 0 \leq t' \leq t. \phi_s(X_0 + \Lambda t')}{(s, X_0) \xrightarrow{t} (s, X_1)}$$

We introduce the following definitions:

A *trajectory* σ is an infinite sequence $q_0 \xrightarrow{t_0} q_1 \xrightarrow{t_1} \dots$ where for all $i \geq 0$ $q_i \xrightarrow{t_i} q_{i+1}$ is such that $q_i \xrightarrow{t_i} q_{i+1}$, or there exists a label a such that $q_i \xrightarrow{a} q_{i+1}$ and $t_i = 0$. A *position* π of σ is a pair (i, t) with $0 \leq t \leq t_i$. We define a total order $\preceq$ on the positions, as $(i, t) \preceq (j, t')$ iff $i < j$ or $i = j \wedge t < t'$. Let $q_i = (s_i, X_{0i})$ and $q_{i+1} = (s_i, X_{0i+1})$ where $X_{0i+1} = X_{0i} + \Lambda_i.t_i$, with $\Lambda_i \in [A_{s_i}, B_{s_i}]$. The *state at position* (i, t) of σ is $\sigma(i, t) = (s_i, X_{0i} + \Lambda_i.t)$. The *time elapsed at position*

(i, t) is $\tau_\sigma(i, t) = t + \sum_{j < i} t_j$, that is, the time elapsed since the beginning of the trajectory. A trajectory σ *diverges* if for all $t \in \mathbb{R}^+$ there exists a position π such that $\tau_\sigma(\pi) > t$. The set of all divergent trajectories is represented by Σ, and the set of all divergent trajectories starting at q by $\Sigma(q)$.

Remark. A different semantics for LHS is proposed in [AHH93] for which the results developed in the paper are also valid.

Example 1. The temperature of a tank grows linearly with time at rates belonging to the interval [4,6]. For safety reasons the temperature must be kept within a minimum of 79 and a maximum of 90 degrees. A controller cools the tank by moving two independent rods. The rates of cooling belong to the intervals [-4,-1] and [-3,-2], respectively. Furthermore, each rod can be used again only if a given time has elapsed since the end of its previous movement. These minimum times are 7 and 5 minutes, respectively. If the temperature cannot be decreased because there is no available rod, a complete shutdown is required. A variant of this example is given in [JLHM91, NOSY93]. Figure 1 shows a linear hybrid system for the controller.

We use three variables: θ measures the temperature of the tank and x_1 and x_2 count the time elapsed since the last use of the respective rod. Initially, the temperature takes its minimum value and both rods can be used.

The equations at locations specify the evolution laws and the predicate specifies the invariant for the location.

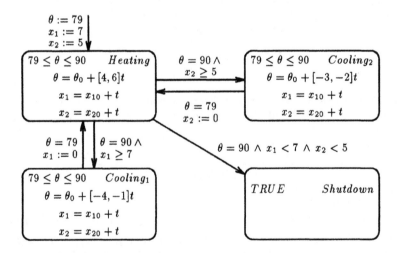

Fig. 1. Temperature control system

3 The Logic TCTL

The real-time temporal logic TCTL is a real-time extension of CTL [CES86]. Many versions of TCTL have been defined [ACD90, Alu91, HNSY92]. We adopt here the one defined in [ACD90] using the temporal operators $\exists \mathcal{U}$ and $\forall \mathcal{U}$ of CTL with subscripts that limit their scope in time.

Intuitively, $f_1 \exists \mathcal{U}_{\#n} f_2$ means that there exists a trajectory such that f_2 holds at some state at a time t satisfying $t \# n$ and f_1 continuously holds at all intermediate states. $f_1 \forall \mathcal{U}_{\#n} f_2$ means that for every trajectory the above property holds. For example, we can write $\mathbf{true} \exists \mathcal{U}_{<5} p$ to say that along some trajectory, p becomes true before time 5.

3.1 Syntax

Let P be a finite set of atomic propositions. The formulas of TCTL are defined by the following grammar:

$$f ::= p \mid l \prec x \prec u \mid \neg f \mid f_1 \vee f_2 \mid f_1 \exists \mathcal{U}_{\#n} f_2 \mid f_1 \forall \mathcal{U}_{\#n} f_2$$

where $p \in P$, $x \in X$, $l, u \in \mathbb{Z} \cup \{-\infty, +\infty\}$, $\prec \in \{<, \leq\}$, $n \in \mathbb{N}$ and $\# \in \{<, \leq , >, \geq, =\}$.

We use standard abbreviations such as $\forall \Diamond_{\#n} f$ for $\mathbf{true} \forall \mathcal{U}_{\#n} f$, $\exists \Diamond_{\#n} f$ for $\mathbf{true} \exists \mathcal{U}_{\#n} f$, $\exists \Box_{\#n} f$ for $\neg \forall \Diamond_{\#n} \neg f$ and $\forall \Box_{\#n} f$ for $\neg \exists \Diamond_{\#n} \neg f$. In Sect. 5 we use the abbreviation $f_1 \forall \mathcal{W}_{\#n} f_2$ for the formula $\neg(\neg f_1 \exists \mathcal{U}_{\#n} \neg f_2)$. The unrestricted temporal operators of CTL correspond to TCTL operators subscripted by ≥ 0.

3.2 Semantics

TCTL formulas are interpreted over the divergent trajectories generated by the model of a linear hybrid system. Let H be a LHS, $\langle Q, \rightarrow \rangle$ its model, and M an interpretation of atomic propositions over the set of states, i.e. $M(p) \subseteq Q$ for all $p \in P$.

The satisfaction relation *the state $q \in Q$ satisfies the formula f under the interpretation M*, denoted $q \models_M f$, is inductively defined as follows:

$q \models_M p$ iff $q \in M(p)$;
$q \models_M l \prec x \prec u$ iff $l \prec q(x) \prec u$;
$q \models_M \neg f$ iff $q \not\models_M f$;
$q \models_M f_1 \vee f_2$ iff $q \models_M f_1$ or $q \models_M f_2$;
$q \models_M f_1 \exists \mathcal{U}_{\#n} f_2$ iff for some sequence $\sigma \in \Sigma(q)$ there exists a position π such that $\sigma(\pi) \models_M f_2$ and $\tau_\sigma(\pi) \# n$, and for all $\pi' \preceq \pi$, $\sigma(\pi') \models_M f_1 \vee f_2 \wedge \tau_\sigma(\pi') \# n$;
$q \models_M f_1 \forall \mathcal{U}_{\#n} f_2$ iff for all sequences $\sigma \in \Sigma(q)$ there exists a position π such that $\sigma(\pi) \models_M f_2$ and $\tau_\sigma(\pi) \# n$, and for all $\pi' \preceq \pi$, $\sigma(\pi') \models_M f_1 \vee f_2 \wedge \tau_\sigma(\pi') \# n$;

H satisfies f under M, denoted $H \models_M f$, if all the states of H satisfy f.

Example 2. We express in TCTL three properties of Example 1:

- The location *Shutdown* is not reachable:

$$(Heating \land \theta = 79 \land x_1 = 7 \land x_2 = 5) \Rightarrow \forall \Box \neg Shutdown \qquad (1)$$

- Rod 1 (Rod 2) can be used again only if at least 7 minutes (5 minutes) have elapsed since the end of its previous movement:

$$(Heating \land \theta = 79 \land x_1 = 0) \Rightarrow \neg \exists \Diamond_{<7} Cooling_1 \qquad (2)$$

$$(Heating \land \theta = 79 \land x_2 = 0) \Rightarrow \neg \exists \Diamond_{<5} Cooling_2 \qquad (3)$$

4 From Constant Slope Hybrid Systems to Timed Graphs

We define the class of Non-Zero Constant Slope Hybrid System, denoted $\text{CSHS}_{\neq 0}$, such that:

- All the rates are different from zero.
- The invariant and guards are boolean combinations of inequalities of the form $l \prec x \prec u$ with $\prec \in \{<, \leq\}$.
- For all edges $e = \langle s, a, \psi, v, s' \rangle \in E$ and for all variables $x \in X$, if the rates of change of x are different at s and s', then $x := v_x$ is in v.
- At any location s any variable with positive (negative) slope has a finite lower (upper) bound in the invariant ϕ_s.

In this section we show how a given $\text{CSHS}_{\neq 0}$ can be transformed into an equivalent TG. To this end, we use two transformations: the K-transformation and the $\frac{1}{K}$-transformation, whose successive application to a $\text{CSHS}_{\neq 0}$ gives a K-timed graph and a timed graph. The key idea is to apply linear transformations of variables which are proper to each location. For the K-transformation the aim is to obtain the same evolution laws for all variables and locations by multiplying by appropriate constants. Consequently, conditions and assignments are modified. The $\frac{1}{K}$-transformation consists in changing the time scale. Both transformations preserve bisimilarity of the underlying models.

4.1 Translating $\text{CSHS}_{\neq 0}$ into K-TG

The K-transformation. Given $H = \langle S, X, E, A, \phi \rangle$ in $\text{CSHS}_{\neq 0}$, we define the set of functions $\kappa = \{\kappa_s \mid s \in S\}$ as follows:

$$\kappa_s(X) = \frac{K}{A_s} X + C_s$$

where:

- $K = \text{lcm}\{\text{abs}(A_s(x)) \mid s \in S, x \in X\}$, that is, K is the least common multiple of the absolute values of the rates appearing in H, and
- C_s is a tuple such that $C_s(x)$ is the maximum of the absolute values of all the constants appearing in assignments to x in edges with target location s or in conditions on x in guards of edges with source location s, multiplied by $\text{abs}(\frac{K}{A_s(x)})$.

For given $H = \langle S, X, E, A, \phi \rangle$ in $\text{CSHS}_{\neq 0}$, we construct a K-timed graph $H_K = \langle S, X', E', \mathbf{K}, \phi' \rangle$ such that:

- X' is in bijection with X.
- For each $e = \langle s, a, \psi, v, s' \rangle \in E$ we construct the edge $e' = \langle s, a, \psi', v', s' \rangle \in E'$ where ψ' is obtained replacing each condition of the form $l \prec x \prec u$ appearing in ψ by a condition $l' = \frac{K}{A_s(x)}l + C_s(x) \prec x' \prec \frac{K}{A_s(x)}u + C_s(x) = u'$ when the rate $A_s(x)$ is positive, or $u' \prec x' \prec l'$ otherwise. The assignment v' is obtained replacing each $x := [\alpha, \beta] \in v$ by $x' := [\frac{K}{A_s(x)}\alpha + C_s(x), \frac{K}{A_s(x)}\beta + C_s(x)] = [\alpha', \beta']$ when the rate $A_s(x)$ is positive, or $x' := [\beta', \alpha']$ otherwise.
- ϕ'_s is obtained from ϕ_s exactly as ψ' is obtained from ψ.

Remark. When the slope of x is positive (negative) and l or α (u or β) are $-\infty$ (∞) we take the conjunction with the invariant ϕ_s before applying the transformation. This operation does not change the semantics of the system.

Notice that H_K is constructed from H so as the relation $X' = \kappa_s(X)$ holds at each location s.

Property Preservation by K-transformation. We want to prove that H and H_K are bisimilar. We recall here the definition of bisimulation.

Definition 1. Let $T_1 = \langle Q_1, \rightarrow_1 \rangle$ and $T_2 = \langle Q_2, \rightarrow_2 \rangle$ be two labeled transition systems over a set $\mathcal{L}$ of labels and a relation $\rho \subseteq Q_1 \times Q_2$.

The relation ρ is a *bisimulation* between T_1 and T_2, if $\rho(q_1, q_2)$ implies for every label $\ell \in \mathcal{L}$:

- If $q_1 \xrightarrow{\ell}_1 q'_1$, then there exists q'_2 such that $q_2 \xrightarrow{\ell}_2 q'_2$ and $\rho(q'_1, q'_2)$.
- If $q_2 \xrightarrow{\ell}_2 q'_2$, then there exists q'_1 such that $q_1 \xrightarrow{\ell}_1 q'_1$ and $\rho(q'_1, q'_2)$.

We say that T_1 and T_2 are *bisimilar*, denoted $T_1 \sim T_2$, if there exists a bisimulation ρ between T_1 and T_2.

For H_1 and H_2 linear hybrid systems, we write $\langle H_1, q_1 \rangle \sim \langle H_2, q_2 \rangle$ if the transition systems generated respectively from q_1 and q_2 are bisimilar. We write $H_1 \sim H_2$ if there exists a bisimulation ρ between the models of H_1 and H_2.

We naturally extend the function κ to a relation on states in the following way:

$$\kappa((s, X), (s', X')) \text{ iff } s = s' \text{ and } X' = \kappa_s(X)$$

Proposition 2. *For all $\text{CSHS}_{\neq 0}$ H, $H \sim H_K$.*

Proof. It is straightforward to prove that the relation κ is a bisimulation between H and H_K. □

Our goal is to prove that H and H_K satisfy the same TCTL formulas modulo the transformation of the predicates over X according to κ. So, for any formula f we define a formula $| f |_K$, such that H satisfies f iff H_K satisfies $| f |_K$. The formula $| f |_K$ is obtained by substitution of any predicate of the form $l \prec x \prec u$ ocurring in f by $\bigvee_{s \in S} p_s \wedge l' \prec x' \prec u'$, where p_s is an atomic proposition which holds only at location s and $l' \prec x' \prec u'$ is the condition defined in Sect. 4.1.

Proposition 3. *For any H in $CSHS_{\neq 0}$ and for any TCTL formula f:*

$$H \models_M f \text{ iff } H_K \models_M f \mid_K$$

Proof. By induction on the structure of the formula based on the fact that $H \sim H_K$. □

Example (continued). We consider the CSHS H obtained from the LHS of Fig. 1 by taking the rate of θ equal to 6 at location $Heating$, to -4 at location $Cooling_1$ and to -3 at location $Cooling_2$. The evolution laws for θ are different at all locations. However, since all conditions are of the form $\theta = c$ (for a constant c) we can add the assignment $\theta := c$ without changing the semantics of the system. Thus, $H \in CSHS_{\neq 0}$ and the K-transformation is applicable to obtain the K-TG of Fig. 2. The set of functions κ is given by:

$$\theta' = 2\theta + 180 \quad \text{at location } Heating$$
$$\theta' = -3\theta + 270 \quad \text{at location } Cooling_1$$
$$\theta' = -4\theta + 360 \quad \text{at location } Cooling_2$$
$$x_1' = 12x_1 + 84 \quad \text{at all locations}$$
$$x_2' = 12x_2 + 60 \quad \text{at all locations}$$

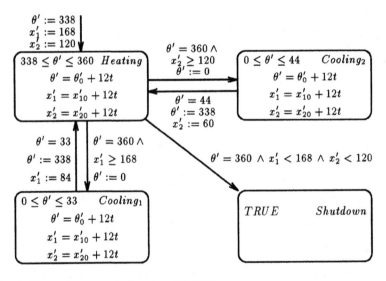

Fig. 2. Temperature control system (K-TG)

The transformation of formulas (1), (2) and (3) according to $\mid \cdots \mid_K$ gives:

$$(Heating \wedge \theta' = 338 \wedge x_1' = 168 \wedge x_2' = 120) \Rightarrow \forall \square \neg Shutdown \quad (4)$$
$$(Heating \wedge \theta' = 338 \wedge x_1' = 84) \Rightarrow \neg \exists \Diamond_{<7} Cooling_1 \quad (5)$$
$$(Heating \wedge \theta' = 338 \wedge x_2' = 60) \Rightarrow \neg \exists \Diamond_{<5} Cooling_2 \quad (6)$$

By Proposition 3, H satisfies formulas (1), (2) and (3) iff H_K satisfies formulas (4), (5) and (6), respectively.

4.2 Translating K-TG into TG

The problem of verifying TCTL formulas on K-timed graphs can be reduced to the problem of verifying TCTL formulas on timed graphs. Given a K-timed graph H_K a timed graph H_1 can be obtained such that the satisfaction of a TCTL formula f on H_K is equivalent to the satisfaction of a formula $| f |_{\frac{1}{K}}$ obtained by transforming f.

The $\frac{1}{K}$-transformation. Given a K-timed graph $H_K = \langle S, X, E, \mathbf{K}, \phi \rangle$, this transformation gives a timed graph $H_1 = \langle S, X, E, \mathbf{1}, \phi \rangle$. That is, an evolution law of the form $x = x_0 + Kt$ in H_K is replaced by the law $x = x_0 + t$ in H_1.

Property Preservation by $\frac{1}{K}$-transformation. Given a formula f, we define a formula $| f |_{\frac{1}{K}}$ by replacing the subscripts $\#n$ by $\#Kn$ in the temporal operators.

Proposition 4. *For any K-timed graph H_K and any* TCTL *formula f:*

$$H_K \models_M f \text{ iff } H_1 \models_M | f |_{\frac{1}{K}}$$

Example (continued). The $\frac{1}{K}$-transformation is applicable to the system H_K of Fig. 2 to obtain a TG H_1. The transformation of formulas (4), (5) and (6) according to $| \cdots |_{\frac{1}{K}}$ gives:

$$(Heating \wedge \theta' = 338 \wedge x_1' = 168 \wedge x_2' = 120) \Rightarrow \forall \Box \neg Shutdown \qquad (7)$$
$$(Heating \wedge \theta' = 338 \wedge x_1' = 84) \Rightarrow \neg \exists \Diamond_{<84} Cooling_1 \qquad (8)$$
$$(Heating \wedge \theta' = 338 \wedge x_2' = 60) \Rightarrow \neg \exists \Diamond_{<60} Cooling_2 \qquad (9)$$

By Proposition 4 H_K satisfies formulas (4), (5) and (6) iff H_1 satisfies formulas (7), (8), and (9), respectively.

The following results were obtained using the tool KRONOS [NSY92, OY92] to verify the formulas (7), (8) and (9) on the timed graph H_1.

formula	time in sec.	iterations	result
(7)	0.050	4	*true*
(8)	0.066	4	*true*
(9)	0.083	4	*true*

Then H_K satisfies formulas (4), (5) and (6) and by Proposition 3 the CSHS$_{\neq 0}$ H satisfies formulas (1), (2) and (3).

Synthesis of the results. By combination of Propositions 3 and 4, we obtain:

Proposition 5. *For any CSHS$_{\neq 0}$ H and any* TCTL *formula f:*

$$H \models_M f \ \textit{iff} \ H_K \models_M | f |_K \ \textit{iff} \ H_1 \models_M \| f |_K |_{\frac{1}{K}}$$

This proposition shows that the model-checking problem for CSHS$_{\neq 0}$ and TCTL is reduced to the problem of model-checking for TG. Now, the model-checking algorithm in [HNSY92] can be adapted for the timed graphs considered here, that is timed graphs having assignments of intervals instead of resets to zero. It can be shown that in this case the meaning of the formulas can be expressed with the same class of state predicates.

5 From Linear Hybrid Systems to Timed Graphs

In this section we extend the translation method proposed for CSHS to LHS. We first define a subclass of LHS, denoted LHS$_{\neq 0}$, such that:
- For all locations $s \in S$ and variables $x \in X$, the end-points of the rate interval $A_s(x)$ and $B_s(x)$ are both positive or both negative, that is $0 \notin [A_s(x), B_s(x)]$.
- The invariant and the guards are boolean combinations of inequalities of the form $l \prec x \prec u$ with $\prec \in \{<, \leq\}$.
- For all edges $e \in E$ and for all variables $x \in X$, if the rates of change of x are different at the source location and at the target location, then x must be in the assignment.
- At any location s any variable with positive (negative) slope has a finite lower (upper) bound in the invariant ϕ_s.

5.1 Translating LHS$_{\neq 0}$ into K-TG

The following proposition shows that unfortunately the result that for any H in CSHS$_{\neq 0}$ there exist a bisimilar K-TG cannot be extended to LHS$_{\neq 0}$.

Proposition 6. *Given a $H = \langle S, X, E, A, B, \phi \rangle$ in LHS$_{\neq 0}$ such that for some $s \in S$*
- *$A_s < B_s$*
- *there exists $e \in E$ such that $e = \langle s, a, [L, U], v, s' \rangle$*

there exists no H_K in K-TG with initial state q_K such that $\langle H, q \rangle \sim \langle H_K, q_K \rangle$ for all the states q from which there exists a trajectory crossing the edge e.

The Generalized K-transformation. Proposition 6 states that it is not possible (in general) to transform a given LHS$_{\neq 0}$ into a bisimilar K-TG. However, we define a transformation of H into an abstract H', in the sense that every transition in the model of H corresponds to a transition in the model of H' (the model of H simulates the model of H').

Given $H = \langle S, X, E, A, B, \phi \rangle$ in LHS$_{\neq 0}$, and a set of variables X' in bijection with X, we define the relation κ as the set $\kappa = \{\kappa_s \mid s \in S\}$ relating X and X' as follows:

$$\kappa_s(X, X') \quad \text{iff} \quad \frac{K}{B_s}X + C_s \leq X' \leq \frac{K}{A_s}X + C_s$$

where:

- $K = \text{lcm}\{\text{abs}(A_s(x)), \text{abs}(B_s(x)) \mid x \in X, s \in S\}$, that is K is the least common multiple of the absolute values of all different end-points of the rate intervals.
- C_s is a tuple such that $C_s(x)$ is the maximum of the absolute values of all the constants appearing in assignments to x in edges with target location s or in conditions on x in guards of edges with source location s, multiplied by $\max(\text{abs}(\frac{K}{A_s(x)}), \text{abs}(\frac{K}{B_s(x)}))$.

From H and κ the K-timed graph $H_K = \langle S, X', E', \mathbf{K}, \phi' \rangle$ is defined as follows:

- X' is in bijection with X.
- For each $e = \langle s, a, \psi, v, s' \rangle \in E$ we construct the edge $e' = \langle s, a, \psi', v', s' \rangle \in E'$ where ψ' is obtained replacing each condition of the form $l \prec x \prec u$ appearing in ψ by a condition $l' = \frac{K}{B_s(x)}l + C_s(x) \prec x' \prec \frac{K}{A_s(x)}u + C_s(x) = u'$ when $A_s(x)$ and $B_s(x)$ are positive, or $u' \prec x' \prec l'$ otherwise. The assignment v' is obtained replacing each $x := [\alpha, \beta] \in v$ by $x' := [\frac{K}{B_s(x)}\alpha + C_s(x), \frac{K}{A_s(x)}\beta + C_s(x)] = [\alpha', \beta']$ when $A_s(x)$ and $B_s(x)$ are positive, or $x' := [\beta', \alpha']$ otherwise.
- ϕ'_s is obtained from ϕ_s exactly as ψ' is obtained from ψ.

The remark of Sect. 4.1 is also applicable in this case. Is important to note that when $A = B$, H is a CSHS$_{\neq 0}$ and the generalized K-transformation coincides with the K-transformation defined in Sect. 4.1.

Simulation. We want to prove that H simulates H_K. We first recall the definition of simulation.

Definition 7. Let $T_1 = \langle Q_1, \to_1 \rangle$ and $T_2 = \langle Q_2, \to_2 \rangle$ be two labeled transition systems over a set $\mathcal{L}$ of labels and a relation $\rho \subseteq Q_1 \times Q_2$.

The relation ρ is a *simulation* from T_1 to T_2, if $\rho(q_1, q_2)$ implies for every label $\ell \in \mathcal{L}$:

- If $q_1 \xrightarrow{\ell}_1 q'_1$, then there exists q'_2 such that $q_2 \xrightarrow{\ell}_2 q'_2$ and $\rho(q'_1, q'_2)$.

We say that T_1 *simulates* T_2, denoted $T_1 \sqsubseteq T_2$, if there exists a relation ρ such that ρ is a simulation from T_1 to T_2.

For H_1 and H_2 linear hybrid systems, we write $H_1 \sqsubseteq H_2$ if there exists a simulation ρ from the model of H_1 to the model of H_2.

We naturally extend the relation κ on states in the following way:

$$\kappa((s, X), (s', X')) \quad \text{iff} \quad s = s' \text{ and } \kappa_s(X, X')$$

Lemma 8. *For all LHS$_{\neq 0}$ H, $H \sim_\kappa H_K$ if only timed transitions are considered.*

Lemma 9. *For all $LHS_{\neq 0}$ H, $H \sqsubseteq H_K$ if only discrete transitions are considered.*

Proposition 10. *For all $LHS_{\neq 0}$ H, $H \sqsubseteq H_K$.*

Proof. By combining Lemmas 8 and 9. □

Property Preservation by K-transformation. We show that an interesting subset of TCTL, called ∀TCTL, is preserved by the relation K from H_K to H, that is, if $f \in$ ∀TCTL is satisfied by H_K then H satisfies a corresponding formula $| f |_K \in$ ∀TCTL. ∀TCTL is the subset of TCTL defined as follows:

$$f ::= p \mid \neg p \mid l \prec x \prec u \mid f_1 \wedge f_2 \mid f_1 \vee f_2 \mid f_1 \forall \mathcal{W}_{\#n} f_2 \mid f_1 \forall \mathcal{U}_{\prec n} f_2$$

where $f_1 \forall \mathcal{W}_{\#n} f_2$ and p, x, l, u, $\prec$, n and $\#$ are as in definition of Sect. 3.1.

For any formula f in ∀TCTL the formula $| f |_K$ is obtained by replacing any predicate of the form $l \prec x \prec u$ by $\frac{l - C_s(x)}{K} A_s(x) \prec x' \prec \frac{u - C_s(x)}{K} B_s(x)$.

Proposition 11. *For any $LHS_{\neq 0}$ H and any ∀TCTL formula f,*

$$\text{if } H_K \models_M f \text{ then } H \models_M | f |_K .$$

Example 3. Consider the example of Sect. 1 modeled by the LHS H of Fig. 1. The relation κ for H is:

$$
\begin{aligned}
2\theta + 270 \leq \theta' \leq 3\theta + 270 &\quad \text{at location } Heating \\
-12\theta + 1080 \leq \theta' \leq -3\theta + 1080 &\quad \text{at location } Cooling_1 \\
-6\theta + 540 \leq \theta' \leq -4\theta + 540 &\quad \text{at location } Cooling_2 \\
x_1' = 12x_1 + 84 &\quad \text{at all locations} \\
x_2' = 12x_2 + 60 &\quad \text{at all locations}
\end{aligned}
$$

Figure 3 shows the K-TG H_K obtained. The $\frac{1}{K}$-transformation is applicable to the system H_K to obtain a TG H_1. Consider the following formulas:

$$(Heating \wedge 428 \leq \theta' \leq 507 \wedge x_1' = 168 \wedge x_2' = 120) \Rightarrow \forall \Box \neg Shutdown \ (10)$$
$$(Heating \wedge 428 \leq \theta' \leq 507 \wedge x_1' = 84) \Rightarrow \neg \exists \Diamond_{<84} Cooling_1 \ (11)$$
$$(Heating \wedge 428 \leq \theta' \leq 507 \wedge x_2' = 60) \Rightarrow \neg \exists \Diamond_{<60} Cooling_2 \ (12)$$

It can be shown that if H_1 satisfies formulas (10), (11) and (12) then the LHS$_{\neq 0}$ H of Fig. 1 satisfies formulas (1), (2) and (3).

The following results were obtained using the tool KRONOS to verify the formulas (10), (11) and (12) on the TG H_1.

formula	time in sec.	iterations	result
(10)	0.084	7	*false*
(11)	0.117	4	*true*
(12)	0.084	4	*true*

Notice that formulas (11) and (12) are verified which implies that formulas (2) and (3) are satisfied by the system of Fig. 1. On the contrary, formula (10) is not verified and nothing can be said about the validity of formula (1).

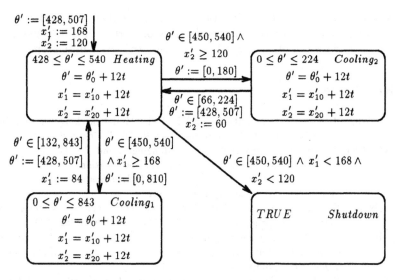

Fig. 3. Temperature control system (H_K)

6 Conclusion

As mentioned in the introduction, the paper contributes in two different directions to the development of a verification methodology for LHS:

- it shows that the model checking method for TCTL on timed graphs can be extended to a non-trivial class of CSHS, the so called Non-Zero Constant Slope Hybrid Systems $CSHS_{\neq 0}$. Even though the latter class is semantically equivalent to timed graphs, we noticed that several examples of LHS in the literature happen to belong to $CSHS_{\neq 0}$.
- it shows that well-known results of property preservation by abstractions can be adapted to the case of $LHS_{\neq 0}$. The generalized K-transformation presented in Sect. 5 maps this class to timed graphs and, consequently, the verification of a large class of properties such as invariance and bounded inevitability, on some element of $LHS_{\neq 0}$, can be reduced to the verification of the same property on the corresponding timed graph. It remains to evaluate through examples the usefulness of this approach, that is how much faithful are the abstractions defined by K.

These results should be related to the results of [PV94] which consider almost the same class of LHS by using two different approaches. The restriction to have assignments for variables whose evolution law changes is essential for both. The differences concern our requirement that at some position the sign of the rate of the variables cannot change while we accept also open intervals for guards and assignments. The restriction to closed intervals is essential for the decidability result presented in [PV94]. This result is certainly an important one and stronger than our results if they are restricted to decidability of the reachability problem. However, a general comparison is difficult for at least two reasons. The result in [PV94] is obtained by digitization which means that it is hardly exploitable in practice and it concerns only reachability properties; it is not obvious that it can be extended to other properties, for instance inevitability properties.

References

[ACD90]   R. Alur, C. Courcoubetis, and D. Dill. Model-checking for real-time systems. In *Proc. 5th LICS*, pages 414–425. IEEE Comp. Soc. Press, 1990.

[ACHH93]   R. Alur, C. Courcoubetis, T. A. Henzinger, and Pei-Hsin Ho. Hybrid automata: an algorithmic approach to the specification and analysis of hybrid systems. In *Workshop on Theory of Hybrid Systems*, Lyngby, Denmark, June 1993. LNCS 736, Springer-Verlag.

[AD90]   R. Alur and D. Dill. Automata for modeling real-time systems. In *Proc. 17th ICALP*, pages 322–335. LNCS 443, Springer-Verlag, 1990.

[AHH93]   R. Alur, T. Henzinger, and P.-H. Ho. Automatic symbolic verification of embedded systems. In *14 th Annual Real-Time Systems Symposium*, pages 2–11. IEEE Comp. Soc. Press, 1993.

[Alu91]   R. Alur. *Techniques for automatic verification of real-time systems*. PhD thesis, Department of Computer Science, Stanford University, August 1991.

[CES86]   E.M. Clarke, E.A. Emerson, and A.P. Sistla. Automatic verification of finite-state concurrent systems using temporal logic specifications. *ACM Transactions on Programming Languages and Systems*, 8(2):244–263, 1986.

[HNSY92]   T. Henzinger, X. Nicollin, J. Sifakis, and S. Yovine. Symbolic model-checking for real-time systems. In *Proc. 7th LICS*. IEEE Comp. Soc. Press, 1992.

[JLHM91]   M. Jaffe, N. Leveson, M. Heimdahl, and B. Melhart. Software requirements analysis for real-time process-control systems. *IEEE Transactions on Software Engineering*, 17(3):241–258, 1991.

[KPSY93]   Y. Kesten, A. Pnueli, J. Sifakis, and S. Yovine. Integration graphs: a class of decidable hybrid systems. In *Workshop on Theory of Hybrid Systems*, Lyngby, Denmark, June 1993. LNCS 736, Springer-Verlag.

[MMP91]   O. Maler, Z. Manna, and A. Pnueli. From timed to hybrid systems. In *Proc. REX Workshop "Real-Time: Theory in Practice"*, the Netherlands, June 1991. LNCS 600, Springer-Verlag.

[NOSY93]   X. Nicollin, A. Olivero, J. Sifakis, and S. Yovine. An approach to the description and analysis of hybrid systems. In *Workshop on Theory of Hybrid Systems*, Lyngby, Denmark, June 1993. LNCS 736, Springer-Verlag.

[NSY91]   X. Nicollin, J. Sifakis, and S. Yovine. From ATP to timed graphs and hybrid systems. In *Proc. REX Workshop "Real-Time: Theory in Practice"*, the Netherlands, June 1991. LNCS 600, Springer-Verlag.

[NSY92]   X. Nicollin, J. Sifakis, and S. Yovine. Compiling real-time specifications into extended automata. *IEEE TSE Special Issue on Real-Time Systems*, 18(9):794–804, September 1992.

[OY92]   A. Olivero and S. Yovine. *Kronos: a Tool for Verifying Real-time Systems. User's Guide and Reference Manual*. VERIMAG, Grenoble, France, 1992.

[PV94]   A. Puri and P. Varaiya. Decidability of hybrid systems with rectangular differential inclusions. This volume.

Decidability of Hybrid Systems with Rectangular Differential Inclusions*

Anuj Puri and Pravin Varaiya

Department of Electrical Engineering and Computer Science,
University of California, Berkeley, CA-94720

Abstract. A hybrid system is modeled with a finite set of locations and a differential inclusion associated with each location. We discuss a subclass of hybrid systems with constant rectangular differential inclusions. The continuous state of the system is $x \in \mathbb{R}^n$ with x_i evolving with differential inclusion $\dot{x}_i \in [L_i, U_i]$ where L_i, U_i are integers (i.e., the slope of trajectory of x_i could be changing, but is restricted to remain within $[L_i, U_i]$). A transition from one location to another can be made provided the state satisfies the enabling condition for the transition. The state can also be initialized to a new value during the transition. The differential inclusion for x_i can be changed when x_i is an integer or when x_i is initialized to a new value. We show that the verification problem for this class of hybrid systems is decidable. With this approach, systems with unsynchronized and drifting clocks can be modeled, a general differential equation can be abstracted by breaking the state space into regions with constant differential inclusions, and many previously presented hybrid system examples can be verified.

1 Introduction

Hybrid systems are modeled as automata with a finite set of locations and continuous state $x \in \mathbb{R}^n$. There is a differential inclusion at each location and the edges between locations have enabling conditions [1, 12, 13, 16]. The hybrid system starts in a specified location with an initial condition $x_0 \in \mathbb{R}^n$. The continuous trajectory evolves according to the differential inclusion associated with that location. At some time t, $x(t)$ satisfies the enabling condition for a transition and a jump is made to a new location. The state x could be initialized to a new value during the jump. At the new location, x begins evolving according to the differential inclusion associated with that location. After some time, another transition is made, and so on.

In this paper, we study hybrid systems with constant differential inclusions of the form $[L_1, U_1] \times \cdots \times [L_n, U_n]$ where L_i, U_i are integers. The continuous state of the system, $x \in \mathbb{R}^n$, evolves according to $\dot{x}_i \in [L_i, U_i], i = 1, \cdots, n$ (i.e., the slope of trajectory x_i could be changing but is restricted to remain within

---

* Research supported by NSF under grants ECS 9111907 and IRI 9120074, and by the PATH program, University of California, Berkeley

$[L_i, U_i])$. These are also called Bounded-Rate Automata in [4]. A transition from one location to another can be made provided the state satisfies the enabling condition for the transition. A transition to a new location with different differential inclusion can be made only when x_i is an integer value or when x_i is initialized. During the transition, the state can be initialized to a new value. We show that under these conditions, interesting verification problems for the hybrid system are decidable. In particular, we show that the languages generated by our hybrid automata are regular. Our model does not include integration graphs [10], since we permit the differential inclusion for x_i to change only when x_i is initialized, or when x_i is an integer.

Systems with clocks [6, 2, 3, 9, 8, 15], where $\dot{x}_i = 1$, are special cases of the hybrid systems we consider. With our approach, systems with unsynchronized or drifting clocks can be modeled, systems with differential equations can be abstracted by breaking the state space into regions with constant differential inclusions [16], and it follows that for many hybrid system examples [1, 13], there is a decision procedure that will terminate.

In Sect. 2, we introduce the sub-class of hybrid systems. In Sect. 3, we present the decidability results.

2 Hybrid Automata

2.1 Preliminaries

Notation $\mathbb{R}$ is the set of reals and $\mathbb{Z}$ is the set of integers. $\mathbb{R}^+$ is the set of non-negative reals and $\mathbb{Z}^+$ is the set of non-negative integers. For $X \subset \mathbb{Z}$, define $X\Delta = \{k\Delta | k \in X\}$. For an interval $[a, b]$, where $a, b \in \mathbb{Z}$ and $\frac{1}{\Delta} \in \mathbb{Z}^+$, define $[a, b]_\Delta = \{a, a + \Delta, \ldots, b\}$. For $x \in \mathbb{R}$, define $\lfloor x \rfloor_\Delta = k\Delta$ where k is the largest integer for which $k\Delta \leq x$. In this paper, we always take the floor with respect to Δ, so we write $\lfloor x \rfloor$ instead of $\lfloor x \rfloor_\Delta$.

Differential Inclusion A differential equation is $\dot{x} = f(x)$ where $x \in \mathbb{R}^n$ and $f : \mathbb{R}^n \to \mathbb{R}^n$. A solution to the differential equation with initial condition $x_0 \in \mathbb{R}^n$ is any differentiable function $\phi(t)$, where $\phi : \mathbb{R}^+ \to \mathbb{R}^n$ such that $\phi(0) = x_0$ and $\dot{\phi}(t) = f(\phi(t))$.

A differential inclusion is written as $\dot{x} \in f(x)$ where $x \in \mathbb{R}^n$ and f is a set-valued map from $\mathbb{R}^n$ to $\mathbb{R}^n$ (i.e., $f(x) \subset \mathbb{R}^n$). A solution to the differential inclusion with initial condition $x_0 \in \mathbb{R}^n$ is any differentiable function $\phi(t)$, where $\phi : \mathbb{R}^+ \to \mathbb{R}^n$ such that $\phi(0) = x_0$ and $\dot{\phi}(t) \in f(\phi(t))$.

A differential equation with a given initial condition has a unique solution (under Lipshitz conditions), whereas a differential inclusion has a family of solutions.

In this paper we consider constant differential inclusions $\dot{x} \in \beta$, $\beta = [L_1, U_1] \times \cdots \times [L_n, U_n]$ where $L_i, U_i \in \mathbb{Z}$ (i.e., $\dot{x}_i \in [L_i, U_i]$). We define $\mathcal{B}$ to be the set of all such constant differential inclusions.

Enabling Conditions Enabling conditions will be associated with edges between locations in the hybrid automaton. Similar to [2],we define Φ inductively to be the set of all enabling conditions:

$\delta := x_i \le c | x_i \ge c | \delta_1 \wedge \delta_2 | \delta_1 \vee \delta_2$ where $c \in \mathbb{Z}$.

Enabling conditions are closed subsets of $\mathbb{R}^n$.

Setting the Initial State During a transition, some components of the state may be initialized to a new value. We associate an initialization relation $\lambda = (\lambda_1, \cdots, \lambda_n)$ with an edge where $\lambda_i = [l_i, u_i]$ or $\lambda_i = id$, and $l_i, u_i \in \mathbb{Z}$. When $\lambda_i = id$ (the identity relation), the value of x_i does not change during the transiton. But for $\lambda_i = [l_i, u_i]$, x_i is initialized non-deterministically to a value in $[l_i, u_i]$. For $x \in \mathbb{R}^n$, define

$$\lambda[x] = \{x' \in \mathbb{R}^n | x_i' = x_i \text{ for } \lambda_i = id, \text{ and } x_i' \in [l_i, u_i] \text{ for } \lambda_i = [l_i, u_i]\}.$$

We define S to be the set of all initialization relations.

2.2 Syntax

A hybrid automaton $H = (L, \Sigma, D, \mathcal{I}, \psi)$ where L is a finite set of locations, Σ is a finite set of events, $D : L \rightarrow \mathcal{B}$ associates a differential inclusion with each location, $\mathcal{I} \subset L$ is a set of initial locations, and $\psi \subset L \times L \times \Sigma \times \Phi \times S$ are the edge labels ($(l, l', \sigma, \delta, \lambda) \in \psi$ labels edge (l, l') with event σ, enabling condition δ, and initialization relation λ). We further require that the differential inclusion for x_i is changed only when x_i is initialized. That is, for edge label $(l, l', \sigma, \delta, \lambda)$, $d = D(l)$, $d' = D(l')$, $d_i = d_i'$ when $\lambda_i = id$.

Fig. 1 is an example of the kind of hybrid automaton we consider in this paper. Note that the differential inclusion for y is changed when making the transition from location C to location D because y is equal to -4 at the transition. That is the same as checking y is equal to -4, followed by initializing y to -4. However, the differential inclusion for x cannot change when going from location C to location D.

2.3 Semantics

The hybrid automaton starts at an initial location with state $x = 0$. At location l, the state x moves according to the differential inclusion $D(l)$. It can make a transition from location l to location l' with edge label $(l, l', \sigma, \delta, \lambda)$ provided $x \in \delta$. After the transition, the state is $x' \in \lambda[x]$ and the new differential inclusion is $D(l')$.

The state trajectory x moves in two phase steps [14]. In the first phase, time progresses and x changes continuously. In the second phase, a sequence of transitions is made instantaneously (Fig. 2). Formally, x is a multiple-valued function. It is defined on $[0, T_0], [T_0', T_1], [T_1', T_2] \ldots$ with $T_i' = T_i$. For $T_i' < T_{i+1}$, x is differentiable on $[T_i', T_{i+1}]$. A transition is made at time T_i, with the state

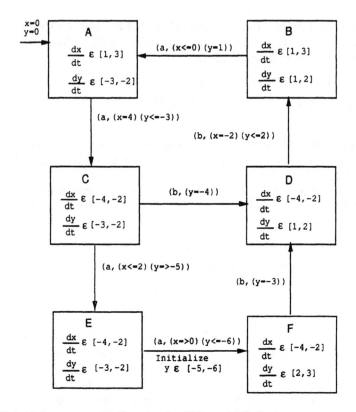

Fig. 1. Hybrid Automata with Rectangular Differential Inclusions

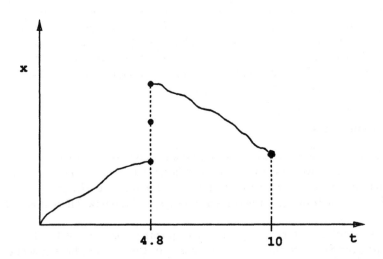

Fig. 2. Sample Trajectory

being $x(T_i)$ before the transition, and $x(T_i')$ after the transition. In Fig. 2, x is defined on $[0, 4.8], [4.8, 4.8], [4.8, 10]$, and it makes two successive transitions at time 4.8.

We associate a language $\mathcal{L}(H)$ with our hybrid automaton. Given $\sigma \in \Sigma^\omega$, we say $\sigma \in \mathcal{L}(H)$ provided there is a sequence of edges $< l_0, l_1, \sigma_0, \delta_0, \lambda_0 >$ $< l_1, l_2, \sigma_1, \delta_1, \lambda_1 > \cdots$ where $l_0 \in \mathcal{I}$ and $< D(l_0), \delta_0, \lambda_0 >< D(l_1), \delta_1, \lambda_1 > \cdots$ is consistent. We say $< D(l_0), \delta_0, \lambda_0 >< D(l_1), \delta_1, \lambda_1 > \cdots$ is consistent when there is a trajectory x for which this sequence of transitions is feasible.

Definition 1. A sequence $< D(l_0), \delta_0, \lambda_0 >< D(l_1), \delta_1, \lambda_1 > \ldots$ is continuous-time consistent provided there is a multiple-valued function $x : \mathbb{R}^+ \to \mathbb{R}^n$ and a sequence of intervals $[0, T_0], [T_0', T_1], [T_1', T_2] \ldots$ with $T_i' = T_i$ such that
1) $x(0) = 0$
2) $x(T_i) \in \delta_i$
3) $x(T_i') \in \lambda_i[x(T_i)]$
4) For $T_i' < T_{i+1}$, $\dot{x}(t) \in D(l_i)$ for $t \in [T_i', T_{i+1}]$

We similarly define the hybrid system which operates in discrete time according to a difference equation. Define

$$\frac{1}{\Delta} = LCM\{L_i, U_i | D_i(l) = [L_i, U_i] \text{ for } l \in L \text{ and } 1 \leq i \leq n\}$$

where LCM is the least common multiple of the set. For the example in Fig. 1, $\frac{1}{\Delta} = LCM\{1, 2, 3, 4\} = 12$.

Definition 2. A sequence $< D(l_0), \delta_0, \lambda_0 >< D(l_1), \delta_1, \lambda_1 > \ldots$ is discrete-time consistent provided there is a multiple-valued function $x : (\mathbb{Z}^+ \Delta) \to (\mathbb{Z}\Delta)^n$ and a sequence of intervals $[0, T_0], [T_0', T_1], [T_1', T_2] \ldots$ where $T_i \in (\mathbb{Z}^+ \Delta)$ and $T_i' = T_i$ such that
1) $x(0) = 0$
2) $x(T_i) \in \delta_i$
3) $x(T_i') \in (\lambda_i[x(T_i)])_\Delta$ (i.e., for $(\lambda_i)_j = [a, b]$, $x_j(T_i') \in [a, b]_\Delta$, and for $(\lambda_i)_j = id$, $x_j(T_i') = x_j(T_i)$)
4) For $T_i' < T_{i+1}$, $x_j((n+1)\Delta) - x_j(n\Delta) \in D(l_i)_j \Delta$ for $T_i' \leq n\Delta, (n+1)\Delta \leq T_{i+1}$

We call H_C the hybrid system which operates in continuous time and H_D the system which operates in discrete time. For $\sigma \in \mathcal{L}(H_C)$, there is a sequence $< D(l_0), \delta_0, \lambda_0 >< D(l_1), \delta_1, \lambda_1 > \ldots$ which is continuous-time consistent. Similarly for $\sigma \in \mathcal{L}(H_D)$, there is a sequence $< D(l_0), \delta_0, \lambda_0 >< D(l_1), \delta_1, \lambda_1 > \ldots$ which is discrete-time consistent.

3 Decidability Results

Our main result is that $\mathcal{L}(H_C)$ is a regular language. To prove this, we follow an approach similar to [8, 15]. We first show that $\mathcal{L}(H_C) = \mathcal{L}(H_D)$ and then prove that $\mathcal{L}(H_D)$ is regular.

We show $\mathcal{L}(H_C) = \mathcal{L}(H_D)$ by proving a sequence $\rho = < D(l_0), \delta_0, \lambda_0 > <$
$D(l_1), \delta_1, \lambda_1 > \ldots$ is discrete time-consistent iff it is continuous-time consistent. A discrete-time consistent sequence ρ has a discrete trajectory x_d. The continuous trajectory x_c obtained by linear interpolation from x_d also satisfies the constraints of ρ. Therefore, every discrete-time consistent sequence is also continuous-time consistent. The converse, that a continuous-time consistent sequence is also discrete-time consistent, is more difficult to prove. Lemma 3 states that a sequence ρ with continuous trajectory will also have a piece-wise linear trajectory satisfying it (Fig. 3). Lemma 4 shows that continuous trajectory x_i on $[t_j, t_{j+1}]$ with integer end points can be made into a trajectory on $[\lfloor t_j \rfloor, \lfloor t_{j+1} \rfloor]$ with same integer end points (Fig. 4). In Lemma 5 and Lemma 6, we show that for a sequence ρ and a continuous trajectory satisfying it at $T_0 \leq T_1 \leq T_2 \ldots$, there is another continuous trajectory which satisfies it at $\lfloor T_0 \rfloor, \lfloor T_1 \rfloor, \lfloor T_2 \rfloor, \ldots$. In Lemma 7, we finally show that there is a discrete trajectory x_d which also satisfies ρ.

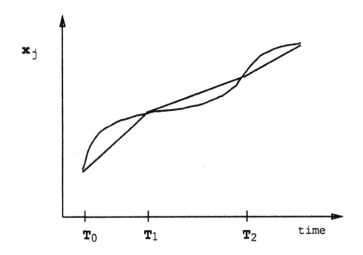

Fig. 3. Mean-Value Theorem

Lemma 3. *If a sequence $\rho = < D(l_0), \delta_0, \lambda_0 > < D(l_1), \delta_1, \lambda_1 > \ldots$ has a continuous trajectory, then it also has a piecewise linear trajectory (Fig. 3).*

Proof: Suppose x satisfies ρ at $T_0 \leq T_1 \leq T_2 \ldots$, then form x' by linear interpolation between points $x(0), x(T_0), x(T_1), \ldots$. From the Mean-Value theorem, it follows x' satisfies ρ.

Lemma 4. *Given a differentiable function x_i on $[t_j, t_{j+1}]$ with $\dot{x}_i \in [L, U]$ and $x_i(t_j) = k, x_i(t_{j+1}) = l$ where $k, l \in \mathbb{Z}$. We define x'_i on $[\lfloor t_j \rfloor, \lfloor t_{j+1} \rfloor]$ by linear interpolation between $x'_i(\lfloor t_j \rfloor) = k$ and $x'_i(\lfloor t_{j+1} \rfloor) = l$ (Fig. 4). Then $\dot{x}'_i \in [L, U]$.*

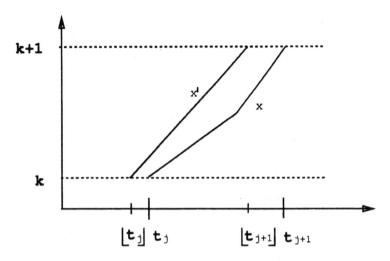

Fig. 4. Discretized Times for Trajectory with Integer End-Points

Proof: Since

$$L \leq \frac{x_i(t_{j+1}) - x_i(t_j)}{t_{j+1} - t_j} \leq U$$

for $U > 0$, we get

$$\frac{x_i(t_{j+1}) - x_i(t_j)}{U} = \frac{l - k}{U} = m\Delta \leq \lfloor t_{j+1} \rfloor - \lfloor t_j \rfloor$$

where $m \in \mathbb{Z}$. Since $x'_i(\lfloor t_{j+1} \rfloor) = x_i(t_{j+1})$ and $x'_i(\lfloor t_j \rfloor) = x_i(t_j)$, we get

$$\frac{x'_i(\lfloor t_{j+1} \rfloor) - x'_i(\lfloor t_j \rfloor)}{\lfloor t_{j+1} \rfloor - \lfloor t_j \rfloor} \leq U$$

Similar proof holds when $U \leq 0$. Similarly

$$L \leq \frac{x'_i(\lfloor t_{j+1} \rfloor) - x'_i(\lfloor t_j \rfloor)}{\lfloor t_{j+1} \rfloor - \lfloor t_j \rfloor}$$

Lemma 5. *Given a differentiable function x_i on $[b, c]$ with $\dot{x}_i \in [L, U]$, there is a function x'_i on $[\lfloor b \rfloor, \lfloor c \rfloor]$ with $\dot{x}'_i \in [L, U]$ such that for $j, k \in \mathbb{Z}$, $j \leq x_i(t) \leq k$ implies $j \leq x'_i(\lfloor t \rfloor) \leq k$ (Fig. 5).*

Proof: We look at the integer crossing points of x_i (Fig. 5). Using consecutive integer crossing points t_1, t_2, from Lemma 4, we obtain x'_i on $[\lfloor t_1 \rfloor, \lfloor t_2 \rfloor]$ where $\dot{x}'_i \in [L, U]$. For $t \in [t_1, t_2]$ and $j, k \in \mathbb{Z}$, $j \leq x_i(t) \leq k$ implies $j \leq x'_i(\lfloor t \rfloor) \leq k$. In case $x_i(b)$ is not an integer, we extend x_i to $a < b$ (keeping $\dot{x}_i \in [L, U]$) so that $x_i(a)$ is an integer and then reason as above. Similar reasoning applies if $x_i(c)$ is not an integer. After obtaining x'_i, we restrict it to $[\lfloor b \rfloor, \lfloor c \rfloor]$.

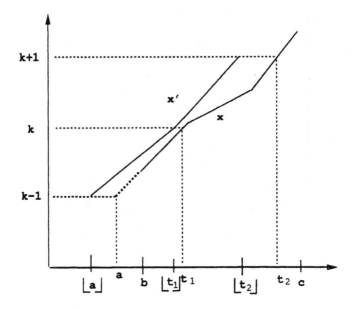

Fig. 5. Creating a Trajectory with Discretized Times

Lemma 6. *If a trajectory x satisfies ρ at times $T_0, T_1, T_2, \ldots$, then there is a trajectory x' which satisfies ρ at times $\lfloor T_0 \rfloor, \lfloor T_1 \rfloor, \lfloor T_2 \rfloor, \ldots$.*

Proof: Given x_i on $[0, T_{k_1}][T_{k_1}, T_{k_2}], \ldots$ where T_{k_j} are times at which x_i is initialized and x_i is continuous on interval $[T_{k_j}, T_{k_{j+1}}]$ with $\dot{x}_i \in [L_j, U_j]$ (x_i maybe discontinuous because it could get initialized to a new value at T_{k_j}). The x'_i on $[0, \lfloor T_{k_1} \rfloor][\lfloor T_{k_1} \rfloor, \lfloor T_{k_2} \rfloor], \ldots$ from Lemma 5 satisfies ρ at times $\lfloor T_0 \rfloor, \lfloor T_1 \rfloor, \lfloor T_2 \rfloor, \ldots$. Since for $i = 1, \ldots, n$, x'_i satisfies ρ at $\lfloor T_0 \rfloor, \lfloor T_1 \rfloor, \lfloor T_2 \rfloor, \ldots$, we get x' satisfies ρ.

Lemma 7. *If a continuous trajectory x satisfies ρ at times $\lfloor T_0 \rfloor, \lfloor T_1 \rfloor, \lfloor T_2 \rfloor, \cdots$ then the discrete trajectory x_d where $(x_d)_i(k\Delta) = \lfloor x_i(k\Delta) \rfloor$ also satisfies ρ at times $\lfloor T_0 \rfloor, \lfloor T_1 \rfloor, \lfloor T_2 \rfloor \ldots$.*

Proof: x_d satisfies the enabling conditions since $k \leq x_i(\lfloor T_j \rfloor) \leq l$ implies $k \leq \lfloor x_i(\lfloor T_j \rfloor) \rfloor \leq l$ for $k, l \in \mathbb{Z}$. Furthermore $(x_d)_i$ also satisfies the difference inclusion constraints because

$$L \leq \frac{x_i((k+1)\Delta) - x_i(k\Delta)}{\Delta} \leq U$$

implies

$$L \leq \frac{\lfloor x_i((k+1)\Delta) \rfloor - \lfloor x_i(k\Delta) \rfloor}{\Delta} \leq U$$

Theorem 8. $\mathcal{L}(H_C) = \mathcal{L}(H_D)$

Proof: $\sigma \in \mathcal{L}(H_D)$ $(\sigma \in \mathcal{L}(H_C))$ provided there is a sequence $\rho = < D(l_0), \delta_0, \lambda_0 > < D(l_1), \delta_1, \lambda_1 > \ldots$ which is discrete-time (continuous-time) consistent. As discussed at beginning of Sect. 3, every discrete-time consistent sequence is continuous-time consistent. From Lemma 3 - Lemma 7, every continuous-time consistent sequence is also discrete-time consistent. Thus $\sigma \in \mathcal{L}(H_D)$ iff $\sigma \in \mathcal{L}(H_C)$.

Theorem 9. $\mathcal{L}(H_D)$ *is a regular language.*

Proof: We will construct a finite state automaton which generates $\mathcal{L}(H_D)$. Let M_j be the largest integer with which x_j is compared or initialized, and m_j, the smallest such integer. Define $\Gamma_j = \{<\} \cup \{m_j, m_j + \Delta, \ldots, M_j\} \cup \{>\}$, $\Gamma = \Gamma_1 \times \cdots \times \Gamma_n$, and the finite set of states $Q = L \times \Gamma$. The finite state automaton $H_D = (Q, \Sigma, \rightarrow)$ where the transition relation $\rightarrow \subset Q \times \Sigma \times Q$ is defined as:

- $(l, v) \xrightarrow{\epsilon} (l, v')$ where $v'_j = "<"$ when $v_j = "<"$ or $v_j + w < m_j$ for some $w \in D(l)_j \Delta$; $v'_j = ">"$ when $v_j = ">"$ or $v_j + w > M_j$ for some $w \in D(l)_j \Delta$; $v'_j \in \{m_j, m_j + \Delta, \ldots, M_j\}$ when $v'_j - v_j \in D(l)_j \Delta$.
- $(l, v) \xrightarrow{\sigma} (l', v')$ provided $(l, l', \sigma, \delta, \lambda) \in \psi$, $v \in \delta$, and $v'_j \in (\lambda_j[v_j])_\Delta$.

The first part of the definition represents passage of Δ time and the second is an instantaneous transition.

It is not necessary to keep track of x_j when it exceeds M_j. This is clear when $\dot{x}_j \in [L_j, U_j]$ where $L_j, U_j \geq 0$ because once x_j exceeds M_j, it can become less than M_j only by being initialized. But it remains true even when $L_j \leq 0$ and $U_j \geq 0$ because any trajectory which exceeds M_j and then falls below it can be replaced by another which stays at M_j. Similar reasoning applies to m_j.

Theorem 10. $\mathcal{L}(H_C)$ *is a regular language*

Proof: From Theorem 8 and Theorem 9.

References

1. R. Alur, C. Courcoubetis, T.A. Henzinger and P.-H. Ho, Hybrid automata: an algorithmic approach to the specification and analysis of hybrid systems, *Workshop on Theory of Hybrid Systems*, Lyngby, Denmark, October 1992.
2. R. Alur and D. Dill, Automata for modeling real-time systems, *Proc. 17th ICALP*, Lecture Notes in Computer Science 443, Springer-Verlag, 1990.
3. R. Alur and T. Henzinger, Logics and models of real-time: A survey, *Proc. REX Workshop Real-Time: Theory in Practice*, Lecture Notes in Computer Science 600, Springer-Verlag, The Netherlands, June 1991.
4. R. Alur, T.A. Henzinger and P.-H. Ho, Automatic symbolic verification of embedded systems, *Proc. of the 14th Annual Real-time Systems Symposium*, IEEE Computer Society Press, 1993.
5. R. Alur, A. Itai, R. Kurshan and M. Yannakakis, Timing Verification by Successive Approximation, *Proc. 4th Workshop Computer-Aided Verification*, Lecture Notes in Computer Science 663, Springer-Verlag, 1992.
6. D. Dill, Timing Assumptions and verification of finite-state concurrent systems, *Automatic Verification Methods for Finite-State Systems*, Lecture Notes in Computer Science 407, Springer-Verlag, 1989.

7. F.Balarin and A.Sangiovanni-Vincentelli, A Verification Strategy for Timing-Constrained Systems, *Proc. 4th Workshop Computer-Aided Verification*, Lecture Notes in Computer Science 663, Springer-Verlag, 1992.

8. T. Henzinger, Z. Manna and A. Pnueli, What Good are Digital Clocks ?, *Proc. 19th ICALP*, Lecture Notes in Computer Science, Springer-Verlag, 1992.

9. T.A. Henzinger, X. Nicollin, J. Sifakis and S.Yovine, Symbolic model-checking for real-time systems, *Proc. 7th Symp. on Logics in Computer Science*, IEEE Computer Society Press, 1992.

10. Y. Kesten, A. Pnueli, J. Sifakis and S. Yovine, Integration graphs: a class of decidable hybrid systems, *Workshop on Theory of Hybrid Systems*, Lyngby, Denmark, October 1992.

11. O. Maler, Z. Manna and A. Pnueli, From Timed to Hybrid Systems, *Proc. REX Workshop on Real-Time: Theory in Practice*, Lecture Notes in Computer Science 600, Springer-Verlag, The Netherlands, June 1991.

12. J. A. Mcmanis, Verification and Control of Real-Time Discrete Event Dynamical Systems, Ph.D Thesis, Department of Electrical Engineering and Computer Science, University of California, Berkeley, 1993.

13. X.Nicollin, A.Olivero, J. Sifakis and S. Yovine, An Approach to the Description and Analysis of Hybrid Systems, *Workshop on Theory of Hybrid Systems*, Lyngby, Denmark, October 1992.

14. X. Nicollin and J. Sifakis, An overview and synthesis on timed process algebras, *Proc. 3rd Workshop Computer-Aided Verification*, Denmark, July 1991.

15. A. Puri, Real-Time Systems: Discrete Time vs. Dense Time, Unpublished, May 1993.

16. A. Puri and P. Varaiya, Modeling and Verification of Hybrid Systems, Preprint.

17. *Proc. REX Workshop on Real-Time: Theory in Practice*, Lecture Notes in Computer Science 600, Springer-Verlag, The Netherlands, June 1991.

Suspension Automata: A Decidable Class of Hybrid Automata*

Jennifer McManis and Pravin Varaiya

University of California at Berkeley

Abstract. A hybrid automaton consists of a discrete state component represented by a finite automaton, coupled with a (vector) continuous state component governed by a differential equation. For hybrid automata it is possible to reduce certain verification problems to those of checking language containment or language emptiness. Here we present a class of hybrid automata called suspension automata for which conditions can be given under which these problems are decidable.

1 Introduction

The state of a hybrid system has two components. There is a discrete component which evolves as in a finite automaton, and a (vector) continuous component which is governed by a differential equation. The two components interact: the differential equation in force at any time is determined by the state of the automaton, and the occurrence of transitions of the automaton is determined by the value of the continuous component. Details of the modeling formalism depend on tradition and on the purpose at hand as seen in the recent volume [1]. The model used here is a modification of the 'hybrid automaton' of [2]. It is called the 'suspension automaton.'

Verification problems for hybrid automata are not always decidable. The most important decidable special case is the timed automaton [3, 4, 5]. A timed automaton comprises a finite automaton and a collection of 'timers.' A timer has a continuous positive component which increases at rate one. The value of the timer states enables or disables discrete transitions of the system. At such transitions, a timer's value may be reset to zero. Timed automata are used to impart real-time constraints on the transitions of finite state machines. Important digital circuit timing analysis and (bounded) timed Petri nets can be formulated in this way. Software implementations of decision procedures for these automata are being developed [6, 7].

For many other hybrid systems the timed automaton is not a sufficiently descriptive model. Among these are common scheduling, synchronization, and communication policies. These systems require that, at the minimum, one should be able to assign timers a rate of 0 as well as 1. The following is a simple example of this sort of system.

---

* Research supported by National Science Foundation under grants ECS111907 and IRI9120074.

Example 1. Scheduling Jobs Under Preemptive Least Time Remaining Policy

Suppose a single processor must perform several different types of tasks upon request. For each task type there is a lower bound on the time between requests and a fixed time needed to perform the task. If more than one task has been requested at a given time then the processor must have some policy for determining which task is performed first. One common policy is the *least time remaining* policy. Under this policy the processor performs the task requiring the least amount of time to complete. As a concrete example, suppose we have two task types with the following characteristics.

Task 1: interarrival time ≥ 8 and processing time $= 2$

Task 2: interarrival time ≥ 4 and processing time $= 3$

One question that is commonly asked is: 'Will the processor complete a task of a given type before the next task of that type is requested?' In this case it is easy to see that the answer to the question will be 'no.' If Task 2 is preempted for service of Task 1, then a second request for Task 2 may arrive before completion of service to the originally requested task.

Formally, problems such as those posed for Ex. 1 may be addressed by examining the logical sequencing of events possible for the system. One common technique is verification through language containment:

> Given a specification represented by a language (set of event sequences) $\mathcal{L}^{spec}$ and system behavior represented by a language $\mathcal{L}^{sys}$ check whether $\mathcal{L}^{sys} \subseteq \mathcal{L}^{spec}$?

This problem is decidable if both $\mathcal{L}^{sys}$ and $\mathcal{L}^{spec}$ are ω-regular, that is finite state representations exist for the languages. In general, the language $\mathcal{L}^{sys}$ for a hybrid system need not be ω-regular. This paper is devoted to defining conditions under which a finite state representation may be derived.

A suspension automaton is an extension of a timed automaton in which a timer can be 'suspended,' i.e., its rate of increase can be set to 0. A suspended timer can later be unsuspended to resume increasing at rate 1. A 'rate assignment' specifies which timers are suspended or unsuspended as a function of the discrete components of the state. Suspension automata can be used to represent scheduling, synchronization, and communication policies in a very natural fashion. Unfortunately, the untimed language of a suspension automata is not guaranteed to be ω-regular. In order to avoid this problem, additional restrictions must be made to the form of the automata. In this paper, a trick is introduced by which one may replace the suspension of a timer by the decrementation of its value. In Sect. 2 the suspension automaton is defined. It is shown in Sect. 3 that the system of Ex. 1 may be modeled using the suspension automaton. In Sect. 4 a condition is given under which the untimed language is guaranteed to be finite state. It is shown that the system in Ex.1 may be represented by an automaton satisfying the conditions. However, two problems remain. First, the conditions given in Sect. 4 are not easily checked. Second, the automata representing the systems of interest which satisfy the conditions are not easily derived. In Sect. 5 a systematic method is given for transforming a class of suspension automata (including those such as the one given in Sect. 3) into automata guaranteed to satisfy the conditions given in Sect. 4. Together these results give a reasonable procedure for verifying an interesting class of hybrid systems.

2 The Suspension Automaton

The suspension automaton defined below is similar to the hybrid automaton of [2]. It limits the dynamic behavior of the timers to a rate of increase of either one or zero. In addition, it expands the timer reset operation to allow for the decrementation of timer values by integer amounts. A *suspension automaton* is a tuple $(\mathcal{A}, L_0, \mathcal{F})$ where $\mathcal{A}$ is a suspension transition system denoting the causal behavior of the system, $L_0 \subseteq L$ is a restriction on the initial conditions of the system, and $\mathcal{F} \subseteq \mathcal{P}(L)$ (where $\mathcal{P}(\cdot)$ represents the power set) is a restriction on the set of locations visited infinitely often.

Formally, $\mathcal{A}$ is a tuple $(L, \Sigma, T, R, V_{inc}, Edge)$ where:

- L is a finite set of locations representing the discrete state of the system.
- Σ is a finite event set.
- T is a collection of timers $\{T_i\}_{i=1}^{N}$. Each timer has associated with it $v_i \in \mathbb{R}$, the value of the timer. Let v be a vector of dimension N representing the timer values.
- $R = \{R^l : l \in L\}$ is the rate assignment where $R^l \subseteq \{0,1\}^N$ defines the possible rates of timers in location l. For any $r \in R^l$, r_i gives the rate of T_i. If R^l has a single element for each $l \in L$ say that the automaton is *rate deterministic*.
- $V_{inc} = \{V^l : l \in L\}$ where each $V^l \subseteq \mathbb{R}^N$ is an inclusion condition. While in any given location l, the timer values must remain in V^l.
- $Edge \subseteq L \times \Sigma \times L \times \mathcal{P}(\mathbb{R}^N) \times Reset$ is an edge relation. Each $e \in Edge$ is of the form $e = (l, \sigma, l', V^e, reset^e)$ where:
    - l is the current location, σ an event, and l' the next location. This corresponds to the discrete edge transition.
    - V^e is an enabling condition. In order to make a transition, the current timer values must be in V^e.
    - $reset^e$ is a collection of conditions each of which either resets the timer value to 0 (denoted by $v_i := 0$) or decrements the timer by some fixed integer constant (denoted by $v_i := v_i - c$).

Only very specific forms of V^e and V^l will be considered. These will be called *simple regions*. Say that $V \subseteq \mathbb{R}^N$ is a simple region if there exists a set of constraint equations each of the form; $v_i \sim c$ or $v_i - v_j \sim c$ where $\sim$ is one of

$<, >, \leq, \geq$ or $=$, and $c \in \mathbb{Z}$, such that V is the set of all $v \in \mathbb{R}$ satisfying the constraints.

The suspension automaton may be depicted by a labeled directed graph. Ex. 2 shows this representation. Here, each location is given as a node with a location label, rate assignment, and inclusion set denoted as a collection of constraint equations. Edges are represented as directed arcs between nodes labeled with an event, a set of constraint equations representing V^e, and $reset^e$.

Example 2. The Representation of the Task 1 Component of the System

There are two events – the arrival of a request denoted by $\mathbf{r_1}$ and the completion of task service denoted by $\mathbf{c_1}$. The request interarrival time is regulated by an interarrival timer T_2 and the service time by a process timer T_1.

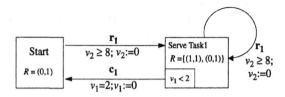

The constraint $v_2 \geq 8$ ensures that the task request event $\mathbf{r_1}$ occurs no more often that every 8 time units. The nondeterministic rate assignment indicates that the rate is not under the control of the task, but subject to external control. The inclusion condition $v_1 < 2$ together with the enabling condition $v_1 = 2$ ensures that the completion event $\mathbf{c_1}$ will occur when the task has received the required amount of service.

2.1 Behavior of Suspension Systems

Formally, the behavior of the system will be represented in terms of runs. The run describes the behavior of the system in terms of timer behavior as the system moves from location to location. Formally, given a sequence ρ drawn from Σ, a *run* for ρ with respect to $\mathcal{A}$ is an object of the following form:

$$_{v(0)}(l(0), t(0))_{v'(0)} \xrightarrow{\rho(1)} _{v(1)} (l(1), t(1))_{v'(1)} \xrightarrow{\rho(2)} _{v(2)} (l(2), t(2))_{v'(2)} \cdots$$

where:

- $v(0) = 0$.
- for all $n \geq 0$:
  - $t(n) \geq 0$.
  - There exists $r \in R^l$ where $l = l(n)$ such that $v'(n) = v(n) + rt(n)$.
  - For all $0 \leq t < t(n)$, $v(n) + rt \in V^l$ where $l = l(n)$.
  - There exists an edge $e = (l, \sigma, l', V^e, reset^e)$ where:

* $l = l(n)$, $\sigma = \rho(n+1)$, and $l' = l(n+1)$.
* The enabling conditions are satisfied – $v'(n) \in V^e$.
* The reset conditions are satisfied.
  · If $v_i := 0 \in reset^e$, then $v_i(n+1) = 0$.
  · Else if $v_i := v_i - c \in reset^e$, then $v_i(n+1) = v_i'(n) - c$.
  · Else, $v_i(n+1) = v_i'(n)$.

A run is said to be an *accepting run* if $l(0) \in L_0$ and $inf(\{l\}) \in \mathcal{F}$ where $inf(\{l\})$ is the set of all locations visited infinitely often. Let $\mathcal{L}^{seq}(\mathcal{A}, L_0, \mathcal{F})$ be the set of all ρ with accepting runs.

3 Representation of Scheduling Using Suspension Automata

Scheduling may be represented quite naturally through rate assignment where a process timer is assigned rate 1 if its task is of highest priority and rate zero otherwise. Fig. 1 shows this for Ex. 1. For the sake of simplicity, the representation of interarrival times is omitted. In addition, edges representing requests arriving before completion of already requested tasks of the same type are not shown. The locations represent the status of the tasks in the system – waiting for service, receiving service, or not requested. There are only two timers, process timers T_1 for Task 1 and T_2 for Task 2. The timer values have the interpretation of processor time received. Thus the priority decision (and hence the rate assignment) may be formulated as a property of the timer values.

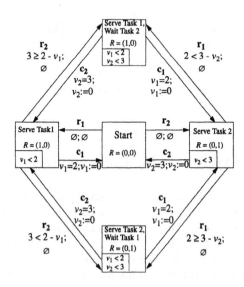

Fig. 1. Representation of least time remaining policy using timer suspension

The automaton in Fig. 2 represents an alternative way of representing scheduling. This automaton is not as straightforward to interpret, but can be show to have a finite state representation for its untimed language. In this case, preempted tasks have their process timers decremented by the service time of the preempting task. Although the automaton is similar to that of Fig. 1 several complications are introduced. Now a distinction must be made between a task request arriving and having to wait and a task which is receiving service being preempted. Also, the timer values when decremented no longer have the interpretation of processor time received. Although this does not affect transitions for two task systems, in general, this will make the priority decision harder to formulate in terms of timer values.

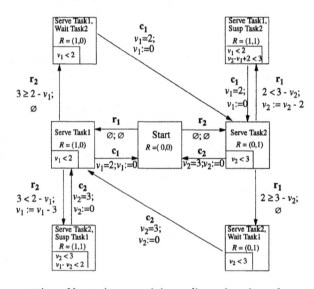

Fig. 2. Representation of least time remaining policy using timer decrementation

4 A Decidability Result for Decrementation Automata

In general, the untimed language of a suspension automaton need not be finite state. In this section one set of conditions is given under which a finite state representation is guaranteed. The result presented here is a simple extension of the result for timed automata [4, 5]. The reasoning is as follows. The full state of any suspension automaton is given by the location, current rate, and values of the timers. This state evolves continuously with time. However, for a given location and rate assignment, the evolution of the timer values may be fully predicted for the duration of the location knowing their values when the system first entered that location. Thus one may define a sampled transition system on the state space $L \times \{0,1\}^N \times \mathbb{R}^N$ where only the values of the timers upon initially entering a location are tracked. This transition system is by no means finite state. However, one may combine states which are indistinguishable in terms of past and future logical behavior. The resulting automaton is called the *timer region automaton*.

The above approach is the one taken in [4] to show that timed automata have ω-regular languages. The proposed state equivalence groups timer values into *timer regions* based on the following two observations. First, in terms of the transitions taken, it is enough to know the integer components of the timer values and the relative ordering of the fractional components of the timer values. Second, there is some upper bound past which the value of the timer no longer matters at all.

This result may be extended by making two more observations:

- If $v_i = 0$, then the suspension of T_i is represented by keeping $v_i = 0$.
- If v_i is decremented by an integer amount, the relative ordering of the fractional parts of the timers does not change, and the change in the integer value is predictable.

Thus, under the above two conditions it should still be sufficient to track just the relative ordering of the fractional values and the integer values of the timers. What remains is to give conditions under which suitable upper and lower bounds on the region of interest for the timer values may be derived. The following theorem gives one such set of conditions. This set of conditions is not in itself checkable, but in Sect. 5 it is shown how to derive automata that are guaranteed to satisfy the conditions and model interesting systems such scheduling policies.

Theorem 1. *Given an accepting run for a rate deterministic automaton $(\mathcal{A}, L_0, \mathcal{F})$ let $dec(i,n)$ be the amount v_i was decremented by upon the nth transition.*

Suppose for all accepting runs for each $T_i \in T$ there exists K_i^1 and K_i^2 such that:

- *If $r_i = 0$, then $v_i = 0$.*
- *For all M, $\sum_{m=1}^{M} dec(i,m) - t(m) \leq K_i^1$*
- *For all M, whenever $v_i(M) \geq K_i^2$, v_i is no longer decremented and for any edge taken, the constraints representing V^e do not involve the comparison of v_i to another timer.*

If the above conditions are satisfied, then $\mathcal{L}^{seq}(\mathcal{A}, L_0, \mathcal{F})$ is ω-regular.

Intuitively, the first condition ensures that timers may be suspended only if their value is fixed. The second condition gives the lower bound and the third condition gives the upper bound. A timer may never have a value less than K_i^1 and once it is greater than K_i^2 its actual value ceases to be of importance.

Note that the automaton of Fig. 2 satisfies the condition of Theorem 1. That is, $r_i = 0$ only when $v_i = 0$, and the values $K_1^1 = 3$, $K_1^2 = 2$, $K_2^1 = 2$, and $K_2^2 = 3$ satisfy the remaining conditions.

4.1 Defining A Timer Region Automaton

In order to prove Theorem 1, we will first define the timer regions and then specify the transition function between regions as a function of the edges of the suspension automaton. This is sufficient to define a finite state machine accepting $\mathcal{L}^{seq}(\mathcal{A}, L_0, \mathcal{F})$. The appropriate equivalence classes are similar to those in [4, 5]. For each timer T_i let C_i be the largest constant appearing in a constraint equation involving v_i defining either an enabling or inclusion condition. Let $U_i = max(K_i^2, C_i)$ and $L_i = -K_i^1$. Let $int(\cdot)$ be the integer component of a number and $fract(\cdot)$ the fractional component. Say that $v \approx v'$ if either:

- for some i, j, $v_i < L_i$ and $v_j' < L_j$
- or for all i, $v_i \geq L_i$ and for all i, j
  - either $int(v_i) = int(v_i')$ or both $v_i > U_i$ and $v_i' > U_i$
  - and whenever $v_i \leq U_i$ and $v_j \leq U_j$
    $fract(v_i) \leq fract(v_j)$ if and only if $fract(v_i') \leq fract(v_j')$.

Let $[v]$ indicate the equivalence class containing v and $[\mathbb{R}^N]_\approx$ the set of all equivalence classes induced by $\approx$. These equivalence classes will be the timer regions. Note that there are only a finite number of timer regions. The key to constructing an automaton accepting $\mathcal{L}^{seq}(\mathcal{A}, L_0, \mathcal{F})$ is to describe the transition from one timer region to the next. Below, the succession of timer regions is described. Two sorts of transitions are possible for timer regions – an ϵ-transition denoting the passage of time and a σ transition denoting the occurrence of a discrete event. Assume the system is in location l.

- **Time Succession:** Say that $[v'] \neq [v]$ is a time successor of $[v]$ if for some $t > 0$, $v + tr^l \in [v']$, for all $0 \leq t' \leq t$, $v + t'r^l \in [v] \cup [v']$, and $[v'] \subseteq \overline{V^l}$ where $\overline{V^l}$ is the closure of V^l. That is, $[v']$ is the first timer region reachable from any point in $[v]$ by the passage of time.
- **Discrete Transition:** Say that $[v']$ is a σ successor of $[v]$ if there exists an edge $e = (l, \sigma, l', V^e, reset^e)$ such that:
  - $[v] \subseteq V^e$.
  - $\exists v'' \in [v']$ such that:
    * If $v_i := 0 \in reset^e$, then $v_i'' = 0$.
    * Else if $v_i := v_i - c \in reset^e$, then $v_i'' = v_i - c$.
    * Else $v_i'' = v_i$.
  - $[v'] \subseteq V^{l'}$.

Lemma 2. *Given $(\mathcal{A}, L_0, \mathcal{F})$ satisfying the conditions of Theorem 1, the language accepted by the timer region automaton is exactly $\mathcal{L}^{seq}(\mathcal{A}, L_0, \mathcal{F})$.*

Proof. Let $(\mathcal{A}, L_0, \mathcal{F})$ be a suspension automaton satisfying the conditions of Theorem 1.

Suppose we are given a run for ρ with respect to $(\mathcal{A}, L_0, \mathcal{F})$. Note that for this run it is never the case that either $v_i < L_i$ or $v_i' < L_i$. It can be seen that $[v'(n)]$ is reachable from $[v(n)]$ by a series of ϵ-transitions denoting time succession. Similarly, it can be seen that $[v(n+1)]$ may be reached from $[v'(n)]$ by a $\rho(n+1)$ transition.

Now suppose that we have a run for ρ through the timer region automaton. A run for $(\mathcal{A}, L_0, \mathcal{F})$ may be constructed as follows. Note that the timer region $[v]$ where for some T_i, $v_i < L_i$, is not reachable. Let $\{[v]\}$ be the sequence of timer regions visited. Define a subsequence $\{[v](k_n)\}$ where $[v](k_0) = [v](0)$ and $[v](k_{n+1})$ is the next timer region reached from $[v](k_n)$ after a series of time transitions by a σ-transition (specifically $\rho(n+1)$). Next define the subsequence $\{[v](k'_n)\}$ by $[v](k'_n) = [v](k_{n+1} - 1)$. In other words, $[v](k'_n)$ is the last timer region reached from $[v](k_n)$ by a series of timer transitions and for which $[v](k_{n+1})$ is a $\rho(n+1)$ successor. Let $v(0) \in [v](k_0)$. Recursively construct $\{v\}$ and $\{v'\}$ as follows:

- Given $[v](k_n)$ and $[v](k'_n)$ we can choose a $t(n)$ such that
  $v'(n) = v(n) + t(n)r^l \in [v](k'_n)$ where r^l is the current rate assignment.
- Given $v'(n)$ applying reset rules will lead to $v(n+1) \in [v](k_{n+1})$.

Now, since $v(n)$ and $v'(n)$ are in the appropriate equivalence classes it is ensured that the enabling and inclusion conditions are satisfied. □

5 Translation of Suspension To Decrementation

In this section, a procedure is given for transforming a suspension based automaton into a decrementation based automaton. Provided that certain conditions are met, the decrementation based automaton is guaranteed to satisfy the conditions of Theorem 1. This procedure is designed with the structure of scheduling, communication, and synchronization policies in mind. As a basic step it requires the definition of a precedence or priority relation among timers as a function of location. In general, the priority assignment can be linked to system concepts such as the notion of task priority.

To gain intuition, consider the evolution of v_2 for the automata of Figs. 1 and 2 in the case that Task 2 is requested at time 8 and Task 1 is requested at time 8.5. This evolution is shown in Fig. 3.

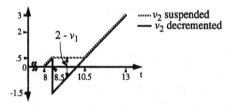

Fig. 3. A comparison of runs

Here, the suspension and decrementation of T_2 is closely linked to the behavior of T_1. That is, between times 8.5 and 10.5 we can think of T_1 as being in an 'active state.' T_2 is suspended while T_1 is in this state. Furthermore, the time T_1 is in this state is predictable (it is the time of service for Task 1) and this predictability may be used to determine the amount v_2 should be decremented by. Finally, note that given v_1, the decremented value of v_2 may be deduced from the suspended value and vice versa.

5.1 Precedence System

The precedence system formalizes the notion of timer behavior and interaction shown in Fig. 3. Suppose we are given an automaton $\mathcal{A}$. Say that a timer is p_i-*valued* for $\mathcal{A}$ if its behavior can be characterized in the following way. The timer has two states – an idle state where its rate and value are both zero, and an active state where its value and rate may be non-zero. The timer transitions from the active to idle state if and only if its value reaches p_i and the timer may not be reset or decremented while in the active state.

A *precedence assignment* for $\mathcal{A}$ is a collection of relations $Prec = \{Pr^l : l \in L\}$ with $Pr^l \subseteq T_\perp \times T$ (where $T_\perp = T \cup \{\perp\}$) denoting the relative priorities of the timers while in location l. $(\perp, T_i)$ is used to indicate timer T_i being of high precedence without there being a timer of lower precedence.

Fig. 4 gives a precedence assignment for the least time remaining policy. Only the top half is given, the bottom half may be defined in a similar fashion. Note that the precedence assignment corresponds to the task priority assignment as it is given in Fig. 1.

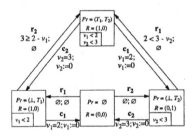

Fig. 4. Precedence for least time remaining policy

A *precedence system* provides a link between precedence assignment and rate assignment for p_i-valued timers. Let $T = T^\prec \cup T^{\not\prec}$ where $T^\prec$ is the set of timers involved in the precedence assignment. Require that all timers in $T^\prec$ be p_i-valued and that timers in $T^{\not\prec}$ are never decremented. Explicitly represent the state of the timers in $T^\prec$ as follows. Let $T0 \subseteq T^\prec$ be the set of idle timers. Require that a timer leave $T0$ the first time a location l is reached where $(T_i, T_j) \in Pr^l$ and there does not exist $(T_j, T_k) \in Pr^l$. Call the new automaton including the explicit representation of timer state the *precedence automaton*. Say that $Prec$ is *consistent* with R if the following condition holds:

$\forall l \in L$, $r_i^l = 0$ if and only if either $T_i \in T0$ or $\exists (T_i, T_j) \in Pr^l$.

A *precedence system* is a precedence automaton $\mathcal{A}$ where $Prec$ is consistent with R.

In order to define a precedence system for the automaton in Fig. 4 the only thing to do is make $T0$ an explicit part of the location. This is done in Fig. 5. For this automaton $Prec$ is consistent with R. In terms of discrete transitions, this brings the representation closer to that of Fig. 2. However, scheduling is still represented through the suspension of timers.

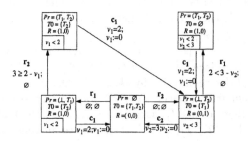

Fig. 5. Precedence system for least time remaining policy

In order to provide the final link between timer behavior and decrementation a few more conditions must be satisfied. Most importantly, it must be possible to identify which timer is suspending which at any time, and to ensure that this suspension relation doesn't change arbitrarily. This may be achieved by restricting the form of $Prec$. Say that $Prec$ is *prioritizing* if for all ρ the following condition holds for any run:

> Given a transition from $(l, T0)$ to $(l', T0')$, if $T_i, T_j \notin T0$ and $(T_i, T_j) \in Pr^l$, then either $(T_i, T_j) \in Pr^{l'}$ or $T_j \in T0'$.

Thus, during the active period of any two timers their precedence relation to each other remains fixed. The prioritizing condition may either be ensured by design or checked as a language property. Say that Pr^l is *tree-like* if whenever $(T_i, T_j) \in Pr^l$ and $(T_i, T_k) \in Pr^l$ then either $(T_j, T_k) \in Pr^l$ or $(T_k, T_j) \in Pr^l$. $Prec$ is tree-like if for all $l \in L$, Pr^l is tree-like. Finally, say that Pr^l is *complete* if whenever $(T_i, T_j) \in Pr^l$ and $(T_j, T_k) \in Pr^l$, then $(T_i, T_k) \in Pr^l$. Note that any precedence assignment may be completed.

It is also necessary to restrict the form of the enabling and inclusion conditions. Say that $V \subseteq \mathrm{I\!R}^N$ is rectangular if there exists a set of constraint equations all of the form $v_i \sim c$ describing it. Note that the precedence system given in Fig. 5 satisfies the tree-like and prioritizing conditions and has rectangular enabling and inclusion regions.

5.2 Translation of Precedence Automaton to Decrementation

Suppose that $\mathcal{A}$ is a precedence system where $Prec$ is prioritizing, tree-like, and complete, and for all l and e, V^l and V^e are rectangular. An automaton $Dec(\mathcal{A})$ may be produced as follows.

- **Decrementation:** Let e be an edge transitioning from $(l, T0)$ to $(l', T0')$. Suppose that $T_i \in T0$, but $T_i \notin T0'$. For all $T_j \notin T0'$ such that $(T_j, T_i) \in Pr^{l'}$, add $v_j := v_j - p_i$ to $reset^e$.
- **Rate Assignment:** Set the rate assignment so that timers are rate zero if and only if they are in $T0$.
- **Skewing V^l and V^e:** Given v, $T0$, and Pr^l, define $\bar{v} = skew(v, T0, Pr^l)$ as follows. If $T_i \in T0$ then $\bar{v}_i = 0$. Otherwise, $\bar{v}_i = v_i - \sum_{j \in J_i}(p_j - v_j)$ where $J_i = \{j : (T_i, T_j) \in Pr^l\}$.
  Let $skew(V, T0, Pr^l) = \{\bar{v} : \exists v \in V \text{ s.t. } \bar{v} = skew(v, T0, Pr^l)\}$.
  - For all $l \in L$ replace V^l by $skew(V^l, T0, Pr^l)$.
  - For all $e \in Edge$ replace V^e by $skew(V^e, T0, Pr^l)$.
  For any rectangular V, $skew(V, T0, Pr^l)$ may be represented as follows. For all $T_i \notin T0$ if there exists T_{j_i} such that $(T_i, T_{j_i}) \in Pr^l$ and there does not exist T_j such that $(T_{j_i}, T_j) \in Pr^l$, then replace all constraint equations of the form $v_i \sim c$ by $v_i - v_{j_i} + p_{j_i} \sim c$.

$Dec(\mathcal{A})$ for the precedence system of Fig. 5 is shown in Fig. 6. Two changes are made to $\mathcal{A}$ – in the case that Task 2 is preempted, its process timer is decremented and the inclusion condition is skewed.

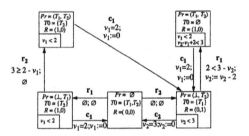

Fig. 6. $Dec(\mathcal{A})$ for least time remaining policy

Theorem 3. *Suppose $\mathcal{A}$ is a precedence system where $Prec$ is prioritizing, tree-like, and complete, and for all l and e, V^l and V^e are rectangular. In this case, $\mathcal{L}^{seq}(\mathcal{A}) = \mathcal{L}^{seq}(Dec(\mathcal{A}))$.*

Proof. Given a run with respect to one of the automata, it is enough to construct the appropriate timer values to prove there exists a corresponding run with respect to the other. Let $\bar{v}$ denote the timer values for $Dec(\mathcal{A})$ and v denote the timer values for $\mathcal{A}$.

Given a run with respect to $\mathcal{A}$, let $\bar{v}(n) = skew(v(n), Pr^{l(n)}, T0(n))$ and $\bar{v}'(n) = skew(v'(n), Pr^{l(n)}, T0(n))$. It is easy to verify that this indeed defines a run with respect to $Dec(\mathcal{A})$.

Similarly, suppose we are given a run with respect to $Dec(\mathcal{A})$. Let $v_i(n) = \bar{v}_i(n) - \bar{v}_{j_i}(n) + p_{j_i}$ where j_i satisfies $(T_i, T_{j_i}) \in Pr^{l(n)}$ and there does not exist T_j such that $(T_{j_i}, T_j) \in Pr^{l(n)}$. Define $v_i'(n)$ in a similar fashion. It can be verified that this defines a run with respect to $\mathcal{A}$. $\qquad\square$

Theorem 4. *Suppose $\mathcal{A}$ is a precedence system where Prec is prioritizing, tree-like, and complete, and for all l and e, V^l and V^e are rectangular. In this case, $Dec(\mathcal{A})$ satisfies the conditions of Theorem 1.*

Proof. By the definition of the precedence system and the construction of $Dec(\mathcal{A})$ it is clear that $r_i = 0$ only if $v_i = 0$. Let $J_i = \{j : \exists l \text{ s.t. } (T_i, T_j) \in Pr^l\}$. Then $K_i^1 = \sum_{j \in J_i} p_j$ satisfies the second condition of Theorem 1. $K_i^2 = p_i$ satisfies the final condition of Theorem 1. $\qquad\square$

6 Discussion

In this paper, a class of hybrid automata for which language containment is decidable is defined. The conditions presented are somewhat restrictive, but they do allow for the representation of interesting systems that could not be represented using rate 1 automata. Future work should seek to relax these conditions, for instance, a relaxation of the requirement that timers are suspended for a fixed amount of time. In addition, techniques should be explored to minimize the state explosion problem inherent in all the current timer-based hybrid automata.

References

1. R. Grossman, A. Nerode, A Ravn, and H. Rischel, editors. *Hybrid Systems*, volume LNCS 736. Springer, 1993.
2. R. Alur, C. Courcoubetis, T. Henzinger, and P. Ho. Hybrid automata: An algorithmic approach to the specification and verification of hybrid systems. In *Hybrid Systems*, volume LNCS 736, pages 209–229. Springer, 1993.
3. D. Dill. Timing assumptions and verification of finite-state concurrent systems. In *Automatic Verification Methods for Finite State Systems*, volume LNCS 407. Springer, 1989.
4. R. Alur and D. Dill. Automata for modeling real-time systems. In *Proceedings of the 17th Annual Colloquium on Automata, Languages, and Programming*, 1990.
5. R. Alur, C. Courcoubetis, and D. Dill. Model-checking for real-time systems. In *IEEE Proceedings of the Fifth Annual Symposium on Logic and Computer Science*, 1990.
6. A. Olivero and S. Yovine. Kronos: A tool for verifying real-time systems. user's guide and reference manual. draft 0.0. Preprint. VERIMAG, B.P. 53X, 38401 Grenoble, France, May 1993.
7. P. Tzounakis. Verification of real time systems: The extension of COSPAN in dense time. Master's thesis, University of Crete, 1992.

Verification of Context-Free Timed Systems Using Linear Hybrid Observers[*]

Ahmed Bouajjani[** †]   Rachid Echahed[*** †]   Riadh Robbana[** †]

Abstract. We address the verification problem of infinite timed systems. We consider context-free timed systems defined as a generalization of the (regular) timed graphs [ACD90]. Then, we propose decision procedures for the verification of invariance properties of these systems, expressed by means of observation variables. These variables record relevant informations about the computations of the observed system. They are permanently updated along these computations without any interference with the behaviour of the system. Observation variables are either additional clocks (timers), nonbounded integer variables (accumulators), or constant slope continuous (real valued) variables (integrators).

1 Introduction

During the last decade, the framework of specification and verification of systems has been extensively developed. The first works in this area consider the case of systems modeled by *finite-state* automata. For such systems, several specification and verification methods have been elaborated, using behavioural equivalence (or preorder) testing [Mil80, Par81, KS83] or model-checking techniques for logic-based specifications [QS82, CES83, VW86, EL86, CS91]. Recent investigations aim at extending these techniques in order to deal with two kinds of systems: *timed* and/or *hybrid systems*, in particular with a dense time domain (positive reals) [ACD90, Cer92, HNSY92, ACHH93, BES93, KPSY93], and separately, systems having (discrete) infinite state spaces, in particular *context-free systems* (or equivalently systems supplied with a stack) [BBK87, BS92, CHS92]. This paper presents a unification of these two directions of investigations, by proposing automatic verification methods for *context-free timed systems*. First, let us give a brief overview of the existing works on the verification of timed systems and context-free systems.

Concerning timed systems, several models and description formalisms have been proposed [ACD90, NRSV90, Wan90, CGL93]. One of the most adopted models for these systems is certainly the powerful model of finite *timed graphs* introduced in [ACD90]. Model-checking algorithms have been proposed for timed graphs w.r.t. formulas of the timed temporal logic TCTL [ACD90, HNSY92]. Recently, the scope of the investigations around timed systems has been broadened by considering specification languages allowing to reason about new interesting notions as *durations*, and by extending the models and the verification techniques for timed systems to the case of hybrid systems

---

[*] Partially supported by the ESPRIT-BRA project REACT.

[**] VERIMAG-SPECTRE, Miniparc-Zirst, Rue Lavoisier, 38330 Montbonnot St-Martin, France. VERIMAG is a joint laboratory of CNRS, INPG, UJF and VERILOG SA., SPECTRE is a project of INRIA.

[***] LGI-IMAG, BP53, 38041 Grenoble cedex, France.

[†] Ahmed.Bouajjani@imag.fr, Rachid.Echahed@imag.fr, Riadh.Robbana@imag.fr

[CHR91, MMP92, MP93, ACHH93, ACH93, BES93, KPSY93, NOSY93]. The notion of duration of a state property has been introduced in [CHR91] and corresponds to the accumulated time spent in some given computation sequence while the considered property holds. So, the notion of duration allows to reason about several times corresponding to durations of some relevant properties. For instance, we may require in the specification of a communication node that

1. *in every session, the time consecrated to recover lost messages is less than 10 % of the global time.*

Some automatic verification techniques have been proposed in order to decide whether a finite timed system meets some kinds of duration properties as invariance in [KPSY93, ACH93] and nested invariance and eventuality properties in [BES93]. These techniques allow to verify for instance the property (1) above.

Concerning the works on context-free systems, important results have been obtained in the last few years about the (un)decidability of behavioural equivalences and preorders [BBK87, GH91, CHS92]. Mainly, it has been shown that bisimulation equivalence is decidable for these systems [CHS92]. There are less works concerning the extension of the logic-based specification and verification approach to the case of context-free systems. The first work on this topic proposes a model-checking algorithm for *regular* properties of these systems, expressed in the propositional μ-calculus [BS92]. However, a wide class of relevant properties of infinite-state systems (in particular, the context-free ones) are nonregular properties and cannot be expressed in the μ-calculus neither by finite-state ω-automata. For instance, in the specification of a communication node, we may need to express the fact that

2. *between the beginning and the ending of every session, there are exactly the same numbers of receptions and transmissions.*

In addition, we have also to require that

3. *during every session, the number of transmissions never exceeds the number of receptions.*

Actually, as these examples show, significant properties of infinite-state systems are essentially properties involving constraints on *numbers of occurrences* of some events (or number of states satisfying some state property). In [BER93] we have proposed a temporal logic that allows to express such constraints using formulas in Presburger arithmetic, and we have proposed automatic verification procedures for a wide class of nonregular properties, including (2) and (3), for (untimed) context-free systems.

In this paper, we address the verification problem for *context-free timed systems*, i.e., timed systems such that the underlying untimed structure is a context-free system. First, we propose a definition of these systems which is a generalization of the definition of (finite) timed graphs given in [ACD90]. Then, we consider the verification problem of context-free timed systems w.r.t. properties involving constraints on delays, durations, and on numbers of occurrences of some (events) state properties.

The properties we tackle in this paper are *invariance properties*, and our approach of verification is based on the use of *linear hybrid observers*. In general, given a system to verify, an observer is another system running in parallel, which maintains a set of *observation variables*, where informations about the execution of the observed system are recorded. The observer does not interfere with the execution of the system and its aim is only to check permanently the truth of some invariance property on its observation variables.

We define a linear hybrid observer as an observer that disposes of three kinds of observation variables: timers that can be reset at some transitions of the observed system, discrete (integer valued) variables, called *accumulators*, that can be incremented or decremented at some transitions of the system, and finally, continuous (real valued) variables, called *integrators*, that change continuously with a constant slope (which may be different) at each location of the observed system. The accumulators allow to count linear combinations of numbers of occurrences of some events whereas the integrators allow to count linear combinations of durations.

The remainder of this paper is organized as follows. In Section 2, we introduce some notations about sequences and context-free grammars. In Section 3, we define context-free timed systems, in Section 4, we define their linear hybrid observers. In Section 5, we define *invariance formulas* and dually *reachability formulas*, expressed on observation variables of linear hybrid observers. In Section 6, we present decidability results for the verification problem of such invariance formulas on context-free timed systems. For this, we consider two different cases: The first case (see Section 6.1) corresponds to the consideration of pure timed constraints, i.e., the observer has only timers and no accumulators nor integrators. We give for this case a decision procedure for the verification of invariance formulas based on a reduction to the emptiness problem of context-free languages. In the second case (see Section 6.2) we prove the decidability of the verification problem of reachability formulas that involve constraints on timers and accumulators but at most one integration constraint. This result is proven for CFTS's whose transition guards are non-strict, by reducing the considered problem to the satisfiability problem of integer linear constraints.

2 Preliminaries

We introduce in this section some notations and recall some well-known notions about sequences, languages and grammars.

Let Σ be a finite alphabet. We denote by Σ^* the set of finite sequences over Σ. Given a sequence $\sigma \in \Sigma^*$, $|\sigma|$ denotes the length of σ. Let λ denote the empty sequence, i.e., the sequence of lenght 0. Let $\Sigma^+ = \Sigma - \{\lambda\}$. For every $a \in \Sigma$, $|\sigma|_a$ is the number of occurrences of a in σ. In the sequel, we write $a \in \sigma$ to denote the fact that a appears in the sequence σ. For every $i \in \{1, \ldots, |\sigma|\}$, $\sigma(i)$ is the ith element of σ. Consider a nonempty sequence $\sigma \in \Sigma^*$. For every $i \in \{1, \ldots, |\sigma|\}$, we denote by σ_i the prefix of σ ending at position i, i.e., the subsequence $\sigma(1) \cdots \sigma(i)$. For $i = 0$, we consider that $\sigma_i = \lambda$.

A context-free grammar G over Σ is a tuple $(\Sigma, N, Prod, S)$ where N is a set of nonterminals, $Prod$ is a set of productions of the form $A \to \alpha$ where $A \in N$ and $\alpha \in (\Sigma \cup N)^*$, and S is the starting symbol. Given a production $p = ``A \to \alpha" \in Prod$, we denote by $lhs(p)$ the left hand side of p (i.e., A) and by $rhs(p)$ its right hand side (i.e., α). We adopt standard notations for the derivation relation ($\Longrightarrow$) and its reflexive-transitive closure ($\stackrel{*}{\Longrightarrow}$). We use subscripts to precise the set of productions or the sequences of productions used in the derivation. We denote by $L(G)$ the language generated by the grammar G (i.e., the set of sequences $\sigma \in \Sigma^*$ such that $S \stackrel{*}{\Longrightarrow} \sigma$). We use CFG to abbreviate *context-free grammar* and CFL for *context-free language*. For more details concerning the theory of formal languages, see for instance [Har78].

3 Context-Free Timed Systems

Finite timed graphs have been introduced in [ACD90] as a powerful model for real-time systems. A timed graph consists in a finite location graph augmented by a set of timers (clocks) that can be reset and/or tested at each transition. A computation of the so modeled system can visit some location and stay in it by letting time progress (during the visit, the timers run continuously), and then moves to another location by taking some transition in the graph, provided that the enabling guard (condition on the values of the timers) is satisfied ; some of the timers may be reset after taking the transition.

In this paper, our aim is to deal with a class of infinite timed graphs. We consider timed graphs having a context-free structure, i.e., they are defined by a context-free set of rules (similar to productions in CFG's). Actually, we generalize the definition of timed graphs by replacing their binary transition relation between locations with a relation between locations and sequences of locations. Let us now give the formal definitions.

Let $\mathcal{P}rop$ be a set of atomic propositions. Let $\mathcal{C}$ be a set of timers (clocks). A *time guard* on $\mathcal{C}$ is any boolean combination of constraints of the form $x \prec n$ where $x \in \mathcal{C}$, $\prec \in \{<, \leq\}$ and $n \in I\!N$. Let $\Gamma_{\mathcal{C}}$ be the set of time guards on $\mathcal{C}$.

A *context-free timed system* (CFTS for short) is a tuple $\mathcal{M} = (\Sigma, \mathcal{V}ar, \mathcal{C}, \delta, \gamma, \rho, \Pi)$ where $\Sigma = 2^{\mathcal{P}rop}$, $\mathcal{V}ar$ is a set of location variables, $\mathcal{C}$ is a set of timers, $\delta \subseteq \mathcal{V}ar \times \mathcal{V}ar^*$ is a set of derivation rules, $\gamma : \delta \to \Gamma_{\mathcal{C}}$, $\rho : \delta \to 2^{\mathcal{C}}$ and $\Pi : \mathcal{V}ar \to \Sigma$. The function γ associates with each derivation rule $d \in \delta$ a time guard that should be satisfied by the values of the timers when d is applied, the function ρ associates with each rule d the set of timers that should be reset at each application of d. Finally, the function Π associates with each location variable $X \in \mathcal{V}ar$ the set of atomic propositions that hold at X.

An alternative definition of context-free timed systems can be given using pushdown automata instead of sets of derivation rules. We can translate one definition to the other by extending the standard transformations between context-free grammars and pushdown automata. Notice that a particular case of such systems are 1-counter systems, i.e., systems with one nonbounded integer variable.

Now, we give the *operational semantics* of the CFTS $\mathcal{M}$. A state of the system $\mathcal{M}$ is constituted by a *nonempty* sequence of location variables and a valuation that assigns to each timer a positive real value, i.e., a state is a pair $\langle \alpha, E \rangle$ where $\alpha \in \mathcal{V}ar^+$ and $E : \mathcal{C} \to I\!R^+$. Let $\mathcal{S}_{\mathcal{M}}$ be the set of states of the system $\mathcal{M}$.

We define two transition relations $\to$ and $\rhd$ between states of $\mathcal{M}$. The relation $\to$ corresponds to transitions due to time progress at some location whereas $\rhd$ corresponds to moves between locations using derivation rules in δ. Before giving the formal definition of these relations, let us first introduce some notations.

Given a derivation rule $d = (X, \alpha)$, we denote by $fst(d)$ (resp. $snd(d)$) the first (resp. second) component of the rule, i.e. the location variable X (resp. the sequences of location variables α). Given a valuation $E : \mathcal{C} \to I\!R^+$, a timer $x \in \mathcal{C}$ and a time value $t \in I\!R^+$, we denote by $E[x \leftarrow t]$ the new valuation which assigns t to x and coincides with E for all the other timers. Moreover, for any $t \in I\!R^+$, we denote by $E + t$ the valuation E' such that for every $x \in \mathcal{C}$, $E'(x) = E(x) + t$. Finally, given a valuation E and a constraint g, we denote by $E \vdash g$ the fact that the evaluation of g under the valuation E is true.

Now, we define two families of relations $\xrightarrow{t}$ and $\triangleright_d$ with $t \in I\!R^+$ and $d \in \delta$. For every $t \in I\!R^+$ and every $d \in \delta$, these relations are defined as the smallest relations included in $\mathcal{S}_\mathcal{M} \times \mathcal{S}_\mathcal{M}$ such that:

- $\langle \alpha, E \rangle \xrightarrow{t} \langle \alpha, E + t \rangle$,
- $fst(d) = X$ and $E \vdash \gamma(d)$ implies $\langle X \cdot \alpha, E \rangle \triangleright_d \langle snd(d) \cdot \alpha, E[x \leftarrow 0]_{x \in \rho(d)} \rangle$

Notice that in the definition of $\triangleright_d$, we suppose that $snd(d) \cdot \alpha \neq \lambda$. Nevertheless, this assumption is not a restriction, since in case λ is derivable via the relation δ from some $\beta \in Var^+$, we can add to the system a fresh variable $\sharp$ which represents a termination location, and for any state $\langle \beta, E \rangle$, consider rather the state $\langle \beta \cdot \sharp, E \rangle$.

We define $\rightarrow = \bigcup_{t \geq 0} \xrightarrow{t}$ and $\triangleright = \bigcup_{d \in \delta} \triangleright_d$, and we consider the relation $\hookrightarrow = \rightarrow \cup \triangleright$. We denote by $\overset{*}{\hookrightarrow}$ the reflexive-transitive closure of $\hookrightarrow$. For any pair of states s and s', $s \overset{*}{\hookrightarrow} s'$ means that s' is *reachable* from s in $\mathcal{M}$.

Given a state s, a *computation sequence* of $\mathcal{M}$ starting from s is a sequence

$$\langle s_0, \tau_0 \rangle \langle s_1, \tau_1 \rangle \cdots \langle s_n, \tau_n \rangle \in \mathcal{S}_\mathcal{M}^*$$

such that $s_0 = s$, $\tau_0 = 0$, and for every i such that $0 \leq i < n$, either $s_i \triangleright s_{i+1}$ and $\tau_i = \tau_{i+1}$, or there exists an amount of time $t \in I\!R^+$ such that $s_i \xrightarrow{t} s_{i+1}$ and $\tau_{i+1} = \tau_i + t$. We denote by $\mathcal{C}omput(\mathcal{M}, s)$ the set of computation sequences of $\mathcal{M}$ starting from s.

4 Linear Hybrid Observers

Let $\mathcal{M} = (\Sigma, Var, \mathcal{C}, \delta, \gamma, \rho, \Pi)$ be a CFTS. A *linear hybrid observer* (LHO) for the system $\mathcal{M}$ is a tuple $\mathcal{O} = (\mathcal{X}, \mathcal{A}, \mathcal{I}, \zeta, \kappa, \partial)$ where $\mathcal{X}$ is a set of timers, $\mathcal{A}$ is a set of accumulators (discrete integer variables), $\mathcal{I}$ is a set of integrators (constant slope continuous real variables), $\zeta : \delta \to 2^{\mathcal{X}}$, $\kappa : \delta \to (\mathcal{A} \to \mathbb{Z})$ and $\partial : Var \to (\mathcal{I} \to \mathbb{Z})$. The function ζ associates with each derivation rule $d \in \delta$ the set of timers in $\mathcal{X}$ that should be reset at each application of d. The function κ associates with each rule d and accumulator $u \in \mathcal{A}$, an integer that should be added to u whenever d is applied. Finally, the function ∂ associates with each location variable $X \in Var$ and integrator $u \in \mathcal{I}$, an integer rate at which u changes continuously while the computation is at X. This means that if the computation stays t amount of time at X, the variation of u is $\partial(X)(u) \cdot t$.

The composition of the system $\mathcal{M}$ with the observer $\mathcal{O}$ is a context-free hybrid system $\mathcal{M}_\mathcal{O}$. A state of this system is an enrichment of a state of $\mathcal{M}$ by valuations for the new timers, accumulators and integrators of $\mathcal{O}$. Let $\mathcal{T} = \mathcal{C} \cup \mathcal{X}$. Formally, a state of $\mathcal{M}_\mathcal{O}$ is a tuple $\langle \alpha, E_T, E_\mathcal{A}, E_\mathcal{I} \rangle$ where $\alpha \in Var^+$, $E_T : \mathcal{T} \to I\!R^+$, $E_\mathcal{A} : \mathcal{A} \to \mathbb{Z}$ and $E_\mathcal{I} : \mathcal{I} \to I\!R$. Let $\mathcal{S}_{\mathcal{M}_\mathcal{O}}$ be the set of states of the hybrid system $\mathcal{M}_\mathcal{O}$.

The computations of $\mathcal{M}_\mathcal{O}$ are obtained as well by enriching the computations of $\mathcal{M}$. We extend the definitions of the transition relations $\xrightarrow{t}$ and $\triangleright_d$ to states of $\mathcal{M}_\mathcal{O}$. Let us first introduce some notations. Given a valuation of the accumulators $E : \mathcal{A} \to \mathbb{Z}$ and a mapping $D : \mathcal{A} \to \mathbb{Z}$, we denote by $E + D$ the new valuation E' such that for every $u \in \mathcal{A}$, $E'(u) = E(u) + D(u)$. Moreover, given a valuation of the integrators $E : \mathcal{I} \to I\!R$, a mapping $R : \mathcal{I} \to \mathbb{Z}$ and an amount of time $t \in I\!R^+$, we denote by $E + R \cdot t$ the new valuation E' such that, for every $u \in \mathcal{I}$, $E'(u) = E(u) + R(u) \cdot t$.

Now, for every amount of time $t \in I\!R^+$ and every derivation rule $d \in \delta$, we consider that $\xrightarrow{t}$ and $\triangleright_d$ are the smallest relations included in $\mathcal{S}_{\mathcal{M}_\mathcal{O}} \times \mathcal{S}_{\mathcal{M}_\mathcal{O}}$ such that:

- $\langle X \cdot \alpha, E_T, E_A, E_I \rangle \xrightarrow{t} \langle X \cdot \alpha, E_T + t, E_A, E_I + \partial(X) \cdot t \rangle$,
- $fst(d) = X$ and $E \vdash \gamma(d)$ implies
$$\langle X \cdot \alpha, E_T, E_A, E_I \rangle \rhd_d \langle snd(d) \cdot \alpha, E_T[x \leftarrow 0]_{x \in \rho(d) \cup \zeta(d)}, E_A + \kappa(d), E_I \rangle$$

We extend also the relations $\rightarrow$ and $\rhd$, as well as the relation $\hookrightarrow = \rightarrow \cup \rhd$, to states of $\mathcal{M}_{\mathcal{O}}$ and define the computation sequences of $\mathcal{M}_{\mathcal{O}}$ accordingly. Given a state $s \in \mathcal{S}_{\mathcal{M}_{\mathcal{O}}}$, we denote by $Comput(\mathcal{M}_{\mathcal{O}}, s)$ the set of computation sequences of $\mathcal{M}_{\mathcal{O}}$ starting from s.

5 Invariance Properties

Given a CFTS $\mathcal{M}$ and a LHO $\mathcal{O}$, we consider the set of *invariance formulas* φ defined by:

$$\varphi ::= \forall \Box \bigwedge \bigvee \psi$$
$$\psi ::= \pi \mid \xi$$
$$\pi ::= P \mid \neg \pi \mid \pi \vee \pi$$
$$\xi ::= u \sim k \quad \text{with } u \in \mathcal{X} \cup \mathcal{A} \cup \mathcal{I}, \sim \in \{<, >, \le, \ge\}, k \in \mathbb{Z}$$

A constraint $\xi = u \sim k$ is called either a *time*, *accumulation* or *integration* constraint according to the fact that u is respectively a timer, an accumulator or an integrator.

In order to give the semantics of invariance formulas, we define a satisfaction relation $\models$ between states of $\mathcal{M}_{\mathcal{O}}$ and these formulas.

Let $s = \langle \alpha, E_T, E_A, E_I \rangle$ be a state of $\mathcal{M}_{\mathcal{O}}$. Suppose that the sets T, A and I have no common elements and let E be the union of the functions E_T, E_A and E_I. The satisfaction relation $\models$ is inductively defined by:

$$s \models \forall \Box \phi \quad \text{iff } \forall s'. \, s \xrightarrow{*} s'. \, s' \models \phi$$
$$s \models P \quad \text{iff } P \in \Pi(\alpha(1))$$
$$s \models \neg \phi \quad \text{iff } s \not\models \phi$$
$$s \models \phi_1 \vee \phi_2 \text{ iff } s \models \phi_1 \text{ or } s \models \phi_2$$
$$s \models u \sim k \quad \text{iff } E \vdash u \sim k$$

Let us introduce the operator $\exists \Diamond = \neg \forall \Box \neg$. It is easy to see from the definition of the relation $\models$ that $s \models \exists \Diamond \phi$ if and only if there exists some state s' that is reachable from s and satisfies ϕ. Formulas of the form $\exists \Diamond \phi$ are called *reachability formulas*.

Using standard laws for the boolean connectives together with the fact that any formula $\exists \Diamond (\phi_1 \vee \phi_2)$ is equivalent to $(\exists \Diamond \phi_1) \vee (\exists \Diamond \phi_2)$, it can easily be shown that every invariance formula is equivalent to the negation of a formula of the form

$$\bigvee_{i=1}^{n} \exists \Diamond (\pi_i \wedge \bigwedge_{j=1}^{m_i} \xi_i^j)$$

where we assume without loss of generality that if ξ_i^j is a time constraint, then it is of the form $u \sim n$ with $n \in \mathbb{N}$, if ξ_i^j is an accumulation constraint, then it is of the form $u \le k$, and finally, if ξ_i^j is an integration constraint, it is of the form $u \prec k$ with $\prec \in \{<, \le\}$.

6 Decidability Results

We present gradually decidability results concerning the the satisfaction relation $\models$ between states of CFTS's and invariance formulas. First, we consider the case when the observer uses only timers (clocks). Thus, the invariance formulas in this case contain only (pure) time constraints and correspond to a subclass of TCTL formulas [ACD90]. Then, we consider the case when the observer uses in addition accumulators and integrators. The consideration of integration constraints imposes some restrictions as in [KPSY93].

Let us fix for the remainder of this section a CFTS $\mathcal{M} = (\Sigma, Var, \mathcal{C}, \delta, \gamma, \rho, \Pi)$ and a LHO $\mathcal{O} = (\mathcal{X}, \mathcal{A}, \mathcal{I}, \zeta, \kappa, \partial)$, and let $T = \mathcal{C} \cup \mathcal{X}$. We suppose without loss of generality that the sets $\mathcal{C}$, $\mathcal{X}$, $\mathcal{A}$ and $\mathcal{I}$ are disjoint.

6.1 Pure Time Constraints

In this subsection we tackle the verification problem of invariance (resp. reachability) formulas involving only time constraints. We call these formulas pure time constraints (PTC) invariance (resp. reachability) formulas. Actually, we can consider equivalently that the observer $\mathcal{O}$ is such that $\mathcal{A} = \mathcal{I} = \emptyset$. We show that in this case, the verification problem of reachability formulas (hence, invariance formulas) is decidable by reducing it to the nonemptiness problem of context-free languages, using and extending the notion of *region graphs* introduced in [ACD90].

In the remainder of this subsection, valuations of accumulators and integrators are omitted in representations of states since they are undefined. So, we suppose that a state is a pair $\langle \alpha, E \rangle$ where E is a valuation of timers. Let $\mathcal{E} = [T \to I\!\!R^+]$ be the set of valuations of timers. In [ACD90], an equivalence relation $\cong$ on $\mathcal{E}$ is defined so that any pair of states of a (finite) timed graph with equivalent time valuations satisfies the same reachability formulas (actually, the same TCTL formulas). It can be easily verified that this result holds also for context-free timed systems.

Lemma 6.1 *Let $\alpha \in Var^+$ and φ be a PTC reachability formula. Then, $\forall E, E' \in \mathcal{E}$, $E \cong E'$ implies $\langle \alpha, E \rangle \models \varphi$ iff $\langle \alpha, E' \rangle \models \varphi$.*

The Lemma above allows to analyse the satisfaction of reachability formulas on a countable structure instead of the noncountable state graph $(\mathcal{S}_{\mathcal{M}_{\mathcal{O}}}, \hookrightarrow)$. This countable structure is what is called *region graph* in [ACD90]. It corresponds to the quotient graph of $(\mathcal{S}_{\mathcal{M}_{\mathcal{O}}}, \hookrightarrow)$ w.r.t. $\cong$. Let us see how this structure is obtained.

First of all, it is shown in [ACD90] that given a formula φ, the relation $\cong$ induces a *finite* partition of $\mathcal{E}$, depending on the constants that are compared with timers both in the time guards of $\mathcal{M}$ and in time constraints of φ. Then, consider a PTC reachability formula $\varphi = \exists \Diamond(\pi \wedge \bigwedge_{i=1}^{n} \xi_n)$, and let $[\mathcal{E}]$ be the quotient set of $\mathcal{E}$ under the relation $\cong$. We denote by $[E]$ the equivalence class of $E \in \mathcal{E}$.

Also, from [ACD90], the set $[\mathcal{E}]$ can be supplied by a successor function $succ$ between equivalence classes which captures time progress. The function $succ$ is defined in the following manner: For every $E, E' \in \mathcal{E}$, $succ([E]) = [E']$ if and only if $E \not\cong E'$, and $\exists t \in I\!\!R^+$ such that $E' = E + t$, and $\forall t' \in I\!\!R^+$, $0 \leq t' < t$, either $E + t' \cong E$ or $E + t' \cong E'$.

Now, we define a *region* as a pair $\langle \alpha, [E] \rangle$, where $\alpha \in Var^+$ and $[E] \in [\mathcal{E}]$. Let Reg be the set of such regions. The region graph $\mathcal{R}(\mathcal{M}_{\mathcal{O}}, \varphi)$ defined by $\mathcal{M}_{\mathcal{O}}$ and φ, is defined as the graph $\langle Reg, Edg \rangle$ where Reg is the vertex set and the edge set Edg is the smallest set such that:

- Each vertex (region) $\langle \alpha, [E] \rangle$ has an edge to $\langle \alpha, succ([E]) \rangle$,

- Each vertex $\langle X \cdot \alpha, [E] \rangle$ has an edge to $\langle \beta \cdot \alpha, [E[x \leftarrow 0]_{x \in T}] \rangle$, for every rule $d = (X, \beta) \in \delta$ such that $E \vdash \gamma(d)$

The region graph can be seen as a kind of product of the context-free timed system $\mathcal{M}_{\mathcal{O}}$ and the finite state graph $([\mathcal{E}], succ)$. Notice that the region graphs considered in [ACD90] are finite since they are obtained from a finite (regular) timed graphs. Here, a region graph is in general infinite but has a *context-free structure*.

Now, let $s = \langle \alpha, E \rangle$ be a state and suppose that we are interested in the problem of checking whether $s \models \varphi$. By Lemma 6.1, this problem reduces to decide whether there exists some path in the region graph $\mathcal{R}(\mathcal{M}_{\mathcal{O}}, \varphi)$ which starts from $\langle \alpha, [E] \rangle$ and reaches some region $\langle \alpha', [E'] \rangle$ such that $\langle \alpha', E' \rangle \models (\pi \wedge \bigwedge_{i=1}^{n} \xi_i)$. We show that this problem is actually decidable by reducing it to the nonemptiness problem of a context-free language $\mathcal{G}_{(s, \varphi)}$. This grammar has an empty alphabet (so, its language is either empty or equal to $\{\lambda\}$) and it is defined in such a manner that all its successful derivations correspond to paths in the region graph starting from $\langle \alpha, [E] \rangle$ and reaching some region $\langle \alpha', [E'] \rangle$ such that $\langle \alpha', E' \rangle \models (\pi \wedge \bigwedge_{i=1}^{n} \xi_i)$.

In the sequel, given a location variable X and a state formula π, we write $\Pi(X) \models \pi$ if $(\bigwedge_{P \in \Pi(X)} P) \wedge (\bigwedge_{P \notin \Pi(X)} \neg P) \Rightarrow \pi$. Now, let $\mathcal{F} = \{[E] \in [\mathcal{E}] \ : \ E \vdash \bigwedge_{i=1}^{n} \xi_i \}$. We define the CFG $\mathcal{G}_{(s, \varphi)}$ by $(\emptyset, N, Prod, S)$ where $N = \{ \langle\!| [F], X, [F'] |\!\rangle \ : \ X \in Var$ and $[F], [F'] \in [\mathcal{E}] \} \cup \{ [\![[F], X, [F']]\!] \ : \ X \in Var$ and $[F], [F'] \in [\mathcal{E}] \} \cup \{S\}$ and the set of productions $Prod$ is the smallest set containing the following productions:

(P1) $S \to \langle\!| [E_0], \alpha(1), [E_1] |\!\rangle \cdots \langle\!| [E_{i-1}], \alpha(i), [E_i] |\!\rangle [\![[E_i], \alpha(i+1), [E_{i+1}]]\!]$
    if $E_0 = E$, $i \in \{0, \ldots, |\alpha| - 1\}$, $\forall j \in \{1, \ldots, i\}, [E_j] \in [\mathcal{E}]$, and $[E_{i+1}] \in \mathcal{F}$,
(P2) $\langle\!| [E_1], X, [E_2] |\!\rangle \to \langle\!| succ([E_1]), X, [E_2] |\!\rangle$,
(P3) $\langle\!| [E_1], X, [E_2] |\!\rangle \to$
        $\langle\!| [F_0], \beta(1), [F_1] |\!\rangle \cdots \langle\!| [F_{m-2}], \beta(m-1), [F_{m-1}] |\!\rangle \langle\!| [F_{m-1}], \beta(m), [E_2] |\!\rangle$
    if $d = (X, \beta) \in \delta$ with $|\beta| = m > 0$, $E_1 \vdash \gamma(d)$,
        $F_0 = E_1[x \leftarrow 0]_{x \in \rho(d) \cup \zeta(d)}$, and $[F_1], \ldots, [F_{m-1}] \in [\mathcal{E}]$,
(P4) $\langle\!| [E_1], X, [E_2] |\!\rangle \to \lambda$
    if $d = (X, \lambda) \in \delta$, $E_1 \vdash \gamma(d)$, and $E_2 = E_1[x \leftarrow 0]_{x \in \rho(d) \cup \zeta(d)}$,
(P5) $[\![[E_1], X, [E_2]]\!] \to [\![succ([E_1]), X, [E_2]]\!]$,
(P6) $[\![[E_1], X, [E_2]]\!] \to \langle\!| [F_0], \beta(1), [F_1] |\!\rangle \cdots \langle\!| [F_{i-1}, \beta(i), [F_i] |\!\rangle [\![[F_i], \beta(i+1), [E_2]]\!]$
    if $d = (X, \beta) \in \delta$ with $|\beta| = m > 0$, $i \in \{0, \ldots, m-1\}$,
        $E_1 \vdash \gamma(d)$, $F_0 = E_1[x \leftarrow 0]_{x \in \rho(d) \cup \zeta(d)}$, and $[F_1], \ldots, [F_i] \in [\mathcal{E}]$,
(P7) $[\![[E_1], X, [E_1]]\!] \to \lambda$ if $\Pi(X) \models \pi$.

It can be verified that using the productions (P2), (P3) and (P4), every successful derivation (that generates λ) starting from some nonterminal $\langle\!| [F], X, [F'] |\!\rangle$ corresponds to paths in $\mathcal{R}(\mathcal{M}_{\mathcal{O}}, \varphi)$ that are of the form:

$$\langle X \cdot \beta, [F] \rangle \langle \mu_1 \cdot \beta, [F_1] \rangle \cdots \langle \mu_m \cdot \beta, [F_m] \rangle \langle \beta, [F'] \rangle$$

for some $\beta \in Var^+$, and where, for every $i \in \{1, \ldots, m\}$, $\mu_i \in Var^+$ and $[F_i] \in [\mathcal{E}]$. Indeed, the productions (P2) correspond to transitions in the region graph due to time progress, whereas the productions (P3) correspond to transitions due to applications of derivation rules in δ; the productions (P4) ensure in addition the continuity of the paths, i.e., the fact that segments generated from successive location variables constitute effectively an existing path in the region graph.

Furthermore, it can also be verified that using the productions (P5), (P6), and (P7), together with the productions (P2), (P3) and (P4), every successful derivation starting from some nonterminal $[[F], X, [F']]$ corresponds to paths in the region graph that are of the form:

$$\langle X \cdot \beta, [F] \rangle \langle \mu_1 \cdot \beta, [F_1] \rangle \cdots \langle \mu_m \cdot \beta, [F_m] \rangle \langle \mu_{m+1} \cdot \beta, [F'] \rangle$$

for some $\beta \in Var^+$, and where, for every $i \in \{1, \ldots, m\}$, $\mu_i \in Var^+$ and $[F_i] \in [\mathcal{E}]$, and $\mu_{m+1} \in Var^+$ with $\Pi(\mu_{m+1}(1)) \models \pi$.

Then, it can be deduced that, there exists some path in the region graph starting from $\langle \alpha, [E] \rangle$, and reaching some region $\langle \alpha', [E'] \rangle$ where $(\pi \wedge \bigwedge_{i=1}^n \xi_i)$ is satisfied, if and only if there exists some rank $i \in \{0, \ldots, |\alpha| - 1\}$, and some equivalence classes $[E_0], \ldots, [E_{i+1}]$, such that $E_0 = E$ and $E_{i+1} \in \mathcal{F}$, and

$$\langle [E_0], \alpha(1), [E_1] \rangle \cdots \langle [E_{i-1}], \alpha(i), [E_i] \rangle [[E_i], \alpha(i+1), [E_{i+1}]] \overset{+}{\Longrightarrow} \lambda$$

This is exactly what is expressed by the productions concerning the starting symbol (P1). Then, we obtain the following result:

Proposition 6.1 $s \models \varphi$ iff $L(\mathcal{G}_{(s, \varphi)}) \neq \emptyset$.

Finally, it is well known that the emptiness problem for context-free languages is decidable (see for instance [Har78]). Then, we obtain the following decidability result:

Theorem 6.1 *The verification problem of PTC invariance formulas on CFTS's is decidable.*

6.2 Single Integration Constraint

We consider now the case where the observer $\mathcal{O}$ contains accumulators and integrators. The consideration of integration constraints can be done only under some restrictions on the CFTS's and the reachability formulas.

We say that a CFTS is *closed* if all its time guards are conjunctions of non-strict constraints, i.e., of the form $x \leq n$ or $x \geq n$. A single integration constraint (SIC) reachability formula is of the form

$$\exists \Diamond (\pi \wedge \bigwedge_{i=1}^n \xi_i)$$

where among the ξ_i's, there is at most one integration constraint and all the time constraints are non-strict (of the form $x \leq n$ or $x \geq n$).

We show in this section that the satisfaction relation between closed CFTS's and SIC reachability formulas is decidable. For this, we show first that in this case, we can check the satisfaction relation by considering only the computations where time takes integer values. Then, we reduce the verification problem to the satisfiability problem of integer linear constraints.

Discrete satisfaction relation

Let $\mathcal{M}$ be a CFTS and $\mathcal{O}$ a LHO of $\mathcal{M}$. We introduce a discrete satisfaction relation between states of $\mathcal{M}_{\mathcal{O}}$ and formulas. This relation is a specialization of $\models$ which takes into account only states where time has an integer value. Let us give the formal definitions.

A state $s = \langle \alpha, E_T, E_A, E_I \rangle \in \mathcal{S}_{\mathcal{M}_{\mathcal{O}}}$ is called *integer state* if $\forall u \in T$, $E_T(u) \in \mathbb{N}$ and $\forall u \in I$, $E_I(u) \in \mathbb{Z}$. Let $\mathcal{Z}(\mathcal{S}_{\mathcal{M}_{\mathcal{O}}})$ be the set of integer states of $\mathcal{M}_{\mathcal{O}}$.

We can modify in the obvious manner the definition of the relation $\rightarrow$ to obtain the definition of the transition relation $\rightarrow_{\mathcal{Z}} \subseteq \mathcal{Z}(\mathcal{S}_{\mathcal{M}_{\mathcal{O}}}) \times \mathcal{Z}(\mathcal{S}_{\mathcal{M}_{\mathcal{O}}})$ corresponding to discrete (integer) progress of time, and consider the relation $\hookrightarrow_{\mathcal{Z}} = \rightarrow_{\mathcal{Z}} \cup \triangleright$. Then, we can define accordingly *integer computation sequences*.

Now, we introduce the discrete satisfaction relation $\models_{\mathcal{Z}}$ between integer states and invariance formulas (and consequently, reachability formulas). For this, it is sufficient to consider that, for every integer state $s \in \mathcal{Z}(\mathcal{S}_{\mathcal{M}_{\mathcal{O}}})$, $s \models_{\mathcal{Z}} \forall \Box \phi$ iff $\forall s'. s \stackrel{*}{\hookrightarrow}_{\mathcal{Z}} s'. s' \models_{\mathcal{Z}} \phi$.

Obviously, for every integer state and formula φ, $s \models_{\mathcal{Z}} \varphi$ implies $s \models \varphi$. The reverse implication is far less trivial. We prove this implication by using digitization techniques as in [HMP92, KPSY93] (see the full paper [BER94]). Thus, we have

Proposition 6.2 *Let $s \in \mathcal{Z}(\mathcal{S}_{\mathcal{M}_{\mathcal{O}}})$ and φ a SIC reachability formula. Then, $s \models \varphi$ iff $s \models_{\mathcal{Z}} \varphi$.*

From Reachability to Satisfiability

We give now the reduction of the verification problem of SIC reachability formulas to the satisfiability problem of integer linear constraints.

Let $s = \langle \alpha, E_T, E_A, E_I \rangle$ be an integer state, and consider a SIC reachability formula

$$\varphi = \exists \Diamond (\pi \wedge (\bigwedge_{i=1}^{n_1} \xi_i) \wedge (\bigwedge_{i=1}^{n_2} \varrho_i) \wedge \vartheta)$$

where the ξ_i's and the ϱ_i's are respectively time and accumulation constraints, and ϑ is an integration constraint. Suppose that we are interested in the problem of deciding whether $s \models \varphi$.

First of all, recall that by Proposition 6.2, this verification can be done by considering only the integer computations starting from s. Let us focus for the moment on time constraints only, and consider the PTC reachability formula

$$\varphi' = \exists \Diamond (\pi \wedge \bigwedge_{i=1}^{n_1} \xi_i).$$

Then, clearly we can use the verification method shown in Section 6.1 to decide whether $s \models \varphi'$. Recall that it consists in solving a reachability problem in an infinite region graph via a reduction to the nonemptiness problem of a context-free language. However, in the present case, we can consider only regions corresponding to integer valuations of timers (by Proposition 6.2 and since φ' is also a SIC reachability formula).

So, let $\mathcal{E}_{\mathcal{Z}}$ be the set of time valuations E such that, for every $x \in T$, $E(x) \in \mathbb{N}$ and consider the integer successor function $succ_{\mathcal{Z}}$ defined by $succ_{\mathcal{Z}}([E]) = [E + 1]$, for every $E \in \mathcal{E}_{\mathcal{Z}}$. Notice that $succ_{\mathcal{Z}}$ is always defined and that in the graph $([\mathcal{E}_{\mathcal{Z}}], succ_{\mathcal{Z}})$, every path reaches eventually some self-looping equivalence class, when the value of each timer

x becomes greater than the maximal constant compared with x in either time guards of $\mathcal{M}$ or time constraints ξ_i of φ'.

Following the same lines as in Section 6.1, we can define straightforwardly the integer region graph $\mathcal{R}_Z(\mathcal{M}_\mathcal{O}, \varphi')$ where the regions involve only equivalence classes in $[\mathcal{E}_Z]$ and the edges are defined by means of the successor function $succ_Z$ instead of $succ$.

Then, it is clear that $s \models \varphi'$ if and only if there exists some path in $\mathcal{R}_Z(\mathcal{M}_\mathcal{O}, \varphi')$ which starts from $\langle \alpha, [\mathcal{E}_T] \rangle$ and reaches some region that satisfies $(\pi \wedge \bigwedge_{i=1}^{n_1} \xi_i)$. In order to solve this reachability problem, we proceed as in Section 6.1 by reducing it to the nonemptiness problem of a context-free language. Indeed, consider the CFG $\mathcal{G}_{(s,\varphi')}^Z$ defined exactly as in Section 6.1 where the function $succ_Z$ is used instead of the function $succ$. It is clear that $s \models \varphi'$ if and only if $L(\mathcal{G}_{(s,\varphi')}^Z) \neq \emptyset$.

Now, let us see how to take into account the accumulation and integration constraints. Consider the following derivation in $\mathcal{G}_{(s,\varphi')}^Z$:

$$\sigma = \text{``} \omega \Longrightarrow_{p_1} \omega_1 \cdots \Longrightarrow_{p_n} \omega_n \Longrightarrow_{p_{n+1}} \lambda \text{''} \tag{1}$$

where ω and the ω_i's are nonempty sequences of nonterminals (in N^+). From the definition of $\mathcal{G}_{(s,\varphi')}^Z$, necessarily ω_n is constituted by one nonterminal of the form $[[F], X, [F]]$ for some location variable X such that $\Pi(X) \models \pi$ and some valuation F such that $F \vdash \bigwedge_{i=1}^{n_1} \xi_i$.

Actually, the verification problem $s \models \varphi$ consists in deciding whether the grammar $\mathcal{G}_{(s,\varphi')}^Z$ has a successful derivation (generating λ) σ (as in 1) such that $\omega = S$ (S is the starting symbol), and furthermore this derivation σ must correspond to some computation sequence which validates the accumulation and integration constraints, given the initial valuations $E_\mathcal{A}$ and $E_\mathcal{I}$.

We show in the sequel that this decision problem is reducible to solving a set of integer linear constraints $\Omega_{(s,\varphi)}$. Let us define the sets of variables that are involved in this set of constaints. First, with every accumulator $u \in \mathcal{A}$ we associate a variable a_u, and for the unique integrator v appearing in ϑ we associate the variable ι_v. These variables stand for the values of the corresponding accumulators and integrator at the end of the computation represented by σ. Moreover, with every production $p \in Prod$ we associate a variable w_p which stands for the number of applications of p in σ.

Before giving the constraints that constitute $\Omega_{(s,\varphi)}$, let us introduce some notations. Let $Prod_\tau$ be the set of productions (P2) and (P5), and $Prod_\delta$ the set of productions (P3), (P4) and (P6) in the definition of $\mathcal{G}_{(s,\varphi')}^Z$. The productions in $Prod_\tau$ correspond to time progress, whereas every production p in $Prod_\delta$ corresponds to the application of some derivation rule we denote by d_p. Also, for every production p, we denote by X_p the location variable X if $lhs(p)$ is of the form $[[F], X, [F']]$ or $\langle\!\langle [F], X, [F'] \rangle\!\rangle$.

Now, we say that a sequence of productions $\mathcal{P} \in Prod^*$ is *elementary* if all its productions apply to different nonterminals, i.e., $\forall p \in Prod.\ |\mathcal{P}|_p \leq 1$. Given a nonterminal A and a sequence $\omega \in N^+$, we define Π_ω^A to be the set of elementary sequences $\mathcal{P}$ on $Prod$ such that $\exists \nu \in N^*,\ \omega \overset{*}{\Longrightarrow}_\mathcal{P} A\nu$. Notice that the set Π_ω^A is finite.

We are able now to define $\Omega_{(s,\varphi)}$. First of all, let us define the a_u's and ι_v in terms of the w_p's. For every accumulator u, we define the linear constraint ACC_u:

$$(0 \leq a_u) \ \wedge \ \Big(a_u = E_\mathcal{A}(u) + \sum_{p \in Prod_\delta} \kappa(d_p)(u) \cdot w_p\Big)$$

and concerning the integrator v, we define the constraint INT_v:

$$(0 \leq \iota_v) \wedge (\iota_v = E_{\mathcal{I}}(v) + \sum_{p \in Prod_r} \partial(X_p)(v) \cdot w_p)$$

Now, let us define the constraints on the w_p's. These constraints must be satisfiable by some valuation W of the w_p's, if and only if W corresponds to numbers of applications of productions p's in some existing derivation σ (1) of the grammar.

So, first of all, the constraints on the w_p's must express the fact that any occurrence of a nonterminal appearing along σ must be reduced so that we get the empty sequence λ. Thus, in the derivation σ, for any nonterminal A, the number of the A-reductions, i.e., applications of some productions p such that $lhs(p) = A$ (A-productions), must be equal to the number of the A-introductions, i.e., the number of the occurrences of A in ω and in the right-hand sides of the applied productions. Given a nonterminal A and a sequence $\omega \in N^+$, this fact is expressed by the linear constraint REDUCT_ω^A defined by:

$$\sum_{p \in Prod} |lhs(p)|_A \cdot w_p = |\omega|_A + \sum_{p \in Prod} |rhs(p)|_A \cdot w_p$$

However, some solutions of $\bigwedge_{A \in N} \mathsf{REDUCT}_\omega^A$ may assign to some variable w_p a value which is non null while p is not necessarily involved in some derivation σ.

Indeed, consider some valuation W that validates $\bigwedge_{A \in N} \mathsf{REDUCT}_\omega^A$ and suppose that it corresponds to some derivation σ as in (1). Consider also some nonterminal B which does not appear in ω neither in any ω_i in the considered derivation σ. Now, assume that there is some production $p = B \rightarrow B$ in $Prod$. We can define another valuation W' which assigns to w_p any strictly positive integer and coincides with W on the other variables. Clearly, the new valuation W' is also a solution of $\bigwedge_{A \in N} \mathsf{REDUCT}_\omega^A$. This, solution must be discarded since the values of the accumulators u's and the integrator v, which are calculated from W' using ACC_u and INT_v, do not correspond necessarily to values that can be obtained from some existing derivation of the grammar $\mathcal{G}^Z_{(s,\varphi')}$.

Thus, we must express in addition, the fact that for any nonterminal A, there exists some A-production p with $w_p > 0$ if and only if A appears in ω or in the ω_i's. This is done by the constraint REACH_ω^A:

$$\sum_{p \in Prod} |lhs(p)|_A \cdot w_p > 0 \Leftrightarrow \bigvee_{\mathcal{P} \in \Pi_\omega^A} \bigwedge_{p \in \mathcal{P}} w_p > 0.$$

Now, the linear system $\Omega_{(s,\varphi)}$ is defined by:

$$\Omega_{(s,\varphi)} = (\bigwedge_{A \in Prod}(\mathsf{REDUCT}_S^A \wedge \mathsf{REACH}_S^A)) \wedge (\bigwedge_{u \in \mathcal{A}} \mathsf{ACC}_u) \wedge \mathsf{INT}_v \wedge$$
$$(\bigwedge_{i=1}^{n_2} \varrho_i[a_u/u]_{u \in \mathcal{A}}) \wedge \vartheta[\iota_v/v]$$

where, for any constraint ξ, $\xi[y/x]$ denotes the constraint obtained by substituting in ξ each occurrence of x by y. Then, we can prove in a similar way as in [BER93] that:

Proposition 6.3 $s \models \varphi$ iff $\Omega_{(s,\varphi)}$ is satisfiable.

Since the satisfiability of integer linear constraints is decidable, we obtain the following decidability result:

Theorem 6.2 *The verification problem of SIC reachability problem on closed CFTS's is decidable.*

7 Conclusion

We propose in this paper automatic verification methods of general invariance properties for infinite timed systems with a context-free underlying structure.

Our specification approach is expressively powerful. It consists in expressing invariance properties by means of observation variables (timers, accumulators and integrators) that are maintained during the execution of the system, i.e., the observation variables can be seen as additional variables that can be modified but never tested by the system.

We have obtained strong decidability results for the verification of such invariance properties on CFTS's. Indeed, when we consider time and accumulation constraints, we have shown that the verification problem of invariance properties on CFTS's is decidable without any restriction. This result generalizes (concerning invariance properties) the one presented in [ACD90] to the case of context-free systems, and the one presented in [BER93] to the case of timed systems. When integration constraints are taken into account, we have shown that the verification problem of *single integration constraints* is decidable for closed CFTS's. In the full paper [BER94], we also show that for CFTS's with one timer that is reset at each derivation, the verification problem of invariance properties is decidable without any restriction on the number of integration constraints. The results concerning integration constraints generalize those given in [KPSY93].

Several extensions of the work presented in this paper can be considered. Among them, it would be interesting to consider the verification problem for other classes of properties like eventuality properties, and more generally, properties expressible in some temporal logic allowing duration as well as occurrence constraints, obtained for instance as a combination of the logics DTL (Duration Temporal Logic) [BES93] and PCTL (Presburger CTL) [BER93].

References

[ACD90]   R. Alur, C. Courcoubetis, and D. Dill. Model-Checking for Real-Time Systems. In *LICS'90*. IEEE, 1990.

[ACH93]   R. Alur, C. Courcoubetis, and T. A. Henzinger. Computing Accumulated Delays in Real-time Systems. In *Hybrid Systems*, 1993. LNCS 736.

[ACHH93] R. Alur, C. Courcoubetis, T. Henzinger, and P-H. Ho. Hybrid Automata: An Algorithmic Approach to the Specification and Verification of Hybrid Systems. In *Hybrid Systems*, 1993. LNCS 736.

[BBK87]   J.C.M. Baeten, J.A. Bergstra, and J.W. Klop. Decidability of Bisimulation Equivalence for Processes Generating Context-Free Languages. Tech. Rep. CS-R8632, 1987. CWI.

[BER93]   A. Bouajjani, R. Echahed, and R. Robbana. Verification of Nonregular Temporal Properties for Context-Free Processes. submitted for publication, 1993.

[BER94]   A. Bouajjani, R. Echahed, and R. Robbana. Verification of Context-Free Timed Systems using Linear Hybrid Observers. Tech. Rep. Spectre-94-4, Verimag, Grenoble, January 1994.

[BES93]   A. Bouajjani, R. Echahed, and J. Sifakis. On Model Checking for Real-Time Properties with Durations. In *LICS'93*. IEEE, 1993.

[BS92]   O. Burkart and B. Steffen. Model Checking for Context-Free Processes. In *CONCUR'92*, 1992. LNCS 630.

[Cer92]   K. Cerans. Decidability of Bisimulation Equivalence for Parallel Timer Processes. In *CAV'92*, 1992. LNCS 663.

[CES83]   E.M. Clarke, E.A. Emerson, and E. Sistla. Automatic Verification of Finite State
           Concurrent Systems using Temporal Logic Specifications: A Practical Approach. In
           POPL'83. ACM, 1983.
[CGL93]   K. Cerans, J. Godskesen, and K. Larsen. Timed Modal Specification: Theory and
           Tools. In *CAV'93*. LNCS 697, 1993.
[CHR91]   Z. Chaochen, C.A.R. Hoare, and A.P. Ravn. A Calculus of Durations. *Information
           Processing Letters*, 40:269–276, 1991.
[CHS92]   S. Christensen, H. Hüttel, and C. Stirling. Bisimulation Equivalence is Decidable for
           all Context-Free Processes. In *CONCUR'92*, 1992. LNCS 630.
[CS91]   R. Cleaveland and B. Steffen. A Linear-Time Model-Checking Algorithm for
           the Alternation-Free Modal Mu-Calculus. In *Proc. Computer-Aided Verification
           (CAV'91)*, 1991. LNCS 575.
[EL86]   E.A. Emerson and C.L. Lei. Efficient Model-Checking in Fragments of the Proposi-
           tional μ-Calculus. In *LICS'86*, 1986.
[GH91]   J.F. Groote and H. Hüttel. Undecidable Equivalences for Basic Process Algebra.
           Tech. Rep. ECS-LFCS-91-169, 1991. Dep. of Computer Science, Univ. of Edinburgh.
[Har78]   M.A. Harrison. *Introduction to Formal Language Theory*. Addison-Wesley Pub.
           Comp., 1978.
[HMP92]   T. Henzinger, Z. Manna, and A. Pnueli. What Good are Digital Clocks? In
           ICALP'92, 1992. LNCS 623.
[HNSY92]   T.A. Henzinger, X. Nicollin, J. Sifakis, and S. Yovine. Symbolic Model-Checking for
           Real-Time Systems. In *LICS'92*. IEEE, 1992.
[KPSY93]   Y. Kesten, A. Pnueli, J. Sifakis, and S. Yovine. Integration Graphs: A Class of De-
           cidable Hybrid System s. In *Hybrid Systems*, 1993. LNCS 736.
[KS83]   P. Kanellakis and S.A. Smolka. CCS Expressions, Finite State Processes, and Three
           Problems of Equivalence. In *PODC'83*. ACM, 1983.
[Mil80]   R. Milner. A Calculus of Communication Systems. 1980. LNCS 92.
[MMP92]   O. Maler, Z. Manna, and A. Pnueli. A Formal Approach to Hybrid Systems. In *REX
           workshop on Real-Time: Theory and Practice*, 1992. LNCS 600.
[MP93]   Z. Manna and A. Pnueli. Verifying Hybrid Systems. In *Hybrid Systems*, 1993. LNCS
           736.
[NOSY93]   X. Nicollin, A. Olivero, J. Sifakis, and S. Yovine. An Approach to the Description
           and Analysis of Hybrid Systems. In *Hybrid Systems*, 1993. LNCS 736.
[NRSV90]   X. Nicollin, J.-L. Richier, J. Sifakis, and J. Voiron. ATP: an Algebra for Timed
           Processes. In *IFIP TC2 Working Conf. on Prog. Concepts and Methods*, 1990. Israel.
[Par81]   D. Park. Concurrency and Automata on Infinite Sequences. In *5th GI-Conference
           on Theoretical Computer Science*. 1981. LNCS 104.
[QS82]   J-P. Queille and J. Sifakis. Specification and Verification of Concurrent Systems in
           CESAR. In *Intern. Symp. on Programming, LNCS 137*, 1982.
[VW86]   M.Y. Vardi and P. Wolper. An Automata-Theoretic Approach to Automatic Pro-
           gram Verification. In *LICS'86*. IEEE, 1986.
[Wan90]   Y. Wang. Real Time Behaviour of Asynchronous Agents. In *CONCUR'90*, 1990.
           LNCS 458.

On the Random Walk Method for Protocol Testing

Milena Mihail
Bellcore

Christos H. Papadimitriou*
University of California, San Diego

Abstract

An important method for testing large and complex protocols repeatedly generates and tests a part of the reachable state space by following a random walk; the main advantage of this method is that it has minimal memory requirements. We use the coupling technique from Markov chain theory to show that short trajectories of the random walk sample accurately the reachable state space of a nontrivial family of protocols, namely, the symmetric dyadic flip-flops. This is the first evidence that the random walk method is amenable to rigorous treatment.

Following West's original reasoning, efficient sampling of the reachable state space by random walk suffices to ensure effectiveness of testing. Is, however, efficient sampling of the random walk necessary for the effectiveness of the random walk method? In the context of Markov chain theory, "small cover time" can be thought of as a simpler justification for the effectiveness of testing by random walk; all symmetric (reversible) protocols possess the small cover time property.

Thus the conclusions of our work are that (i)the random walk method can be understood in the context of known Markov chain theory, and (ii)symmetry (reversibility) is a general protocol style that supports testing by random simulation.

1   Introduction

Testing complex protocols for conformance with their specifications is an ever-increasing technological challenge —see Holzmann's book for a detailed exposition [13]. According to Holzmann, traditional testing techniques based on full or controlled partial state exploration become very complex (perhaps prohibitively so) when the reachable state space becomes larger than, say, 10^8 [13] [4] [17]. For larger protocols, an alternative testing method is what could be called the *random walk method* (a.k.a. *random partial state exploration*, or *random simulation*) [13] [19] [18]: Beginning at the starting protocol state, a random applicable action of the protocol is chosen and applied, another random step is repeated at the resulting state, and so on, up to some sufficiently large and yet computationally feasible number of steps. At each step the next action is chosen among the applicable actions either according to a distribution that reflects the protocol's operation, or uniformly at random —the latter technique is good for exploring the less traveled paths in the state space. The main advantage of this random simulation method is that it has minimal space (memory) requirements: to simulate a step of the random walk we need to store only the current state and a description of the actions (which is a natural lower bound on memory for any state exploration process).

When and why is the random walk testing technique sound? Engineers argue that testing is effective if the random walk looses memory of its starting state after a relatively short trajectory, thus the final point of the walk simulates accurate observation of the system in steady state, or, preferably, fair sampling of the protocol's reachable state space. Under such *rapid mixing* conditions, errors are recovered in time proportional to their probability of occurrence in stationarity, and the whole reachable state space can be potentially visited in time roughly proportional to its size (which is again a natural lower bound on time for any exhaustive search). In his influential paper introducing and advocating the random walk technique [19], West argued that this is indeed the behavior of certain validation experiments over the OSI Session Layer, and, in the formal sense, in the case of a very simple hypothetical *totally decoupled* system (a system without communication between its entities).

---

*Research supported by the National Science Foundation.

There are several mathematical tools at the disposal of the full and controlled partial state exploration approach to protocol testing, such as temporal logic algorithms for model verification (eg. see [20] [8] [9] for early references), symbolic methods, compression methods, projection methods, and techniques for testing and minimizing finite state machines (eg. see [15] [21] for recent references). In contrast, there is very little rigorous work on the random walk method —besides West's observation concerning the decoupled protocol. In Markov chain theory, it is now well understood that many common random walks *do not* converge in any satisfactory way —and a theory of those that do is emerging [10] [1] [2] [14] [16] [6] [11] [12]. In this context, West's example of a decoupled system is easily identified as the *hypercube*, the most simple case of a rapidly mixing random walk.

The aim therefore is to use Markov chain theory to separate those protocols which can be effectively tested by random walk from those which cannot. Naturally, circular characterizations such as "protocols whose reachable state space is an expander" are completely useless. For a proposed family of protocols to be interesting, it should be definable in terms of "local," "syntactic" properties of the protocols. Ideally, one would like syntactic guidelines, perhaps "styles" of protocol writing, that guarantee convergence of the associated random walk; for example, "reversibility of actions" could be such a property. A possible analogy here is the "structured programming" paradigm in software engineering, which guarantees positive debugging and maintainability properties of the resulting code. The question therefore is, *what are examples of broad families of protocols whose state space can betested by random walks?*

In this paper we identify a family of protocols, namely *systems of symmetric dyadic flip-flops* (SSDF's), and prove that the random walk always converges to the uniform distribution over the state space of such protocols in time polylogarithmic in the size of the state space (Theorem 3.1). Thus, by West's statistical arguments, this class can be tested by random walk. SSDF's are systems of communicating finite automata (see also Figure 1) with two states per entity (hence the name flip-flop); each action involves only two entities (hence the attribute "dyadic"); and *symmetry* (*reversibility*), that is, whenever an action that affects two coordinated transitions of two entities is present, the reverse action is also present. Although the expressibility of SSDF's is rather restricted, in the sense that their reachable state spaces have fairly regular characterizations, the transition graph over the reachable state space is highly irregular, and consequently our rapid mixing argument demonstrates nontrivial technique (and is a giant step beyond the mixing argument for the hypercube decoupled protocol, the only one heretofore analyzed in this way [19]). Therefore our result provides the first substantial evidence that the random walk method is amenable to formal Markov chain analysis and understanding.

The technique employed to prove our main rapid mixing result uses the *coupling* argument from Markov chain theory [1] [10] [5] [6]. To show that the random walk from some starting state converges within polynomial time to its stationary distribution, we imagine side-by-side two such walks, one from the starting state and one from an arbitrary state of the reachable state space, and coordinate their transitions so that each walk is simulated fairly, but in small expected time the two states coincide and thus the two walks are indistinguishable; since the second random walk can be assumed to start from the stationary distribution over all reachable states, fast convergence is established.

Can arguments for the effectiveness of the random walk method extend beyond the class of SSDF's? This question is twofold. Firstly, can rapid mixing arguments for the justification of the random walk method extend beyond SDFF's? Secondly, are rapid mixing conditions necessary for the random walk testing method to be effective?

For the first question, there are counter-examples showing that, if any of the conditions defining SDFF's fails, rapid mixing also fails [7]. But perhaps characterizations somewhat different to the features of SDFF's may reveal further rapid mixing cases (experiments suggest that "random" symmetric protocols mix rapidly, thus, capturing features of randomness that ensure rapid mixing might be a way to proceed).

For the second question, we point out that *small cover time* may be a simpler justification for the effectiveness of the random walk method; small cover time is a strictly weaker property than rapid mixing. The cover time of a graph is the expected time for a random walk to visit all the vertices of the graph, and it fits naturally in the context of state space exploration. All symmetric systems posses the small cover time property. Thus a possible heuristic for testing by random simulation could be to design systems as reversible as possible, and/or impose artificial reversibility for simulation purposes.

2 Preliminaries

A *protocol* is a system of communicating finite automata, the *entities*. Each entity is driven by a set of *actions*, and for a ⟨state, action⟩ pair, the *transition function* determines (i)whether the action is applicable on the particular state (not all actions are applicable on all states), and if so, (ii)which is the new state after the action is applied. The system has *communication* in that an action may simultaneously involve several entities. When the system evolves as a whole, an action is applicable iff it is applicable for all entities that it involves. This determines the *transition function* of the *combined automaton*, with combined states consisting of vectors of size n (storing the state of each entity) and actions consisting of the union of actions over all entities. Finally there is a specified *starting combined state*, and a *reachable state space* consisting of combined states to which the initial state can be driven by a sequence of applicable actions; realize that the size N of the combined state space is exponential in n. The aim of *testing* is to explore the reachable state space of the reachable state space of the combined automaton. The aim of *efficient testing* is to explore the reachable state space using resources that match the natural lower bounds (up to polynomial factors, that is, time poly$N$ and space poly n, and for practical efficiency as close to linear as possible).

We shall also consider a protocol as a *Markov chain* (or *random walk on the underlying graph of the combined automaton*). Without loss of generality, we consider a slightly modified version of the natural random walk: Beginning at the starting combined state, at each step, some action is chosen uniformly at random among all actions of the system, and if the action is applicable to the current combined state, then, with probability $1/2$, it is applied; in all other cases the combined state does not change. How can such a scheme support efficient testing? Firstly realize that it satisfies the memory requirement for efficiency: to simulate each step, we need to store only the current state and the set of actions. To argue about time requirements we need additional conditions.

A protocol is *symmetric* if it is it is symmetric as a combined automaton and as a Markov chain. This amounts to the protocol being *reversible* as a system: for each action a that drives

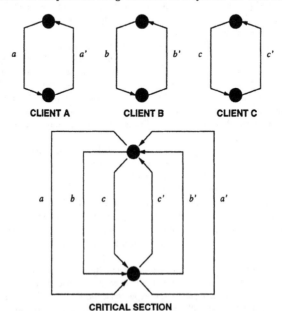

Figure 1: A system of Symmetric Dyadic Flip-Flops implementing Critical Section with Three Clients.

combined state $\vec{q}$ to combined state $\vec{q'}$, there exists a reverse action a^{-1} that drives $\vec{q'}$ to $\vec{q}$. Well known Markov chain theory suggests that the random walk on symmetric protocols *converges to the uniform distribution* over the whole reachable state space. Now for testing purposes, convergence to uniformity is quite favorable. We can use a point of the random walk *after convergence has (nearly) occurred* as a uniformly random point of the state space. For a reachable state space of size N, $O(N \log N)$ uniform samples, in expectation, generate the whole space (in addition, for any fraction ϵ of the reachable state space, a representative state is generated by ϵ^{-1} samples in expectation). But can we ensure *efficient convergence* to uniformity? We focus on a class of symmetric protocols that we call *systems of symmetric dyadic flip-flops* (SSDF's), and show that their convergence to uniformity is efficient, in the sense that it is ensured after poly log N steps (Theorem 3.1).

Three features determine symmetric dyadic flip-flops: They are symmetric in the way discussed above. They are flip-flops, meaning that all entities have two states; we can assume that these are 0 and 1. They are dyadic, meaning that each action involves at most two automata; without loss of generality we assume that each action involves exactly two automata. We impose the additional technical condition that there are no self-loops, that is actions that leave the state of some involved automaton unchanged (self loops are treated in [7]). Thus, there are two kinds of actions: the so called *bar* actions that change a pair of 0's to 1's and their inverses changing a pair of 1's to 0's, and the so called *cross* actions changing a pair of a 1 and a 0 to a pair of a 0 and a 1. Given a system of symmetric dyadic flip-flops, we define its *communication graph*, with one vertex for each automaton, and with an edge $[u,v]$ present if and only if there is an action (and its inverse) involving both u and v; again, without loss of generality, we assume that if there is a bar (resp. cross) action involving u and v, then there is no cross (resp. bar) action involving u and v. Now realize that the transitions of the protocol can be completely specified by its communication graph, once its edges have been labeled as bar or cross; we call this the *labeled communication graph*.

For each combined state $\vec{q}$ we also consider its *action communication graph*: if $[u,v]$ is a bar edge and the bits u and v are similar, or if $[u,v]$ is a cross edge and the bits u and v are opposite, then one of actions a and a^{-1} that correspond to $[u,v]$ is applicable on $\vec{q}$, thus we mark $[u,v]$ as *active*; otherwise we mark $[u,v]$ as *inactive*. Realize that if a state $\vec{q'}$ differs from a state $\vec{q}$ on exactly one bit, say v, then the action communication graphs of $\vec{q}$ and $\vec{q'}$ are identical on edges not incident to v, and exactly opposite on edges incident to u. In the next section, we shall repeatedly think of combined states in terms of their action communication graphs and furthermore, of transitions as follows (see also Figure 2): for some active edge $[u,v]$ with corresponding actions a and a^{-1} exactly one of which, say a, is applicable to $\vec{q}$, the effect of applying a on $\vec{q}$ can be thought of as activating the edge $[u,v]$ on the action communication graph of $\vec{q}$ thus changing bits u and v, and, in the resulting action communication graph, changing all edges incident to u or v from active to inactive and vice versa, except for $[u,v]$ itself which remains unchanged.

Clearly the labeled communication graph is an invariant of the protocol, while the action communication graph changes from state to state. However, the action graph inherits certain invariants of the labeled graph. We shall need several such invariants in the next section; here we start by demonstrating the simplest one in Lemma 2.1 below. In particular, we say that a *cycle of the labeled graph* has *odd parity* iff it contains an odd number of bar edges, and we say that a *cycle of the action graph of a specific state* has *odd parity* iff it contains an odd number of active edges (here we mean generalized cycle: a closed trail with arbitrary partial overlaps).

Lemma 2.1 (Cycle invariant) (See also Figure 2). *For any cycle of the communication graph, the parity of the cycle in action graphs is invariant over all combined states, and it is the same as the parity of the cycle in the labeled graph.*

PROOF. The parity of cycles in labeled and action graphs are trivially identical for the combined state $\vec{q} = 00 \ldots 0$, and changing the bits of $\vec{q}$ one at a time preserves the parity of cycles on action communication graphs.

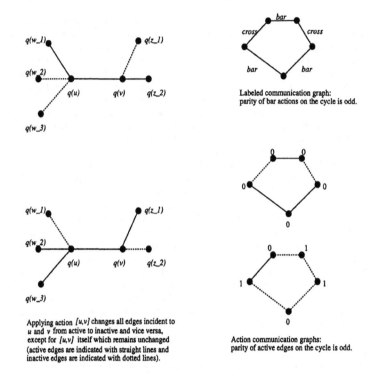

Applying action $[u,v]$ changes all edges incident to u and v from active to inactive and vice versa, except for $[u,v]$ itself which remains unchanged (active edges are indicated with straight lines and inactive edges are indicated with dotted lines).

Labeled communication graph: parity of bar actions on the cycle is odd.

Action communication graphs: parity of active edges on the cycle is odd.

Figure 2: Transitions and cycle parities on action and labeled communication graphs.

3 SDFF's, Coupling, and Rapid Mixing

In this section we prove that SDFF's, viewed as Markov chains, posses the rapid mixing property, that is, the random walk is arbitrarily close to uniformity over N reachable states after poly $\log N$ number of steps. To establish rapid mixing we use the *coupling* method from Markov chain theory [1] [10] [5] [6]. Consider two random walks on the protocol, random walk $\vec{q}(t)$ starting from some starting state $\vec{q}(0)$, and random walk $\vec{q'}(t)$ starting from some arbitrary state $\vec{q'}(0)$ reachable from $\vec{q}(0)$; $\vec{q'}(t)$ can be thought of as representing the uniform stationary distribution. Intuitively, we must show that $\vec{q}(t)$ and $\vec{q'}(t)$ will soon be indistinguishable. To do this, we run $\vec{q}(t)$ and $\vec{q'}(t)$ side-by-side, by applying subtle coordination in their transitions. The coordination is such that both random walks still obey the appropriate transition probabilities, but within expected time poly $n =$ poly $\log N$ the two walks converge to the same state. More formally, where the *coupling time* T is the first time that $\vec{q}(t)$ and $\vec{q'}(t)$ match: $\vec{q}(T) = \vec{q'}(T)$, and $E[T]$ is its worst case expectation (over all initial configurations $\vec{q}(0)$ and $\vec{q'}(0)$ that are reachable from each other), the *variation distance* of $\vec{q}(t)$ from stationarity can be bounded by (see [1] [10] [5] for a proof):

$$d(t) = \max_Q \left(\Pr[\vec{q}(t) \in Q] - \frac{|Q|}{N} \right) \leq 1/2^{\frac{t}{2E[T]}}$$

For the coupling, we keep pebbles on the vertices of the communication graph where $\vec{q}(t)$ and $\vec{q'}(t)$ differ and measure progress in terms of the number of pebbles left. In particular, where $[u,v]$ is an edge of the communication graph and there are pebbles on vertices here $\vec{q}(t)$ and $\vec{q'}(t)$ differ, realize that the following hold: (i)If there are pebbles on both u and v, then either $[u,v]$ is active for both $\vec{q}(t)$ and $\vec{q'}(t)$, or $[u,v]$ is inactive for both $\vec{q}(t)$ and $\vec{q'}(t)$. The case where $[u,v]$ is active for both processes is particularly favorable: if $[u,v]$ is activated on exactly one of the processes, then this process will change bits u and v thus matching them with the other process, and the pebbles can be removed from vertices u and v. (ii)If there is a pebble on exactly one of u and v, say u, then $[u, v]$ is active on exactly one of $\vec{q}(t)$ and $\vec{q'}(t)$ and inactive on the other. This case is indifferent: if $[u, v]$ is activated for the process for which it is active, then the two processes will match the bit u but will unmatch the bit v, thus the pebble should be moved from u to v. (iii)If there are pebbles on neither u nor v, then the two processes match on these two bits and, as far as $[u, v]$ is concerned, they can proceed in full coordination. The above three observations lead us to the following coupling (which is convenient to think of as it evolves on action communication graphs, in the way explained in Section 2):

> Let $\vec{q}(t)$ and $\vec{q'}(t)$ be the states at time t;
> Pick $[u, v]$ uniformly at random among all edges of the communication graph;
> Toss a fair coin;

Case ia   IF there are pebbles on both u and v and $[u, v]$ is active in both processes

THEN $\begin{cases} \text{if heads then } [u, v] \text{ is activated on } \vec{q}(t) \text{ only;} \\ \text{if tails then } [u, v] \text{ is activated on } \vec{q'}(t) \text{ only;} \\ \text{in either case the pebbles are removed from both } u \text{ and } v; \end{cases}$

Case ib   IF there are pebbles on both u and v and $[u, v]$ is inactive in both processes
THEN nothing happens;

Case ii   IF there is a pebble on u but not on v, thus $[u, v]$ is active for exactly one of the processes
THEN, if heads, $[u, v]$ is activated on the process for which it is active, and the pebble is moved from u to v;
ELSE nothing happens;

Case iii   IF there are no pebbles on either u or v, thus $[u, v]$ is either active or inactive for both processes,
THEN, if heads and $[u, v]$ is active, $[u, v]$ is activated on both processes;
ELSE nothing happens;

Theorem 3.1 (Main Theorem) *For any SDFF with n entities and m actions, for any initial configurations $\vec{q}(0)$ and $\vec{q'}(0)$ that belong to the reachable state space, the expectation of the coupling time T (when all pebbles are removed and $\vec{q}(T) = \vec{q'}(T)$) is $E[T] = O(n^3 m^2)$, and hence the variation distance decreases as $2^{-\Omega(t/n^3 m^2)}$.*

PROOF (outline). We wish to establish that the favorable *Case ia* with two pebbles placed at the endpoints of an active edge occurs often enough. What causes difficulty in the pebble game is that evolution is determined by the action graphs, and these graphs change. We therefore go through a sequence of reductions to reformulate a pebble game on the time invariant labeled communication graph. By the end of Lemma 3.4 the pebble game is played only on the labeled communication graph.

Firstly, the pebble game can be reduced to one which starts with only two pebbles, where the first time that *Case ia* occurs the game is finished. Let τ be the time for *Case ia* to occur when we start with two pebbles, and let $E[\tau]$ be its worst case expectation. It can be argued (and is intuitive to understand) that

$$E[T] \leq nE[\tau] \tag{1}$$

Now for the two-pebble game, say a red and a blue, there may be many possibilities by which *Case ia* occurs; we focus on a single one. In Lemma 3.2 we isolate a "target" trail on the action communication graphs with the property that this trail always contains an odd number of active edges in both processes. In Lemma 3.3(i) we reduce removal of the pebbles to traversal of the target trail: when this odd trail is of length one, the pebbles are neighboring by an active edge and thus *Case ia* has occurred. However, straightforward traversal of the target trail is probabilistically very unlikely, thus in Lemmas 3.3(ii) and 3.4 we account for the likely case of repeated "circular" efforts until the target trail is finally traversed.

Lemma 3.2 (Target trail). *If $\vec{q'}$ is reachable from $\vec{q}$ and they differ in exactly bits u and v (the red and blue pebbles), then there is a "target" trail from u to v on the communication graph such* *that the trail contains an odd number of active edges in the action communication graphs of both $\vec{q}$ and $\vec{q'}$.*

PROOF (outline). Any trail from u to v has the same number of active edges in both $\vec{q}$ and $\vec{q'}$ (all edges incident to neither u nor v are similar —and so is $[u, v]$, if present, and all edges incident to exactly one of u and v are opposite). Thus we need to find an odd active edge trail (*odd trail*, for short) only for $\vec{q}$.

If the action communication graph of $\vec{q}$ contains a cycle with an odd number of active edges (and by Lemma 2.1 the labeled graph contains and odd cycle), then an odd trail can be found trivially: start with a path from u to some vertex of the cycle, end with a path from some vertex of the cycle to v, and use the odd parity of the cycle to insert a trail that ensures correct parity of active edges. If all cycles in the action communication graph of $\vec{q}$ contain an even number of active edges, then it can be shown that all trails from u to v are odd (this reasoning is somewhat more detailed and is omitted here).

Note that the specific target trail constructed in this proof is of length at most m.

Remark: As a byproduct of Lemma 3.2 we obtain regular characterizations of the reachable state spaces. For protocols whose labeled graph contains an odd cycle, all vectors with the same parity of bits can be reached, and a similar (but somewhat finer) characterization holds for protocols whose labeled graph contains no odd cycle. Such observation draw limitations on the expressibility of SDDF's in terms of their reachable state spaces, however, it should be obvious that the transitions over these state spaces are highly irregular.

Lemma 3.3 *Let $u = v_0, v_1, \ldots, v$ be a target trail of Lemma 3.2. (i)If edge $[u, v_1]$, which is active in exactly one of $\vec{q}$ and $\vec{q'}$, is applied, thus moving the red pebble to v_1, then the suffix $v_1, \ldots, v$ is a target trail for the new configurations. (ii)If edge $[u, u']$, $u' \neq v_1$, which is also active in exactly one of $\vec{q}$ and $\vec{q'}$, is applied, thus moving the red pebble to u', then the concatenation $u', u, v_1, \ldots, v$ is a target trail for the new configurations.*

Lemma 3.4 *Let $u = v_0, v_1, \ldots, v$ be a target trail of Lemma 3.2. Let $v_i, \ldots, v_j = v_i$ be a cycle that has even number of bar edges in the labeled graph, and by Lemma 2.1, the cycle contains an even number of active edges in the action graphs of both $\vec{q}$ and $\vec{q'}$. Then $u = v_0, v_1 \ldots v_i, v_j, \ldots, v$ is also a target trail for $\vec{q}$ and $\vec{q'}$.*

The proofs of Lemmas 3.3 and 3.4 are straightforward and omitted.

We proceed to finish the proof of Main Theorem 3.1. In particular, we are now ready to formulate a time invariant pebble game.

Let $u = v_0, v_1, \ldots, v_l = v$ be a target trail for $\vec{q}(0)$ and $\vec{q'}(0)$. Lemmas 3.3 and 3.4 suggest that when, for some (so called "return") time τ_r, the red pebble returns on u' after having traveled a cycle $u = y_0, y_1 \neq v_1, \ldots, u$ that has an even number of bar edges in the labeled graph, and, for the same τ_r, the blue pebble returns on v after having traversed a cycle $v = z_0, z_1 \neq v_{l-1}, \ldots, v$, then $u = v_0, v_1, \ldots, v_l = v$ is still a target trail for $\vec{q}(\tau_r)$ and $\vec{q'}(\tau_r)$.

And if the first time that a pebble moves after τ_r, it is either the red pebble moving along $[u, v_1]$, or the blue pebble moving along $[v, v_{l-1}]$, then the target trail shortens by one. The probability that this happens is at least $1/n$, thus in expected n attempts it will indeed occur.

Now it follows that the expected time for a target trail of initial length l to become a single edge is

$$E[\tau] \leq l \cdot n \cdot E[\tau_r] \leq n^2 E[\tau_r] \tag{2}$$

For the return time τ_r, combining arguments of [3] and averaging principles (about odd and even cycles), it can be shown that

$$E[\tau_r] \leq O(m^2) \tag{3}$$

Now (1), (2), and (3) complete the proof of the Main Theorem 3.1.

4 More General Protocols and Cover Times

Three main restrictions define the class of SDFF's: symmetry, two states per enity, and two entities per action. In Figures 3 and 4 we outline examples where if any of the above conditions fail, rapid mixing fails (more extensive discussions are in [7]; the counter-like counter-examples were pointed out to us by Ernie Cohen). Thus, in the strict sense, rapid mixing cannot extend beyond SDFF's, unless we obtain some characterization somewhat different to the features involved in SDFF's.

On the positive side, experiments with random graphs indicate that random symmetric protocols mix rapidly [7]. Therefore, perhaps there are features of randomness that can be captured to establish rapid mixing beyond SDFF's (in graph theory, the process of isolating features of random graphs, and constructing explicit graphs that posses these features thus inheriting certain behaviors of random graphs, is well explored). For example, the combined automata and the communication graphs (hypergraphs) of random symmetric protocols have small diameter —a feature strongly violated by counter-like counter-examples (where not only wost case points are in long distances, but also average case points are in long distances).

However, is rapid mixing essential for the effectiveness of testing by random walk?

If the requirement of testing is that, for any fraction ϵ of the reachable state space of size N, we visit a representative element from this set in expected time roughly $\epsilon^{-1}\text{poly}\log N$ (which is the natural lower bound), then rapid mixing (which achieves efficient testing almost by definition) appears necessary.

If the requirement of testing is weakened to exhaustive state space exploration, then perhaps "small cover time" for the combined automaton is a simpler justification for effectiveness of testing.

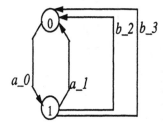

A system of assymetric dyadic flip–flops that does not mix rapidly: Starting from the 00...0 combined state, it takes 2^n attempts in expectation for a 1 to propagate to the n–th bit. This is because, at each step, with probability 1/2, some "b" action will bring the system to its starting combined state.

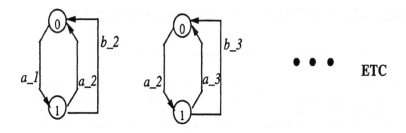

Figure 3: Assymmetric counter-example to rapid mixing.

For a random walk on a graph, the *cover time* is the time by which all vertices have been visited at least once. All symmetric (undirected) graphs are known to have cover times at most cubic in their sizes [3], and most graphs have cover times slightly bigger than linear. Thus all symmetric protocols posses the small cover-time property, and, in the sense of exhaustive search of the reachable state space, are amenable to effective testing by random walk.

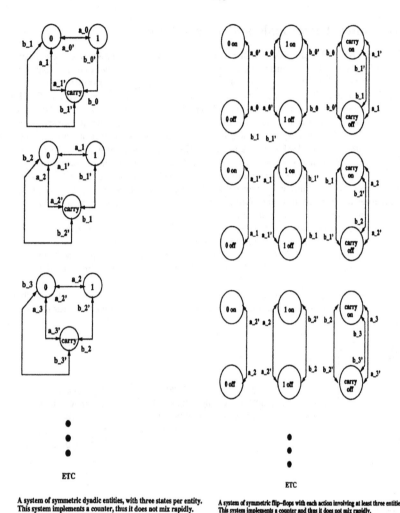

A system of symmetric dyadic entities, with three states per entity. This system implements a counter, thus it does not mix rapidly.

A system of symmetric flip-flops with each action involving at least three entities. This system implements a counter and thus it does not mix rapidly.

Figure 4: Counter-like counter-examples to rapid mixing.

For general assymmetric protocols (and graphs) little is known about bounds on their cover times —besides the fact that, in the worst case, these bounds are exponential in the size of the graphs. However, experiments suggest more refined behavior (in particular, that under suitable adaptations of the random walk, only a small fraction of the states remains unvisited after what seems to be superpolynomial effort).

References

[1] Aldous, D., "Random Walks on finite groups and rapidly mixing Markov chains", *Seminaire de Probabilites XVII*, Lecture Notes in Mathematics, Vol. 986, Springer Ferlag, Berlin, 1983.

[2] Aldous, D., "On the Markov Chain Simulation Method for uniform combinatorial distributions and simulated annealing", *Probability in Eng. and Inf. Sci. 1*, 1987, pp 33-46.

[3] Aleliunas, R., Karp, R.M., Lipton, J.R., Lovasz, L., and Rackoff, C., "Random walks, universal traversal sequences, and the complexity of maze problems", *Proc. 20th IEEE Symp. on Foundations of Computer Science*, 1979, 218-233.

[4] Apt, K.R., and Kozen, D.Z., "Limits for automatic verification of finite state concurrent systems", *Inf. Processing Letters*, Vol. 22. No. 6, 1986, pp. 307-309.

[5] Broder, A.Z., "How hard is it to marry at random? Approximating the Permanent", *Proc. 18th ACM Symp. on the Theory of Computing*, 1986, pp. 55-58.

[6] Broder, A.Z., "Generating random Spanning Trees", *Proc. 30th IEEE Symp. on Foundations of Computer Science*, 1989, pp 4422-447.

[7] Cohen, E., Mihail, M., Papadimitriou, C.H., and Tsantilas, T., "Testing Protocols by Random Walk: Mixing and Cover Times", Bellcore TM ARA-4-94, 1994.

[8] Browne, M.C., Clarke, E.M., Dill, D.L., and Mishra, B., "Automatic verification of sequential circuits using temporal logic", *IEEE Trans. on Computers*, Vol. C-35, No. 12, 1986, pp. 1035-1043.

[9] Clarke, E.M., Emerson, E.A., and Sisla, A.P., "Automatic verification of finite state concurrent systems using temporal logic specifications: a practical approach", *Proc. 10th ACM Symposium on Principles of Programming Languages*, 1983.

[10] Diaconis, P., "Group Theory and Statistics", Lecture Notes from a course taught at Harvard, Spring 1982.

[11] Dyer, M., Frieze, A., and Kannan, R., "A random polynomial time algorithm for estimating volumes of convex bodies", *Proc. 21st ACM Symp. on the Theory of Computing*, 1989, pp 375-381.

[12] Feder, T., and Mihail, M., "Balanced Matroids", *Proc. 24th ACM Symp. on the Theory of Computing*, 1992, pp. 26-37.

[13] Holzmann, G.J., *Design and Validation of Computer Protocols*, Prentice Hall Software Series, 1991.

[14] Jerrum, M.R., and Sinclair, A., "Approximating the Permanent", *Proc. 20th ACM Symp. on the Theory of Computing*, 1988, pp 235-243.

[15] Lee, D., and Yannakakis, M., "Online minimization of transition systems", *Proc. 24th ACM Symp. on the Theory of Computing*, 1992, pp. 264-274.

[16] Mihail, M., "Conductance and Convergence of Markov Chains, A Combinatorial Treatment of Expanders", *Proc. 30th IEEE Symp. on Foundations of Computer Science*, 1989.

[17] Reif, J.H., and Smolka, S.A., "The complexity of reachability in distributed communicating processes", *Acta Informatica*, Vol. 25, 1988, pp. 333-354.

[18] Rudin, H., "Protocol development success stories: Part 1", *Protocol Specification, Testing, and Verification*, XII, Elsevier Science Publishers, (North-Holland), 1992.

[19] West, C.H., "Protocol Validation in Complex Systems", *Proc.. 8th ACM Symposium on Principles of Distributed Computing*, 1989, pp. 303-312.

[20] Wolper, P., "Specifying interesting properties of programs in propositional temporal logic", *Proc. 13th ACM Symposium on Principles of Programming Languages*, 1986, pp. 148-193.

[21] Yannakakis, M., and Lee, D., "Testing Finite State Machines", *Proc. of the 23rd ACM Symp. on the Theory of Computing*, 1991, pp. 476-485.

An Automata-Theoretic Approach to Branching-Time Model Checking (Extended Abstract)

Orna Bernholtz[1] and Moshe Y. Vardi[2] and Pierre Wolper[3]*

[1] The Technion, Department of Computer Science, Haifa 32000, Israel.
Email: ornab@cs.technion.ac.il
[2] Rice University, Department of Computer Science, P.O. Box 1892, Houston,
TX 77251-1892, U.S.A. Email: vardi@cs.rice.edu
[3] Université de Liège, Institut Montefiore, B28, 4000 Liège Sart-Tilman, Belgium.
Email: pw@montefiore.ulg.ac.be

Abstract. Translating linear temporal logic formulas to automata has proven to be an effective approach for implementing linear-time model-checking, and for obtaining many extensions and improvements to this verification method. On the other hand, for branching temporal logic, automata-theoretic techniques have long been thought to introduce an exponential penalty, making them essentially useless for model-checking. Recently, Bernholtz and Grumberg have shown that this exponential penalty can be avoided, though they did not match the linear complexity of non-automata-theoretic algorithms. In this paper we show that *alternating tree automata* are the key to a comprehensive automata-theoretic framework for branching temporal logics. Not only, as was shown by Muller et al., can they be used to obtain optimal decision procedures, but, as we show here, they also make it possible to derive optimal model-checking algorithms. Moreover, the simple combinatorial structure that emerges from the automata-theoretic approach opens up new possibilities for the implementation of branching-time model checking, and has enabled us to derive improved space complexity bounds for this long-standing problem.

1 Introduction

Temporal logics, which are modal logics geared towards the description of the temporal ordering of events, have been adopted as a powerful tool for specifying and verifying concurrent programs [Pnu81]. One of the most significant developments in this area is the discovery of algorithmic methods for verifying temporal logic properties of *finite-state* programs [CES86, QS81]. This derives its significance both from the fact that many synchronization and communication protocols can be modeled as finite-state programs, as well as from the great ease of use of fully algorithmic methods. Finite-state programs

---

* The work of this author was supported by the Esprit BRA action REACT and by the Belgian Incentive Program "Information Technology" - Computer Science of the future, initiated by the Belgian State - Prime Minister's Office - Science Policy Office. Scientific responsibility is assumed by its authors.

can be modeled by transition systems where each state has a bounded description, and hence can be characterized by a fixed number of Boolean atomic propositions. This means that a finite-state program can be viewed as a finite *propositional Kripke structure* and that its properties can be specified using *propositional* temporal logic. Thus, to verify the correctness of the program with respect to a desired behavior, one only has to check that the program, modeled as a finite Kripke structure, satisfies (is a model of) the propositional temporal logic formula that specifies that behavior. Hence the name *model checking* for the verification methods derived from this viewpoint. A survey can be found in [Wol89].

We distinguish between two types of temporal logics: linear and branching [Lam80]. In linear temporal logics, each moment in time has a unique possible future, while in branching temporal logics, each moment in time may split into several possible futures. For linear temporal logics, a close and fruitful connection with the theory of automata on infinite words has been developed [VW86a, VW94]. The basic idea is to associate with each linear temporal logic formula a finite automaton on infinite words that accepts exactly all the computations that satisfy the formula. This enables the reduction of linear temporal logic problems, such as satisfiability and model-checking, to known automata-theoretic problems, yielding clean and asymptotically optimal algorithms. Furthermore, these reductions are very helpful for implementing temporal-logic based verification methods, and are the key to techniques such as *on-the-fly* verification [VW86a, JJ89, CVWY92] that help coping with the "state-explosion" problem.

For branching temporal logics, the automata-theoretic counterpart are automata on infinite trees. By reducing the satisfiability to the nonemptiness problem for these automata, optimal decision procedures have been obtained for various branching temporal logics [Eme85, EJ88, ES84, SE84, VW86b]. Unfortunately, the automata-theoretic approach does not seem to be applicable to branching-time model checking. Indeed, model checking can be done in linear running time for CTL [CES86, QS81] and the alternation-free fragment of the μ-calculus [Cle93], and is in NP$\cap$co-NP for the general μ-calculus [EJS93], whereas there is an exponential blow-up involved in going from formulas to automata. Thus, using the construction of a tree automaton as a step in a model-checking algorithm seems a non-starter, which can only yield algorithms with exponential running time. (Indeed, the proof in [EJS93] avoids the construction of tree automata that correspond to μ-calculus formulas.)

A different automata-theoretic approach to branching-time model checking, based on the concepts of *amorphous automata* and *simultaneous trees*, was suggested by Bernholtz and Grumberg in [BG93]. Amorphous automata have a flexible transition relation that can adapt to trees with varying branching degree. Simultaneous trees are trees in which each sub-tree is duplicated twice as the two leftmost successors of its root. Simultaneous trees thus enable the automaton to visit different nodes of the same path simultaneously. Bernholtz and Grumberg showed that CTL model checking is *linearly* reducible to the acceptance of a simultaneous tree by an amorphous automaton and that the latter problem can be solved in quadratic running time.

While this constitutes a meaningful first step towards applying automata-theoretic techniques to branching-time model checking, it is not quite satisfactory. First, unlike the situation with linear temporal logic, different automata are required to solve model checking and satisfiability and thus, we do not get a uniform automata-theoretic treatment for the two problems. Second, and more crucial, the complexity of the resulting algorithm is quadratic in both the size of the specification and the size of the program, which makes this algorithm impractical; after all, most of the current research in this area is attempting to develop methods to cope with *linear* complexity.

In this paper, we argue that *alternating tree automata* are the key to a comprehensive and satisfactory automata-theoretic framework for branching temporal logics. Alternating tree automata generalize the standard notion of nondeterministic tree automata by allowing several successor states to go down along the same branch of the tree. It is known that, while the translation from branching temporal logic formulas to nondeterministic tree automata is exponential, the translation to alternating tree automata is linear [MSS88, EJ91]. In fact, Emerson stated that "μ-calculus formulas are simply alternating tree automata" [Eme94]. Muller et al. showed that this explains the exponential decidability of satisfiability for various branching temporal logics. We show here that this also explains the efficiency of model checking for those logics. The crucial observation is that for model checking, one does not need to solve the nonemptiness problem, but rather the *1-letter* nonemptiness problem. This problem (testing the nonemptiness of an alternating automaton that is defined on trees labeled with a singleton alphabet) is substantially simpler. Thus, alternating tree automata provide a unifying and optimal framework for both satisfiability and model-checking problems for branching temporal logic.

We first show how our automata-theoretic approach unifies previously known results about model checking for branching temporal logics. The alternating automata used by Muller et al. in [MSS88] are of a restricted type called *weak alternating automata*. To obtain an exponential decision procedure for the satisfiability of CTL and related branching temporal logics, Muller et al. used the fact that the nonemptiness problem for these automata is in exponential time [MSS86]. We prove that their 1-letter nonemptiness is decidable in linear running time, which yields an automata-based model checking algorithm of linear running time for CTL. The same technique can also be used to show that model-checking for the alternation-free μ-calculus can be done in linear running time. For the general μ-calculus, it follows from the results in [EJ91] that μ-calculus formulas can be linearly translated to alternating *Rabin* automata. We prove here that the 1-letter nonemptiness of alternating Rabin automata is in NP, which entails that model checking of μ-calculus formulas is in NP∩co-NP.

As the algorithms obtained by our approach match known complexity bounds for CTL [CES86] and the μ-calculus [Cle93, EJS93], what are the advantages offered by our approach? The first advantage is that it immediately broadens the scope of efficient model checking to other, and more expressive, branching temporal logics. For example, the dynamic logic considered in [MSS88] allows, in the spirit of [Wol83], nondeterministic tree automata as operators. Since this logic has a linear translation to weak alternating automata, it follows directly from our results that it also has a linear model-checking algorithm.

The second advantage comes from the fact that our approach combines the Kripke structure and the formula into a single automaton before checking this automaton for nonemptiness. This facilitates the use of a number of implementation heuristics. For instance, the automaton combining the Kripke structure and the formula can be computed *on-the-fly* and limited to its reachable states. This avoids exploring the parts of the Kripke structure which are irrelevant for the formula to be checked, and hence addresses the issue raised in the work on *local* model checking [SW91, VL93], while preserving optimal complexity and ease of implementation.

The third advantage of the automata-theoretic approach is that it offers new and significant insights into the space complexity of CTL model checking. It comes from the observation that the weak alternating automata that are obtained from CTL formulas have a special structure: they have *bounded alternation*. A careful analysis of the 1-letter nonemptiness problem for weak alternating automata with this property yields a top-down model-checking algorithm for CTL that is in NLOGSPACE in the size of the Kripke structure. A similar result holds for CTL*. This is very significant since it implies that, for concurrent programs, model checking can be done in space polynomial in the size of the program description, rather than requiring space of the order of the exponentially larger expansion of the program, as is the case with standard bottom-up model-checking algorithms.

2 Preliminaries

2.1 Temporal Logics and μ-Calculi

The temporal logic CTL (Computation Tree Logic) provides temporal operators that are composed of a path quantifier immediately followed by a single linear-time operator. The path quantifiers are A ("for all paths") and E ("for some path"). The allowed linear-time operators are X ("next time") and U ("until"). A *positive normal form* CTL formula is a CTL formula in which negations are applied only to atomic propositions. It can be obtained by pushing negations inward as far as possible, using De Morgan's laws and dualities. For technical convenience, we use the linear-time operator $\tilde{U}$ as a dual of the U operator ($p\tilde{U}q \equiv \neg((\neg p)U(\neg q))$), and write all CTL formulas in positive normal form. The *closure*, $cl(\psi)$, of a CTL formula ψ is the set of all CTL subformulas of ψ (including ψ). It is easy to see that for every ψ, $|cl(\psi)| \leq |\psi|$. The logic CTL* is defined similarly to CTL except that arbitrary linear-time formulas can appear in the scope of a path quantifier. See [Eme90] for more details on CTL and CTL*.

The propositional μ-calculus is a propositional modal logic augmented with least and greatest fixpoint operators. Specifically, we consider a μ-calculus where formulas are constructed from Boolean propositions with Boolean connectives, the temporal operators EX and AX as well as least (μ) and greatest (ν) fixpoint operators. For example, the μ-calculus formula $\mu y(q \vee (p \wedge EXy))$ is equivalent to the CTL formula $EpUq$. Assuming μ-calculus formulas are written in positive normal form (negation only applied to atomic propositions), a formula is *alternation free* if there are no occurrences of ν (μ) on any syntactic paths from an occurrence of μy (νy) to an occurrence of y. For more details, see [Koz83, EJ88, Var88].

The semantics of the logics described above is defined with respect to a *Kripke structure*, $K = \langle W, R, w^0, L \rangle$, where W is a set of states, $R \subseteq W \times W$ is a transition relation that must be total (i.e., for every $w \in W$ there exists $w' \in W$ such that $\langle w, w' \rangle \in R$), w^0 is an initial state, and $L : W \rightarrow 2^{AP}$ maps each state to a set of atomic propositions true in this state. The notation $K, w \models \varphi$ indicates that a formula φ holds at a state w of the structure K. Also, $K \models \varphi$ iff $K, w^0 \models \varphi$.

2.2 Alternating Tree Automata

For an introduction to the theory of automata on infinite trees see [Tho90]. *Alternating automata* on infinite trees generalize nondeterministic tree automata and were first introduced in [MS87]. For simplicity, we refer first to automata over infinite binary trees. Consider a nondeterministic tree automaton $A = \langle \Sigma, Q, \delta, q_0, F \rangle$. The transition relation δ maps an automaton state $q \in Q$ and an input letter $\sigma \in \Sigma$ to a set of pairs of states. Each such pair suggests a nondeterministic choice for the automaton's next configuration. When the automaton is in a state q and is reading a node x labeled by a letter σ, it proceeds by first choosing a pair $\langle q_1, q_2 \rangle \in \delta(q, \sigma)$ and then splitting into two copies. One copy enters the state q_1 and proceeds to the node $x \cdot 0$ (the left successor of x), and the other copy enters the state q_2 and proceeds to the node $x \cdot 1$ (the right successor of x).

For a given set D, let $\mathcal{B}^+(D \times Q)$ be the set of positive Boolean formulas over $D \times Q$ (i.e., Boolean formulas built from elements in $D \times Q$ using $\wedge$ and $\vee$), where we also allow the formulas true and false and, as usual, $\wedge$ has precedence over $\vee$. We can represent δ using $\mathcal{B}^+(\{0, 1\} \times Q)$. For example, $\delta(q, \sigma) = \{\langle q_1, q_2 \rangle, \langle q_3, q_1 \rangle\}$ can be written as $\delta(q, \sigma) = (0, q_1) \wedge (1, q_2) \vee (0, q_3) \wedge (1, q_1)$.

In nondeterministic tree automata, each conjunction in δ has exactly one element associated with each direction. In alternating automata on binary trees, $\delta(q, \sigma)$ can be an arbitrary formula from $\mathcal{B}^+(\{0, 1\} \times Q)$. We can have, for instance, a transition

$$\delta(q, \sigma) = (0, q_1) \wedge (0, q_2) \vee (0, q_2) \wedge (1, q_2) \wedge (1, q_3).$$

The above transition illustrates that several copies may go to the same direction and that the automaton is not required to send copies to all the directions. Formally, an *alternating tree automaton* is a tuple $A = \langle \Sigma, Q, \delta, q_0, F \rangle$ where Σ is the input alphabet, Q is a finite set of states, $q_0 \in Q$ is an initial state, F specifies the acceptance condition, and $\delta : Q \times \Sigma \to \mathbb{N}^+ \times \mathcal{B}^+(\mathbb{N} \times Q)$ is the transition function. We require that if $\delta(q, \sigma) = \langle k, \theta \rangle$, then $\theta \in \mathcal{B}^+(\{0, \ldots, k-1\} \times Q)$. In other words, a transition specifies a branching degree and a matching Boolean transition. A transition can only be applied to a node of a tree with a branching degree equal to the one specified by the transition. A run r of an alternating automaton A on a tree T is a tree where the root is labeled by q_0 and every other node is labeled by an element of $\mathbb{N} \times Q$. Each node of r corresponds to a node of T. Suppose that the path to a node y in r is labeled by $q_0, (c_1, q_1), \ldots, (c_m, q_m)$, then y corresponds to the node $x = c_1 \cdots c_m$ of T. Intuitively, the node y of r describes the automaton reading the node x of T. Note that many nodes of r can correspond to the same node of T; in contrast, in a run of a nondeterministic automaton on T there is a one-to-one correspondence between the nodes of the run and the nodes of the tree. The labels of a node and its successors have to satisfy the transition function. The run is accepting if all its infinite paths satisfy the acceptance condition. For formal details see [MS87] (our definition here extends the definition in [MS87] by allowing trees with varying branching degrees).

Amorphous automata extend conventional tree automata in that they can handle trees with both varying and unspecified branching degrees. Their amorphous nature enables them to be adjusted during their run to any branching degree. Amorphous automata were first introduced in [BG93], where they extend nondeterministic Büchi tree automata. We introduce here an amorphous version of alternating automata. In a standard alternating automaton, each transition is a pair consisting of a branching degree k and a Boolean formula in $\mathcal{B}^+(\{0, \ldots, k-1\} \times Q)$. In amorphous alternating automata, the transition depends on the (unknown in advance) branching degrees of the nodes being read by the automaton. Formally, the transition is a function $\delta : Q \times \Sigma \times \mathbf{N}^+ \to \mathcal{B}^+(\mathbf{N} \times Q)$, such that $\delta(q, \sigma, k) \in \mathcal{B}^+(\{0, \ldots, k-1\} \times Q)$. The notion of runs of amorphous alternating automata is a natural extension of the notion of runs of alternating automata. When the automaton is in a state q as it reads a node x that is labeled by a letter σ and has k successors, it applies the transition $\delta(q, \sigma, k)$.

In [MSS86], Muller et al. introduce *weak alternating automata* (WAA). The definition applies also to amorphous alternating automata. In an (amorphous) WAA, $F \subseteq Q$ and there exists a partition of Q into disjoint sets, Q_i, such that for each set Q_i, either $Q_i \subseteq F$, in which case Q_i is an *accepting set*, or $Q_i \cap F = \phi$, in which case Q_i is a *rejecting set*. In addition, there exists a partial order $\leq$ on the collection of the Q_i's such that for every $q \in Q_i$ and $q' \in Q_j$ for which q' occurs in $\delta(q, \sigma, k)$, for some $\sigma \in \Sigma$ and $k \in \mathbf{N}^+$, $Q_j \leq Q_i$. Thus, transitions from a state in Q_i lead to states in either the same Q_i or a lower one. It follows that every infinite path of a run of a WAA, ultimately gets "trapped" within some Q_i. The path then satisfies the acceptance condition iff Q_i is an accepting set.

3 Alternating Automata and Model Checking

In this section we introduce an automata-theoretic approach to model checking for branching temporal logic. The model-checking problem for a branching temporal logic is as follows. Given a Kripke structure K and a branching temporal formula ψ, determine whether $K \models \psi$. We solve this problem as follows. A Kripke structure K can be viewed as a tree, T_K, that corresponds to the unwinding of K from w^0. Let ψ be a branching temporal formula. Suppose that A_ψ is an amorphous alternating automaton that accepts exactly all the trees that satisfy ψ (amorphousness is used to handle the unspecified branching degrees of the nodes in the models of ψ and alternation reduces the size of the state set of A_ψ from exponential to linear in the length of ψ). The product of K and A_ψ either contains a single tree, T_K, in which case $K \models \psi$, or is empty, in which case $K \not\models \psi$. The model-checking problem can thus be solved as follows:

(1) Construct the amorphous alternating automaton A_ψ.
(2) Construct an alternating automaton $A_{K,\psi} = K \times A_\psi$ by taking the product of K and A_ψ. This automaton simulates a run of A_ψ on T_K.
(3) Output "Yes" if $\mathcal{L}(A_{K,\psi}) \neq \phi$, and "No", otherwise.

The type of A_ψ (i.e., its acceptance condition, its weakness, etc.) as well as the type of $A_{K,\psi}$ and consequently the complexity of the nonemptiness test depend on the logic in which ψ is specified. The crucial point in our approach is that the automaton $A_{K,\psi}$ is an automaton over a 1-letter alphabet; this reduces the complexity of the nonemptiness test. Note that in general, unlike the case for nondeterministic automata, the nonemptiness problem for alternating automata cannot be reduced to the 1-letter emptiness problem. It is taking the product with K that yields here an automaton over a 1-letter alphabet.

Let ψ be a branching temporal formula. We associate with ψ an amorphous alternating automaton $A_\psi = \langle 2^{AP}, cl(\psi), \delta_\psi, \psi, F_\psi \rangle$, which accepts precisely all the tree models of ψ. The details of the construction depends on the logic under consideration; we will see some examples in the next section. Let $K = \langle W, R, w^0, L \rangle$ be a Kripke structure. It is convenient to assume that the nodes of W are ordered. Thus, for every node w, let $succ_R(w) = \langle w_0, \ldots, w_{k-1} \rangle$ be the ordered list of w's R-successors.

We now define the product automaton. $A_{K,\psi} = \langle \{a\}, W \times cl(\psi), \langle w^0, \psi \rangle, \delta, F \rangle$ where δ and F are defined as follows.

- Let $\varphi \in cl(\psi)$, $w \in W$, $succ_R(w) = \langle w_0, \ldots, w_{k-1} \rangle$, and $\delta_\psi(\varphi, L(w), k) = \theta$. Then $\delta(\langle w, \varphi \rangle, a) = \langle k, \theta' \rangle$, where θ' is obtained from θ by replacing each atom (c, η) in θ by the atom $(c, \langle w_c, \eta \rangle)$.
- F is defined according to the acceptance condition F_ψ of A_ψ. For example, if $F_\psi \subseteq cl(\psi)$ is a Büchi condition, then $F = W \times F_\psi$ is also a Büchi condition. If $F_\psi = \{\langle L_1, U_1 \rangle, \ldots, \langle L_m, U_m \rangle\}$ is a Rabin condition, then $F = \{\langle W \times L_1, W \times U_1 \rangle, \ldots, \langle W \times L_m, W \times U_m \rangle\}$ is also a Rabin condition.

It is easy to see that $A_{K,\psi}$ is of the same type as A_ψ. In particular, if A_ψ is a WAA, then so is $A_{K,\psi}$.

Proposition 1. (1) $|A_{K,\psi}| = O(|K| * |A_\psi|)$.
(2) $\mathcal{L}(A_{K,\psi})$ is nonempty iff $K \models \psi$.

Proposition 1 can be viewed as an automata-theoretic generalization of Theorem 4.1 in [EJS93].

In conclusion, given a branching temporal formula ψ for which there exists an automaton A_ψ such that A_ψ accepts exactly all the trees that satisfy ψ, model checking of a Kripke structure K with respect to ψ, is reducible to checking the 1-letter emptiness of an automaton of the same type as A_ψ and of size $O(|K| * |A_\psi|)$. In the following sections, we show how this approach can be used to derive in a uniform way known coplexity bounds for model checking of CTL and μ-calculus formulas, as well as to obtain new space complexity bounds.

4 Applications

The efficiency of the method we presented in the previous section depends on the efficiency of the translation of formulas to automata, as well as the efficiency of the 1-letter nonemptiness test.

4.1 Model checking for CTL

Vardi and Wolper showed how to solve the satisfiability problem for CTL via an exponential translation of CTL formulas to Büchi automata on infinite trees [VW86a]. Müller et al. provided a simpler proof, via a linear translation of branching dynamic logic formulas to WAA [MSS88]. We extend here the ideas of Muller et al. by demonstrating a linear translation from CTL formulas to amorphous WAA.

Theorem 2. *Given a CTL formula ψ, we can construct in linear running time an amorphous WAA $A_\psi = \langle 2^{AP}, cl(\psi), \rho, \psi, F \rangle$ such that $\mathcal{L}(A_\psi)$ is exactly the set of tree models satisfying ψ.*

Proof. The set F of accepting states consists of all formulas in $cl(\psi)$ of the form $A\varphi_1 \tilde{U} \varphi_2$ or $E\varphi_1 \tilde{U} \varphi_2$. It remains to define the transition function ρ.

- $\rho(p, \sigma, k) = \text{true if } p \in \sigma.$ - $\rho(p, \sigma, k) = \text{false if } p \notin \sigma.$
- $\rho(\neg p, \sigma, k) = \text{true if } p \notin \sigma.$ - $\rho(\neg p, \sigma, k) = \text{false if } p \in \sigma.$
- $\rho(\varphi_1 \wedge \varphi_2, \sigma, k) = \rho(\varphi_1, \sigma, k) \wedge \rho(\varphi_2, \sigma, k).$
- $\rho(\varphi_1 \vee \varphi_2, \sigma, k) = \rho(\varphi_1, \sigma, k) \vee \rho(\varphi_2, \sigma, k).$
- $\rho(AX\varphi_2, \sigma, k) = \bigwedge_{c=0}^{k-1}(c, \varphi_2).$
- $\rho(EX\varphi_2, \sigma, k) = \bigvee_{c=0}^{k-1}(c, \varphi_2).$
- $\rho(A\varphi_1 U \varphi_2, \sigma, k) = \rho(\varphi_2, \sigma, k) \vee (\rho(\varphi_1, \sigma, k) \wedge \bigwedge_{c=0}^{k-1}(c, A\varphi_1 U \varphi_2)).$
- $\rho(E\varphi_1 U \varphi_2, \sigma, k) = \rho(\varphi_2, \sigma, k) \vee (\rho(\varphi_1, \sigma, k) \wedge \bigvee_{c=0}^{k-1}(c, E\varphi_1 U \varphi_2)).$
- $\rho(A\varphi_1 \tilde{U} \varphi_2, \sigma, k) = \rho(\varphi_2, \sigma, k) \wedge (\rho(\varphi_1, \sigma, k) \vee \bigwedge_{c=0}^{k-1}(c, A\varphi_1 \tilde{U} \varphi_2)).$
- $\rho(E\varphi_1 \tilde{U} \varphi_2, \sigma, k) = \rho(\varphi_2, \sigma, k) \wedge (\rho(\varphi_1, \sigma, k) \vee \bigvee_{c=0}^{k-1}(c, E\varphi_1 \tilde{U} \varphi_2)).$

To show that A_ψ is a WAA, we define a partition of Q into disjoint sets and a partial order over the sets. Each formula $\varphi \in cl(\psi)$, constitutes a (singleton) set $\{\varphi\}$ in the partition. The partial order is then defined by $\{\varphi_1\} \leq \{\varphi_2\}$ iff $\varphi_1 \in cl(\varphi_2)$. Since each transition of the automaton from a state φ leads to states associated with formulas in $cl(\varphi)$, the weakness conditions hold. In particular, each set is either contained in F or disjoint from F.

We now describe an efficient algorithm to test 1-letter nonemptiness of WAA.

Theorem 3. *The 1-letter nonemptiness problem for weak alternating automata is decidable in linear running time.*

Proof. See Appendix A.1

Theorems 2 and 3 yield a model-checking algorithm for CTL with linear (in the size of the input structure and in the size of the input formula) running time. The bottom-up labeling of the algorithm used in the proof of Theorem 3 is clearly reminiscent of the bottom-up labeling that takes place in the standard algorithm for CTL model checking [CES86]. Thus, the automata-theoretic approach seems to capture the combinatorial essence of CTL model checking.

4.2 Model Checking for the μ-Calculus

The intimate connection between the μ-calculus and alternating automata has been noted in [EJ91, Eme94]. We show here that our automata-theoretic approach provides a clean proof that model checking for the μ-calculus is in NP∩co-NP. The key steps in the proof are in showing that μ-calculus formulas can be efficiently translated to amorphous alternating Rabin automata, and that the 1-letter nonemptiness problem for alternating Rabin automata is in NP.

Theorem 4. *Given a μ-calculus formula ψ, we can construct in linear running-time an amorphous alternating Rabin automaton $A_\psi = \langle 2^{AP}, cl(\psi), \rho, \psi, F \rangle$ such that $\mathcal{L}(A_\psi)$ is exactly the set of tree models satisfying ψ.*

Proof. Emerson and Jutla showed how to translate μ-calculus formulas to alternating *Streett* automata [EJ91]. The extension to amorphous automata is straightforward. By constructing an amorphous alternating Streett automaton for $\neg\psi$ and then complementing it (it is easy to complement alternating automata [MS87]), we obtain an amorphous alternating Rabin automaton.

Theorem 5. *The 1-letter nonemptiness problem for alternating Rabin automata is decidable in nondeterministic polynomial running time.*

Proof. Without loss of generality we can assume that we are dealing with automata on trees of fixed branching degree, say k. Since the 1-letter k-ary tree is homogeneous (i.e., all subtrees are the same), we can pretend that successor states which are going down the same branch of the tree, are actually going down separate branches. Thus, we can apply techniques from the theory of nondeterministic Rabin automata, developed in [Eme85, VS85], to show that the 1-letter nonemptiness problem is in NP.

Combining Theorems 4 and 5, Proposition 1, and the observation in [EJS93] that checking for satisfaction of a formula ψ and a formula $\neg\psi$ has the same complexity, we get that the model-checking problem for the μ-calculus is in NP∩co-NP.

For the alternation-free μ-calculus, we can prove an analogue to Theorem 2. It follows then from Theorem 3, that model checking for the alternation-free μ-calculus can be done in linear running time.

5   The Space Complexity of Model Checking

Pnueli and Lichtenstein argued that, when analyzing the complexity of model checking, a distinction should be made between complexity in the size of the input structure and complexity in the size of the input formula; it is the complexity in size of the structure that is typically the computational bottleneck [LP85]. The Kripke structures to which model-checking is applied are often obtained by constructing the reachability graph of concurrent programs, and can thus be very large. So, even linear complexity (in terms of the input structure) can be excessive, especially as far as space is concerned. The question is then whether it is possible to perform model-checking without ever holding the whole structure to be checked in memory at any one time. For linear temporal formulas, the answer has long been known to be positive [VW86a]. Indeed, this problem reduces to checking the emptiness of a Büchi automaton on words which is NLOGSPACE-complete. Thus, if the Büchi automaton whose emptiness has to be checked is obtained as the product of the components of a concurrent program (as is usually the case), the space required will be polynomial in the size of these components rather than of the order of the exponentially larger Büchi automaton. Pragmatically, this is very significant and is, to some extent, exploited in the "on the fly" approaches to model checking and in related memory saving techniques [CVWY92].

Is the same true of CTL model-checking? The answer to this question was long thought to be negative. Indeed, the bottom-up nature of the known model-checking algorithms seemed to imply that storing the whole structure was required. Using our automata-theoretic approach to CTL model-checking, we are able to show that this is not so. Technically, this means that we will now prove that model-checking for CTL is NLOGSPACE-complete in the size of the Kripke structure. To prove this result, we will first show that the 1-letter WAA we construct for CTL model-checking have a special property (bounded alternation). Then, we will present an alternative algorithm for checking emptiness of WAA with this property.

Consider the product automaton $A_{K,\psi} = K \times A_\psi$ for a Kripke structure K and CTL formula ψ. The states of this automaton are elements of $W \times cl(\psi)$ and are partitioned into subsets Q_i according to their second component (two states are in the same Q_i if their second components are identical). Thus the number of Q_i's is bounded by the size of $cl(\psi)$ and is independent of the size of the Kripke structure. If one examines the Q_i's closely, one notices that they all fall into one of the following three categories:

1. Sets from which all transitions lead exclusively to states in lower Q_i's. These are the Q_i's corresponding to all elements of $cl(\psi)$ except U-formulas and $\tilde{U}$-formulas. We call these *transient* Q_i's.
2. Sets Q_i such that, for all $q \in Q_i$, the transition $\delta(q, a, k)$ only contain conjunctively related elements of Q_i, i.e. if the transition is rewritten in conjunctive normal form, there is at most one element of Q_i in each conjunct. These are the Q_i's corresponding to the $A\varphi_1 U\varphi_2$ and $A\varphi_1 \tilde{U}\varphi_2$ elements of $cl(\psi)$. We call these *universal* Q_i's.
3. Sets Q_i such that, for all $q \in Q_i$, the transition $\delta(q, a, k)$ only contain disjunctively related elements of Q_i. These are the Q_i's corresponding to the $E\varphi_1 U\varphi_2$ and $E\varphi_1 \tilde{U}\varphi_2$ elements of $cl(\psi)$. We call these *existential* Q_i's.

This means that it is only when moving from one Q_i to the next, that alternation actually occurs (alternation is moving from a state that is conjunctively related to its siblings to a state that is disjunctively related to its siblings, or vice-versa). If the number of Q_i's is fixed and if the depth of transitions is bounded (i.e., if their parse tree has bounded depth), we call a WAA that satisfies this property a *bounded-alternation* WAA.

Let us now turn to the nonemptiness problem for bounded-alternation WAA. Theorem 3 shows that the problem can be solved in linear running time. Notice that the algorithm used there is essentially a bottom-up labeling of the Boolean graph of the automaton. We will now show that by using a top-down exploration of this Boolean graph, we can get a space efficient 1-letter nonemptiness algorithm for bounded-alternation WAA.

Theorem 6. *The 1-letter nonemptiness problem for bounded-alternation WAA is NLOGSPACE-complete.*

Proof. See Appendix A.2

We note that for general WAA the 1-letter nonemptiness problem is P-complete.

Now, let us define the structure complexity of model-checking as the complexity of this problem in terms of the size of the input Kripke structure, i.e. assuming the formula fixed (this was called *program complexity* in [VW86a]). The following is then a direct consequence of Theorem 6.

Theorem 7. *The structure complexity of CTL model-checking is NLOGSPACE-complete.*

Theorem 7 can be extended to ECTL* [VW84]. The alternating automata that correspond to ECTL* formula are not in general weak. Nevertheless, a careful analysis shows that these automata do have a special structure and Theorem 6 can be extended to such automata.

If the Kripke structure is obtained as the product of the components of a concurrent program, this implies that CTL (and ECTL*) model-checking can be done in polynomial space with respect to the size of this program. It is also interesting to note that a less space-efficient deterministic version of the algorithm given in the proof of Theorem 6 can be viewed as the automata-theoretic counterpart of the algorithm presented in [VL93].

References

[Bee80]   C. Beeri. On the membership problem for functional and multivalued dependencies in relational databases. *ACM Trans. on Database Systems*, 5:241–259, 1980.

[BG93]   O. Bernholtz and O. Grumberg. Branching time temporal logic and amorphous tree automata. In *Proc. 4th Conferance on Concurrency Theory*, volume 715 of *Lecture Notes in Computer Science*, pages 262–277, Hildesheim, August 1993. Springer-Verlag.

[CES86]   E.M. Clarke, E.A. Emerson, and A.P. Sistla. Automatic verification of finite-state concurrent systems using temporal logic specifications. *ACM Transactions on Programming Languages and Systems*, 8(2):244–263, January 1986.

[Cle93]   R. Cleaveland. A linear-time model-checking algorithm for the alternation-free modal μ-calculus. *Formal Methods in System Design*, 2:121–147, 1993.

[CVWY92] C. Courcoubetis, M.Y. Vardi, P. Wolper, and M. Yannakakis. Memory efficient algorithms for the verification of temporal properties. *Formal Methods in System Design*, 1:275–288, 1992.

[EJ88]   E. A. Emerson and C. Jutla. The complexity of tree automata and logics of programs. In *Proceedings of the 29th IEEE Symposium on Foundations of Computer Science*, White Plains, oct 1988.

[EJ91]   E. A. Emerson and C. Jutla. Tree automata, mu-calculus and determinacy. In *Proceedings of the 32nd IEEE Symposium on Foundations of Computer Science*, pages 368–377, San Juan, Oct 1991.

[EJS93]   E. A. Emerson, C. Jutla, and A.P. Sistla. On model-checking for fragments of μ-calculus. In *Computer Aided Verification, Proc. 5th Int. Workshop*, volume 697, pages 385–396, Elounda, Crete, June 1993. Lecture Notes in Computer Science, Springer-Verlag.

[Eme85]   E.A. Emerson. Automata, tableaux, and temporal logics. In *Proc. Workshop on Logic of Programs*, volume 193 of *Lecture Notes in Computer Science*, pages 79–87. Springer-Verlag, 1985.

[Eme90]   E.A. Emerson. Temporal and modal logic. *Handbook of theoretical computer science*, pages 997–1072, 1990.

[Eme94]   E.A. Emerson. Automated temporal reasoning about reactive systems. In *VIII-th BANFF Higher Order Workshop*, 1994. unpublished abstract of forthcoming talk.

[ES84]   A.E. Emerson and A.P. Sistla. Deciding full branching time logics. *Information and Control*, 61(3):175–201, 1984.

[JJ89]   C. Jard and T. Jeron. On-line model-checking for finite linear temporal logic specifications. In *Automatic Verification Methods for Finite State Systems, Proc. Int. Workshop, Grenoble*, volume 407, pages 189–196, Grenoble, June 1989. Lecture Notes in Computer Science, Springer-Verlag.

[Koz83]   D. Kozen. Results on the propositional μ-calculus. *Theoretical Computer Science*, 27:333–354, 1983.

[Lam80]   L. Lamport. Sometimes is sometimes "not never" - on the temporal logic of programs. In *Proceedings of the 7th ACM Symposium on Principles of Programming Languages*, pages 174–185, January 1980.

[LP85]   O. Lichtenstein and A. Pnueli. Checking that finite state concurrent programs satisfy their linear specification. In *Proceedings of the Twelfth ACM Symposium on Principles of Programming Languages*, pages 97–107, New Orleans, January 1985.

[MS87]   D.E. Muller and P.E. Schupp. Alternating automata on infinite trees. *Theoretical Computer Science*, 54,:267–276, 1987.

[MSS86]   D.E. Muller, A. Saoudi, and P.E. Schupp. Alternating automata, the weak monadic theory of the tree and its complexity. In *Proc. 13th Int. Colloquium on Automata, Languages and Programming*. Springer-Verlag, 1986.

[MSS88]   D. E. Muller, A. Saoudi, and P. E. Schupp. Weak alternating automata give a simple explanation of why most temporal and dynamic logics are decidable in exponential time. In *Proceedings 3rd IEEE Symposium on Logic in Computer Science*, pages 422–427, Edinburgh, July 1988.

[Pnu81]   A. Pnueli. The temporal semantics of concurrent programs. *Theoretical Computer Science*, 13:45–60, 1981.

[QS81]   J.P. Queille and J. Sifakis. Specification and verification of concurrent systems in Cesar. In *Proc. 5th Int'l Symp. on Programming*, volume 137, pages 337–351. Springer-Verlag, Lecture Notes in Computer Science, 1981.

[SE84]   R. S. Street and E. A. Emerson. An elementary decision procedure for the mu-calculus. In *Proc. 11th Int. Colloquium on Automata, Languages and Programming*, volume 172. Lecture Notes in Computer Science, Springer-Verlag, July 1984.

[SW91]   C. Stirling and D. Walker. Local model checking in the modal mu-calculus. *Theoretical Computer Science*, 89(1):161–177, 1991.

[Tho90]   W. Thomas. Automata on infinite objects. *Handbook of theoretical computer science*, pages 165–191, 1990.

[Var88]   M.Y. Vardi. A temporal fixpoint calculus. In *Proc. 15th ACM Symp. on Principles of Programming Languages*, pages 250–259, San Diego, January 1988.

[VL93]   B. Vergauwen and J. Lewi. A linear local model checking algorithm for ctl. In *Proc. CONCUR '93*, volume 715 of *Lecture Notes in Computer Science*, pages 447–461, Hildesheim, August 1993. Springer-Verlag.

[VS85]   M.Y. Vardi and L. Stockmeyer. Improved upper and lower bounds for modal logics of programs. In *Proc 17th ACM Symp. on Theory of Computing*, pages 240–251, 1985.

[VW84]   M.Y. Vardi and P. Wolper. Yet another process logic. In *Logics of Programs*, volume 398, pages 501–512. Lecture Notes in Computer Science, Springer-Verlag, 1984.

[VW86a]   M.Y. Vardi and P. Wolper. An automata-theoretic approach to automatic program verification. In *Proceedings of the First Symposium on Logic in Computer Science*, pages 322–331, Cambridge, June 1986.

[VW86b]   M.Y. Vardi and P. Wolper. Automata-theoretic techniques for modal logics of programs. *Journal of Computer and System Science*, 32(2):182–21, April 1986.

[VW94]   M.Y. Vardi and P. Wolper. Reasoning about infinite computations. *Information and Computation*, 110(2), May 1994. (To appear).

[Wol83] P. Wolper. Temporal logic can be more expressive. *Information and Control*, 56(1–2):72–99, 1983.

[Wol89] P. Wolper. On the relation of programs and computations to models of temporal logic. In B. Banieqbal, H. Barringer, and A. Pnueli, editors, *Proc. Temporal Logic in Specification*, volume 398, pages 75–123. Lecture Notes in Computer Science, Springer-Verlag, 1989.

A   Proofs

A.1   Theorem 3

We present an algorithm with linear running time for checking the nonemptiness of the language of a WAA $A = \langle \{a\}, Q, \delta, q_0, F \rangle$.

As A is weak, there exists a partition of Q into disjoint sets Q_i such that there exists a partial order $\leq$ on the collection of the Q_i's and such that for every $q \in Q_i$ and $q' \in Q_j$ for which q' occurs in $\delta(q, a)$, $Q_j \leq Q_i$. Thus, transitions from a state in Q_i lead to states in either the same Q_i or a lower one. In addition, each set Q_i is classified as accepting, if $Q_i \subseteq F$, or rejecting, if $Q_i \cap F = \phi$.

The algorithm labels the states of A with either 'T', standing for **true**, or 'F', standing for **false**. Intuitively, states $q \in Q$ for which the language of A^q (i.e., the language of A with q as the initial state) is nonempty are labeled with 'T' and states q for which the language of A^q is empty are labeled with 'F'. The language of A is thus nonempty iff the initial state q_0 is labeled with 'T'. The algorithm works in phases and proceeds up the partial order. Let $Q_1 \leq \ldots \leq Q_n$ be an extension of the partial order to a total order. In each phase i, the algorithm handles states from the minimal set Q_i which still has not been labeled.

States that belong to a set Q_i that is minimal in the partial order, are labeled according to the classification of Q_i. Thus, they are labeled with 'T' if Q_i is an accepting set, and with 'F' if it is rejecting. Once a state $q \in Q_i$ is labeled with 'T' or 'F', transition functions in which q occurs are simplified accordingly, i.e., a conjunction with a conjunct 'F' is simplified to 'F' and a disjunction with a disjunct 'T' is simplified to 'T'. Consequently, a transition function $\delta(q', a)$ for some q', (not necessarily from Q_i) can obtain its truth value. q' is then labeled, and evaluation proceeds further.

Since the algorithm proceeds up the total order, when it reaches a state $q \in Q_i$ that is still not labeled, it is guaranteed that all the states in all Q_j for which $Q_j < Q_i$, have already been labeled. Hence, all the states that occur in $\delta(q, a)$ have the same status as q. That is, they belong to Q_i and are still not labeled. The algorithm then labels q and all the states in $\delta(q, a)$ according to the classification of Q_i. They are labeled 'T' if Q_i is accepting and are labeled 'F' otherwise.

Using an AND/OR graph, as suggested in [Bee80], the algorithm can be implemented in linear running time. Typically, the graph, induced by the transition function, keeps the labeling performed during the algorithm execution. Simplification of each transition function $\delta(q, a)$ for all $q \in Q$, then costs $O(|\delta(q, a)|)$.

A.2 Theorem 6

The property of bounded-alternation WAA we use is that, from a state of a Q_i, it is possible to search for another reachable state of the same Q_i in NLOGSPACE. For transient Q_i, there are no such states. For universal and existential Q_i, the exact notion of reachability we use is the transitive closure of the following notion of immediate reachability. Assume, we have a Boolean value for all states in sets lower than Q_i. Then a state q' is immediately reachable from a state q, if it appears in the transition from q when this transition has been simplified using the know Boolean values for states in lower Q_i. Note that the simplified transition is always a conjunction for a state of a universal Q_i, and a disjunction for a state of an existential Q_i.

The following procedure labels the states of the automaton with 'T' (accepts) or 'F' (does not accept).

1. One starts at the initial state.
2. At a transient state q, one applies the procedure to the successor states. The labels that are obtained for these successor states are then substituted in the transition from q, and q is labeled with the Boolean value that is thus obtained for the transition.
3. At a state q of a universal Q_i, one proceeds as follows. We call a state q' of Q_i *provably true* if, when the procedure is applied to the successors of q' that are not in Q_i, and the Boolean expression for the transition from q' is simplified, it is identically true. States that are *provably false* are defined analogously.
   (a) One searches in NLOGSPACE for a reachable state q' of the same Q_i that is provably false (note that this requires applying the procedure recursively to all states from lower Q_i's that are touched by the search). If such a state q' is found, the state q is labeled 'F'.
   (b) If no such state exists, one searches in NLOGSPACE for a state q' of Q_i that is reachable from q and from itself. If such a state is found, q is labeled according to the classification of the Q_i.
   (c) if none of the first two cases apply, q is labeled 'T'.
4. At a state q of an existential Q_i, one proceeds as follows.
   (a) One searches in NLOGSPACE for a reachable state q' of the same Q_i that is provably true. If such a state q' is found, the state q is labeled 'T'.
   (b) If no such state exists, one searches in NLOGSPACE for a state q' of Q_i that is reachable from q and from itself. If such a state is found, q is labeled according to the classification of the Q_i.
   (c) if none of the first two cases apply, q is labeled 'F'.

The procedure is recursive, but as the depth of the transitions is bounded by the number of Q_i's, so does the depth of recursion. Since each invocation of the procedure can be executed in NLOGSPACE, the whole procedure is thus NLOGSPACE. Completeness in NLOGSPACE is immediate by reduction from the corresponding problem for nondeterministic sequential automata.

Realizability and Synthesis of Reactive Modules

Anuchit Anuchitanukul   Zohar Manna

Computer Science Department
Stanford University
Stanford, CA 94305
anuchit@cs.stanford.edu

April 4, 1994

Abstract

We present two algorithms: a realizability-checking algorithm and a synthesis algorithm. Given a specification of reactive asynchronous modules expressed in propositional ETL (Extended Temporal Logic), the realizability-checking algorithm decides whether the specification has an actual implementation, under the assumptions of a random environment and fair execution. It also creates a structure which can then be transformed by the synthesis algorithm into a program, represented as a labeled finite automaton. Unlike previous approaches, the realizability-checking algorithm can handle fairness assumptions. The realizability-checking algorithm is incremental and it directly manipulates formulas in linear temporal logic without having to transform into a branching-time logic or other representations.

1   Introduction

The problem of automatic program synthesis has been previously studied in many different frameworks. For functional programs, the specification is a first-order formula expressing the desired relationship between inputs and outputs, where the synthesized program can be extracted from a constructive proof of the formula [MW80].

Later, [EC82] and [MW84] extended the approach to reactive programs. The synthesized program is extracted from a proof of the satisfiability of the specification. However, the reactive programs considered in the approach do not have any interaction with the environment, that is, they are *closed* systems.

The effort to synthesize reactive modules, i.e., *open* systems, was first reported in [PR89a]. In that paper, the synthesis of reactive *synchronous* modules from a specification in linear-time temporal logic is linked to the problem of checking the validity of a branching-time temporal formula obtained by transforming the original specification.

The restriction to synchronous systems (or the game of perfect information) was removed in [PR89b] where the problem of synthesizing asynchronous systems is considered. In that work, a linear-time temporal specification is transformed into a formula in branching-time temporal logic by introducing read and write variables, and by adding constraints on the variables.

Several notions of realizability were introduced and studied in [ALW89]. For the finite case, the approach taken is similar to the automata approach of [PR89a,b]. Because of the choice of the specification, the method can check realizability in a more general sense, that is, when the behavior of the environment is restricted. However, [ALW89] only considered synchronous systems.

In [WD91], the approach of [PR89b] was extended to handle shared variables and the restriction on read and write sequences was relaxed. The paper also generalizes [ALW89] to include the asynchronous and real-time cases. The main technique used is still by transformations from one representation (automata) to another.

The assumption shared by all of the works mentioned previously is that the problem of realizability can be solved simply by transforming the specification (under some representation) into an automata representation (tree automaton) and then checking for non-emptiness. We argue that this assumption is not valid when we want to solve a more general realizability problem. There are properties which cannot be encoded into a specification, such as the assumption of fairness. We would like to be able to determine whether a specification is realizable, assuming that the execution of the system and the environment is fair. Since the synthesized program does not exist at the time the specification is written, there is no location or transition in the program to refer to in the specification. We cannot encode in the specification the assumption that the synthesized program is to be executed fairly. This problem arises regardless of the choice of the specification language. Therefore, any realizability-checking and synthesis algorithms must handle this explicitly. Although [ALW89] did define a notion of realizability under fairness (and a theorem relating it to realizability without fairness), no solution was provided.

In this paper, we present two algorithms, one for realizability checking and the other for synthesis. The specification language we study is the Extended Temporal Logic (ETL) described in [Wo83]. We also introduce a scheduling variable μ following the approach in [BKP84]. Although useful for expressing specifications and for extending the algorithm to handle sequential composition, μ is not essential to the algorithms. The realizability-checking algorithm is based on the tableau decision procedure described in [Wo85]. Given a specification, the first algorithm checks for *strong* realizability under fairness and random environment assumptions and generates a structure called a *realizability graph*. The synthesis algorithm takes the generated realizability graph and produces a program which satisfies the specification. Unlike other approaches, the synthesis algorithm will generate a more general class of programs which may have some disabled transitions. The transitions in the generated program may be labeled weakly or strongly fair as necessary. Since the realizability-checking algorithm is a tableau-based algorithm, it manipulates only formulas of linear temporal logic, which are subformulas of the original specification.

2 Definitions

A [infinite] *behavior* σ over a state space Σ is a pair $\langle \sigma_v, \sigma_s \rangle$ of two equal-length [infinite] sequences: a sequence of *states* $\sigma_v = s_0 s_1 s_2 \ldots$ where $s_i \in \Sigma$ and a *scheduling* sequence $\sigma_s = a_0 a_1 a_2 \ldots$ where $a_i \in \{0, 1\}$. We denote the set of all infinite behaviors from Σ by $Bhv(\Sigma)$, or Bhv if Σ is clear from the context, and the set of all finite behaviors by $Bhv_{fin}(\Sigma)$.

We can represent a behavior $\langle s_0 s_1 s_2 \ldots, a_0 a_1 a_2 \ldots \rangle$ pictorially as

$$\xrightarrow{a_0} s_0 \xrightarrow{a_1} s_1 \xrightarrow{a_2} s_2 \ldots$$

The intended meaning is that the move from s_i to s_{i+1} is caused by the environment if $a_{i+1} = 0$, and by the system if $a_{i+1} = 1$. Since we always assume that the environment chooses the initial state, we require that the scheduling sequence always begins with 0, i.e., $a_0 = 0$.

Given a behavior $\sigma = \langle s_0 s_1 s_2 \ldots, a_0 a_1 a_2 \ldots \rangle$, we write $State(i, \sigma)$ to denote s_i and $Sched(i, \sigma)$ to denote a_i. If $\sigma = \langle \sigma_v, \sigma_s \rangle$, then $\sigma|_i$ denotes a behavior $\langle \sigma_v|_i, \sigma_s|_i \rangle$ where $\sigma_v|_i$ [$\sigma_s|_i$] is the prefix of σ_v [σ_s] of length i.

Let $\Pi^0 : Bhv_{fin} \mapsto \Sigma^*$ be a function that maps a finite behavior σ to a subsequence of states which are caused by the system, namely, all $State(i, \sigma)$ where $Sched(i, \sigma) = 1$. Let $\Pi^1 : Bhv_{fin} \mapsto \Sigma^*$ be a function that maps a finite behavior σ to a subsequence of states which are observed by the system (precede a system state), that is, all $State(i, \sigma)$ where $Sched(i + 1, \sigma)$ is defined and $Sched(i + 1, \sigma) = 1$, or $i + 1$ is the length of σ.

A *computer* $f : \Sigma^* \times \Sigma^* \mapsto \Sigma$ is a partial function which takes a history of all the states the system caused and all the states the system observed and selects a state as the next move of the system. A *run* of a computer f is an infinite behavior such that for all i, if $Sched(i, \sigma) = 1$ then $f(\Pi^0(\sigma|_i), \Pi^1(\sigma|_i))$ is defined and equal to $State(i, \sigma)$. Therefore, a behavior is a run of a computer if every system move is the result of f computed with the information regarding the system's own moves and all the moves the system has observed in the past.

A run σ of f is *weakly fair* iff for all j, if $f(\Pi^0(\sigma|_i), \Pi^1(\sigma|_i))$ is defined for all $i \geq j$, then $Sched(k, \sigma) = 1$ for some $k \geq j$, i.e., if f is continuously enabled beyond a certain point, it has to be taken eventually. Similarly, a run σ is *strongly fair* iff for all j, if for all $j' \geq j$ there exists $i \geq j'$ such that $f(\Pi^0(\sigma|_i), \Pi^1(\sigma|_i))$ is defined, then $Sched(k, \sigma) = 1$ for some $k \geq j$.

Let $Run_{sf}(f)$ be all possible strongly fair runs and $Run_{wf}(f)$ be all possible weakly fair runs of the computer f. A set B of behaviors is *realizable* (under fairness and random environment assumptions) iff there exists a computer f such that $Run_{sf}(f) \subseteq B$. If only weak fairness is assumed, B is realizable iff there exists a computer f such that $Run_{wf}(f) \subseteq B$.

3 Preliminaries

3.1 Specification Language

The specification language studied here is Extended Temporal Logic (ETL) augmented with a special predicate μ. The use of ETL and μ is not necessary for the realizability-checking and synthesis algorithms. Clearly, the algorithms can handle any subset of the language, including ordinary propositional temporal logic specifications without μ. Adding μ to the language is necessary, however, to express some common forms of specifications such as mutual exclusion. Without μ, we would have to separate the environment assumption and the system property. With μ, the whole specification can be expressed in a single formula.

In ETL, there are infinitely many temporal operators. Each corresponds to a non-terminal symbol of a right-linear grammar. A right-linear grammar G is a tuple (V_N, V_T, P) such that

- $V_N = \{\mathcal{G}_1, \ldots, \mathcal{G}_m\}$ is a finite set of non-terminal symbols.

- $V_T = \{t_1, \ldots, t_n\}$ is a finite set of terminal symbols.

- P is a finite set of production rules of the forms $\mathcal{G}_i \to t_j$ or $\mathcal{G}_i \to t_j \mathcal{G}_k$ where $\mathcal{G}_i, \mathcal{G}_k \in V_N$ and $t_j \in V_T$.

For each non-terminal symbol $\mathcal{G}_i$, the corresponding temporal operator $\mathcal{G}_i(\phi_1, \ldots, \phi_n)$ has exactly n arguments (n is the number of terminal symbols).

Given a set $\mathcal{P}$ of propositions and a truth-value assignment function $\pi : \Sigma \mapsto 2^{\mathcal{P}}$, the semantics of a formula on an infinite behavior σ is defined as follows:

- $\sigma \models \phi$ iff $\langle \sigma, 0 \rangle \models \phi$.

- $\langle \sigma, i \rangle \models p$ iff $p \in \pi(State(i, \sigma))$, for any proposition $p \in \mathcal{P}$.

- $\langle \sigma, i \rangle \models \mu$ iff $Sched(i, \sigma) = 1$.

- $\langle \sigma, i \rangle \models \bigcirc \phi$ iff $\langle \sigma, i+1 \rangle \models \phi$.

- $\langle \sigma, i \rangle \models \mathcal{G}(\phi_1, \ldots, \phi_n)$ iff there is a word (finite or infinite) $w = t_{n_0} t_{n_1} t_{n_2} \cdots$ (each $t_{n_j} \in V_T$), generated by $\mathcal{G}$, and for all $j \geq 0$, $\langle \sigma, i+j \rangle \models \phi_{n_j}$.

- Other cases ($\phi_1 \vee \phi_2$, $\phi_1 \wedge \phi_2$, and $\neg \phi$) are standard.

Clearly, any formula ϕ defines a set of infinite behaviors B which satisfy the formula, i.e., $\sigma \models \phi$ iff $\sigma \in B$. Therefore, we define a specification to be realizable if the corresponding set of behaviors is realizable.

3.2 Elementary Formulas

A formula is called *elementary* if it is either

- an atomic formula, i.e., an atomic proposition (including μ) or its negation, or

- a next formula, i.e., a formula that has $\bigcirc$ as its main connective.

3.3 Decomposition Rules

The following decomposition rules are used in the tableau graph construction algorithm to decompose non-elementary formulas. The meaning of a decomposition rule is that in order to satisfy the formula on the left hand side, one of the sets on the right hand side must be satisfied.

- $(\phi_1 \vee \phi_2) \Longrightarrow \{\{\phi_1\}, \{\phi_2\}\}$

- $(\phi_1 \wedge \phi_2) \Longrightarrow \{\{\phi_1, \phi_2\}\}$

- $\neg(\phi_1 \vee \phi_2) \Longrightarrow \{\{\neg\phi_1, \neg\phi_2\}\}$

- $\neg(\phi_1 \wedge \phi_2) \Longrightarrow \{\{\neg\phi_1\}, \{\neg\phi_2\}\}$

- $(\neg\neg\phi) \Longrightarrow \{\{\phi\}\}$

- $(\neg \bigcirc \phi) \Longrightarrow \{\{\bigcirc\neg\phi\}\}$

- For an ETL grammar operator $\mathcal{G}(\phi_1, \ldots, \phi_n)$ with grammar productions of the form: $\mathcal{G} \rightarrow t_{a_i} \mathcal{G}_{b_i}$ where $1 \leq i \leq l$ is the index of the production rules of $\mathcal{G}$, $t_{a_i} \in V_T$ and $\mathcal{G}_{b_i} \in V_N$ (which may or may not be present), we have the following decomposition rules:

$$\mathcal{G}(\phi_1, \ldots, \phi_n) \Longrightarrow \bigcup_{1 \leq i \leq l} \{\{\phi_{a_i}, \bigcirc\mathcal{G}_{b_i}(\phi_1, \ldots, \phi_n)\}\}$$

$$\neg\mathcal{G}(\phi_1, \ldots, \phi_n) \Longrightarrow \{\bigcup_{1 \leq i \leq l} \{\neg\phi_{a_i} \vee \bigcirc\neg\mathcal{G}_{b_i}(\phi_1, \ldots, \phi_n)\}\}$$

3.4 Tableau Graph

Before we proceed to describe the realizability-checking algorithm, we will briefly explain a tableau graph construction similar to that in [Wo85]. A tableau graph is a directed graph in which each node n is labeled with a set of formulas, denoted by $\Phi(n)$.

- A node n in a tableau graph is called a *state* node iff $\Phi(n)$ contains only elementary formulas.

- A node n is *environment-compatible* iff $\mu \notin \Phi(n)$.

- Similarly, a node n is *system-compatible* iff $\neg\mu \notin \Phi(n)$.

Given a formula $\hat{\psi}$ to be checked for satisfiability, the tableau graph for $\hat{\psi}$ is created as follows:

First,

1. create a node *(root)* and label it with $\{\hat{\psi}\}$.

Repeatedly apply steps 2 and 3.

2. If a node n, with no successor, contains a non-elementary formula ϕ in its label $\Phi(n)$, and if the decomposition rule for ϕ is $\phi \implies \{S_1, \ldots, S_t\}$, then for each set of formulas S_i, create a successor of n and label it with $(\Phi(n) - \{\phi\}) \cup S_i$. However, if there is a node with the same label already, then just connect n to the existing node.

3. For a state node n with label $\Phi(n)$, create (if no duplication occurs) a successor of n and label it with $\{\phi \mid \bigcirc\phi \in \Phi(n)\}$.

Finally,

4. Remove all inconsistent nodes (the nodes containing a proposition p and its negation $\neg p$).

A loop in a tableau graph is called a *self-supporting* loop if for any state node n in the loop, there is a finite path in the loop starting from n such that all formulas of the form $\bigcirc\mathcal{G}(\ldots)$ in $\Phi(n)$ are *fulfilled* on the path. A formula $\bigcirc\neg\mathcal{G}(\ldots)$ with the decomposition rule,

$$\neg\mathcal{G}(\phi_1, \ldots, \phi_n) \implies \{ \bigcup_{1 \leq i \leq l} \{\neg\phi_{a_i} \vee \bigcirc\neg\mathcal{G}_{b_i}(\phi_1, \ldots, \phi_n)\}\}$$

is *fulfilled* at a state n if the next state n' on the path contains a term from each of the disjunctions of the decomposition rule and if the term is $\bigcirc\neg\mathcal{G}_{b_i}(\ldots)$ then it is also fulfilled at n' (i.e. at the next state down the path).

3.5 Maximally Consistent Subsets

Given a set of state nodes N, a subset $N_{mcs} \subseteq N$ is *maximally consistent* if both of the following conditions are satisfied:

- **Consistent:** It is not the case that for some proposition p other than μ and for some nodes $n_1, n_2 \in N_{mcs}$, both $p \in \Phi(n_1)$ and $\neg p \in \Phi(n_2)$. In other words, the union of all the observable atomic formulas (which are all atomic formulas except μ and $\neg\mu$) in the labels of the nodes in N_{mcs} is consistent.

- **Maximal:** There is no other subset $N' \subseteq N$ such that N' satisfies the above condition (consistent) and $N_{mcs} \subset N'$.

3.6 Maximally Negation-Consistent Subsets

For a set of state nodes N, a subset $N_{mncs} \subseteq N$ is *maximally negation-consistent* if both of the following conditions are satisfied:

- **Negation-consistent:** There exists a function f which maps each node $n \in N_{mncs}$ to an atomic formula $f(n) \in \Phi(n)$ which is not μ or $\neg\mu$, and the set $P = \{\neg f(n) \mid n \in N_{mncs}\}$ is consistent. The set P is called the *falsifying* set for N_{mncs}.

- **Maximal:** There is no other subset $N' \subseteq N$ such that N' satisfies the above condition (negation-consistent) and $N_{mncs} \subset N'$.

3.7 Realizability Graph

The structure created by the realizability-checking algorithm is called a *realizability graph*. A realizability graph is a directed bipartite graph $(V_s, V_n, E_{sn}, E_{ns})$ where

- V_s is a set of nodes called *R-state* nodes and labeled by a *node-label* which is a set of tableau graph nodes and a *write-label* which is a set of atomic formulas.

- V_n is a set of nodes called *R-non-state* nodes and labeled by a node-label.

- V_{sn} is a set of links from R-state nodes to R-non-state nodes.

- V_{ns} is a set of links from R-non-state nodes to R-state nodes.

3.8 Embedding

An increasing sequence $d_0 \ldots d_l$ of integers is an *embedding* of a path [loop] $n_0 \ldots n_k$ in a tableau graph into a path [loop] $v_0 \ldots v_l$ in a realizability graph if both of the following conditions hold:

- for all $0 \leq i \leq l$, n_{d_i} is in the node-label of v_i.

- for all n_j, if $j \neq d_i$ for all $0 \leq i \leq l$, then n_j is environment-compatible.

It is straightforward to extend the definition to allow the embedding of an infinite path in a tableau graph into a (finite or infinite) path in a realizability graph.

4   Realizability-Checking Algorithm

The key idea in the algorithm is that the realizability graph represents a game between the system and the environment in which the environment can make any finite number of moves after a system's move. This is represented by the alternate levels of R-state and R-non-state nodes. Given a formula ψ to be tested for realizability, the algorithm construct a tableau graph for the negation of ψ. To "win the game", the environment must try to force the execution to stay on a path in the tableau graph which falsifies ψ; whereas the system must try to push the execution out of such path.

We start constructing the realizability graph from an R-state node which contains the root node n_{root} of the tableau graph. Since the environment can make any number of moves, it may try to follow any path in the tableau graph from n_{root}. Without the complete knowledge of all the moves the environment makes, the system cannot determine which path the environment has taken. It can only use the information from the state it observes when it is scheduled to run, to determine a *set* of all state nodes accessible from n_{root} the path might have led into. In the worst case, such s set will be a maximally consistent subset of all accessible state nodes. Therefore, we construct an R-non-state successor of the R-state node, for each maximally consistent subset. For its own move, the system must try to push the execution out of any path which the environment might follow (and win) afterward. The best move that the system can possibly make is to falsify as many successor nodes of the nodes in the node-label of the R-non-state node and in essence, to limit the possible paths left for the environment to follow. This is the reason why we compute the maximally negation-consistent subsets and the falsifying set of atomic formulas. The remaining nodes which are not falsified can be computed by subtracting the maximally negation-consistent subsets from the set of all successors of the nodes in the R-non-state node. For each best move possible, we create an R-state successor of the R-non-state successor, put the remaining nodes in its node-label and continue expanding the realizability graph from the new R-state node.

In the algorithm, at each R-state node v_s, we compute a set *Disabled* by collecting all state nodes in the labels of every deleted R-non-state successor of v_s. When an R-non-state successor v_{ns} is deleted, it means the system will not be able to satisfy the specification by making a transition from v_s through v_{ns}. Therefore, we should consider such a transition "disabled". As a result, we put every state node in the deleted R-non-state node into the set *Disabled* because the environment can choose to move into some states in which the transition through the deleted R-non-state node is disabled.

Finally, we also have to check at each R-state node that the environment cannot win by remaining in a loop containing disabled state nodes.

4.1 Main procedure

1. First, create a tableau graph Glb for the formula $\neg\psi$ where ψ is the formula to be tested for realizability.

2. Create an R-state node *(root)* and label it with the set $\{n_{root}\}$ where n_{root} is the root node of Glb.

3. Call the subroutine *Expand*, passing the root node as its parameter, to expand the realizability graph in a depth-first fashion.

4. Finally, check if the root node of the final realizability graph is deleted. If it is not deleted, then the formula ψ is realizable. Otherwise, it is unrealizable.

4.2 Subroutine *Expand* (Realizability Graph Construction)

Given an R-state node v_s with a node-label $L(v_s)$, expand the realizability graph as follows:

1. If $L(v_s)$ is empty, then do nothing and return.

2. Let N_{acc} be the set of all state nodes n_k accessible from some $n_0 \in L(v_s)$ through some path $n_0 \ldots n_k$ in Glb such that for all $0 < i \leq k$, n_i is an environment-compatible node.

3. If there is a node in N_{acc} which contains only atomic formulas, then delete v_s and return from *Expand*.

4. Set *Disabled* to be the empty set.

5. For each maximally consistent subset N_{mcs} of N_{acc},

   (a) Create an R-non-state node v_{ns} as a successor of v_s and label v_{ns} by N_{mcs}.

   (b) Let N' be the set of all system-compatible state nodes n_k accessible from some $n_0 \in N_{mcs}$ through a path $n_0 \ldots n_k$ where for all $0 < i < k$, n_i is not a state node.

   (c) For each maximally negation-consistent subset N_{mncs} of N' and the corresponding falsifying set P of atomic formulas,

      i. Create an R-state node as a successor of v_{ns} and label it by a node-label $N' - N_{mncs}$ and a write-label P. Then, recursively call *Expand* on the new node.

      ii. However, if there is an R-state node v'_s with the same node-label and write-label, and if, in addition, the node v'_s itself is marked, *"satisfied"*, then connect v_{ns} to v'_s. If v'_s is not marked *"satisfied"*, then check whether there exists a self-supporting loop in *Glb* that can be embedded into the loop $v_s \ldots v'_s$. If there is no such loop in *Glb*, connect v_{ns} to v'_s.

(d) If there is no successor to v_{ns}, delete v_{ns} and add all the nodes in N_{mcs} (the node-label of v_{ns}) to the set variable *Disabled*.

6. Check if there is a self-supporting loop in *Glb* which is accessible from a node in $L(v_s)$ through a path consisting only of environment-compatible nodes, and all state nodes in the loop are environment-compatible and in the set *Disabled*. If there is, then delete v_s. Otherwise, mark v_s "satisfied" and return.

If only weak fairness is allowed in the definition of realizability that we are checking, we only have to look for a self-supporting loop with *at least one* state node in *Disabled*.

5 Synthesis Algorithm

To simplify the presentation, we choose to represent the synthesized module by a labeled finite automaton. A *module automaton* is a tuple $\langle S, \delta, s_0, l \rangle$ where S is a finite set of states, $\delta : S \times 2^{\mathcal{P}} \mapsto 2^S$ the transition relation, $s_0 \in S$ the initial state and $l : S \mapsto 2^{\mathcal{P}}$ the labeling function. A run r is a sequence (finite or infinite) of states from S starting with s_0. We will write $r[k]$ to denote the k-th state in the sequence r and $|r|$ to denote the length of r. A behavior σ with a truth-value assignment $\pi : \Sigma \mapsto 2^{\mathcal{P}}$ is accepted by the automaton iff there is a run r such that for every k, if $r[k+1]$ is defined then $\pi((\Pi^0(\sigma))[k]) = l(r[k+1])$ and $r[k+1] \in \delta(r[k], \pi((\Pi^1(\sigma))[k]))$. With weak fairness, a behavior σ is accepted iff in addition to the previous conditions, if r is finite then for some j, there exist infinitely many $i \geq j$, such that $\delta(r[|r|], \pi(State(i, \sigma))) = \emptyset$. A similar acceptance condition can be defined for the case of strong fairness.

Given a realizability graph, we will synthesize a module automaton which implements the specification. First, for each R-state node v, create a state $s_v \in S$ for the automaton. The initial state s_0 corresponds to the root node of the realizability graph. For the labeling function l, let $l(s_v)$ be the write-label of v.

For each s_v and each $x \in 2^{\mathcal{P}}$, recall the set N_{acc} of all state nodes accessible from the nodes in the node-label $L(v)$ of the R-state node v. Find the largest subset $N \subseteq L(v)$ such that for every state node $n \in N$, all the propositions $p \in \Phi(n)$ are in x and there is no $p \in x$ such that $\neg p \in \Phi(n)$. An R-non-state successor v_{ns} of v is said to *cover* x iff $N \subseteq L(v_{ns})$ where $L(v_{ns})$ is the node-label of v_{ns}. Let V be the set of all R-state successors of the R-non-state successor v_{ns} of v which covers x. Then $\delta(s_v, x) = \{s_{v_s} \in S \mid v_s \in V\}$.

6 Correctness and Completeness

Proposition 6.1 *(Correctness) If A is the module automaton synthesized after checking the realizability of the specification formula ψ under strong [weak] fairness, then for all behaviors σ accepted by A under strong [weak] fairness, $\sigma \models \psi$.*

Proof Outline: Suppose there were a behavior σ accepted by A under strong fairness but $\sigma \not\models \psi$. We will prove that this leads to a contradiction. First, from $\sigma \not\models \psi$, then $\sigma \models \neg\psi$ and we can show that σ can be embedded into a path in the tableau graph Glb. The path starts from the root node of Glb and may be either finite or infinite. A behavior σ can be embedded into a path $n_0 n_1 n_2 \ldots$ in the tableau graph iff for all state nodes n_i, if n_i is the j-th state node in the path, then for all $\phi \in \Phi(n_i)$, $\langle \sigma, j \rangle \models \phi$. Next, we can show that the path $n_0 n_1 n_2 \ldots$ can be embedded into a path $v_0 v_1 v_2 \ldots$ in the realizability graph, using the assumption that σ is accepted by A. We use the fact that we compute the maximally negation-consistent subset in the realizability-checking algorithm to show (by induction) the existence of the part of the embedding from an R-non-state node to an R-state node and the fact that we compute the largest subset $N \subseteq L(v)$ for each R-state node v in the synthesis algorithm for the part of the embedding from an R-state node to an R-non-state node. If the path $n_0 n_1 \ldots$ in Glb is finite, then it must be the case that the last state node n_t in the path must contain only atomic formulas, because $\sigma \models \neg\psi$. It implies that n_t is accessible (in N_{acc}) from some node in the label of the last R-state node v_l of the path $v_0 v_1 \ldots$. If that is the case, v_l would have been deleted in step 3 of the realizability-checking algorithm, a contradiction.

If $n_0 n_1 \ldots$ is infinite but $v_0 v_1 \ldots v_l$ is finite, then we can also derive a contradiction by showing that for the case of strong [weak] fairness, there must be a self-supporting loop within $n_0 n_1 \ldots$ such that all [some] state nodes in the loop are environment-compatible and in the set *Disabled*. The essential step is to use the fact that σ is accepted under strong [weak] fairness and to show from the properties of maximally consistent subsets that all state nodes n_i in the loop must be in the set *Disabled* if $\delta(s_{v_l}, \pi(n_i)) = \emptyset$. However, if such a self-supporting loop exists, then v_l would have been deleted in step 6.

Similarly, we can also derive a contradiction in the case when both $n_0 n_1 \ldots$ and $v_0 v_1 \ldots$ are infinite, by showing that the loops in $v_0 v_1 \ldots$ would have been eliminated in step 5.(c).ii.

Proposition 6.2 *(Termination) The realizability-checking algorithm always terminates.*

Proof Outline: There are only finitely many possible R-state and R-non-state nodes. Therefore, it is not possible to keep expanding the realizability graph forever. It is also clear that there are only finitely many maximally consistent and maximally negation-consistent subsets at any time in the algorithm.

Proposition 6.3 *(Completeness) The formula ψ is realizable iff the root node of the realizability graph is not deleted.*

Proof Outline: One direction of the proof, showing that if the root node of the realizability graph is not deleted then ψ is realizable, is straightforward from the proposition 1 (correctness).

In the other direction, we assume that ψ is realizable and show that the root node of the realizability graph is not deleted. Since ψ is realizable, there exists a function f which realizes it.

We have to define an embedding of a behavior into the realizability graph. A behavior σ can be embedded into a finite or infinite path $v_0 v_1 \ldots$ starting from the root in the realizability graph iff there exists an increasing sequence of integers $d_0 d_1 \ldots$ such that all of the following conditions are true:

- for all $i \geq 0$ and $n \in L(v_{2i})$, there is a formula $\phi \in \Phi(n)$ such that $\langle \sigma, d_i \rangle \models \neg\phi$,

- for all $i > 0$, $\langle \sigma, d_i \rangle \models \mu$,

- for all $i > 0$ and $n \in L(v_{2i+1})$, there is a formula $\phi \in \Phi(n)$ such that $\langle \sigma, d_i - 1 \rangle \models \neg\phi$.

An R-state node is called *reachable* iff there is a behavior of f which can be embedded into some path passing through the node. It is easy to see that the root node must be reachable. We want to show that some of the reachable nodes including the root are not deleted.

First, we can show that a reachable R-state node must not be deleted in step 3; otherwise, we can easily construct a behavior of f which falsifies ψ.

Next, we can show that for every reachable R-state node v and every self-supporting environment-compatible loops accessible from a node in the node-label $L(v)$ of v, there exists a node n in the loop such that for every R-non-state successor v_{ns} of v which contains n in the node-label, there is an R-state successor v' of v_{ns} which is also reachable. Again, we can prove this by showing that if such n does not exist, we can construct a fair behavior of f which falsifies ψ. We also use the fact that the label of v_{ns} is a maximally consistent set to show the existence of the embedding into a path through v_{ns}.

Finally, we can show that for any loop of reachable R-state nodes, if there is a self-supporting loop in *Glb* which can be embedded into it as in step 5.(c).ii, there is a reachable R-state node in the loop which is not deleted as the result of breaking the loop of R-state nodes in step 5.(c).ii. We prove this by considering a behavior of f which can be embedded into a path passing through this loop of reachable R-state nodes. Clearly, the path cannot remain within the loop forever or the behavior will not satisfy ψ.

Acknowledgements

We thank Allen Emerson for his very useful comments and Howard Wong-Toi, Eddie Chang, Nikolaj Bjorner and Henny Sipma for fruitful discussions and for carefully reading the drafts of this paper.

References

[ALW89]   M. Abadi, L. Lamport, and P. Wolper. Realizable and unrealizable concurrent program specifications. *Proc. 16th Int. Colloq. Aut. Lang. and Prog.* Lec. Notes in Comp. Sci. 372, Springer-Verlag, Berlin, 1–17, 1989.

[BKP84]   H. Barringer, R. Kuiper, and A. Pnueli. Now you may compose temporal logic specifications. *Proc. 16th ACM Symp. Theory of Comp.*, 51–63, 1984.

[EC82]   E.A. Emerson and E.M. Clarke. Using branching time temporal logic to synthesize synchronization skeletons. *Sci. Comp. Prog.*, 2(3):241–266, 1982.

[MW80]   Z. Manna and R. Waldinger. A deductive approach to program synthesis. *ACM Trans. Prog. of Lang. and Sys.*, 2(1):90–121, 1980.

[MW84]   Z. Manna and P. Wolper. Synthesis of communicating processes from temporal-logic specifications. *ACM Trans. on Prog. Lang. and Sys.*, 6(1):68–93, 1984.

[PR89a]   A. Pnueli and R. Rosner. On the synthesis of a reactive module. *Proc. 16th ACM Symp. Princ. of Prog. Lang.*, 179–190, 1989.

[PR89b]   A. Pnueli and R. Rosner. On the synthesis of an asynchronous reactive module. *Proc. 16th Int. Colloq. Aut. Lang. Prog.* Lec. Notes in Comp. Sci. 372, Springer-Verlag, Berlin, 652–671, 1989.

[WD91]   H. Wong-Toi and D.L. Dill. Synthesizing processes and schedulers from temporal specifications, *Computer-Aided Verification (Proc. CAV90 Workshop)*, DIMACS Series in Discrete Mathematics and Theoretical Computer Science Vol. 3 (American Mathematical Society, 1991).

[Wo83]   P. Wolper. Temporal logic can be more expressive. *Info. and Cont.*, 56:72–99, 1983.

[Wo85]   P. Wolper. The tableau method for temporal logic: An overview. *Logique et Anal.*, 28:119–136, 1985.

Model Checking of Macro Processes

Hardi Hungar

Computer Science Dept.,
University Oldenburg,
D-26111 Oldenburg, Germany
hungar@informatik.uni-oldenburg.de

Abstract.
Decidability of modal logics is not limited to finite systems. The alternation-free modal mu-calculus has already been shown to be decidable for context-free processes, with a worst case complexity which is linear in the size of the system description and exponential in the size of the formula. Like context-free processes correspond to the concept of procedures without parameters, *macro* processes correspond to procedures with procedure parameters. They too allow deciding mu-calculus formulae, as is shown in this paper by presenting both global (iterative) and local (tableaux-based) procedures. These decision procedures handle correctly also process systems which are defined by unguarded recursion. As expected, the worst case complexity depends on the highest type level in the process description, and it is k-exponential in the size of the formula for a system description with type level k.

1 Introduction

Model checking provides a powerful tool for the automatic verification of behavioral systems. The corresponding standard algorithms fall into two classes: the iterative algorithms (cf. [15, 8, 11, 12]) and the tableaux-based algorithms (cf., e.g. [3, 4, 9, 30, 33, 35]). Whereas the former class usually yields higher efficiency in the worst case, the latter allow *local* model checking (cf. [33]), which avoids the investigation of for the verification irrelevant parts of the process being verified.

At first sight both kinds of algorithms seem to be restricted to finite systems, although Bradfield and Stirling [3, 4] constructed a sound and complete tableau system for the full mu-calculus [28], which can deal with *infinite* transition systems: In general, neither the assertions in the tableaux are purely syntactic nor is well-formedness of a tableau decidable. But by using second-order assertions Burkart and Steffen [5] were able to handle infinite systems given in the form of *context-free* process systems with an iterative algorithm, and a local model checking procedure has been developed by Hungar and Steffen [24]. Also, an automata-based approach has been presented by Iyer [26]. This one does not decide mu-calculus formulae but checks whether the process has a computation which is accepted by a given Büchi automaton.

Context-free processes (CFPs) are essentially equivalent to BPA processes [2]. They are labeled transition systems generated by edge replacement systems: edges labeled with a nonatomic action (a procedure) are to be replaced by the transition system defining the action (the body of the procedure). Because the defining systems may contain nonatomic actions as well, the resulting system might be infinite. But for model checking purposes, only the effect of an action to the validity of subformulae of the formula in question needs to be observed. This allows the reduction of the problem to a finite mutual recursive equation system, which was the central idea of [5, 24].

CFPs can model some interesting infinite structures, for example a stack. They can not, however, model a queue. This should not come as a surprise, because a queue would give the power of a Turing-machine, destroying decidability. Also, two stacks in parallel would yield undecidability. Therefore, CFPs are not closed under parallel composition [6]. But there are other ways in which the expressibility can be increased without destroying decidability of mu-calculus formulae. And this is done within this paper.

Just like context-free grammars can be generalized to macro grammars [17], and further to higher-typed production systems, capable of generating more and more languages [13, 16], one can also study higher-typed expansion rules for the generation of processes, resulting in what I call *macro processes*. In programming language terms, macro processes (MPs) correspond to programs with higher-order procedures, whereas context-free processes correspond to programs with parameterless procedures. Another way of introducing macro processes would be to enrich BPA by process abstraction and application.

Recalling the results of classical program verification [18, 19], one could expect to be able to lift the algorithms working for CFPs to the macro case. And indeed this is the main result of this paper. Already the automata-based approach from [26] can cope with macro processes of second order (only parameterless processes are allowed as parameters). Here, arbitrary finite types are allowed. The price when dealing with MPs is that the upper bound for the worst-case complexity of the algorithms is k-exponential in the size of the formula, where k is the maximal type level in the process system definition. This hyperexponential growth is again in accordance with similar results for higher-order programming languages [22, 23, 29]. It is a worst-case complexity, though, and an efficient implementation should usually behave much better.

The new algorithms also cope with unguarded recursion, which allows to define processes with infinite branching. In that respect they generalize the known algorithms even for unparameterized processes. For parameterized processes, this is of particular interest, because it is much harder to check finite branching for parameterized processes than for unparameterized (context-free) processes.

The next section introduces the technical definitions and demonstrates the modeling power of MPs by giving an example. Then the model checking algorithms are presented, and in the final section extensions and open questions are discussed.

2 Processes and Formulae

In this section I introduce *process graphs* as the basic structure for modeling behavior, and more specifically, *macro process systems* as finite representations of infinite process graphs, as well as the (alternation-free) modal mu-calculus as a logic for specification.

2.1 Macro Process Systems

Definition 1 (Process Graphs). A *process graph* (short *PG*) is a labeled transition system G with distinguished start and end state (s_G and e_G, $s_G \neq e_G$), where no transition originates at e_G. I.e. it consists of a set (of states), a binary relation on the states (transitions), and a labeling of the transitions.

Intuitively, a process graph encodes the operational behavior of a process. That the end state of a process graph must not be the origin of a transition has its reason in the use which will be made later of process graphs: They will replace edges in transition systems, and that an inserted process graph may be entered via its end state has to be avoided. Leaving the inserted process graph via the start state is avoided by using a specific form of insertion, called *embedding*.

Definition 2 (Embedding). Let G and G' be PGs and let s_1 and s_2 be states in G'. *Embedding G into G' between s_1 and s_2* produces a PG where a copy of G is inserted into G' with

- all states in the copy are new except the end state, for which s_2 is used,
- for all edges leaving s_G, an edge with the same label and destination originates at s_1.

The copy of s_G is called the *initial state* of the embedding and s_2 is called its *return state*.

I want to remark that the construction ensuring that an embedded PG is not left through the start state could also have been used for the end states, to eliminate the restriction on transitions leaving end states in Definition 1. This choice is a matter of taste and of technical significance only.

To generalize the notion of a context-free process system, I first introduce *finite types*.

Definition 3 (Finite Types). The set of *finite types* is given by the production rules

$$\tau ::= \beta \mid (\tau \to \tau).$$

β is the base type.

The *level* of a type is inductively defined by:

$$\text{lev}(\beta) = 0, \quad \text{lev}(\tau \to \rho) = max(\text{lev}(\tau) + 1, \text{lev}(\rho)).$$

For every type τ, an infinite set of identifiers X_τ is available, with $x_\tau, p_\tau \in X_\tau$. *Typed terms* are built up from identifiers by type-respecting application, i.e.

$$t_\tau ::= x_\tau \mid (t_{\tau \to \tau} \, t_\tau)$$

Simple finite types are generated from $\beta \to \beta$ instead of β. Accordingly, *simple identifiers* are those of simple types, and *simple terms* do only have subterms of simple type.

Apart from the notions of *simple* types and terms, the definition is fairly standard. Note that types of level 1 are just curried forms of the usual first-order types $\beta \times \cdots \times \beta \to \beta$. I use an ML-style way of writing applications, i.e. $(p\ x\ y)$ means that the result of p applied to x is applied to y.

Simple terms of type $\beta \to \beta$ will be used as nonatomic actions in the following definition of macro processes. Unrestricted finite types are considered in the discussion of extension in the final section.

Definition 4 (Macro Processes). Let *Act* be a set of *atomic actions*.

A *macro process system* (MPS) has the form

$$(p_1\ x_{11}\ \ldots\ x_{1\,n_1}) = G_{p_1}$$

$$\cdots$$

$$(p_k\ x_{k1}\ \ldots\ x_{k\,n_k}) = G_{p_k}$$

$$G_{main}\ .$$

It consists of a set of *process definitions* defining process identifiers $p_1, \ldots, p_k$ and a *main process graph* G_{main}. The process graphs $G_{p_1}, \ldots, G_{p_k}$ and G_{main} have start states $s_{p_1}, \ldots, s_{p_k}$ and s_{main} and end states $e_{p_1}, \ldots, e_{p_k}$ and e_{main}. Labels in the graphs are either atomic actions or simple terms of type $\beta \to \beta$. In addition to the defined identifiers, which may occur in any of the process graphs, the terms in G_{p_i} may contain $x_{i1}, \ldots, x_{i\,n_i}$, i.e. the formal parameters of the definition. The *level* of an MPS is the highest occuring type level.

The level of an MPS is the maximum of the levels of its defined identifiers. Context-free process systems coincide with the set of MPSs of level one.

I will give a simple *operational* semantics for an MPS, by defining its *complete expansion*. This results from the main PG by repetetively replacing all nonatomic actions by their definitions. The complete expansion may be infinite or even infinitely branching.

Definition 5 (Complete Expansion). *Expanding* a nonatomic transition $s_1 \xrightarrow{(p_i\ t_1\ \ldots\ t_n)} s_2$ in the main PG of an MPS embeds G_{p_i}, the defining PG of p_i, with the formal identifiers $x_{i\,j}$ in transition labels replaced by the t_j, between s_1 and s_2 and removes the transition $s_1 \xrightarrow{(p_i\ t_1\ \ldots\ t_n)} s_2$.

An *expansion step* replaces all nonatomic edges in parallel. The *complete expansion* of an MPS results by repeated application of expansion steps. I.e. the complete expansion is the union of a (countable) chain of approximants, where an approximant is the PG of the MPS after a finite number of expansion steps with all nonatomic edges deleted.

The *finite language* of an MPS consists of those strings of atomic actions which label finite paths from the start to the end state of the complete expansion.

Note that expanding an edge in an MPS always yields a well-formed MPS. The result of an expansion step is unique up to renaming of states, thus also the complete expansion is uniquely determined.

2.2   An Example of an MPS

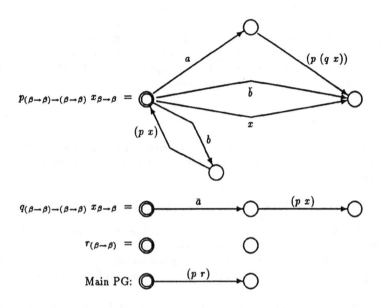

Fig. 1. MPS for a 2-stack.

Figure 1 shows the MPS for a 2-stack. There are several different, but equivalent versions of n-stacks [27]. Here, a 2-stack is a stack of ordinary stacks over a set of basic elements with (partial) operations

- $push_1(d)$: push d on the top ordinary stack
- pop_1: pop the top ordinary stack
- $push_2$: duplicate the top stack
- pop_2: remove the top stack ,

which initially contains an empty ordinary stack.

In the MPS of Figure 1, a and $\bar{a}$ are the actions for $push_1(d)$ and pop_1, and b and $\bar{b}$ are used for $push_2$ and pop_2. This MPS abstracts from the differences between the basic elements and just keeps their count, but it is obvious how to change it in order to model a 2-stack over a given finite set precisely (For each basic element, an atomic action for pushing and popping would be introduced). Also nested stacks of higher degree can be modeled (by MPSs of higher degrees).

A formal argument that the given MPS indeed models a 2-stack would not be hard to give. One can establish a bisimulation between a 2-stack and the MPS. Roughly, the correspondence is as follows. During execution, the parameter of the actual instance of p represents the top 1-stack. It is always of the form $(q^n r)$, which allows n consecutive a actions and thus corresponds to a 1-stack with n elements. A b starts a new recursive call with the same parameter, after which the current call is continued (i.e. it duplicates the top 1-stack), whereas a $\bar{b}$ returns to the previous incarnation of p (i.e. it removes the top 1-stack).

Nested stacks are outside the scope of context-free processes. One could use the pumping lemma to show that the finite language of a nested stack (of nesting depth greater than one) is not context-free. Indeed, in several respects an n-stack comprises the essence of computation with procedures of type level n, cf. [14, 29], which in their computation power form an infinite hierarchy.

2.3 Mu Calculus

The following negation–free syntax defines a sublanguage of the mu-calculus, which in spite of being as expressive as the full mu-calculus allows a simpler technical development.

$$\phi ::= f\!f \mid tt \mid X \mid \phi \wedge \phi \mid \phi \vee \phi \mid [a]\phi \mid \langle a\rangle\phi \mid \nu X.\phi \mid \mu Y.\phi$$

In the above, $a \in Act$, and $X, Y \in Var$, where Var is a set of variables. Only *closed*, *alternation free* formulae will be used, i.e. every variable is bound by a ν or μ, and no ν-subformula has a free variable which, in the context of the whole formula, is bound by a μ, and vice versa. The set of closed alternation-free formulae is denoted by $\mathcal{F}$.

Given a PG, the semantics of a formula is a subset of its states. I do not give the formals of the standard definition. The *closure* $CL(\phi)$ of a formula ϕ is the set of all its subformulae with each free variable replaced (iteratively) by the corresponding fixpoint subformula, e.g. $CL(\nu X.\phi) = \{\nu X.\phi\} \cup CL(\phi[\nu X.\phi/X])$.

2.4 Higher-Order Formulae

Definition 6 (Higher-Order Formulae).
Let the set of *typed higher-order formulae* be defined by

$$\mathcal{F}_\beta =_{df} \mathcal{F}$$
$$\mathcal{F}_{\tau\to\tau} =_{df} \mathcal{P}(\mathcal{F}_\tau) \times \mathcal{F}_\tau \ ,$$

where $\mathcal{P}$ denotes the powerset of a set. A higher-order formula (HOF) is *finite* iff its components are finite. The *basis* of a higher-order formula is the set of all its constituent mu-calculus formulae, i.e. $(b\ \phi_\beta) = \phi_\beta$, $(b\ (\Theta,\phi)) = \{(b\ \theta) \mid \theta \in \Theta\} \cup \{(b\ \phi)\}$.

The relation $\preceq$, defined by

$$\phi_\beta \preceq \psi_\beta =_{df} \phi_\beta = \psi_\beta$$
$$\Phi_\tau \preceq \Psi_\tau =_{df} \forall\phi_\tau \in \Phi_\tau.\exists\psi_\tau \in \Psi_\tau.\phi_\tau \preceq \psi_\tau$$
$$(\Phi_\tau,\phi_\tau) \preceq (\Psi_\tau,\psi_\tau) =_{df} \Psi_\tau \preceq \Phi_\tau \wedge \phi_\tau \preceq \psi_\tau \ ,$$

gives a syntactic approximation of implication on higher-order formulae.

Higher-order formulae will be denoted by lowercase greek letters, whereas uppercase letters will be used for sets of formulae.

The objects denoted by higher-order formulae are terms over the defined identifiers of an MPS. The base case of the definition concerns the validity of second-order assertions for arbitrary PGs, and this is explained as ordinary validity for embeddings of the PG.

Definition 7 (Semantics of Higher-Order Formulae). Let G be a PG. Then $G \models (\Theta_\beta, \phi_\beta)$ iff the initial state of every embedding of G satisfies ϕ_β, whenever the return state satisfies all formulae in Θ_β.

Let $t_{\beta \to \beta}$ be a term of defined identifiers in an MPS. $t_{\beta \to \beta} \models \phi_{\beta \to \beta}$ iff the complete expansion of $s \overset{t_{\beta \to \beta}}{\to} e$ satisfies $\phi_{\beta \to \beta}$.

For higher simple types $\bar{\tau} \to \tau$, $t_{\tau \to \tau} \models (\Theta_\tau, \phi_\tau)$ iff, for all t_τ with $t_\tau \models \Theta_\tau$, it holds that $(t_{\tau \to \tau}\ t_\tau) \models \phi_\tau$ (the MPS may be extended to define identifiers in t_τ).

The usefulness of higher-order formulae for the verification relies on two observations:

- For verification purposes, defined processes are completely determined by the set of formulae they satisfy, which is important for the completeness of compositional reasoning.
- If the validity of a finite formula is concerned, only finite formulae need to be considered, which will make the method effective.

In other words: a finite set of formulae is an adequate abstraction of the true semantics of a macro process. This is formalized in the following proposition.

Proposition 8 (Adequacy).

1. *Let G be a PG, and let s_i resp. s_r be the initial resp. return state of an embedding of G. Then*

$$s_i \models \phi_\beta \quad \text{iff} \quad G \models (\{\theta \in CL(\phi_\beta) \mid s_r \models \theta\}, \phi_\beta) \ .$$

2. *Let $t_{\tau \to \tau}$ and t_τ be terms of defined identifiers. Then*

$$(t_{\tau \to \tau}\ t_\tau) \models \phi_\tau \text{ iff } t_{\tau \to \tau} \models (\{\theta_\tau \in \mathcal{F}_\tau \mid t_\tau \models \theta_\tau \text{ and } (b\ \theta_\tau) \subseteq CL(b\ \phi_{\tau \to \tau})\}, \phi_\tau) \ .$$

Formulae of type $\beta \to \beta$ are essentially the second-order formulae from [24]. Intuitively, $(\Theta_\beta, \phi_\beta)$ is true if the start state always satisfies ϕ_β whenever at the end state a PG is added, so that the end state now satisfies each formula in Θ_β. Higher-order formulae impose this on the PG if all parameters satisfy their specification.

3 Model Checking of Macro Processes

I will formulate the algorithms in this section w.r.t. a given MPS with defined processes $p_1, \ldots, p_k$.

3.1 Local Model Checking

The local method is tableaux-based. I give a set of tableau rules. Other than the iterative algorithm to be presented in the following subsection, local model checking allows to ignore irrelevant parts of the MPS.

Intermediate formulae in the proofs concern either states of defining PGs with appropriate assumptions about the parameters and the return state, or terms which might be called or passed as parameters in calls. Such an intermediate formula is called a *sequent*.

Definition 9 (Sequents). A *(higher-order) assertion* is either a *state assertion* s sat ϕ_β or a *term assertion* t_τ sat ϕ_τ for a simple type τ, with finite ϕ in both cases.

A *hypotheses set* Γ_{p_i} for a defined identifier p_i is a set of assertions of the form

$$\{e_{p_i} \text{ sat } \phi_0, \ x_{i1} \text{ sat } \phi_1, \ \dots \ , x_{in_i} \text{ sat } \phi_{n_i}\} \ ,$$

i.e. a collection of assertions containing one assertion for each parameter and one for the end state of the definition.

A *sequent* is either $\Gamma_{p_i} \vdash s$ sat ϕ_β where s is one of the states of p_i or $\Gamma_{p_i} \vdash t_\tau$ sat ϕ_τ where t_τ only involves defined identifiers and formal parameters of p_i .

The validity of sequents is defined similar to the validity of formulae: The assertion must hold whenever the formal parameters are instantiated with defined identifiers satisfying the hypotheses.

The rule set is an adaptation of the rule set from [24], by adding higher-type reasoning like in [19].

Definition 10 (Successful Tableau, Derivability). A tableau is *successful* if it is finite and all *leaves* are *successful*. *Successful leaves* are of the form

1. $\Gamma \vdash s$ sat tt, or
2. $\{\dots x_\tau \text{ sat } \Theta_\tau \ \dots\} \vdash x_\tau$ sat θ_τ where $\{\theta_\tau\} \preceq \Theta_\tau$, or
3. $\Gamma \vdash s$ sat $\phi(\nu X.\psi)$, where $\phi(\nu X.\psi) \in CL(\nu X.\psi)$, there is another node on the path from the root of the tableau to this node labeled with the same sequent, and the maximal fixpoint gets unfolded between these two nodes, or
4. $\Gamma \vdash p$ sat $(\Theta_{\tau_1}, \dots, (\Theta_\beta, [a]\phi_\beta) \dots)$, where the assertion recurs along a path where the last component of the formula is always $[a]\phi_\beta$.

A sequent is *derivable* if it has a successful tableau.

A node in a tableau may also have no successor without being a leaf. This applies to sequents of the form $\Gamma \vdash s$ sat $[a]\phi_\beta$, if neither a-transitions nor nonatomic transitions originate at s, for example if $s = e_{main}$.

Some remarks in order to explain the mechanism of tableau construction: At any time, reasoning is restricted to one process definition (resp., the main process), and the hypotheses contain a set of assertions about each parameter and the end state of the definition. A subtableau for a process call is entered only when necessary, i.e. when evaluating a modality at the origin of a nonatomic transition. Then, ultimately

Main

$$\frac{H \text{ sat } \phi_\beta}{\{\} \vdash s_{main} \text{ sat } \phi_\beta}$$

Basic

$$\frac{\Gamma \vdash s \text{ sat } \phi_\beta \wedge \psi_\beta}{\Gamma \vdash s \text{ sat } \phi_\beta \quad \Gamma \vdash s \text{ sat } \psi_\beta} \qquad \frac{\Gamma \vdash s \text{ sat } \phi_\beta \vee \psi_\beta}{\Gamma \vdash s \text{ sat } \phi_\beta} \qquad \frac{\Gamma \vdash s \text{ sat } \phi_\beta \vee \psi_\beta}{\Gamma \vdash s \text{ sat } \psi_\beta}$$

$$\frac{\Gamma \vdash s \text{ sat } [a]\phi_\beta}{\Gamma \vdash s' \text{ sat } \phi_\beta \quad \dots \quad \Gamma \vdash s'' \text{ sat } \Theta_\beta \quad \Gamma \vdash t_{\beta \to \beta} \text{ sat } (\Theta_\beta, [a]\phi_\beta)}$$

(All s' where $s \xrightarrow{a} s'$ and all $t_{\beta \to \beta}$, s'' where $s \xrightarrow{t_{\beta \to \beta}} s''$. $s \notin \{ s_p \mid p \text{ defined identifier} \}$).

$$\frac{\Gamma \vdash s \text{ sat } \langle a \rangle \phi_\beta}{\Gamma \vdash s' \text{ sat } \phi_\beta} \ (s \xrightarrow{a} s') \qquad \frac{\Gamma \vdash s \text{ sat } \langle a \rangle \phi_\beta}{\Gamma \vdash s'' \text{ sat } \Theta_\beta \quad \Gamma \vdash t_{\beta \to \beta} \text{ sat } (\Theta_\beta, \langle a \rangle \phi_\beta)} \ (s \xrightarrow{t_{\beta \to \beta}} s'')$$

$$\frac{\Gamma \vdash s \text{ sat } \nu X.\phi_\beta}{\Gamma \vdash s \text{ sat } \phi_\beta[\nu X.\phi_\beta / X]} \qquad \frac{\Gamma \vdash s \text{ sat } \mu Y.\phi_\beta}{\Gamma \vdash s \text{ sat } \phi_\beta[\mu Y.\phi_\beta / Y]}$$

Higher Types

$$\frac{\Gamma \vdash (t_{\tau \to \tau} \ t_\tau) \text{ sat } \phi_\tau}{\Gamma \vdash t_{\tau \to \tau} \text{ sat } (\Theta_\tau, \phi_\tau) \quad \Gamma \vdash t_\tau \text{ sat } \theta_\tau, \theta_\tau \in \Theta_\tau}$$

$$\frac{\Gamma \vdash p \text{ sat } (\Theta_{\tau_1}, (\Theta_{\tau_2}, \dots (\Theta_\beta, \phi_\beta) \dots))}{\{x_1 \text{ sat } \Theta_{\tau_1}, x_2 \text{ sat } \Theta_{\tau_2}, \dots, e_p \text{ sat } \Theta_\beta\} \vdash s_p \text{ sat } \phi_\beta}$$

Fig. 2. Tableau Rules

a modality formula about the initial state of another process has to be proven. This depends only on the successors of the initial state, which makes this reasoning sound, considering the construction of the complete expansion.

Since all rules are (backwards) sound, the soundness of the method follows from the validity of each successful leaf. Both forms of recurring leaves require an argument. The recurrence of a maximal formula by unfolding the maximal fixpoint on the connecting path suffices (in the alternation-free mu-calculus) for an application of the maximal fixpoint characterization. Note that the maximal fixpoint formula need not recur itself, because it might only appear in disjoint subtableaux handling process calls (compare the example in [24]). If a p sat $(\Theta_{\tau_1}, \dots, (\Theta_\beta, [a]\phi_\beta) \dots)$ recurs without the box being removed on the path, all possibilities of invalidating the assertion by doing an a-step are examined in subtableaux along the path. The minimality of the complete expansion guarantees that this is sufficient.[1]

The completeness relies on the finiteness of the set of relevant assertions and sequents, and on the adequacy of the higher-order semantics. To prove a formula ϕ_β,

---

[1] Box recurrence is an effect of allowing unguarded recursion in MPSs.

the basis of intermediately needed higher-order formulae is in $CL(\phi_\beta)$ (finiteness). For every valid ν-formula, some sequent with an element of its closure must recur, if the rules are applied with care (adequacy). And a valid μ-formula must have a finite justification.

The finiteness accounts also for the completeness of an exhaustive search. Summing up, we have:

Theorem 11. *The tableau system provides an effective, sound and complete model checking procedure for macro processes and alternation-free formulae.*

3.2   Iterative Model Checking

The iterative algorithm follows the idea of [5], only that it is extended to macro processes and that it copes with infinitely branching processes (generated by unguarded recursion).

Since the formulae are alternation-free, minimal and maximal fixpoints can be treated separately. To check the validity of a formula ϕ for the start state (of the complete expansion), the iterative algorithm essentially computes all valid sequents. To be more precise, if s is a state of p_i and a complete semantic description of the parameters and the end state of p_i is given, the subset of $CL(\phi)$ which is valid at s is computed. The semantic description of an end state is a subset of $CL(\phi)$ (the subformula which are assumed to be valid). For a parameter, a monotonic function over $\mathcal{P}(CL(\phi))$ of appropriate type is taken as description. For identifiers of $\beta \to \beta$, this is just the second-order semantics of [5]: The function yielding the set of formulae valid at the start state, given the set of formulae at the end[2]. For higher types, it is a higher-order monotonic function. The function (of parameter and end state descriptions) computed for the s_{p_i} gives the semantic description for p_i, the result for s_{main} gives the set of formulae valid for the MPS.

I will sketch the computation step for minimal fixpoint operators. It is best explained for the case of one fixpoint only. Every fact about subformulae not containing the fixpoint variable can be assumed to be known. Let Ψ be the set of closure formulae containing the fixpoint formula. The computation iteratively approximates from below the subsets of Ψ valid at the states of the definitions for all possible parameter descriptions. It starts with the end states of process definitions initialized according to the respective description values, and everything else initialized to the empty set. Then, modalities and conjunctions and disjunctions are evaluated, and fixpoints may be unfolded. Transitions labeled with atomic actions are easy to cope with, transitions labeled with terms need an evaluation of the term according to the parameter descriptions (for parameters of the definition) respectively the current aproximation of the higher-order semantics of a defined identifier. This computation is monotonic, i.e. the sets of valid formulae do increase[3]. Therefore, it will reach a fixpoint, which itself gives a monotonic function for each state.

---

[2] This description of the algorithm is still oversimplified. E.g. in case that the argument set is inconsistent, the result of the computation of the algorithm need not be the full set $CL(\phi)$. In general, only a subset is computed. But the result is correct if the argument set consists of all subformulae valid for some process graph. A similar remark applies to parameter specifications.

[3] This relies on the monotonicity of the functions for parameter descriptions.

If recursion was always guarded, this fixpoint would be the correct approximation of the semantics of the definitions (as in [5]). In the presence of unguarded recursion, it has to be checked whether each box formula $[a]\theta$ which is computed as ff is properly invalidated by some other formula, i.e. whether according to the parameter descriptions there is a transition labeled with a to a state where θ is false. This will always be the case except when a defined process recursively calls itself (with semantically equivalent parameters) without any atomic action occuring before the second call. If some box formulae are set to tt because they do not pass the test, the iteration continues until finally a fixpoint is reached where all false box-formulae are properly invalidated.

The computation process for maximal fixpoint formulae is the dual procedure. It approximates the result from above, and has a special check for the validity of diamond formulae.

Once the higher-order descriptions of the defined identifiers are computed, it is easy to decide whether the formula in question holds at the start state of the main process graph.

The complexity of this iterative computation is dominated by length of representations of the descriptions of defined identifiers. This representation is k-exponential in $O(lev(\tau))$ for an identifier of type τ. I.e. for context-free processes the algorithm is exponential, for second-order processes it is double exponential, and so on.

4 Extensions and Open Problems

The presented model checking procedures are restricted to MPSs with simple types, i.e. the type hierarchy in fact starts with $\beta \to \beta$. It is, however, quite straightforward to extend them to arbitrary finite types. But what should be a process of type $\beta \to (\beta \to \beta)$? The answer is simple: The argument of type β of a simple process is instantiated with the return state of the call, so an extra argument of this type should be regarded as a second possible return state. A call to this process would have to be represented by a *hyperedge* with one origin and two destinations, and the defining process graph would have *two* end states.

Hyperedge replacement systems [21] are, anyhow, a more adequate generalization of context-free string grammars to graph grammars than edge replacement systems. They can generate all [7] graphs of pushdown automata [32] (exactly the same set of finitely branching graphs). In [6], the iterative model checking procedure for CFPs is already extended to pushdown processes, and this set is shown to be closed under parallel composition with finite processes.

The same holds here. All results of this paper carry over to the more general case. The tableau rules do not even have to be changed much, only the modality rules have to take care of a tuple of end states of nonatomic transitions. And the set of processes at each type level will also be closed under parallel composition with finite processes, in the same sense as pushdown processes are.

Several open questions concern the complexity of the methods. The k-exponential worst-case complexity is an upper bound, but I conjecture that it can also be established as a *lower* bound - for the worst case. In many cases, a clever algorithm should perform much better. The tableau system gives the basis of an incremental algorithm, and the behavior of a program in the style of [34] would be interesting to observe. Very often, only a small subset of all parameter values is relevant, so one could save a lot.

Of more theoretical interest are two other questions. First, whether it is possible to give a *denotational* semantics to MPSs. And second (perhaps after solving the first problem), whether monadic second-order logic, which is more expressive than the mu-calculus, is decidable for macro processes. Also, of course, to extend the decision procedure to handle the full mu-calculus is something worth trying.

References

1. Andersen, H., *Model checking on boolean graphs*. ESOP '92, LNCS 582 (1992), 1–19.
2. Bergstra, J.A., and Klop, J.W., *Process theory based on bisimulation semantics*. LNCS 354 (eds de Bakker, de Roever, Rozenberg) (1988), 50–122.
3. Bradfield, J.C., *Verifying temporal properties of systems*. Birkhäuser, Boston (1992).
4. Bradfield, J.C., and Stirling, C. P., *Verifying temporal properties of processes*. Proc. CONCUR '90, LNCS 458 (1990), 115-125.
5. Burkart, O., and Steffen, B., *Model checking for context-free processes*. CONCUR '92, LNCS 630 (1992), 123–137.
6. Burkart, O., and Steffen, B., *Pushdown processes: Parallel composition and model checking*. Tech. Rep. Aachen/Passau (1994), 17 p. (1992), 123–137.
7. Caucal, D., and Monfort, R., *On the transition graphs of automata and languages*. WG 90, LNCS 484 (1990), 311–337.
8. Clarke, E.M., Emerson, E.A., and Sistla, A.P., *Automatic verification of finite state concurrent systems using temporal logic specifications*. ACM TOPLAS 8 (1986), 244–263.
9. Cleaveland, R., *Tableau-based model checking in the propositional mu-calculus*. Acta Inf. 27 (1990), 725-747.
10. Cleaveland, R., Parrow, J., and Steffen, B., *The concurrency workbench*. Workshop Automatic Verification Methods for Finite-State Systems, LNCS 407 (1989), 24–37.
11. Cleaveland, R., and Steffen, B., *Computing behavioral relations, logically*. ICALP '91, LNCS 510 (1991).
12. Cleaveland, R., and Steffen, B., *A linear-time model-checking algorithm for the alternation-free modal mu-calculus*. CAV 91, LNCS 575 (1992), 48–58.
13. Damm, W., *The IO- and OI-hierarchies*, TCS 20 (1982), 95–205.
14. Damm, W., and Goerdt, A., *An automata-theoretic characterization of the OI-hierarchy*, Inf. and Cont. 71 (1986), 1–32.
15. Emerson, E.A., and Lei, C.-L., *Efficient model checking in fragments of the propositional mu-calculus*. 1st LiCS (1986), 267–278.
16. Engelfriet, J., and Schmidt, E.M., *IO and OI*, JCSS 15 (1977), 328–353, and JCSS 16 (1978), 67–99.
17. Fischer, M.J., *Grammars with macro-like productions*, 9th Conf. Switching and Automata Theory, IEEE (1968), 131–142.
18. German, S.M., Clarke, E.M. and Halpern, J.Y., *Reasoning about procedures as parameters in the language L4*, Inf. and Comp. 83 (1989) 265–359. (Earlier version: 1st LiCS (1986) 11–25)
19. Goerdt, A., *A Hoare calculus for functions defined by recursion on higher types*, Logics of Programs 1985, LNCS 193, 106–117.

20. Habel, A., *Hyperedge replacement: Grammars and languages.* PhD thesis, Bremen (1989), 193 p.

21. Habel, A., and Kreowski, H.-J., *May we introduce to you: Hyperedge replacement,* Graph-grammars and their application to computer science 1986, LNCS 291 (1987), 15–26.

22. Hungar, H., *Complexity of proving program correctness,* TACS '91, LNCS 526 (1991), 459–474.

23. Hungar, H. *The complexity of verifying functional programs,* STACS '93, LNCS 665 (1993), 428–439.

24. Hungar, H., and Steffen, B., *Local model checking for context-free processes.* ICALP '93, LNCS 700 (1993), 593–605.

25. Huynh, D.T., and Tian, L., *Deciding bisimilarity of normed context-free processes is in* Σ_2^p. Tech. Rep. UTDCS-1-92, Univ. Texas Dallas (1992).

26. Iyer, S.P., *A note on model checking context-free processes,* North American Process Algebra Workshop '93 (ed. Bard Bloom).

27. Kowalczyk, W., Niwinski, D., and Tiuryn, J. *A generalization of Cook's auxiliary-pushdown-automata theorem,* Fund. Inf. 12 (1989) 497–506.

28. Kozen, D., *Results on the propositional μ-calculus.* TCS 27 (1983), 333–354.

29. Kfoury, A.J., Tiuryn, J. and Urzyczyn, P., *The hierarchie of finitely typed functions,* 2nd LiCS (1987) 225–235.

30. Larsen, K. G., *Proof systems for satisfiability in Hennessy-Milner logic with recursion.* TCS 72 (1990), 265–288.

31. Larsen, K.G., *Efficient local correctness checking.* CAV '92.

32. Muller, D.E., and Schupp, P.E., *The theory of ends, pushdown automata, and second-order logic.* TCS 37 (1985), 51–75.

33. Stirling, C. P., and Walker, D. J., *Local model checking in the modal mu-calculus.* TAPSOFT '89, LNCS 351 (1989), 369-383.

34. Vergauwen, B., and Lewi, J., *A linear local model checking algorithm for CTL.* CONCUR '93, LNCS 715 (1993), 447–461.

35. Winskel, G., *A note on model checking the modal mu-calculus.* ICALP '89, LNCS 372 (1989), 761–772.

Methodology and System for Practical Formal Verification of Reactive Hardware

Ilan Beer, Shoham Ben-David, Daniel Geist, Raanan Gewirtzman and Michael Yoeli

IBM Science & Technology, Haifa, Israel

Abstract. Making formal verification a practicality in industrial environments is still difficult. The capacity of most verification tools is too small, their integration in a design process is difficult and the methodology that should guide their usage is unclear.

This paper describes a step-by-step methodology which was developed for the practical application of formal verification. The methodology was successfully realized in a production environment of hardware design. The realization involved the development of a system consisting of several tools, while using the SMV [McM93] verification tool as the system core.

This system was used in the verification of eight designs. We specifically elaborate on the verification of a bus-bridge design, which was particularly successful in uncovering and eliminating many hardware design errors.

1   Introduction

Most commercial formal verification packages are used for either verification of combinational logic or state machine comparison. Other packages employ theorem-proving methods to reason about systems, but their practical usage requires extensive user intervention. An important type of verification, which has attracted much attention in the academy, but is still under a limited use in the industry, is proving that a design satisfies certain properties which describe its desired *ongoing* behavior. This kind of verification is particularly applicable to *reactive systems* (cf. [Pnu86, MP91]), i.e., systems which continuously interact with their environment, such as process controllers.

Recently we have developed an industry-oriented formal verification methodology and a system which are especially suitable for the verification of reactive hardware designs. The system, which is based on advanced academic research tools, employs symbolic model-checking to verify behavioral properties of designs. It is interoperable with several industrial design environments and supports various hardware description languages. It also incorporates user-friendly means for error diagnostics. So far the system has been successfully applied in the verification of several hardware designs developed at IBM. It has detected many design errors which have been otherwise difficult to find. As a result, formal verification has become an integral part of the VLSI chip design methodology in the Haifa Design Group. We believe that the cooperation with several design teams has matured the system by focusing on industry-oriented considerations rather then research aspects.

This paper provides an outline of the formal verification activity at the IBM Haifa Research Laboratory. The rest of this section briefly reviews related work and background. Section 2 describes the verification methodology and Section 3 presents the main features of the system developed. Section 4 illustrates the application of the methodology and system to a real-life example. Section 5 concludes with more results, problems, and plans.

1.1 Related Work

In this section we survey two works which are related to our work in two main aspects: both are suitable for formal verification of reactive systems and both have been exercised in industrial environments.

Much of the theory relevant to this kind of verification, as well as its application, has been developed at Carnegie-Mellon University. These works are mainly based on temporal logic and model-checking (e.g. [CES83]). Our work was influenced by that of K. McMillan [McM93], who took a major part in the development of the theory of symbolic model checking and, based on it, implemented the SMV model checker. He used SMV to verify a protocol of an industrial computer, but did it within an academic framework.

A relevant methodology of considerable interest is presented in [Kur87]. An implementation of this methodology is provided by the verification tool called COSPAN (COordination-SPecification ANalyzer), which forms a part of the industrial verification system used at AT&T Bell Labs. COSPAN facilitates the development and formal verification of reactive systems, such as communication protocols. Within this methodology, system development starts with a high-level model and proceeds by successive refinements, until the final implementation is reached. Each refinement is accompanied by a formal proof. The COSPAN methodology is applicable to verification during development, while we are concerned with a posteriori verification.

1.2 Background

Temporal logic is a branch of formal logic, applicable to the verification of time-dependent systems. A survey of temporal logics can be found in [Eme89]. In this work we refer to **CTL** [CE81], a branching-time temporal logic. CTL is interpreted over infinite computation trees, representing all feasible execution sequences of a finite state machine. It allows formulation of a rich class of temporal properties, including safety and liveness. Examples of CTL formulas can be found in Section 4. ACTL [SG90, GL91] is a subset of CTL. It reasons about properties which are valid only if they hold on every path of the computation tree. As will be seen later, ACTL is useful in the context of abstraction.

Model checking is the process of verifying that a temporal logic formula holds w.r.t. a suitably represented finite state machine. CTL model checking complexity is linear in the formula length and exponential in the number of state variables. It is efficient compared to other relevant temporal logics such as LTL and CTL*, whose complexity is exponential in both formula length and number of state variables. Early CTL model checkers required an explicit representation of the complete state transition graph, which stressed the well-known state explosion problem. **Symbolic model checking** [McM93] provides efficient model checking facilities, without the need for explicit state enumeration. An improvement of the CTL model-checking algorithm provides for facilities to take into account only suitably selected paths by applying **fairness constraints**.

Reduction and Abstraction

Many real-life hardware designs are too large to be verified by straightforward model checking; even symbolic model checking may fail, due to the size of the design to be verified. Reduction and abstraction are used to alleviate this difficulty. Assume we are given a description of a hardware design M, and that we wish to verify that M satisfies a set F of formulas. **Reduction** is a construction of a simplified version of M, say M', such that $M' \models f \Leftrightarrow M \models f$ for every f in F. M' is called an exact reduction of M w.r.t. F; we also say that this reduction "strongly" preserves the properties of F. **Abstraction** is yet another way of simplifying a design. Simplification takes place by hiding internal details which are irrelevant to the verified property. Any ACTL formula valid in the abstracted design is also valid in the original one [GL91]. (Note that this property does not hold for CTL formulas in general.) Abstraction is a powerful tool because applying it may yield much smaller designs than when applying reduction. However, we are faced with a difficulty when the property of interest is not valid in the abstracted design.

2 Methodology

The methodology employed by our verification system is described in this section. It is based on concepts and methods described in [Bee92] and depicted in Figure 1. First, an overview is presented, and then we discuss some aspects which we find interesting.

2.1 Overview

The methodology suggests to follow these steps:

1. **Study of Expected Behavior**
   Unfortunately, it is not yet a common practice to supply formal specifications for hardware designs. Specification of the design, as well as the expected behavior of its environment, is usually provided by means of an informal description. Specifications can be obtained from several possible sources, such as published communication protocols, design documents and personal communication with designers.

2. **Partitioning**
   Formal verification systems have a limited capacity. The state explosion problem prevents verification of designs as a whole. Our solution is to partition the design and verify one part at a time. Usually designers develop their designs in a modular fashion and their partitioning is used for verification. When verifying one part,

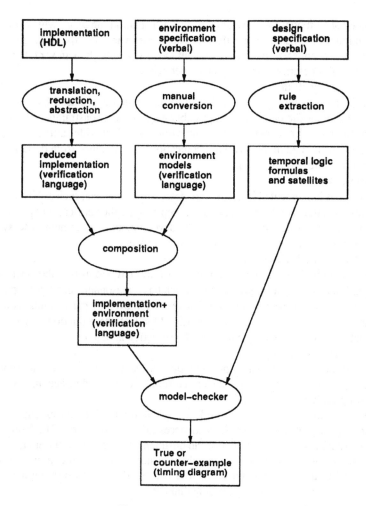

Fig. 1. Verification flow chart

abstract models of the others are treated as an environment. A specification of the interface between parts is obtained from the designers. In the sequel, when referring to a design, we mean a design partition. For more about partitioning see Section 2.3.

3. Environment Modeling

A hardware design should usually be verified in conjunction with its environment, which imposes restrictions on the feasible behavior of the design. Verifying the design on its own, may lead to finding errors in execution sequences which are not possible in reality. Out of the informal information provided about the environment, a formal model is generated. This model represents all possible interactions between the environment and the design, along their common interface. All other irrelevant activities of the environment are hidden by using abstraction techniques. We found it beneficial to model the environment in a verification language which provides abstraction mechanisms, rather than a general hardware description language. To increase the confidence in the environment models, we check them w.r.t. temporal-logic formulas describing their expected behavior.

4. **Formal Specification**

   The informal specification of the design to be verified is replaced by a finite set of CTL formulas (hereafter: rules), possibly with fairness constraints and satellites (see Section 2.2). Implementation-oriented rules, which provide additional require-ments not mentioned in the written specifications, are obtained from designers and transformed into CTL formulas.

5. **Translation**

   The design is implemented in a hardware description language, such as VHDL. This description is translated into a language acceptable by the verification system. The translation process depends on the actual development environment. The system described in Section 3 is an example of such a process.

6. **Reduction and Abstraction**

   Often, the design size after partitioning is still too large to fit into the verification system. Usually, only a fragment of it, which has a reasonable size, participates in the verification of a specific formula. The rest is removed by per-formula reduction and abstraction methods (see Section 1.2). Much of the reduction can be done automatically. If it is insufficient, manual abstraction is required.

7. **Actual Verification**

   The translated and reduced implementation is composed with the environment models and verified w.r.t. the specification formulas, using model checking facilities.

8. **Handling Failures**

   In case of a failure, the system produces, whenever feasible, an execution trace illustrating the problem. The cause is not necessarily a design error. The failure may be due to a mistaken rule formulation, an unsuitable abstraction of the environment, or the application of an inadmissible reduction step. Only after the failure has been pinpointed to an actual design error, it is presented to the design engineer in a suitably familiar form, such as a timing diagram.

Verification Modes

Two verification modes are defined: development and maintenance. During develop-ment, models are established, most of the rules are formulated, and the main reduction steps are performed. Model checking is applied until the design is successfully verified w.r.t. most rules. At this stage the design is relatively stable and maintenance mode begins. New versions of the implementation, obtained from designers, are translated, reduced and verified automatically. New rules are added if necessary. Development mode is the responsibility of the verification persons, while the maintenance mode can often be operated by designers.

2.2 Satellites

CTL formulas are not powerful enough to express all state machine properties which one would wish to verify. Furthermore, some properties expressible in CTL require formulas which are cumbersome to construct and difficult to understand. In many cases linear-time formulas are more adequate than branching-time ones. Although more expressive logics have been established (e.g. CTL* [Eme89]), relevant model checkers are inherently less efficient, and practical realizations are not available. To overcome some of these CTL limitations, the concept of a **satellite**, an extension of history variables [Cli73], was introduced in [Bee92]. A satellite is a small finite state machine connected externally to the design to be verified. It can read the design output at any moment but, to prevent the satellite from affecting the design behavior, the design does not read the satellite's state. A satellite records particular events of interest, to be retrieved later by CTL formulas. A concept similar to satellites has been introduced in [Lon93] as *observer processes*. Satellites resemble PLTL (linear time logic of the past [Eme89]), but their expressive powers are incomparable: satellites cannot capture the fact that an event occurred infinitely often (PLTL can), while they can easily capture that a property holds in time t *modulo* k (PLTL cannot). Using CTL formulas combined with satellites, one can reason about branching-time future and linear-time past. Satellites do not come for free. Their main disadvantage is the increase in the number of state variables, causing an earlier explosion of the state-space.

2.3 Partitioning and Modular Verification

Partitioning can be a problem if the verification of a property requires the presence of more than one partition. Modular Verification [GL91] is a powerful method dealing with such situations. Suppose that the design is a composition of the form: $M = M_1 \| M_2 \| ... \| M_n$. If we construct abstract models $A_1, A_2, ..., A_n$, such that A_i is an abstraction of partition M_i, then $A = A_1 \| A_2 \| ... \| A_n$ is an abstraction of M. If A is relatively small we can check it w.r.t ACTL formulas. If a formula is valid in A it is also valid in M (the opposite is not necessarily true).

2.4 More about Environment Abstraction

The informal description of the environment is frequently either partial or parametric, and it presents a variety of actions from which an actual environment is free to choose. It is essential that the formal model precisely reflects this situation: i.e., the model includes all feasible actions, but no others. Otherwise not all possible execution sequences of the design will be verified or, alternatively, non-feasible ones will be examined. To illustrate this point, assume that the environment informal description contains the statement *"State A always implies the eventual occurrence of state B"*. In this case the environment is free to arbitrarily choose any finite delay. The relationship between states A and B can be represented by the finite state machine M shown in Figure 2, where the transition from A to B is non-deterministic. To prevent an infinite self loop around state A, which means a (forbidden) infinite delay, we impose on M a fairness constraint "$GF(\neg A)$", which means that the environment is infinitely often not in state A.

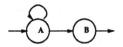

Fig. 2. Environment model example

3 Verification System

A system supporting the methodology of Section 2 is described below. The system is composed of existing tools and of new programs developed by us. Only its main features are described.

3.1 The SMV Model Checker

The core of the verification system is SMV - a symbolic model-checker developed by K.McMillan at CMU [McM93]. SMV is used for checking finite-state systems against specifications written in the temporal logic CTL. Its input language is very similar to other hardware description languages but, in addition to describing design implementations, it also allows to specify abstract, nondeterministic models. SMV employs symbolic methods: it represents logic functions, relations and sets of states as Binary Decision Diagrams (BDD [Bry86]).

To increase the capacity and speed of SMV, we augmented it with an algorithm for ordering the transition-relation partitions [GB94]. We are also using several heuristics to order BDD variables. These improvements allow us to verify larger designs than with the original system.

3.2 Translation

Our system deals with real design implementations, which are given in a Hardware Description Language (HDL). The system can read structural VHDL and IBM-internal HDLs. Existing compilers are used to translate the HDL files into gate level, which is then loaded into an IBM design-automation database for further manipulations, such as reduction. Finally this internal representation is translated into the SMV language for actual verification. The translation process is fully automated.

3.3 Reduction and Abstraction

Usually, the implementation is too big to fit into SMV. We therefore have to reduce it and extract only the parts relevant for verification. Different rules may refer to different parts of the design, thus the reduction made is based on the rule to be verified. The reduction process is semi-automatic: the user assigns attributes to interface signals (sometimes to internal signals also) and the system uses this information to reduce the design automatically. There are several possible attributes of signals. Some important attributes: *essential, constant* and *cutting* are described here.

The *essential* attribute is assigned to signals which either participate in a rule or must be observed by an environment model. The basic reduction algorithm leaves only cones of influence of essential signals, removing design components which are not necessary for the evaluation of these signals.

Signals whose exact contents are not supposed to affect the correctness of the checked properties can be assigned constant values. This is often the situation with some bits of wide input buses. A constant-elimination algorithm propagates the constants into the design, thus simplifying its structure by reducing the number of inputs and memory elements. This operation sometimes over-reduces the design and must be used with care, to avoid the elimination of feasible execution paths.

The reduction described so far is sometimes not sufficient and the reduced design is still too large. In these cases, an internal "surgery" is needed. This is done by assigning the *cutting* attribute to internal signals, thus making them design inputs. This technique can be used for the replacement of sub-blocks, whose internal operation is irrelevant to the verification of a specific property, by simpler abstract blocks with an equivalent external behavior. The cutting signals interface between the abstract blocks and the rest of the design. This operation is an abstraction, thus only ACTL formulas can be used, and a failure might occur even when the property holds in the original design.

Another reduction algorithm is used for detection and merging of memory elements which may become equivalent as a result of the previous reduction and abstraction steps.

3.4 Counter Examples

When the design fails to obey a rule, SMV presents a counter-example which can be viewed as a short test that exhibits the wrong behavior. The system translates the counter-example into a timing diagram which can be browsed interactively using an X-Windows based GUI. While browsing, the designer can inspect values of interface signals as well as internal signals. The timing diagram looks exactly like standard simulation results, thus giving the designer a familiar framework for debugging. A timing diagram example is shown in Figure 3.

The counter-example is also used to produce a simulation test that can be given as an input for the simulator. Designers use this test to reconstruct the error in their environment and to verify that they have corrected it. The test can also be used while checking the integrated design, when formal methods can no longer be applied because of size limits.

4   Project Example

Carmel is a communication bridge chip between two buses, which comply with different protocols, one of which is the PCI bus [PCI93]. The chip's function is to detect transactions in one bus which are addressed to devices on the other bus and to enable these transactions. Carmel was developed by the Haifa Design Group, IBM Israel. The implementation language was VHDL 1076.

Both formal verification and traditional testing by simulation were employed to verify Carmel. Formal verification was applied only after successful simulation of most test cases, and was used to uncover relatively subtle design errors. We were mainly interested in verifying Carmel's control logic, leaving the relatively simple data-path for simulation. Trying to verify the data-path might have caused an immediate state explosion. The control logic as a whole was too large to fit into the verification system. The designers partitioned Carmel, considering modularity, into several modules, and we used this partitioning as a basis for the formal verification process.

Table 1 summarizes relevant details of two of the partitions we verified. To demonstrate the influence of reduction on the speed of verification, we include two versions of each partition, with different degrees of reduction. It can be seen that a small decrease in the number of state variables can increase the speed significantly. In the table, $Before$ is the number of state variables (memory elements and inputs) before reduction, $After$ is the number of state variables after reduction, $Rules$ is the number of rules verified, $Bugs$ is the number of design errors detected, $RunTime$ is the CPU-time of a typical rule (in seconds, on a Risc System 6000 model 550), and BDD is the number of BDD nodes allocated by the model-checker.

Table 1. Partition details

Function	Before	After	Rules	Bugs	RunTime	BDD
PCI slave[1]	663	74	55	30	1600	1820K
PCI slave[2]		70			550	660K
PCI master[1]	748	76	40	16	9500	1840K
PCI master[2]		63			190	243K

The errors we found were about 40% of all errors detected throughout the entire verification of the chip, including data-path errors. Some designers had initially been skeptical towards formal verification, but they became enthusiastic after receiving the first error reports. They also appreciated the debugging aids provided by our system.

Carmel has been fabricated and tested extensively. No bugs have been found yet.

4.1 Rules

Many rules concerning the bus interfaces have been extracted from published bus standards. These rules are implementation-independent and we used them again in other designs which interfaced these types of buses. More rules, mainly implementation-dependent ones, have been added by the designers. Some were refinements of existing rules, such as addition of timing constraints. Others verified the internal behavior of partitions by referring to internal signals. The following examples exhibit interesting aspects of rules:

Long Counter-Example

In some cases, counter-examples which demonstrated design errors were considerably long. Recalling that a counter-example is a near-shortest simulation test which can detect the error, it is clear that the probability that one will find these errors is low, if one is using traditional methods. The following example, involving the PCI protocol, illustrates such a case: a rule states that *"In a case of target-retry, the master must retry the transaction within 3 time cycles"*. (This is an implementation-specific requirement that refines a generic PCI rule.) *Target-retry* is characterized by *Devsel* and *Stop* low while *Trdy* is high. *Frame* transition from high to low denotes transaction beginning. The rule is represented by the formula: $"SPEC\ AG((\neg Devsel \land \neg Stop \land Trdy) \rightarrow ABF\ 1..3\ Fall(Frame))"$. *Fall* is a transition-detecting satellite and *ABF* is a bounded-future operator. The resulting counter-example is displayed in Figure 3. It shows that a very special sequence of events must be generated in order to uncover the problem. (For those who are familiar with PCI: three back-to-back transactions and two target-aborts are required before the problem shows up.) The problem hiding behind this simple symptom was a serious one, and required major design modifications.

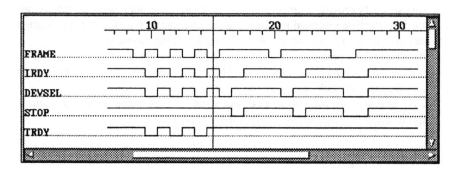

Fig. 3. Counter example timing diagram

Using a Satellite

One of the requirements was that at most K data units are stored in the bridge at any time. Another requirement stated that the bridge cannot write out more data than it reads in. To verify these requirements, we defined a satellite that counted the number of data-read operations minus the number of data-write operations, and then wrote a CTL formula asserting that the counter's value must neither exceed K nor be below zero.

5   Further Experience

In addition to the example of Carmel, discussed in Section 4, we have participated so far (January 94) in the verification of seven other designs, four of them have already been produced. We have found dozens of errors in most of the designs. Many of them, such as deadlocks and other kinds of hard-to-produce situations, were very difficult to find by conventional simulation. After a short experience with the system, even the most skeptical designers gained faith in our methods and cooperated with enthusiasm. Formal verification is now an integral part of the design methodology in the Haifa Design Group.

Observations

- We found the system and methodology particularly useful in the verification of designs with limited influence of the data on the flow of control.
- Experiments with asynchronous designs, or with multiple unsynchronized clocks, resulted in an early state explosion.
- No correlation was found between the complexity of rules and the types of errors they have found: simple rules discovered symptoms of very subtle errors.
- Automated BDD variable ordering is a necessity in an industrial system, because of the frequent design modification and the dependency of the order on the per-formula reduction.

Problems and Plans

- Capacity limitation is a major problem in every formal verification system. Currently we can verify only one partition at a time. To increase our system's capacity, we are investigating several directions, such as improved compositional verification methods, reduction algorithms, and BDD variable ordering methods.
- Another serious problem is the difficulty to decide when a set of CTL formulas, each of them representing a partial specification, covers the whole informal design specification. We seek criteria of adequate coverage.
- CTL formulas are difficult to read and write. We are planning a user friendly specification language on top of CTL and the satellites.

6 Acknowledgements

We thank the designers of the Haifa Design Group, whose cooperation contributed to the maturity of both the methodology and the system. We also thank Israel Berger and Aharon Aharon for supporting this work, and to Yossi Lichtenstein and Orna Grumberg for reviewing this paper.

References

[Bee92] I. Beer, "Formal Verification of Hardware", M.Sc. Thesis, EE Department, Technion, 1992 (in Hebrew).

[Bry86] R.E. Bryant, "Graph-Based Algorithms for Boolean Function Manipulation", IEEE Trans. Computers, Vol. C-35, August 1986.

[CE81] E.M. Clarke and E.A. Emerson, "Design and Synthesis of Synchronization Skeletons using Branching Time Temporal Logic", in: Proceedings of the Workshop on Logic of Programs, LNCS 131, 1981. Temporal Logic Specifications: A Practical Approach", Tenth ACM Symposium on Principles of Programming Languages, Austin, Texas, 1983.

[CES83] E.M. Clarke, E.A. Emerson and A.P. Sistla, "Automatic Verification of Finite-State Concurrent Systems using Temporal Logic Specifications: A Practical Approach", Tenth ACM Symposium on Principles of Programming Languages, Austin, Texas, 1983.

[Cli73] M. Clint, "Program Proving: Coroutines", Acta informatica, 2(1), 50-63, 1973.

[Eme89] E.A. Emerson, "Temporal and Modal Logic", in: Handbook of Theoretical Computer Science, J. van Leeuwen, ed., North-Holland, 1989.

[GB94] D. Geist and I. Beer, "Efficient Model Checking by Automated Ordering of Transition Relation Partitions", submitted for publication, 1994.

[GL91] O. Grumberg and D.E. Long, "Model Checking and Modular Verification", in: LNCS 527, 1991.

[Kur87] R.P. Kurshan, "Reducibility in Analysis of Coordination", in: LNCS 103, 1987.

[Lon93] D.E. Long, "Model Checking, Abstraction and Compositional Verification", Ph.D. Thesis, CMU, 1993.

[McM93] K.L. McMillan, "Symbolic Model Checking", Kluwer Academic Publishers, 1993.

[MP91] Z. Manna and A. Pnueli, "The Temporal Logic of Reactive and Concurrent Systems; Specification", Springer-Verlag, 1991.

[PCI93] "PCI Local Bus Specification, Revision 2.0", PCI Special Interest Group, 1993.

[Pnu86] A. Pnueli, "Applications of Temporal Logic to the Specification and Verification of Recative Systems: A Survey of Current Trends", in: Current Trends in Concurrency, J.W.de Bakker et al., eds., LNCS 224, 1986.

[SG90] G. Shurek and O. Grumberg, "The modular Framework of Computer-Aided Verification: Motivation, Solutions and Evaluation Criteria", CAV90.

Modeling and Verification of a Real Life Protocol Using Symbolic Model Checking

Vivek G. Naik and A. P. Sistla

Department of Electrical Engineering and Computer Science, University of Illinois at Chicago, Chicago, IL 60680

1 Introduction

As the computing systems have grown in size and complexity it has become necessary to develop automated methods for checking the correctness of such systems. *Temporal logic modelchecking* [2] is one of such automated methods for verifying properties of finite state systems. The practical applicability of the original modelchecking system was limited due to the state explosion problem. Recently many techniques have been developed to overcome the state explosion problem. One of the methods that has been finding much application is symbolic modelchecking [1, 8, 3, 6]. The symbolic modelchecking approach, implemented as the SMV system, uses BDDs for symbolically representing sets of states and the transition relation. This approach allowed the possibility of handling systems with extremely large state spaces.

In this paper, we show how symbolic modelchecking has been used to verify a real life protocol. Specifically, we have used SMV tool to model and verify IEEE 802.3 Etherenet CSMA/CD protocol with minimal abstraction. The Etherenet CSMA/CD protocol is a protocol that allows a set of computer systems connected over a local area network to communicate with each other. The major steps involved in using the SMV system for verification of the protocol were to correctly identify the processes within the protocol, to model them in the SMV toolkit, and to specify and verify the required properties of the protocol. Some design issues while modeling such a protocol are also dealt with in the research.

We have verified the protocol under the asynchronous and synchronous models. The major problems encountered in using the SMV system were in modeling of the following aspects associated with the protocol: the *channel, collision detection* and *carrier sensing, delay modeling* (delay is used in successive attempts after a collision using the exponential backoff approach) and *synchronization of transmitters and receivers*. We first modeled the protocol at much detail and checked the properties. Under these two models, we used progressive abstraction to reduce the number of variables in each transmitter and receiver, and thus reduce the time taken for modelchecking. We have verified many properties for different stations, for various values for different values of maximum number attempts and frame sizes.

This paper describes the appraoches employed in the verification purpose. The paper is organized as follows. Section 2 briefly describes the SMV system and

---

[1] This research is partially funded by NSF grant CCR-9212183.

the specification logic CTL. Section 3 gives an introduction to the Ethernet IEEE 802.3 protocol and a formal model of it. Section 4 specifies various problems encountered in modeling in SMV and how they were solved. It describes various components of the protocol. Section 5 lists various correctness properties of the protocol that were specified and verified using SMV, together with the times taken for verification. Section 6 contains concluding remarks.

2 The SMV tool

The inputs to the SMV system consist of the description of transition system modeling the concurrent system and a correctness specification. The correctness specification is a formula of the branching time temporal logic CTL (Computation Tree Logic) [2] . This logic allows the specification of various safety and liveness properties that are of interest to concurrent systems. CTL is a propositional branching time temporal logic.

The input language of SMV allows the description of the state transition system as modules. Each module has a set of parameters that can be instantiated and reused. Thus if the system has many similar components then all of them can be defined as instantiations of a single module definition. It also provides for a hierarchical description of the system. The data types available are Booleans, scalars and fixed arrays. The language allows a parallel assignment syntax. The reader is refered to the [7] for a detailed description of SMV.

3 Ethernet Protocol

3.1 Informal Description

Modern Computer Networks are designed in a highly structured way. A seven layered model was proposed by the International Standards Organization (ISO) as a first step towards international standardization. The layered approach has been taken with the fact that each layer provides some primary service to its upper layer thus making their implementations and design independent of the other layers as long as the services needed are provided. IEEE's 802 standard for local area networks is the key standard for LANs.

Description of the CSMA/CD LAN Protocol

The IEEE 802.3 (from now referred as ethernet) protocol is a Local Area Network(LAN) communication protocol. This standard covers the Physical Layer and Medium Access Control sublayer which is a part of the Data Link Layer. The Data Link Layer sits above the Physical Layer and provides services to the Network layer.

This protocol is based on the concept of ALOHA system developed in 1970s by Norman Abramson and his colleagues at University of Hawaii. The system gives an elegant method for the allocation of a shared channel by multiple users. The users share the single communicating channel. A user sends data in the form of a stream of bits which is called a frame. The basic idea of the original protocol is to let users transmit whenever they have data to send. If collisions occur then users try to transmit the data after a random delay. A much improved version of the above protocol is the CSMA/CD protocol (Carrier Sense Multiple Access wit Collision Detection) protocol. In this protocol, whenever a station wants to send data it first checks if the channel is busy (i.e. if any one else is currently transmitting); If the channel is not busy then it goes ahead with transmission of the data. If the channel is busy, then the station waits until the channel is idle and then transmits the data. Collisions can occur if two stations try to transmit at the same time or with in a short duration of time determined by the propagation delay of the channel. This protocol incorporates a collision detection mechanism. Whenever collisions occur, all transmitting stations are notified of the collision. Rather than finish transmitting their frames, which are irretrievably garbled anyway, the transmitting stations that detect collisions would abruptly stop transmitting and go onto a phase of post collision arbitration. This improves the overall performance.

Post collision operation

The first station to detect the collision aborts transmission and transmits a short noise burst. This noise signal is the jam signal which indicates to all the other stations that there has been collision. The station then waits for a random amount of time and repeats the cycle. After collision the time interval is divided into slots of a period of 2τ, where τ is the one way propagation delay of data transmission. The propagation delay is divided into a slot time of 512 bits (51.2 μsec).

After the first collision each station waits for 0 or 1 slot times before trying again. The number of slots for which a station is going to wait depends upon a random number selected by that station. If two stations pickup the same random number then there will be another collision in their next attempt at transmission. After the second collision the stations wait for a random period of between 0, 1, 2 or 3 slot times and try again. If a third collision occurs then the random number picked will a value between 0 to $2^3 - 1$. In general, after i collisions a random number between 0 to $2^i - 1$ is chosen. After 10 consecutive collisions have been reached then the randomization interval is frozen to a maximum of 1023 slots. After 16 collisions the controller gives up and reports the failure to the upper layer.

This algorithm of dynamically choosing the delay is called *Binary Exponential Backoff*. This mechanism ensures that the delay time adapts to the number of stations involved on collision. This mechanism ensures that collision is resolved in a reasonable interval if many stations collide.

The protocol as such doesn't provide any mechanism for acknowledgement of received frames. Thus the destination station should verify for the checksum and then send the acknowledgement frame if the data received is error free.

3.2 Formal Specification of the protocol

A formal specification of the Ethernet protocol is given in in [11]. In this model each station consists of a set of processes communicating through shared variables. Each process is modeled as a timed transition system with upper and lower bounds on each transition. This specification is much clearer and more readable than the informal specification of Ethernet in IEEE 802.3 [10, 4, 9]. We use this as the basis in our verification.

The various processes at each station and the data flow between these processes is given in figure 1. Each process performs a particular function. The arrows in the figure indicate the communication of variables that are being modified/shared by the processes. The LLC sublayer sends raw data frames to the MAC layer. In the model the actual frame is not sent but the same effect is achieved by the LLC layer just setting a variable to indicate that the MAC layer can start transmission. The MAC layer consists of the following subprocesses.

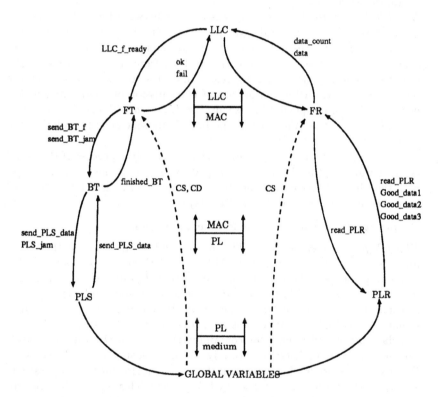

Fig. 1. Basic communication model of the Ethernet Protocol

1. *FT (Frame Transmitter)*: This process takes input from the variable *LLC_f_ready* to indicate that the frame is ready for transmission. It checks for availability of the carrier and if it finds it to be not busy, then it sends a *send_BT_f* signal to the bit transmitter. Then it waits for a *finished_BT* signal from the Bit Transmitter. While waiting, if it encounters a Collision Detection signal then

it sends a *send_BT_jam* signal to the BT and waits for *finished_BT* signal. If there is no collision detect signal till it receives the *finished_BT* signal then it sends an *ok* signal to the LLC indicating that the data transmission has been successful. If there had been a collision the process increments the variable indicating the number of attempts, picks up a random number according to the binary exponential backoff algorithm. It then sets the delay variable to the random number. After waiting for the amonut of time given by the delay, the process returns to the state in which it checks if the carrier is free. If collision occurs again then the process executes the above described steps till the number of attempts is less than maximum. Once the maximum limit is reached then the process terminates in the fail state indicating failure of transmission of the frame.

2. *BT (Bit Transmitter)* : This process is the link between the MAC layer and the processes in the Physical layer processes. The BT, upon receiving the signal *send_BT_f*, sends the *send_PLS_data* signal to the Physical Layer Sender (PLS) process to transmit the next bit from the data frame. If BT receives a *send_BT_jam* signal, then it indicates to the PLS to send the jam sequence bits. After completion of transmission of all the bits in the frame or the jam sequence, BT sends the *finished_BT* signal to the FT. This processes keeps track of the number of bits sent and the index of the next bit to be sent.

3. *FR (Frame Receiver)* : This process is the data receiver process of the MAC layer. It waits for the Carrier Sense signal to be true. It then communicates with the Physical Layer Receiver(PLR) and receives the data bits. The data frame received is then sent to the LLC layer.

The physical layer consists of the following processes:

1. *PLS (Physical Layer Sender)* : This process is the link between the transmission medium and the BT of the MAC Layer. Each station has a variable containing the next data bit to be transmitted. Upon receiving the signal from the BT to transmit data bit, PLS checks if the data to be transmitted is a jam bit or a data bit (this depends on the status of the *send_BT_jam* signal). Accordingly the data bit is set to one of $\{0,1,id\}$ or to $\{J\}$. After the last data bit, an additional bit containing the value ND(no data) is sent; this provides spacing between successive frames transmitted on the channel. Other variables needed for synchronization of the Read and Write operations among all the stations are also set by this process. This topic is discussed in more detail in the Section 5.

2. *PLR (Physical Layer Receiver)* : This process is the receiver section of the Physical Layer. It receives data from the medium (we model them as global variables). When it receives a signal from the FR to read the data in, it reads it and stores it in the buffer. This process reads the data from the variable called the *chnl_data*. This variable is set using all the data bits of the transmitting stations in the model and its value represents the data on the channel. Extra variables are used to synchronize reading of the channel data with the sending of the data by the transmitting station.

The processes FT, BT and PLS form the transmitter part of a station. The processes PLR and FR form the receiver part of a station.

4 Modeling the protocol in SMV

The SMV uses for its input a transition based model of processes communicating to each other through shared and global variables. Thus, it is particularly suited for verifying the Ethernet protocol. However, the following major problems were encountered in modeling the protocol in SMV. Many of these problems are due to the lack of representation of time in SMV.

4.1 Issues in Modeling the protocol in SMV

1. *Channel representation*: The channel through which the stations communicate as we have seen in the previous sections consists of a stream of bits. These bits move at the speed of transmission that is 10 megabits per second. The propagation delay on the channel results in the stream of data bits to be unavailable at the same instant to all the stations on the network. To avoid modeling the propagation delay the data channel is assumed to be one bit long. This part of specification as given in [11] is a real time specification, and it needs to be modified so that it can be handled by SMV.

2. *Simulation of Transmission*: The rate at which the data is transmitted in the protocol is fixed, and the model in [11] achieves this by introducing a fixed time delay between the consecutive bits that are transmitted. The time delay between transmission of successive bits ensures that the receiving stations read the data before sending of the next bit. We achieve this by synchronizing the transmission and reception by using variable arrays.

3. *Processes with time constraints on their transitions*: Some of the processes have time bounds on their transitions. As in the case of transmission of data, the time bounds on the transitions of processes ensure that, for a shared variable between two processes, the previous value of the variable would have been read by the second process before the variable is updated again (most of the processes communicate data in a pattern in which the writing action of one process is followed by a reading action of the other). To get this same effect in the model without time bounds on the transitions it is necessary to validate that the reader has read the previous value before the writer writes again. This involves additional message passing. The technique used is similar to the transmission of data between stations and is discussed in the next section.

4. *The computation of Random Delay*: In the Binary exponential backoff delay mechanism after collision each station waits for a random amount of time slots. On the network it is imperative that if only one station picks up the smallest random number of all the stations involved in the collision, then it is bound to transmit successfully. To ensure that this behavior is preserved in the SMV model, it becomes necessary to have a centralized delay control

mechanism. This would take care of the post collision arbitration.

The following sections describe the basic modeling approach and the processes that effectively model the protocol.

4.2 Modeling the different Aspects in SMV

As indicated before, we considered two different method of modeling a set of stations using the Ethernet protocol. The first method is the asynchronous method where all stations and all processes in each station are modeled as asynchronous processes. The second method is where all stations are modeled as synchronous processes. In this method, the whole station is a single process.

Asynchronous Model

To enable communication between the processes belonging to a station module within themselves and with the other station modules in the model, we define global variables which are shared by all the processes. The variables include those which represent the status of data channel and those which are used for synchronization between the processes to ensure correct behavior.

Various components of the protocol are modeled as follows.

– *Data Channel modeling:* The data Channel was one of the complex aspects to model in the transition system. As mentioned in the section 4.1, some assumptions are made in modeling the channel. The most significant one is that the channel is assumed to be just one bit length, as opposed to that in real life as a stream of bits whose bit length depends upon the physical length of the cable. The data channel is assumed as a set of data bit variables with one data bit for each station on the network. Each Data variable is defined as follows:

$DATA : \{0, 1, 2, J, ND\}$;

These values represent the values that this variable takes. A value of { 0, 1, 2 } indicates good data bits. If the value of the data variable is J then it indicates that the station is writing jam sequence. The value of ND indicates that there is no data. The number of data variables depends on the number of stations in the system. Each station writes into the data variable which corresponds to its id number. The actual data on the channel is given by the variable *chnl_data* which is the composition of all the data variables. In a system with two stations the data on the channel is the composite of the data variables DATA1 and DATA2. The definition of the *chnl_data* is as given below.

```
chnl_data :=
    case
            DATA1 = ND  : DATA2;
            DATA2 = ND  : DATA1;
            1           : J;
    esac;
```

The above DEFINE takes care of the fact that the *chnl_data* has the valid data depending upon the value of the individual data variables. The third assignment sets it to J if the both of the data variables are not ND. Thus the *chnl_data* will always represent the composite of the two stations data bits.

— *Carrier Sense (CS) and Collision Detect (CD) variables*: The Carrier Sense (CS) variable which represents the status of the Channel is defined as follows:
CS := !(DATA1 = ND & DATA2 = ND);
This sets the variable CS to 1 if it is not the case that both the data variables are not ND. Thus a busy channel status is indicated by this variable when it is set.
The Collision Detect (CD) variable which is set to 1 if there is collision on the channel. This is achieved by the following definition:
CD := ! (DATA1 = ND | DATA2 = ND) | (DATA1 = J | DATA2 = J);

— *The Read Array to synchronize the Transmitters and the Receivers*: The data variables, mentioned above, are set by the transmitting side of the station, and the receiving side reads from the *chnl_data* variable. In the real life systems there is a time bound on each of the transitions which occur in the transmitting section and the receiving section. The rate at which the Receiver reads is same as the rate at which the Transmitter sends. Thus the Receiver will never read the same data bit twice from the channel and the Transmitter will not overwrite the last written data. Since we are modeling all processes as completely asynchronous and since we don not have time in our model, we achieve the synchronization between transmitters and receivers by using extra variables.

We use an $n \times n$ array *Read* of binary variables to synchronize the transmitters and receivers. Transmitter i sets the bits in the ith row to zero after writing on the channel and will not write again until all these bits are set to 1 by the receivers indicating that they have read it. Receiver set these bits to 1 after they read the data. Since all receivers run asynchronously we need one bit per receiver; Also, since the transmitters can try to transmit at the same time we need one row of bits per each transmitters. We also have used a model where only a one dimensional array of n bits are used. However, this model does not accurately depict the complete parallelism among the transmitters.

— *Modeling of Delay after collision:* The ethernet protocol uses a Binary Exponential Backoff algorithm for handling the post collision arbitration. When two stations get into collision then each one of them picks up a random number which depends upon the current attempt for sending data. Each station then waits for an amount of time given by the delay chosen by it, and tries to transmit again. Since the clocks at different stations are synchronized to run more or less at the same rate, the station to pick up the least delay will attempt to transmit first in the next try. Clearly, in an asynchronous model, we canot implement this by using an obvious count down of the delay variable. We use a simple centralized delay monitoring mechanism for achieving the above effect.

Each station has its local variable *delay* which is set by the Frame Transmitter process. A value of 0 for this variable indicates that the corresponding station is not in an arbitration, i.e. not waiting to transmit. The Frame Transmitter process sets the *delay* variable according to the binary exponential backoff algorithm and waits for its *stn_go* signal to be set to true. The *stn_go* signal is globally defined for each station. This signal for station 1, in a two station configuration, is defined as follows.

stn1_go := !(stn1.FT.delay = 0) & !(stn2.FT.delay = 0) & (stn1.FT.delay <= stn2.FT.delay)

This definition ensures that the *stn1_go* is set to true only if both the *delay* variables are not zero and the third condition is true. The station which gets its *stn_go* signal to be true then goes ahead to transmitting data and sets the *delay* variable to 9 which then allows the other station to proceed. This further ensures that if this is the case only one station gets the *stn_go* signal then, it will transmit successfully next.

Synchronous Model

In the synchronous model, we do not need the *Read* array for synchronization purposes. Each value written by a transmitter in a clock cycle is read by all the receivers in the next clock cycle. No centralized delay monitoring system was needed. Essentially each transmitter decrements its delay counter in successive clock cycles, and when the delay becomes zero it tries to transmit again. Thus, this delay mechanism, in the synchronous model, is closer to reality than in the asynchronous model. However, in the synchronous model, each step of all the transmitters is synchronized which is not exactly the case in real life. These are the only differences between the asynchronous and the synchronous model.

5 Results and Inference

We were able to verify various properties in the asynchronous as well as the synchronous model. While testing the asynchronous model, we found some errors due to the fact that in the real life protocol there is an inherent assumption about the frame level synchronization which we did not model. We had to change our model appropriately to take this into consideration.

In the synchronous testing we used only one receiver since all the receivers are identical. This receiver is used to check that the transmitted frame is correctly received. We tested for different number of transmitters. We checked for the cases when the number of attempts is 2 and 4. We also checked for a frame size of 3 bits and of one bit. We checked for two properties. The first property asserts that whenever the LLC layer requests the transmission of a frame then eventually the MAC layer (i.e. frame transmitter) responds with a success or a failure message. This property for station 1, called property 1, is expressed by the following CTL formula:

$AG(LLCfready1 = 1 \rightarrow AF(ind1 = success \vee ind1 = fail))$.
Here $ind1$ is the variable through which the MAC layer sends a successful or failure message to the LLC layer. The amount of time taken for different parameter values is given in the table 1.

Table 1. Table of Results for synchronous models for checking property 1

Model Configuration			reachable	relation	time sec
transmitters	attempts	frame size	states	nodes	time
3	2	1	10K	13.3K	287
4	2	1	280K	34.9K	5839
3	4	1	77K	13.3K	3905
3	4	3	750K	140K	20,987

The second property that we checked asserts that the frame received by a receiver is the correct frame. In the synchronous model this is asserted as an invariance property. The following formula asserts this. In this, the predicate $trans1.FT.state1 = done$ indicates that transmitter 1 reaches a done state while the predicate $Gooddata1$ indicates that the frame buffer in the receiver denotes a good data frame from station 1.
$AG(trans1.FT.state = done \rightarrow Gooddata1)$
The timing results for property 2 are given in the table 2. The times given in this table include the time taken for computing the number of reachable states (usin the -r option) and this later time completely dominated the over all time taken for checking this assertion. When this option is removed the checking of this property was extremely fast. In fact, for the case of five transmitters the time reduced to 85 seconds when the -r option was not used.

In our asynchronous model, each station has all the processes on the transmitter side and the receiver side. Also each frame has three bits. We checked for the following properties. The first property (property 1) asserts that each station will eventually reaches a done state or a fail state. This property is asserted by the following CTL formula:
$AF(stn1.FT.state = done|stn1.FT.state = fail)$

The second property asserts that if a station reaches a success state all other stations must have received the frame sent by it. This is asserted by a CTL formula of the form $AG(Stn1.FT.state = done \rightarrow AF\ proper\text{-}reception)$
Here *proper-reception* is a state predicate on the receivers asserting that they received correct frame station 1. It is to be noted we needed an AF modality inside due to the asynchrony.
The third property we checked is that in a two station system whenever there

Table 2. Table of Results for synchronous models for checking property 2

Model Configuration			reachable	relation	time sec
transmitters	attempts	frame size	states	BDD nodes	time sec
3	2	1	10K	13.3K	23
4	2	1	280K	34.9K	247
5	2	1	7,500K	93.4K	3724
3	4	1	77K	13.3K	116
4	4	1	3950K	34.9K	2787
3	4	3	750K	140K	618

is a collision, then both stations reach a done state (successful transmission), or both states reach a fail state. This is expressed by a formula of the form $AF(p)$ where p is a state predicate.

The fourth property that we checked is that, in a two station system, in case of a collision the station picking up the lower delay will successfully transmit. This is expressed as a formula of the form
$$AG(station1\text{-}go \rightarrow AF(stn1.FT.state = done)$$

We checked the above properties for the asynchronous system with configuration of two and three stations in which one or two stations are active. A station is active if it is allowed to send messages. The results are given in table 3.

Table 3. Table of Results for the asynchronous model

Model Configuration		property	reachable	relation	time sec
stations	active stns	property	states	BDD nodes	time sec
2	2	1	701K	24K	51,500
2	2	2	701K	24K	52,000
2	2	3	701K	24K	59,466
2	2	4	701K	24K	62,400
3	2	4	1500K	64K	59,200

As can be seen from table 3 that verification of different properties in the asynchronous model, the modelchecking took more time. We believe that this is due to the inherent complexity of the protocol. We were able to reduce the times by 40%, for some of the cases of the table, by changing the variable ordering.

After modelchecking using a detailed model, we deleted lot of detail in the protocol and also removed the receiver part and modelchecked for the reduced system. In this system we checked for the property 1 given in the previous table. The number of reachable states reduced substantially and we were able to check for two station system with four attempts much faster, i.e. in 1600 seconds (approximately, 25 minutes).

6   Conclusion and Future work

In this paper, for the first time, we verified various properties of a detailed model of the Etherenet Protocol using symbolic modelchecking. The model for the protocol has been developed in stages, and the verification process identified some problems in our modeling. The difficulties in modeling were partly due to the absence of real time in the model checker. Solutions to these problems needed use of additional data structures, to preserve the correct behavior of the protocol. As an example, while developing the model it was found that the transmitter could possibly transmit before the receiver has reset its data buffers. This needed modifications in the model to transmit after all the receivers have reset.

The centralized delay monitoring mechanism used in the asynchronous model works for the case when there are two active stations. It can be modified to work for an arbitrary number of stations. In this case, system allows the station with smallest delay to transmit by setting appropriate flag, and decrements the delay variables of other waiting stations by th delay of the chosen process.

A more general approach for verifying properties of timed trantion systems under the discrete time model, is to transform the timed system into an untimed system by using extra time variables to keep track of the times for which each of the transitions has been enabled. In this case, we need to add another process/transition that models the clock and increments the time variables. The lower bounds bounds associated with each transition can be enforced by adding additional conjucts to the enabling conditions of the transitions. The upper bounds on the transitions are enforced by enabling the clock transition only when all the time variables obey the upper bounds of the corresponding transitions. Also, the time variables need to be reset to zero, in the action parts of each transition, whenever the corresponding transition is disabled or whenever the corresponding transition is executed. All of this can be done by a simple syntactic trnasformation of the transition system, and this can be automated. The SMV system can be used on the transformed system.

Our conclusion is that real life protocols can be verified using the SMV system. As indicated in the paper, our verification under the asynchronous model was done using a fairly detailed description of the system. The CTL specifications, which are used for model checking on the model, cover a wide range of properties. We believe that we can make the modelchecking faster by using dynamic variable reordering. This we expect to do in future.

Future work: The model which was developed is symmetric in the sense that there are more than one instances of the same module used in the model. Thus symbolic model checking with symmetry may make the verification faster. Secondly the SMV system presently doesn't have real time in its language. We feel that model checker for real time systems would model the problem more accurately.

References

1. J. R. Burch, E. M. Clarke, K. L. McMillan, D.L. Dill, "Symbolic Modelchecking: 10^{20} states and beyond", In Proceedings of fifth annual Symposium on Logic in Computer Science, June 1990.
2. E. M. Clarke, E. A. Emerson and A. P. Sistla. "Automatic verification of finite-state concurrent systems using temporal logic." *ACM Trans. Program. Lang. Syst. 8*, 2, (April 1986), 244-263.
3. E. M. Clarke, J. R. Burch, O. Grumberg, D. E. Long and K. L. McMillan. "Automatic verification of Sequential Circuit Designs." *Royal Society of London*, October 1991.
4. J. D. Day and H. Zimmerman. "The OSI reference model." In *Proc. of IEEE*, volume 71, pages 1334-1340, December 1983.
5. O. Lichtenstein and A. Pnueli. "Checking that finite state concurrent programs satisfy their linear specification." In *Conference Record of the Twelfth Annual ACM Symposium on Principles of Programming Languages*, 1985.
6. K. McMillan and J. Schawalbe. "Formal verification of the Encore Gigamax cache consistency protocol." In *International Symposium on Shared Memory Multiprocessors.*, 1991.
7. K. L. McMillan. "The SMV system " February 1992.
8. K.L. McMillan. "Symbolic Modelchecking: An approach to state explosion problem" Ph.D. thesis, CMU-CS-92-131, may 1992.
9. A. Tanenbaum. *Computer Networks.* Prentice Hall, 2nd edition, 1989.
10. ANSI/IEEE std. *Information Processing Systems- Local Area Networks- Part3: Carrier sense multiple access with collision detection (CSMA/CD) access method and physical layer specifications.* The IEEE, Inc., NY, October 1991.
11. H. B. Weinberg and L. D. Zuck. "Timed Ethernet: Real-Time Formal Specification of Ethernet" In *Proc. 3rd CONCUR*, August 1992.

Verification of a distributed Cache Memory by using abstractions *

Susanne Graf

VERIMAG **, Rue Lavoisier, F-38330 Monbonnot, e-mail: graf@imag.fr

Abstract. The purpose of this paper is to verify a distributed cache memory system by using the following general verification method: verify the properties characterizing a complex system on some small finite abstraction of it, obtained as a composition of abstractions of each component of the system. For a large class of systems including infinite state systems, the abstractions of the components can be obtained by replacing all operators on concrete domains by abstract operators on some abstract domain. This holds also for the abstraction of the control part of the system as we consider a kind of guarded command programs where all the control is expressed in terms of operations on explicit control variables.

1 Introduction

The purpose of this paper is to show the practical applicability of the verification method proposed in [LGS+92, GL93a, GL93b, CGL91, Lon93] for infinite state systems. This verification method, based on the principle of abstract interpretation [CC77], proposes to verify a program defining some complex system, where the specification must be given in the form of a set of $\forall$CTL* [SG90] formulas, as follows: define an appropriate abstract program, obtained compositionally from the the given program, and verify the required properties on it. Our way of computing abstract programs is very similar to that proposed in [CGL91, Lon93], but our concept of compositionality is different from that proposed in [Lon93] or in [Pnu85]. We construct a global abstraction of the system by composing abstractions of its components, whereas the other method consists in deducing properties of the composed system from properties of its components. Both approaches are useful, but in the example we treat in this paper, the global properties cannot be deduced easily from properties of the components. An abstraction of each component is obtained applying the principle of abstract interpretation by means of a relation ϱ relating the domain of its variables and the domain of the set of some abstract variables.

In [Loi94] is described a tool allowing to verify finite state systems in a fully automatic way by using this method. Here, we show that the same method is also tractable in practice for infinite state systems where a complete automatization is not possible. In fact, if — depending on the formula one wants to verify — for each component P_i one can guess an appropriate abstraction relation ϱ_i verification becomes often a relatively simple task as

- the corresponding finite state abstract program is reasonably easy to obtain,
- the verification of the properties on the obtained abstract program can be done fully automatically.

In Section 2, we recall all the ingredients we need for our verification method:
- a simple program formalism similar to that used e. g., in [Pnu86],

---

* This work was partially supported by ESPRIT Basic Research Action "REACT"

** Verimag is a joint laboratory of CNRS, Institut National Polytechnique de Grenoble, Universitée J. Fourier and Verilog SA associated with IMAG

- a method to compute abstract programs, consisting in defining for each operator on the concrete domains a corresponding *abstract operator*; this is the only step in the proposed method that cannot be fully automated.
- the temporal logic CTL* and its fragments, used for the description of properties,
- the preservation results allowing to deduce the validity of a property on the concrete program from its validity on the abstract program and
- the compositionality results allowing to compute an abstract program by composing abstractions of its components.

We illustrate all the definitions and results on a small buffer example. In Section 3, we verify — by applying in a systematic way our method — a distributed cache memory system defined in [ABM93]. In [DGJ+93] several (complex) correctness proofs are given for this system based on different methods. Using our method, the verification of this system is almost as simple as the verification of the tiny buffer, as we need almost the the same abstract operations.

2 A verification method using abstractions

2.1 A program description formalism

We adopt a simple program formalism which is not meant as a real programming language but which is sufficient to illustrate our method. A complex system is a parallel composition of basic programs P of the following form

Variables :	$x_1 : T_1, ..., x_n : T_n$
Transitions :	(ℓ_1) $action_1(x_1, ..., x_n, x'_1, ..., x'_n)$
	...
	(ℓ_p) $action_p(x_1, ..., x_n, x'_1, ..., x'_n)$
Initial States :	$init(x_1, ..., x_n)$

where x_i are variables of type T_i and $L_P = \{\ell_1, ..., \ell_n\}$ is the set of program labels. Each $action_i$ is an expression with variables in the set of program variables and a set of primed variables which is a copy of the set of state variables; as in [Pnu86, Lam91], $action_i$ represents a transition relation on the domain of the program variables by interpreting the valuations of $X_P = (x_1, ..., x_n)$ as the state *before*, and the valuations of $X'_P = (x'_1, ..., x'_n)$ as the state *after* the transition. We denote the set of valuations of X_P by $Val(X_P)$.

Semantics : Program P defines in an obvious manner a transition system $S_P = (Q_P, R_P)$ where

- $Q_P = Val(X_P)$ is the set of states,
- $R_P \subseteq Q_P \times Q_P$ is a transition relation defined by $R_P = \{(q, q') \mid \exists i . action_i(q, q')\}$.

The predicate $init$ defines the set of initial states. It is used in the formulas specifying the program: properties are in general of the form $init \Rightarrow \phi$ where ϕ expresses the property one wants to verify.

We do not distinguish variables representing inputs as they need not be treated in a particular manner. However, we annotate in the programs the variables which are meant as inputs as this makes programs easier to read.

Labels are used to name "events" or "actions". If ℓ_i is a label and (v, v') a pair of valuations such that $action_i(v, v')$ is true, then the transition from state v to state v' is an event ℓ. If e is the valuation of the "input" variables extracted from v, then we call this event also $\ell_i(e)$. Events are used for the expression of properties.

Example 1 an infinite lossy buffer. The following program represents an unbounded buffer taking as input elements e of some data domain *elem*. The event *push(e)* enters e (if it has never been entered yet) into the buffer or arbitrarily "loses" it, and *pop(e)* takes e out of the buffer if it is its first element.

Variables :	$e : elem$ (Input)
	$E : set$ of $elem$ (already occurred events $push(e)$)
	$B : buffer$ of $elem$
Transitions :	(push(e)) $allowed(e, E, E') \wedge (add(B, e, B') \vee \mathbf{unch}(B))$
	(pop(e)) $first(B, e) \wedge tail(B, e, B') \wedge \mathbf{unch}(E)$
Initial States : $empty(B)$	

E contains the elements e such that $push(e)$ has already occurred, and $\mathbf{allowed}(e, E, E')$ is necessarily *false* if $e \in E$. All other predicates have the intuitive meanings: $\mathbf{add}(B, e, B')$ holds if B' is obtained by adding element e at the end of the buffer B; $\mathbf{tail}(B, e, B')$ holds if B' is obtained by eliminating e from B if e is its first element ($\mathbf{first}(B, e)$ is *true*); $\mathbf{empty}(B)$ is *true* if B is the empty buffer. $\mathbf{unch}(X)$, where $X = x_1, ...x_n$ is a tuple of program variables, represents the transition relation which lets all variables in X unchanged, i. e., $\mathbf{unch}(X) = \bigwedge_i (x'_i = x_i)$.

We use predicates of the form $add(B, e, B')$ instead of $B' = ADD(B, e)$ where ADD is a function, as abstract operations are in general nondeterministic. This is also the way of representing operations which is proposed, e. g. in [CGL91, Lam91].

Composed programs : In [GL93b] we obtain our results for more general parallel composition operators, but here we need only asynchronous composition. If P_1 and P_2 are programs defined on a tuple of state variables X_1, respectively X_2, then $P_1 \parallel P_2$ is the parallel composition of P_1 and P_2 defining the transition system $S = (Val(X_1 \cup X_2), R)$ where

$$R = R_{P_1} \wedge \mathbf{unch}(X_2 - X_1) \quad \vee \quad R_{P_2} \wedge \mathbf{unch}(X_1 - X_2)$$

Each transition of $P_1 \parallel P_2$ is either a transition of P_1 which leaves all variables which are declared in P_2 but not in P_1 unchanged or the other way round.

2.2 Abstract programs

As proposed in [CGL91, LGS+92], given a program *Prog* and a predicate ϱ on the variables of *Prog* and a tuple of *abstract variables* $X^A = (x_1^A, ...x_m^A)$, representing a relation between the concrete and the abstract domain (a function in [CGL91]), then any program $Prog^A$ defined on X^A, such that for each action *action* of *Prog* there exists an action $action^A$ of $Prog^A$ with the same label, such that

$$\exists X \exists X' . \; \varrho(X, X^A) \wedge \varrho(X', X^{A'}) \wedge action(X, X') \quad \Rightarrow \quad action^A(X^A, X^{A'}) \quad (1)$$

and

$$\exists X . \; \varrho(X, X^A) \wedge init(X) \quad \Rightarrow \quad init^A(X^A)$$

is an *abstraction* or more precisely a ϱ-*abstraction* of *Prog*.

When verifying composed programs, it is interesting to compute an abstract program compositionally, i. e., by composing abstract component programs. From a more general result given in [GL93a], we obtain the following result which is sufficient for the verification of the distributed cache memory system in the following section.

Proposition 1. *Let P_1 and P_2 be programs and ϱ_i total functions from the domain of the variables of P_i into some abstract domains such that $\varrho_1 \cap \varrho_2$ is total and P_1^A, P_2^A are ϱ_i-abstractions of P_i, then $P_1^A \parallel P_2^A$ is a $(\varrho_1 \cap \varrho_2)$-abstraction of $P_1 \parallel P_2$.*

Computation of abstract programs in practice : The idea of abstract interpretation [CC77] is to replace every *function* on the concrete domain used in the program by a corresponding *abstract function* on the abstract domain, and then to analyze the so obtained simpler abstract program instead of the concrete one. Consider the program $Prog^A$ obtained by replacing every basic predicate $op(X, X')$ on the concrete variables by a predicate $op^A(X^A, X^{A'})$ on the abstract variables is a ϱ-abstraction of $Prog$ if, instead of (1), for every basic operation

$$\exists X \exists X' . \; \varrho(X, X^A) \wedge \varrho(X', X^{A'}) \wedge op(X, X') \; \Rightarrow \; op^A(X^A, X^{A'}) \qquad (2)$$

holds. If the expressions in $Prog$ are negation free (as in our buffer), then $Prog^A$ is in fact a ϱ-abstraction of $Prog$. The definition of abstract predicates op^A is the only part of our verification method which cannot be fully automatized. But as we will see, we only need a restricted number of such abstract operations in order to verify a whole class of programs. For example, in the domain of protocol verification, the used data structures are "messages" on which no operations are carried out, "memories" or "registers" in which data can be stored, integers which are mostly used as counters and "buffers" with the usual operations *add, tail, first,..*, as in our examples. In [CGL91] a similar method is proposed.

Example 2 An abstract lossy buffer. To illustrate the idea, consider again the buffer of Example 1. In order to show that the buffer has the property of "order preservation" (see Example 3), it is sufficient to show that the order of any pair of elements $(e_1, e_2) \in elem \times elem$ is preserved. All the information we need about the content of the buffer B is, if and in which order, it contains the elements e_1 and e_2. Furthermore, as each element is supposed to be put into the buffer at most once, we need not distinguish amongst the valuations of B containing e_i more than once. Similarly, for the input variable e we only need to distinguish if its value is e_1, e_2 or any other value. Concerning the value of E determinating which events $push(e)$ are still allowed, we only need to know if the event $push(e_1)$, respectively $push(e_2)$ is still possible or not. This leads us naturally to the abstract domain defined by the abstract variables,

$$e_A : elem_A^2 = \{0, 1, 2\}$$
$$E_A^1, E_A^2 : Bool$$
$$B_A : buffer_A^2 = \{\epsilon, e_1, e_2, e_1 \bullet e_2, e_2 \bullet e_1, \bot\}$$

and the following abstraction relation ϱ^2 defining the correspondence between the concrete and the abstract variables

$$\varrho^2(e, E, B, \; e_A, (E_A^1, E_A^2), B_A) = \varrho_{elem}^2(e, e_A) \wedge \varrho_{set\_of\_elem}^2(E, (E_A^1, E_A^2)) \wedge$$
$$\varrho_{buffer}^2(B, B_A)$$

where for $e : elem$ and $e_A : elem_A^2$

$$\varrho_{elem}^2(e, e_A) = ((e_A = 0) \equiv (e \notin \{e_1, e_2\})) \wedge$$
$$((e_A = 1) \equiv (e = e_1)) \wedge$$
$$((e_A = 2) \equiv (e = e_2))$$

for $E : set$ of $elem$ and $E_A^1, E_A^2 : Bool$, E_A^i expresses that e_i has not occurred yet:

$$\varrho^2_{set\_of\_elem}(E, E^1_A, E^2_A) = ((E^1_A = \exists E' . \; allowed(e_1, E, E')) \wedge$$
$$((E^2_A = \exists E' . \; allowed(e_2, E, E'))$$

and for $B : buffer$ of $elem$ and $B_A : buffer^2_A$

$$\varrho^2_{buffer}(B, B_A) =$$
$$((B_A = \epsilon) \equiv empty(B_{|\{e_1, e_2\}})) \wedge ((B_A = e_1 \bullet e_2) \equiv (B_{|\{e_1, e_2\}} = e_1 \bullet e_2)) \wedge$$
$$((B_A = e_1) \equiv (B_{|\{e_1, e_2\}} = e_1)) \wedge ((B_A = e_2 \bullet e_1) \equiv (B_{|\{e_1, e_2\}} = e_2 \bullet e_1)) \wedge$$
$$((B_A = e_2) \equiv (B_{|\{e_1, e_2\}} = e_2)) \wedge ((B_A = \bot) \text{ in all other cases }))$$

where $B_{|\{e_1, e_2\}}$ is the buffer B restricted to the elements e_1 and e_2. In order to construct an abstract program, we have to define abstract predicates for all the basic predicates used in the concrete buffer program, such as *allowed*, *add*, *tail*, **unch**, etc.

In the case that every abstract variable is related to a single concrete variable, the abstract predicate associated with **unch**(v) is obviously **unch**(v_A) for any abstract variable v_A related to v. The following abstract predicates satisfy the condition (2).

$$allowed^2_A(e_A, (E^1_A, E^2_A), (E^{1'}_A, E^{2'}_A)) = (E^{1'}_A \equiv E^1_A) \wedge (E^{2'}_A \equiv E^2_A) \; \wedge \; (e_A = 0) \vee$$
$$(E^1_A \wedge \neg E^{1'}_A) \wedge (E^2_A \equiv E^{2'}_A)) \; \wedge \; (e_A = 1) \vee$$
$$(E^1_A \equiv E^{1'}_A) \wedge (E^2_A \wedge \neg E^{2'}_A))) \; \wedge \; (e_A = 2)$$

$$add^2_A(B_A, e_A, B'_A) = (B_A = B'_A) \qquad\qquad\qquad\qquad\qquad\qquad \wedge (e_A = 0) \vee$$
$$(B_A \in \{\epsilon, e_2\}) \wedge (B'_A = e_1 \bullet B_A) \vee (B_A \notin \{\epsilon, e_2\}) \wedge (B'_A = \bot) \wedge (e_A = 1) \vee$$
$$(B_A \in \{\epsilon, e_1\}) \wedge (B'_A = e_2 \bullet B_A) \vee (B_A \notin \{\epsilon, e_1\}) \wedge (B'_A = \bot)) \wedge (e_A = 2)$$

$$tail^2_A(B_A, e_A, B'_A) = (B_A = B'_A) \qquad\qquad\qquad\qquad\qquad \wedge (e_A = 0) \vee$$
$$((B_A \in \{e_1, e_1 \bullet e_2\}) \; \Rightarrow \; (B_A = B'_A \bullet e_1)) \wedge (e_A = 1) \vee$$
$$((B_A \in \{e_2, e_2 \bullet e_1\}) \; \Rightarrow \; (B_A = B'_A \bullet e_2)) \wedge (e_A = 2)$$

$$empty^2_A(B_A) = (B_A = \epsilon)$$

$$first^2_A(B_A, e_A) = (e_A = 0) \vee$$
$$(B_A \in \{e_1, e_1 \bullet e_2, \bot\}) \wedge (e_A = 1) \vee$$
$$(B_A \in \{e_2, \bot\}) \wedge (e_A = 2)$$

tail is an example of a predicate defining a function on the concrete domain, but which is nondeterministic on the given abstract domain; $tail^2_A(\bot, 1, B'_A)$ necessarily evaluates to *true* for any value of B'_A (the value of the buffer in the next state).

Using these abstract predicates, the definition of a program representing a ϱ-abstraction of the buffer program becomes trivial. We just replace variables by corresponding abstract variables and every occurrence of a predicate by corresponding abstract one. The resulting abstract program looks almost as the concrete program but defines a very small finite transition system.

The useful abstractions are often obtained by using this kind of abstract domains. Here, we gave in detail the more complicated abstraction of a buffer particularizing two different data elements. But often, it is sufficient to particularize in the same way a single data element. The corresponding abstraction relations ϱ^1_{elem}, $\varrho^1_{set\_of\_elem}$, ϱ^1_{buffer} and abstract predicates $allowed^1_A$, add^1_A, $tail^1_A$,... can be defined by simplifying the above definitions in an obvious manner. For the verification of the cache memory we use also existential abstractions of buffers. The corresponding abstract predicates $add^{ex}(e_A)$, $tail^{ex}(e_A)$,... necessarily evaluate to *true* if e_A is an allowed value of the existentially abstracted buffer.

2.3 Temporal Logic

It remains to recall the definition of temporal logic. Here we restrict ourselves to subsets of CTL* [EH83] for the expression of properties. The preservation results in [LGS+92] are given for subsets of the more powerful branching time μ-calculus augmented by past time modalities.

Definition 2. CTL* is the set of state formulas given by the following definition.
1. Let $\mathcal{P}$ be a set of atomic (a) state respectively (b) path formulas.
2. If ϕ and ψ are (a) state respectively (b) path formulas then $\phi \wedge \psi$, $\phi \vee \psi$ and $\neg\phi$ are (a) state respectively (b) path formulas.
3. If ϕ is a path formula then $\mathbf{A}\phi$ and $\mathbf{E}\phi$ are state formulas.
4. If ϕ and ψ are (a) state or (b) path formulas then $\mathbf{X}\phi$, $\phi\mathcal{U}\psi$ and $\phi\mathcal{W}\psi$ are path formulas.

As usual, we also use the abbreviations $\mathbf{F}\phi$ denoting $true\,\mathcal{U}\,\phi$ and $\mathbf{G}\phi$ denoting $\phi\mathcal{W}false$.

$\mathcal{U}$ is a strong and $\mathcal{W}$ a weak "until" operator, a sequence satisfies $\phi\mathcal{W}\psi$ if ϕ holds up to some point in which ψ holds, and $\phi\mathcal{U}\psi$ expresses the same property and moreover the obligation that such a point satisfying ψ exists. That means that $\mathcal{U}$ and $\mathcal{W}$ are duals by inversing the arguments: $\phi\mathcal{W}\psi = \neg(\neg\psi\mathcal{U}\neg\phi)$.

$\forall$CTL* [SG90] is the subset of CTL* obtained by allowing negations only on atomic formulas and restricting Rule 3 by allowing only the universal path quantifier $\mathbf{A}$.

The *semantics* of CTL* is defined over Kripke structures of the form $M = (S, \mathcal{I})$ where $S = (Q, R)$ is a transition system and $\mathcal{I}$ is a interpretation function mapping elements of $\mathcal{P}$ into sets of states of S.

Definition 3. A *path* in a transition system S is an infinite sequence $\pi = q_0 q_1 \ldots$ such that for every $i \in \mathcal{N}$. $R(q_i, q_{i+1})$. We denote by π_n the nth state of path π and by π^n the sub-path of π starting in its nth state.

Definition 4. Let be $M = (S, \mathcal{I})$ a Kripke structure, $q \in Q$ and π a path of M. Then the satisfaction ($\models_M$) of CTL* formulas on M is defined inductively as follows.
1. Let be $p \in \mathcal{P}$. Then $q \models_M p$ iff $q \in \mathcal{I}(p)$ and $\pi \models_M p$ iff $\pi_0 \in \mathcal{I}(p)$.
2. Let ϕ and ψ be (a) state respectively (b) path formulas. Then,
   (a) $q \models_M \neg\phi$ iff $q \not\models_M \phi$, $q \models_M \phi \wedge \psi$ iff $q \models_M \phi$ and $q \models_M \psi$, $q \models_M \phi \vee \psi$ iff $q \models_M \phi$ or $q \models_M \psi$.
   (b) analogous by replacing q by π
3. Let ϕ be a path formula. Then,
   $q \models_M \mathbf{A}\phi$ iff for every path π starting in q, $\pi \models_M \phi$ and
   $q \models_M \mathbf{E}\phi$ iff there exists a path π starting in q such that $\pi \models_M \phi$.
4. Let ϕ and ψ be (a) state respectively (b) path formulas. Then,
   (a) $\pi \models_M \mathbf{X}\phi$ iff $\pi_1 \models_M \phi$,
   $\pi \models_M \phi\mathcal{U}\psi$ iff there exists $n \in \mathcal{N}$ such that $\pi_n \models_M \psi$ and $\forall k < n$. $\pi_k \models_M \phi$,
   $\pi \models_M \phi\mathcal{W}\psi$ iff for all $n \in \mathcal{N}$. $((\forall k < n$. $\pi_k \models_M \neg\psi)$ implies $\pi_n \models_M \phi)$.
   (b) the same definition obtained by replacing in (a) all states π_i by subsequences π^i.

We say that $M \models \phi$ iff $q \models_M \phi$ for all states of M.

From the more general results given in [LGS+92] we obtain the following proposition concerning preservation of properties of $\forall$CTL*.

Proposition 5 Preservation of ∀CTL*. *Let Prog be a program, ρ a total relation from the domain of Prog into some abstract domain, and Prog_A a ρ-abstraction of Prog. Then, for any φ ∈∀CTL*, $\mathcal{P}$ the set of atomic formulas occurring in φ and $\mathcal{I}$ an interpretation function mapping $\mathcal{P}$ into sets of states of S_{Prog}, we have*

$$Im[\varrho^{-1}] \circ Im[\varrho] \circ \mathcal{I} \ (p) \subseteq \mathcal{I} \ (p) \quad (*) \quad for \ all \ p \in \mathcal{P} \ occurring \ positive \ in \ \phi$$

implies

$$(S_{Prog_A}, Im[\varrho] \circ \mathcal{I}) \models \phi \quad \Rightarrow \quad (S_{Prog}, \mathcal{I}) \models \phi$$

where $Im[\varrho]$ is the image function of ρ. Condition (*) is called consistency *of ρ with $\mathcal{I}(p)$.*

This proposition expresses that, if $\phi \in$ ∀CTL* holds on a ρ-abstraction of the program *Prog* by translating the interpretations of all atomic propositions occurring in the formula by $Im(\varrho)$ into predicates on the abstract domain, and if all these predicates are consistent with ρ, then we can deduce that φ holds on *Prog*. Consistency is not needed for predicates that occur only negated in φ as $Im[\varrho^{-1}](\overline{Im[\varrho](\mathcal{I}(p))}) \subseteq \overline{\mathcal{I}(p)}$. We conclude that, if φ holds on *Prog_A* using the abstract interpretation of $\neg p \ (\overline{Im[\varrho](\mathcal{I}(p))})$, then a stronger property than φ using the concrete interpretation of $\neg p \ (\overline{\mathcal{I}(p)})$ holds on *Prog*. In particular, for the verification of a formula of the form $init \Rightarrow \phi$, *init* need not to be consistent with ρ.

Example 3. The property of *order preservation* can be expressed by the following set of formulas on the set of "observable" atomic predicates

$$\mathcal{P} = \{init, enable(push(e)), after(push(e)), enable(pop(e)), after(pop(e)), ...\},$$

where $enable(\ell)$ is interpreted as the set of states in which event ℓ is possible, and $after(\ell)$ those in which ℓ has just occurred — $after(\ell)$ becomes expressible by adding an explicit boolean program variable $after\_\ell$ which is *true* exactly after any event ℓ.

$$\forall e', e \in elem \ : \ init \ \Rightarrow \ \mathbf{A}(\ [\neg after(push(e)) \mathcal{W} after(push(e'))] \ \Rightarrow$$
$$[\neg enable(pop(e)) \mathcal{W} after(pop(e'))] \ )$$

These formulas can be transformed into ∀CTL* formulas in which only the predicates $after(push(e))$ and $after(pop(e'))$ occur non negated. In order to verify that the concrete buffer program has the property of order preservation, it is sufficient to verify the formula obtained by instanciating e_1 for e and e_2 for e' on the abstract program defined in Example 2. In fact, as e_1 and e_2 represent an arbitrary pair of data values, this verification of a single representative of the set of formulas is sufficient. It is easy to obtain the consistency of predicates of the form $after(\ell)$ by not abstracting the variable $after\_\ell$. In the sequel we suppose, without mentioning it explicitly, that for every predicate $after(\ell)$ occurring in the considered formula such a variable is defined.

3 Verification of a distributed cache memory

3.1 Concrete and abstract specification of a sequential consistent memory

Now we use this verification method for the verification of a particular distributed cache memory which has been presented in [ABM93] and verified using different methods in [JPR93, DGJ+93]. The cache memory is a system of the form $P_1 \parallel P_2 ... \parallel P_n$ where each process P_i is defined as in Figure 3.1. the predicates *add, tail, first* and *empty* are as in the Example 1 and *update* is defined by

$$update(m, (a, d), m') \equiv (m'[a] = d) \wedge (\forall b : address \ . \ (b \neq a \ \Rightarrow \ m'[b] = m[b]) \ ).$$

$$
\begin{array}{ll}
\text{Variables : Input :} & a : address, \ d : datum \\
\underline{\text{local}} : & AD_i : set \ of \ address \times datum_i, \ \text{(data already written)} \\
& C_i : array[address] \ of \ datum \cup \{\epsilon\} \ \text{(local cache memory)} \\
& Out_i : buffer \ of \ (address \times datum_i) \\
\underline{\text{shared}} : & M : array[address] \ of \ datum \ \text{(global memory)} \\
& In_k : buffer \ of \ (address \times (datum \cup \{\epsilon\})), \ k : index
\end{array}
$$

Transitions :

$(write_i(a,d))$ $allowed((a,d), AD_i, AD'_i) \wedge add(Out_i, (a,d), Out'_i) \wedge$
     $unch(C_i, M, In_1, ...In_n)$

$(read_i(a,d))$ $(C_i[a] = d) \wedge empty(Out_i) \wedge empty(In_{i \,|address \times datum_i}) \wedge$
     $unch(C_i, Out_i, AD_i, M, In_1, ...In_n)$

$(mw_i(a,d))$ $first(Out_i, (a,d)) \wedge tail(Out_i, (a,d), Out'_i) \wedge update(M, (a,d), M') \wedge$
     $\forall k : index \ . \ add(In_k, (a,d), In'_k) \wedge unch(C_i, AD_i)$

$(cu_i(a,d))$ $first(In_i, (a,d)) \wedge tail(In_i, (a,d), In'_i) \wedge update(C_i, (a,d), C'_i) \wedge$
     $unch(Out_i, AD_i, M, \{In_j, j \neq i\})$

Init : $(\forall b : address \ . \ (C_i[b] = M[b] = \epsilon)) \wedge empty(Out_i) \wedge empty(In_i)$

Fig. 1. A distributed cache memory system

The event $write_i(a,d)$ does not have any immediate effect neither on the local nor on the central memory, but pushes the pair (a,d) into the local buffer Out_i ; from there it is by the internal event $memory\_write_i(a,d)$ written into the central memory and dispatched into all buffers In_i; the internal event $cashupdate_i(a,d)$ takes the first element (a,d) out of In_i and writes datum d into address a of the local cache memory C_i. The event $read_i(a,d)$ is possible only if $C_i[a] = d$ and no local $write$ event is pending, i. e., if $empty(Out_i) \wedge empty(In_{i \,|address \times datum_i})$. The only difference between our system and the one used in [JPR93] concerns the fact that each pair (a,d) can be the parameter of at most one event $write$. The way we obtain this, is by defining the type $datum$ by $datum = \bigcup_i datum_i$, such that each process "signs" the data it writes, and by using in each process a variable AD_i of type set of $address \times datum_i$ which stores the information if the event $write_i(a,d)$ has already occurred or not, as in the example of the buffer.

    The abstract specification that the system must verify is *sequential consistency* [Lam79], which originally is given in the form of an abstract program. In order to apply our method, we give the abstract specification in terms of a set of properties. Under the assumption that every pair of the form (a,d) can occur at most once as the parameter of some *write* event, sequential consistency can be characterized by the following set of properties expressed in terms of observable events:

Safety properties characterizing a sequential consistent memory:

(S1) $\forall (a,d), (a',d') : address \times datum, j,i : index$ such that $(a,d) \neq (a',d')$
     $init \ \Rightarrow \ \mathbf{A}(\ [\neg after(write_j(a,d)) \mathcal{W} after(write_j(a',d'))] \ \Rightarrow$
         $[\neg enable(read_i(a,d)) \mathcal{W}(enable(read_i(a',d')) \vee$
                             $AG(\neg enable(read_i(a',d'))))] \ )$

(S2) $\forall (a,d) : address \times datum, i : index$

$$init \ \Rightarrow \ \mathbf{A}(\neg enable(read_i(a,d))\mathcal{W}\bigvee_{j:index} after(write_j(a,d)) \ )$$

(S3) $\forall(a,d) : address \times datum, i : index$

$$init \ \Rightarrow \ \mathbf{A}G(after(write_i(a,d)) \ \Rightarrow$$
$$\mathbf{A}((enable(read_i(a)) \ \Rightarrow \ enable(read_i(a,d)))\mathcal{W}AG(\neg enable(read_i(a,d)))))$$

(S4) $\forall(a,d),(a,d') : address \times datum, i_1, i_2 : index$ such that $d \neq d'$

$$init \ \Rightarrow \ \mathbf{A}(\ [\neg after(read_{i_1}(a,d))\mathcal{W}after(read_{i_1}(a,d')) \ ] \ \Rightarrow$$
$$[\neg enable(read_{i_2}(a,d))\mathcal{W}AG(\neg enable(read_{i_2}(a,d'))) \ ] \ )$$

(S1) expresses that in every execution sequence the subsequence of $read_i$ events respects the order of $write_j$ events: whenever (a',d') is written before (a,d) by P_j, then $read_i(a,d)$ is not enabled before either $read_i(a',d')$ has already been enabled or is never enabled again (the second clause is necessary because $read_i(a',d')$ may never be enabled in some computation sequences). (S2) expresses that every $read_i(a,d)$ event is preceded (caused) by some $write_j(a,d)$ event. This is slightly stronger than sequential consistency which may allow $write_j(a,d)$ to occur after $read_i(a,d)$. (S3) expresses the fact that $read_i$ and $write_i$ events od the same process P_i must be consistent with a central memory, i. e., after a $write_i(a,d)$, P_i can read nothing different from d in address a until $read_i(a,d)$ is never enabled again. (S4) expresses analogously to (S1) the fact that $read$ events concerning the same memory cell must be consistent in all processes. All these formulas can be translated into $\forall$CTL* formulas.

3.2 Verification of the cache memory

We verify each parameterized set of formulas on a different abstract program. Our aim is not to find the smallest abstract program that can be used for the verification of each formula, but we want to apply, whenever possible, the already predefined abstractions in order to show that the application of the method is simple and can be done systematically. The cache memory uses the data types and operations of the buffer of Example 1; it uses also a data type "memory" = $array[address]$ of $datum$. As for buffers, we use three different types of abstractions of a variable X of type $memory$ depending on the formula to be verified: we may

– completely forget about it (we do this always for the central memory $\mathbf{M}$)
– keep information about a single pair $(\mathbf{a},\mathbf{d})$ by taking an abstract boolean variable X_A and an abstraction relation $\varrho^1_{memory}(X, X_A) = X_A \equiv (X[\mathbf{a}] = \mathbf{d})$.
– keep information about two pairs $(\mathbf{a}_1, \mathbf{d}_1)$ and $(\mathbf{a}_2, \mathbf{d}_2)$ by taking two abstract boolean variables X_A^1 and X_A^2 and an analogous abstraction relation $\varrho^2_{memory}(X, X_A^1, X_A^2)$.

Suppose the type $elem$ to be $address \times datum$ and take an abstract variable e_A of type $elem_A^1 = \{0,1\}$ already used in the buffer example and the abstraction relation

$$\varrho^1_{elem}((a,d), e_A) = (e_A = 0) \wedge ((a,d) \neq (\mathbf{a},\mathbf{d})) \ \vee \ (e_A = 1) \wedge ((a,d) = (\mathbf{a},\mathbf{d})),$$

exactly as in Example 2; then, it is easy to define an abstract predicate $update_A^1$ by

$$update_A^1(X_A, e_A, X_A') = (e_A = 0) \wedge (X_A' \ \Rightarrow \ X_A) \ \vee \ (e_A = 1) \wedge X_A'$$

expressing that if $(a,d) \neq (\mathbf{a},\mathbf{d})$, $X[\mathbf{a}] = \mathbf{d}$ is only possible in the next state if already in the present state $X[\mathbf{a}] = \mathbf{d}$, and if $(a,d) = (\mathbf{a},\mathbf{d})$, then in the next state $X[\mathbf{a}] = \mathbf{d}$, independently of the value of $X[\mathbf{a}]$ in the present state.

Using these definitions (and analogous ones with superscripts ^{ex} and ^{2}) and those already given in Example 1, the definition of appropriate abstract finite state programs of the cache memory becomes simple.

Abstract programs for property (S3): Each instance of property (S3) involves only events of a single process P_i. However, even if we succeed to verify it on P_i we can *not* deduce their satisfaction on the composed system. In fact, if we replace all processes different from P_i by the process *"Chaos"*, (S3) does not hold any more on the composed abstract program. We use here another approach to compositionality: by Proposition 1, we can abstract each P_j individually and build a global model by composing these small abstract programs. We choose the abstraction relation for all P_j with $j \neq i$ in such a way that shared variables are abstracted in the same way as in P_i and we forget about all local variables; this has as effect to avoid adding certain changes of shared variables which are not allowed by the concrete programs P_j.

Intuitively, (S3) expresses that as soon as $write_i(a,d)$ has occurred, only **d** may be read by P_i on address **a** until **d** has been put into C_i and afterwards been replaced by some other value; that means that we are interested in observing what happens on the buffers Out_i and In_i and the cache C_i. The actions (mw_j) should not disturb the behaviour of P_i observed by (S3) because they cannot push (a,d) into In_i. This leads naturally to the following abstraction relation for P_i:

$$\varrho_i^{S3}((a,d), AD_i, C_i, Out_i, M, In_1, ... In_n, e_A, E_A, C_{iA}, Out_{iA}, In_{iA}) =$$
$$\varrho_{elem}^1((a,d), e_A) \qquad \wedge \qquad \varrho_{set\_of\_elem}^1(AD_i, E_A) \qquad \wedge$$
$$\varrho_{memory}^1(C_i, C_{iA}) \qquad \wedge \qquad \varrho_{buffer}^1(Out_i, Out_{iA}) \qquad \wedge$$
$$\varrho_{buffer}^1(In_i, In_{iA})$$

and for P_j, $j \neq i$ we use the same abstraction as in ϱ_i for the shared variables and forget about all local variables

$$\varrho_j^{S3}((a,d), AD_i, C_i, Out_i, M, In_1, ..., In_n, e_A, In_{iA}) =$$
$$\varrho_{elem}^1((a,d), e_A) \qquad \wedge \qquad \varrho_{buffer}^1(In_i, In_{iA})$$

from which we obtain by replacing concrete by corresponding abstract predicates as defined before, the following abstract program P_i^A for index i,

Variables : abstract input : e_A : *Bool*	
local :	E_A, C_{iA} : *Bool*
	Out_{iA} : $buffer_A^1$
shared :	In_{iA} : $buffer_A^1$
Transitions :	
$(write_i(e_A))$	$allowed_A^1(e_A, E_A, E_A') \wedge add_A^1(Out_{iA}, e_A, Out_{iA}') \wedge \mathbf{unch}(C_{iA}, In_{iA}) \wedge$
$(read_i(e_A))$	$(e_A \Rightarrow C_{iA}) \wedge empty_A^1(Out_{iA}) \wedge empty_A^1(In_{iA}) \wedge$
	$\mathbf{unch}(E_A, C_{iA}, Out_{iA}, In_{iA})$
$(mw_i(e_A))$	$first_A^1(Out_{iA}, e_A) \wedge tail_A^1(Out_{iA}, e_A, Out_{iA}') \wedge$
	$add_A^1(In_{iA}, e_A, In_{iA}') \wedge \mathbf{unch}(C_{iA}, E_A)$
$(cu_i(e_A))$	$first_A^1(In_{iA}, e_A) \wedge tail_A^1(In_{iA}, e_A, In_{iA}') \wedge update_A^1(C_{iA}, e_A, C_{iA}') \wedge$
	$\mathbf{unch}(E_A, Out_{iA})$
Init :	$\neg C_{iA} \wedge empty_A^1(Out_{iA}) \wedge empty_A^1(In_{iA})$

and P_j^A for all indices different from i,

Variables : abstract input : $e_A : Bool$
shared: $\qquad In_{iA} : buffer_A^1$
Transitions : $(write_j(e_A), read_j(e_A), cu_j(e_A))$ $\quad$ unch(In_{iA})
$\qquad\qquad\quad (mw_j(e_A))$ $\qquad\qquad\qquad\qquad first_A^{ex}(e_A) \wedge add_A^1(In_{iA}, e_A, In'_{iA})$
Init : $\qquad\qquad empty_A^1(In_{iA})$

in which we have already eliminated all abstract operations that are always *true*, such as add_A^{ex}, $update_A^{ex}$,.... Notice that the event $(mw_j(true))$ is never executed as $first_A^{ex}(true) = false$ because the buffer Out_j cannot contain the pair (a,d) as $d \in datum_i$. Notice also that the size of the composed system $P_1^A \parallel ... \parallel P_i^A \parallel ... \parallel P_n^A$ is the same, whatever the number of composed programs is, as for all $j \neq i$, the programs P_j^A are identical and $P \parallel P$ and P represent the same transition system.

Abstract programs for property (S2): Property (S2) expresses the fact that any event $read_i$(a,d) is preceded by an event $write_j$(a,d) for some j. Thus, we are interested in observing the buffers Out_j and In_i and the cache C_i. This leads to similar abstraction relations as for the verification of (S3), except that we do not need the unicity of the *write* events and can forget about AD_i but we need abstract buffers Out_{jA} for all j. Thus, all abstraction relations ϱ_j^{S2} are the same:

$$\varrho_j^{S2}(A, D, AD_j, C_j, Out_j, M, In_1, ..., In_n, e_A, Out_{jA}, In_{iA}) =$$
$$\varrho_{elem}^1((A, D), e_A) \qquad\qquad \wedge \qquad\qquad \varrho_{buffer}^1(Out_j, Out_{jA}) \qquad\qquad \wedge$$
$$\varrho_{buffer}^1(In_i, In_{iA})$$

For this abstraction, the size of the obtained global abstract transition system does depend on the number n of processes as we have defined n abstract variables Out_{jA}. In order to obtain an abstract transition system such that its size is independent of n, we can define — instead of the set of local abstract buffers Out_{jA} — a *single global* abstract buffer Out_A defined by a relation of the form

$$\varrho_{buffer}^{1,glob}(\bigcup_{j:index} Out_j, Out_A)$$

which obliges however to redefine the abstract operations add_A, $tail_A$,...

Abstract programs for properties (S1) and (S4): For the verification of (S1) we need to observe events with two different parameters (a_1, d_1) and (a_2, d_2), such that $d_1, d_2 \in datum_j$; thus, we use the abstraction relations with superscript [2] as for the verification of order preservation in the preceding section. We define abstract variables E_A^1, E_A^2 (in P_j^A) in order to guarantee uniqueness of the observed $write_j$ events, Out_{jA} (in P_j^A), C_{i_1}, C_{i_2} (in P_i^A) and a global variable In_{iA} and use the predefined abstraction relations and corresponding abstract operations.

The resulting global abstract transition system is again independent of the number of process as all the abstract programs with indices different from i, j are identical. We need only to consider the case where the indices i and the j are different, as the property for $i = j$ is implied by (S3).

Property (S4) expresses that the sequences of *read* events of any two processes P_{i_1} and P_{i_2} on the same address a are compatible, also when they have been written by two different processes P_{j_1} and P_{j_2}. Thus, for its verification we observe two pairs (a_1, d_1) and

$(\mathbf{a}_2, \mathbf{d}_2)$ such that $\mathbf{a}_1 = \mathbf{a}_2 = \mathbf{a}$ and $\mathbf{d}_1 \in datum_{j_1}$ and $\mathbf{d}_2 \in datum_{j_2}$. Consequently, we need abstract variables E_A^1, E_A^2 (in $P_{j_1}^A$ respectively $P_{j_2}^A$), Out_{j_1A}, Out_{j_2A} (in $P_{j_1}^A$ respectively $P_{j_2}^A$), $\mathbf{C}_{i_11}, \mathbf{C}_{i_12}$ (in $P_{i_1}^A$), $\mathbf{C}_{i_21}, \mathbf{C}_{i_22}$ (in $P_{i_2}^A$) and global variables In_{i_1A} and In_{i_2A}. Here, we have to consider different cases, those where the indices i_1 and j_1 (respectively i_2 and j_2) coincide and those where not.

Now, the verification of the distributed cache memory is almost terminated. The actual construction of global abstract transition systems and the verification of the formulas on them could been done automatically by our tool [GL93b, Loi94]. By Proposition 5, it remains to verify the consistency of the atomic propositions with the used abstraction relations. Properties (S2) and (S4) pose no problem, as in the corresponding $\forall\text{CTL}^*$ formulas only predicates of the form $after(\ell)$ occur non negated. For (S3), in principle the consistency of $enable(read_i(\mathbf{a},\mathbf{d}))$ is required; however, it is used only within the predicate $(enable(read_i(\mathbf{a}))\Rightarrow enable(read_i(\mathbf{a},\mathbf{d})))$ which is equivalent to $\mathbf{C}_i[\mathbf{a}] = \mathbf{d}$ and consistent with the abstraction relation used for (S3). For (S1), it is slightly more complicated to show that the consistency of $enable(read_{i_2}(\mathbf{a}_2, \mathbf{d}_2))$ with the considered abstraction relation ϱ is not needed. The predicate obtained by translating $enable(read_{i_2}(\mathbf{a}_2, \mathbf{d}_2))$ forth and back by ϱ is $\mathbf{C}_{i_2}[\mathbf{a}_2] = \mathbf{d}_2 \wedge \dots$ The property obtained from (S1) by replacing $enable(read_{i_2}(\mathbf{a}_2, \mathbf{d}_2))$ by this weaker predicate, implies nevertheless (S1) for our particular system, because $enable(read_{i_2}(\mathbf{a}_1)\Rightarrow((\mathbf{C}_i[\mathbf{a}_2] = \mathbf{d}_2)\Rightarrow enable(read_{i_2}(\mathbf{a}_2, \mathbf{d}_2)))$ holds. Notice also, that this additional condition is in fact necessary in order to obtain sequential consistency of the given system.

4 Discussion

What have we achieved? A first impression could be that our verification of a cache memory looks much like any other handwritten proof. However, it is quite different: starting right from the beginning, it is in fact rather lengthy to define all the abstraction relations and corresponding abstract predicates, even in order to verify some trivial buffer program. However, having done this once, in order to verify the much more complex cache memory system, we only need a few more definitions obtained a long the same line as the already given ones. In fact, there are many examples of systems, for which we have to verify exactly the same type of properties and which use analogous data structures and operations on them, such that the same abstract domains and operations can be used. Thus, we could build a "library" of useful abstract domains and operations in which new definitions can be added when necessary. A similar approach has been followed by P. and R. Cousot and D. Long concerning "standard" abstractions of integers and operations on them.

The fact that for the verification of an individual property a large part of the system can be abstracted existentially is often necessary in order to obtain tractable global models. If the system is too large or the property is "too global" one can often get results by decomposing the property, as this has been proposed, e. g. in [Kur89].

It can also been observed that our verification method is incremental: it is obviously incremental with respect to changes in the abstract specification, like every method based on the fact that abstract specifications are expressed by a set of properties. But also certain changes of the program allow to use the same or at least very similar abstraction relations and abstract operations. That means that exactly the time consuming and difficult part of the verification process need not to be redone. In the case that the obtained abstract program is not identical to the previous one, the reconstruction of a model and the verification of the properties on it by means of some model checker poses no problem.

An important point for the use of our method in practice, is the formalisms used for the description of programs. We either need a formalism allowing to express nondeterminism or we have to use more complex abstract domains allowing to represent certain sets of classes of concrete values. For example, LOTOS is a specification formalism for which this method can be applied: all the data types and operations on them are specified separately from the control part by means of some "abstract data type" language. This means that, by coding whatever should be abstracted in the data part, to construct an abstract program consists simply in replacing the original data type definitions by simpler ones, whereas the control part of the program remains completely unchanged.

Acknowledgements: I would like to thank the referees who pointed out that my initial characterization of sequential consistency was not sufficient.

References

[ABM93] Y. Afek, G. Brown, and M. Meritt. Lazy caching. *ACM Transactions on Programming Languages and Systems*, 15(1), 1993.

[CC77] P. and R. Cousot. Abstract interpretation: a unified lattice model for static analysis of programs by construction or approximation of fixpoints. In *POPL*, 1977.

[CGL91] E. Clarke, O. Grumberg, D. Long. Model checking and abstraction. In *POPL*, 1991.

[DGJ+93] J. Davis, R. Gerth, W. Jannsen, B. Jonsson, S. Katz, G. Lowe, A. Pnueli, and C. Rump. Verifying sequentially consistent memory. Preliminary report, 1993.

[EH83] E. A. Emerson and J. Y. Halpern. 'Sometimes' and 'not never' revisited: On branching versus linear time. In *POPL*, 1983. also in Journal of ACM, 33:151-178.

[GL93a] S. Graf and C. Loiseaux. Program verification using compositional abstraction. In *TAPSOFT 93, joint conference CAAP/FASE*. LNCS 668, Springer Verlag, April 1993.

[GL93b] S. Graf and C. Loiseaux. A tool for symbolic program verification and abstraction. In *CAV, Heraklion Crete*. LNCS 697, Springer Verlag, 1993.

[JPR93] B. Jonsson, A. Pnueli, and C. Rump. Proving refinement using transduction. Technical report, Weizmann Institute, 1993.

[Kur89] R.P. Kurshan. Analysis of discrete event coordination. In *REX Workshop on Stepwise Refinement of Distributed Systems, Mook*. LNCS 430, Springer Verlag, 1989.

[Lam79] L. Lamport. How to make a multiprocessor that correctly executes multiprocess programs. *IEEE Transactions on Computers*, C-28:690–691, 1979.

[Lam91] L. Lamport. The temporal logic of actions. Technical Report 79, DEC Systems Research Center, 1991.

[LGS+92] C. Loiseaux, S. Graf, J. Sifakis, A. Bouajjani, and S. Bensalem. Property preserving abstractions for the verification of concurrent systems. *To appear in Formal Methods in System Design*, also in *CAV'92*.

[Loi94] C. Loiseaux. Vérification symbolique de programmes réactifs à l'aide d'abstractions. Thesis, Verimag, Grenoble, January 1994.

[Lon93] D. E. Long. Model checking, abstraction and compositional verification. Phd thesis, Carnegie Mellon University, July 1993.

[Pnu85] A. Pnueli. In transition from global to modular temporal reasoning about programs. In *Logics and Models for Concurrent Systems*. NATO, ASI Series F, Vol.13, 1985.

[Pnu86] A. Pnueli. Specification and Development of reactive Systems. In *Conference IFIP, Dublin*. North-Holland, 1986.

[SG90] G. Shurek and O. Grumberg. The modular framework of computer-aided verification: motivation, solutions and evaluation criteria. In *CAV*, LNCS 531, 1990.

Beyond Model Checking

Zohar Manna

Department of Computer Science
Stanford University
Stanford, CA 94305

This talk will describe STEP (the Stanford TEmporal Prover), a system to support computer-aided verification of reactive systems based on their temporal specifications. Unlike most systems for temporal verification, STEP does not concentrate solely on finite-state systems. It combines model checking with algorithmic deductive methods (decision procedures) and interactive deductive methods (theorem proving). The user is expected to interact with the system and provide, whenever necessary, top-level guidance in the form of auxiliary invariants for safety properties, and well-founded measures and intermediate assertions for progress properties.

A verification system which combines model checking and deductive methods can offer a number of advantages over purely model checking or purely deductive approaches. Such a system should

- Avoid the state-explosion problem by decomposing the verification of large systems into the verification of smaller system components, where each component may be verified by a different method.

- Use deduction rules, if appropriate, to replace the global verification of a system component by local proofs of verification conditions.

- Free the user from having to deal with low-level details of verification, which are handled through model checking or decision procedures, and

- Allow the verification of infinite-state systems, such as parameterized systems or systems with infinite data domains.

In short, STEP has been designed with the objective:

> To combine the expressiveness of deductive methods with the simplicity of model checking.

Development efforts have been focused, in particular, in the following areas.

First, in addition to the textual language of temporal logic, the system supports a structured visual language of *verification diagrams*[1] for guiding proofs. Verification diagrams allow the user to construct proofs hierarchically, starting from a high-level, intuitive proof sketch and proceeding incrementally, as necessary, through layers of greater detail.

Second, the system implements powerful techniques (algorithmic and heuristic) for automatic *invariant generation*. Deductive verification in the temporal framework almost always relies heavily on finding, for a given program and specification, suitably

---

[1] Z. Manna and A. Pnueli. *Temporal Verification of Reactive Systems*. Springer-Verlag, New York, 1994. To appear.

strong invariants and intermediate assertions. The user can typically provide an intuitive, high-level invariant, from which the system derives stronger, *top-down invariants*. Simultaneously, *bottom-up invariants* are generated automatically by analyzing the program text. By combining these two methods, the system can often deduce sufficiently detailed invariants to carry through the entire verification process.

Finally, the system provides a built-in facility for automatically checking a large class of first-order and temporal formulas, based on simplification methods, term rewriting, and decision procedures. This degree of automated deduction is sufficient to handle most of the verification conditions that arise during the course of deductive verification — and the few conditions that are not solved automatically correspond to the critical steps of manually constructed proofs, where the user is most capable of providing guidance.

Many of the examples in the Manna-Pnueli textbook *Temporal Verification of Reactive Systems*[2] have been automatically verified using STEP. These include resource allocation protocols based on message-passing and on shared variables, a solution to the readers-writers problem, examples using infinite data domains, and a parameterized solution for the N-process dining philosophers problem. We have also automatically verified a substantially more complex parameterized program, Szymanski's N-process mutual exclusion protocol.

---

[2] Z. Manna and A. Pnueli. *Temporal Verification of Reactive Systems.* Springer-Verlag, New York, 1994. To appear.

Models Whose Checks Don't Explode

R. P. Kurshan

AT&T Bell Laboratories, Murray Hill, New Jersey 07974

Abstract. Automata-theoretic verification is based upon the language containment test

$$\mathcal{L}(P_0 \otimes P_1 \otimes \cdots \otimes P_k) \subset \mathcal{L}(T)$$

where the P_i's are automata which together model a system with its fairness constraints, $\otimes$ is a parallel composition for automata and T defines a specification. The complexity of that test typically grows exponentially with k. This growth, often called "state explosion", has been a major impediment to computer-aided verification, and many heuristics which are successful in special cases, have been developed to combat it. While all such heuristics are welcome advances, it often is difficult to quantify benefit in terms of hard upper bounds. This paper gives a general algorithm for that language containment test which has complexity $O(k)$ when most of the P_i's are of a special type, which generalizes strong fairness properties. In particular, the algorithm and bound reduce to the natural generalization for testing the language emptiness of a nondeterministic Streett automaton, in which the normal acceptance condition is generalized to allow an arbitrary Boolean combination of strong fairness constraints (not just a conjunction), expressible in disjunctive normal form with k literals. The algorithm may be implemented either as a BDD-based fixed point routine, or in terms of explicit state enumeration.

1 Introduction

It is well-known that testing emptiness of language intersection

$$\cap_{i=1}^{k} \mathcal{L}(P_i) = \phi$$

for automata P_i, is PSPACE-complete [Koz77], [GJ79]. This is germane to automata-theoretic formal verification based on automata admitting of a parallel composition $\otimes$ which supports the *language intersection property*:

$$\mathcal{L}(P_1 \otimes \ldots \otimes P_k) = \cap_{i=1}^{k} \mathcal{L}(P_i)$$

as then the test

$$(*) \qquad\qquad \mathcal{L}(P_0 \otimes P_1 \otimes \cdots \otimes P_k) = \phi$$

is PSPACE-complete as well. This latter test enters into automata-theoretic verification when the P_i's model the components of a system together with its fairness constraints and the properties which are to be verified. As a result of this complexity barrier, many heuristics have been proposed for this test, including compositional techniques such as [GL91], [Lon93]. Many of these techniques are completely general and very powerful, leading to checks of $(*)$ which empirically seem to grow linearly with k, in many cases. In given problems, such compositional techniques thus often make the difference between computational tractability and intractability. However, for most interesting cases, there is no guarantee of a linear-time check.

A natural form of incremental check is to compute and then reduce each of the successive terms

$$P_0 \otimes P_1 \otimes \cdots \otimes P_i$$

for $i = 1, \ldots, k$, with the hope that internal cancellations will keep these successive terms small [GS91]. However, a commonly observed problem with this approach is that computing the middle terms (for $i \approx k/2$) very often involves an excessively large amount of computation – larger even than required to compute the final term $P_0 \otimes P_1 \otimes \cdots \otimes P_k$ directly (without benefit of successive reductions). The reason for this is that the middle terms model large and highly unconstrained systems which thus generate many states; many of these states, however, are unreachable in the complete, more constrained model.

The algorithm given in Section 3 also is based upon computations involving the successive terms $P_0 \otimes P_1 \otimes \cdots \otimes P_i$. However, each successive step involves computations in a model no larger than $O(|P_0 \otimes M|)$, where M is the largest P_i.

We assume as the underlying semantic model, the fully expressive type of ω-automata known as L-process [Kur90], for which the language-containment problem

$$\mathcal{L}(P_0 \otimes P_1 \otimes \cdots \otimes P_k) \subset \mathcal{L}(T)$$

may be solved in time linear in the number of edges of the specification model T. In [Kur90], this problem is transformed to the language emptiness problem $(*)$ for L-processes, in time linear in the number of edges of T. Here, we assume this transformation, and address the check $(*)$. We give a general algorithm for this check, which has complexity $O(k)$ when most of the L-processes P_i are of a special type, which generalizes strong fairness properties. In particular, the algorithm and bound reduce to the natural generalization for testing the language emptiness of a nondeterministic Streett automaton, in which the normal Streett acceptance condition is generalized to allow an arbitrary Boolean combination of strong fairness constraints (not just a conjunction), where this is expressible in disjunctive normal form with k literals. Moreover, the algorithm (but not the bound) also applies to the case where the P_i's are entirely arbitrary. The algorithm may be implemented either as a BDD-based fixed point routine, or in terms of explicit state enumeration.

More specifically, for a given L-process P, an L-process Q is defined to be P-adic if it possesses two properties of the Streett strong fairness acceptance condition: it is P-faithful, in the sense that the behavior of Q is a function of the state transitions of P, and it is infinitary, in the sense that its acceptance conditions depend only upon eventualities. The $O(k)$ bound applies to any P-adic L-processes. The essence of the algorithm is very simple, and closely related to a natural efficient test for emptiness of the language of a nondeterministic Streett automaton with k strong fairness constraints: each successive fairness constraint is tested against the set of fair behaviors defined by the previous constraints. Each set of fair behaviors corresponds to a set of strongly connected components of P. Therefore, for each successive constraint, it is necessary only to test the corresponding set of strongly connected components defined by the previous constraint. This gives rise to a recursive procedure which finds the strongly connected components of each strongly connected component defined by the previous constraint. Although each successive set of constraints is defined in terms of an L-process with a transition structure of its own, the fact that it is P-adic ensures that these constraints may be pulled back to the transition structure of P itself, without any ensuing blow-up of the state space.

2   Basics

Boolean Algebra

A Boolean algebra [Hal74] is a set L with distinguished elements $0, 1 \in L$, closed under the Boolean operations:

$$* \; - \; \text{AND}$$
$$+ \; - \; \text{OR}$$
$$\sim \; - \; \text{NOT}$$

with universal element 1 and its complement 0. A Boolean algebra $L' \subset L$ is a *subalgebra* of L if L' and L share the same $0, 1$ and their operations agree. Every Boolean algebra contains the trivial 2-element Boolean algebra $\mathbb{B} = \{0, 1\}$ as a subalgebra. For $x, y \in L$, write $x \leq y$ if and only if $x * y = x$. $S(L)$— the *atoms* of L, are the nonzero elements of L, minimal with respect to $\leq$. For an arbitrary set V, define $\mathbb{B}[V]$ to be the Boolean algebra 2^V, with Boolean set operations. For notational simplicity, for $v \in V$, $\{v\} \in \mathbb{B}[V]$ may be denoted by v.

2.1 Definition For $L_1, \ldots, L_k \subset L$ subalgebras, define their (*interior*) *product*

$$\prod_{i=1}^{k} L_i = \left\{ \sum_{j \in J} x_{1j} * \cdots * x_{kj} \; \middle| \; x_{ij} \in L_i, \; J \text{ finite} \right\} .$$

In [Sik69, §13] it is proved that for any Boolean algebras $L_1, \ldots, L_k$, there exists a Boolean algebra L such that (isomorphic copies of) $L_1, \ldots, L_k \subset L$ and $\prod_{i=1}^{k} L_i = L$. This is defined to be the *exterior* product of $L_1, \ldots, L_k$.

Transition Structure

2.2 Definition Let V be a nonempty set, and let M be a map

$$M : V^2 \to L \quad (V^2 = V \times V, \text{ the Cartesian product}).$$

Say M is an *L-matrix* with *vertices* or *state-space* $V(M) = V$, and *edges* or *transitions* $E(M) = \{e \in V^2 | M(e) \neq 0\}$. M provides the (static) transition function for automata. Note that $M(e) = \sum\limits_{\substack{s \in S(L) \\ s \leq M(e)}} s$ (where each s is an "input letter"). For all $v \in V(M)$, define

$$s_M(v) = \sum_{w \in V(M)} M(v, w) \ .$$

2.3 Definition Let M, N be L-matrices with

$$V(M) \cap V(N) = \phi \ .$$

Their *direct sum* $M \oplus N$ is L-matrix with

$$V(M \oplus N) = V(M) \cup V(N) \ ,$$

defined by:

$$(M \oplus N)(v, w) = \begin{cases} M(v, w) \text{ if } v, w \in V(M) \\ N(v, w) \text{ if } v, w \in V(N) \\ 0 \qquad \text{otherwise} \end{cases}$$

Their *tensor product* $M \otimes N$ is L-matrix with

$$V(M \otimes N) = V(M) \times V(N) \ ,$$

where

$$(M \otimes N)((v, v'), (w, w')) = M(v, w) * N(v', w') \ .$$

2.4 Definition A *path* in M is a string $\mathbf{v} = (v_0, \ldots, v_n) \in V(M)^{n+1}$ for $n \geq 1$ such that $(v_i, v_{i+1}) \in E(M)$ for $i = 0, \ldots, n-1$. If $v_n = v_0$, $\mathbf{v}$ is a *cycle*. Say w is *reachable* from $v \in V(M)$ or $I \subset V(M)$ if there is a path $\mathbf{v}$ with $v_0 = v$ (resp., $v_0 \in I$) and $v_n = w$. Say $C \subset V(M)$ is *strongly connected* if for each $v, w \in C$, there is a path in C from v to w. (**NB**: by this definition, $\{v\}$ is strongly connected if and only if $(v, v) \in E(M)$.) A *directed graph* is a $\mathbb{B}$-matrix.

Automata

2.5 Definition An *L-process* P is a 5-tuple

$$P = (L_P, \ M_P, \ I(P), \ R(P), \ Z(P))$$

where L_P is a subalgebra of L (the *output* subalgebra), M_P is an arbitrary L-matrix, and

$$I(P) \subset V(M_P) \quad (\textit{initial} \text{ states})$$
$$R(P) \subset E(M_P) \quad (\textit{recur} \text{ edges})$$
$$Z(P) \subset 2^{V(M_P)} \quad (\textit{cycle} \text{ sets}) \ .$$

For an L-process P, write

$$V(P) \equiv V(M_P), \quad E(P) \equiv E(M_P), \quad P(v, w) \equiv M_P(v, w), \quad s_P(v) \equiv s_{M_P}(v).$$

2.6 Definition The *selections* of an L-process P at $v \in V(P)$ are the elements of the set

$$S_P(v) = \{s \in S(L_P) \mid s * s_P(v) \neq 0\} .$$

The intended interpretation of "selection" is a set of (nondeterministic) outputs as a function of state. (These may be considered to be outputs either of the associated process or of a hidden internal process.) The nondeterministic nature of selection is an important facility for modelling abstraction: abstraction of function, achieved through modelling an algorithm by a nondeterministic choice of its possible outcomes, and abstraction of duration, achieved through modelling a specific sequence of actions by a delay of nondeterministic duration.

2.7 Definition Let M be an L-matrix and let $\mathbf{v} \in V(M)^\omega$. Set

$$\mu(\mathbf{v}) = \{v \in V(M) \mid v_i = v \text{ infinitely often}\},$$
$$\beta(\mathbf{v}) = \{e \in E(M) \mid (v_i, v_{i+1}) = e \text{ infinitely often}\} .$$

2.8 Definition Let P be an L-process. The *language* of P is the set $\mathcal{L}(P)$ of $\mathbf{x} \in S(L)^\omega$ such that for some run $\mathbf{v}$ of $\mathbf{x}$ in P with $v_0 \in I(P)$,

$$\beta(\mathbf{v}) \cap R(P) = \phi \quad \text{and} \quad \mu(\mathbf{v}) \cap (V(P) \setminus C) \neq \phi \ \forall C \in Z(P) .$$

Such a run $\mathbf{v}$ is called an *accepting* run of $\mathbf{x}$.

Note that if P is an L-process and L is a subalgebra of L', then P is an L'-process. However, the language of P as an L'-process is not the same as the language of P as an L-process (unless $L' = L$). In such cases, the context will make clear which language is meant.

2.9 Definition If P is an L-process and $W \subset V(P)$, define the *restriction of P to W* to be the L-process $P|_W$ with $L_{P|_W} = L_P$, $V(P|_W) = W$, $M_{P|_W}(e) = P(e)$ for all $e \in W^2$, $I(P|_W) = I(P) \cap W$, $R(P|_W) = R(P) \cap W^2$ and $Z(P|_W) = \{C \cap W \mid C \in Z(P)\}$.

2.10 Definition For an L-process P, let $W \subset V(P)$ be the states reachable from $I(P)$ through a path which may be extended to an accepting run of P, and set $P^* = P|_W$.

2.11 Lemma P^* *is an L-process and* $\mathcal{L}(P^*) = \mathcal{L}(P)$.

2.12 Definition Let $P_1, \ldots, P_k$ be L-processes. Then

$$\bigoplus_{i=1}^{k} P_i = \left(L_{P_1}, \ \bigoplus_i M_{P_i}, \ \bigcup_i I(P_i), \ \bigcup_i R(P_i), \ \bigcup_i Z(P_i) \right)$$

$$\bigotimes_{i=1}^{k} P_i = \left(\Pi_i L_{P_i}, \ \bigotimes_i M_{P_i}, \ \mathbf{X} I(P_i), \ \bigcup_i \Pi_i^{-1} R(P_i), \ \bigcup_i \Pi_i^{-1} Z(P_i) \right)$$

where $\Pi_i^{-1} Z(P_i) \equiv \{\Pi_i^{-1} C \mid C \in Z(P_i)\}$. Here, $\oplus P_i$ undefined unless $L_{P_1} = \cdots = L_{P_k}$).

2.13 Lemma *Let $P_1, \ldots, P_k$ be L-processes. Then*

$$\mathcal{L}\left(\bigoplus P_i\right) = \bigcup \mathcal{L}(P_i)$$

$$\mathcal{L}\left(\bigotimes P_i\right) = \bigcap \mathcal{L}(P_i) \, .$$

2.14 Lemma *For L-processes P, Q and $v \in V(P)$, $w \in V(Q)$,*

$$s_{P \otimes Q}(v, w) = s_P(v) * s_Q(w) \, .$$

2.15 Definition Let P be an L-process and let $Q = P|_W$ where W is the set of states of P reachable from $I(P)$. Define P^o to be the directed graph with edges $E(Q^*) \setminus R(Q^*)$. Let $\mathcal{B}(P)$ be the set of strongly connected components of P^o contained in no element of $Z(P)$.

3 *P*-adic Processes

The condition known as *strong-fairness*, although the foundation of the Streett automaton acceptance condition, often is conceived in purely logical terms.

3.1 Definition A *strong-fairness* constraint *on* the set S with designated set of *initial* states $I(S)$ is a pair (L, U) of subsets of S. Its *satisfaction* set is

$$\mathbf{SF}(L, U) = \{\mathbf{v} \in S^\omega \mid v_0 \in I(S), \mu(\mathbf{v}) \cap L \neq \phi \Rightarrow \mu(\mathbf{v}) \cap U \neq \phi\} \, .$$

For a strong-fairness constraint on the set of states $V(P)$ of an L-process P, it is to be understood that the designated set of initial states $I(V(P)) = I(P)$.

Suppose it is required to verify that an L-process P has empty language. It may be that this test fails, unless P is subject to a number of strong-fairness constraints (L_i, U_i) on $V(P)$. (This arises naturally if the system model represented here by P is presented as a Streett automaton.) The strong-fairness constraint (L_i, U_i) may be represented by a 4-state L-process, provided P "outputs its state": *i.e.*, provided the output subalgebra L_P of P contains enough information to determine the state of P from its selections. This always can be accomplished by augmenting the output subalgebra L_P so as to contain the state of P as a component, as in Example 3.2.

3.2 Example Suppose P is an L-process whose output subalgebra L_P is an exterior product of the form $L_P = L' \cdot \mathbb{B}[V(P)]$, and $s_P(v) \leq v$ for all $v \in V(P)$. Then the state of P is a component of its selection: every selection of P is of the form $x * v$ where $x \in S(L')$ and $v \in V(P)$. In this case, a strong-fairness constraint (L_i, U_i) on $V(P)$ may be represented by the 4-state L-process $Q_i = Q_i^L \oplus Q_i^U$ where Q_i^L and Q_i^U are defined as follows: for $X = L_i$ or U_i the respective transition matrix of Q_i^L or Q_i^U is

$$
\begin{array}{c}
 & 0 \quad\; 1 \\
\begin{array}{c} 0 \\ 1 \end{array} & \begin{pmatrix} X' & X \\ X' & X \end{pmatrix}
\end{array}
$$

where $X' = V(P) \backslash X$, $I(Q_i^L) = I(Q_i^U) = \{0\}$, $R(Q_i^L) = \{(0,1),(1,1)\}$, $Z(Q_i^L) = \phi$, $R(Q_i^U) = \phi$, $Z(Q_i^U) = \{\{0\}\}$ and each has the trivial output subalgebra $\mathbb{B}$. Then Q_i is a $\mathbb{B}[V(P)]$-process and as such, for each run $\mathbf{v}$ of P,

$$\mathbf{v} \in \mathcal{L}(Q_i^L) \Leftrightarrow \mu(\mathbf{v}) \cap L_i = \phi,$$
$$\mathbf{v} \in \mathcal{L}(Q_i^U) \Leftrightarrow \mu(\mathbf{v}) \cap U_i \neq \phi$$

and

$$\mathcal{L}(Q_i) = \mathcal{L}(Q_i^L) \cup \mathcal{L}(Q_i^U).$$

Thus,

$$\mathbf{v} \in \mathbf{SF}(L_i, U_i) \Leftrightarrow \mathbf{v} \in \mathcal{L}(Q_i).$$

Consequently, for several strong-fairness constraints (L_i, U_i), the subset of $\mathcal{L}(P)$ whose runs all satisfy $\cap \mathbf{SF}(L_i, U_i)$ is precisely $\mathcal{L}(P \otimes Q)$ where $Q = \otimes Q_i$. Hence, to show that P subject to the strong-fairness constraints (L_i, U_i) has empty language, corresponds to showing

3.3 $\mathcal{L}(P \otimes Q) = \phi$.

The size of $P \otimes Q$ grows geometrically with the number of strong-fairness constraints. In fact, it can be shown that in the worst case, if $\mathcal{L}(P') = \mathcal{L}(P \otimes Q)$, then $2^k \leq |V(P')|$ [HSB94]. Thus, it may seem that the computational complexity of testing (3.3) also should grow thus. However, in what follows, it is shown that this is not the case. In fact, for a class of L-processes Q_i which contains as a proper subset those L-processes derived from strong-fairness constraints (as above), the complexity of testing

3.4 $\mathcal{L}(P \otimes Q_1 \otimes \cdots \otimes Q_k) = \phi$

is only *linear* in the size of k. Moreover, we will see that it is not even necessary to test the Q_i's for membership in this special class: the algorithm to test (3.4) will have complexity which is linear in k when the Q_i's are of this class, but will test (3.4) for any L-processes Q_i whatsoever.

The next definition generalizes the context of Example 3.2.

3.5 Definition Let P, Q be L-processes. Say Q is *P-faithful* provided for all $v \in V(P)$, and $w, w' \in V(Q)$,

$$s_P(v) * Q(w, w') \neq 0 \Rightarrow s_P(v) \leq Q(w, w').$$

Thus, Q is P-faithful if whenever some selection of P at v enables a given transition of Q, then *every* selection at v enables that transition. In other words, Q cannot distinguish among the different selections of P at a given state, and the behavior of Q is a function of the state transitions of P.

3.6 Lemma *If Q is P-faithful and* $\mathbf{x}, \mathbf{y} \in \mathcal{L}(P)$ *share the same run of P, then* $\mathbf{x} \in \mathcal{L}(Q) \Leftrightarrow \mathbf{y} \in \mathcal{L}(Q)$.

3.7 Definition Let P be an L-process and let $L' \subset L$ be a subalgebra. Say L' is *P-faithful* provided that for any $x, y \in S(L')$, if $v \in V(P)$, $x * s_P(v) \neq 0$ and $y * s_P(v) \neq 0$, then $x = y$.

The prototype P-faithful subalgebra is the subalgebra $\mathbb{B}[V(P)]$ of Example 3.2. A P-faithful subalgebra L' is "faithful" to the state of P, inasmuch as distinct atoms $x, y \in S(L')$ correspond to distinct states of P. The atom $x \in S(L')$ "corresponds" to the state $v \in V(P)$ if $x * s_P(v) \neq 0$, and for each state v, this is true of exactly one element of $S(L')$.

3.8 Proposition *Let P be an L-process, $L' \subset L$ a P-faithful subalgebra and let Q be an L'-process. Then Q is P-faithful.*

Proof. Let $v \in V(P)$ and $w, w' \in V(Q)$. Suppose $\widehat{x} \equiv s_P(v) * Q(w, w') \neq 0$, and let $\widehat{y} = s_P(v) * \sim Q(w, w')$. By assumption, $\widehat{x} > 0$, so for some $x \in S(L')$, $x * \widehat{x} > 0$. If $\widehat{y} > 0$, then likewise for some $y \in S(L')$, $y * \widehat{y} > 0$. Thus, $x * s_P(v) \neq 0$ and $y * s_P(v) \neq 0$, so $x = y$. However, $x \leq Q(w, w')$ while $y \leq \sim Q(w, w')$, so $x = y = 0$, a contradiction. It follows that $\widehat{y} = 0$, so $s_P(v) \leq Q(w, w')$, that is, Q is P-faithful.

P-faithfulness is one half of a generalization of strong-fairness. If Q_i is the L-process constructed in Example 3.2 to implement the strong-fairness constraint (L_i, U_i), then by Proposition 3.8, Q_i is P-faithful. The other half of the generalization relates to the acceptance condition, as follows.

3.9 Definition An L-ω-language $\mathcal{L}$ is said to be *infinitary* if whenever $\mathbf{a}, \mathbf{a}' \in S(L)^*$ and $\mathbf{b} \in S(L)^\omega$, then

$$\mathbf{ab} \in \mathcal{L} \Rightarrow \mathbf{a}'\mathbf{b} \in \mathcal{L} \ .$$

An L-process P is *infinitary* if $\mathcal{L}(P)$ is.

Thus, a language $\mathcal{L}$ is infinitary if membership in $\mathcal{L}$ depends only upon eventualities. It is easily seen that each Q_i of Example 3.2 is infinitary. Thus, infinitary and P-faithful together generalize strong-fairness, allowing more general acceptance conditions and sequentiality (defined by the transition structure).

3.10 Definition Let P be an L-process. An L-process Q is said to be *P-adic* if Q is infinitary and P-faithful. Set

$$\mathcal{L}_P = \{\mathcal{L}(P \otimes Q) \mid Q \text{ is } P\text{-adic}\} \ ,$$

the *P-adic languages.*

3.11 Lemma *If Q_1, Q_2 are P-faithful (respectively, infinitary), then the same is true for $Q_1 \otimes Q_2$ and $Q_1 \oplus Q_2$. If $\mathcal{L}$ is infinitary, so is the complementary language $\mathcal{L}'$.*

Proof. Suppose Q_1, Q_2 is P-faithful. Obviously, $Q_1 \oplus Q_2$ is P-faithful. Let $(w_1, w_2), (w_1', w_2') \in V(Q_1 \otimes Q_2)$ and suppose

$$s_P(v) * (Q_1 \otimes Q_2)((w_1, w_2), (w_1', w_2')) \neq 0 .$$

Then $s_P(v) * Q_1(w_1, w_1') * Q_2(w_2, w_2') \neq 0$ so

$$s_P(v) \leq Q_1(w_1, w_1') * Q_2(w_2, w_2') = (Q_1 \otimes Q_2)((w_1, w_2), (w_1', w_2')) .$$

Suppose Q_1, Q_2 are infinitary and $ab \in \mathcal{L}(Q_1 \oplus Q_2)$. Then $ab \in \mathcal{L}(Q_i)$ for $i = 1$ or 2, so for any a', $a'b \in \mathcal{L}(Q_i) \subset \mathcal{L}(Q_1 \oplus Q_2)$. If $ab \in \mathcal{L}(Q_1 \otimes Q_2)$ then $ab \in \mathcal{L}(Q_i)$ for $i = 1$ and 2, so likewise $a'b$ for any a'.

If $\mathcal{L}$ is infinitary and $ab \in \mathcal{L}'$, let a' be chosen. If $a'b \in \mathcal{L}$ then also $ab \in \mathcal{L}$, which is impossible. Thus, $a'b \in \mathcal{L}'$.

3.12 Corollary *$\mathcal{L}_P$ is closed under union and intersection.*

3.13 Note Ken McMillan has shown that $\mathcal{L}_P$ is closed under relative complement, as well: that $\mathcal{L} \in \mathcal{L}_P \Rightarrow \mathcal{L}(P) \setminus \mathcal{L} \in \mathcal{L}_P$. For example, for Q_i as in Example 3.2, each Q_i is P-adic (as already observed). Setting $\widehat{Q}_i = \widehat{Q}_i^L \otimes \widehat{Q}_i^U$, where $\widehat{Q}_i^L$ and $\widehat{Q}_i^U$ are formed from Q_i^L and Q_i^U by interchanging the cycle set and recur edges ($Z(Q_i^L) = \{\{0\}\}$, $R(Q_i^U) = \{(0,1),(1,1)\}$, $R(Q_i^L) = Z(Q_i^U) = \phi$), gives $\mathcal{L}(\widehat{Q}_i) = \mathcal{L}(Q_i)'$. By Lemma (3.11), $\widehat{Q}_i$ is P-adic as well. Incidentally, even if P is as in Example 3.2, it is not the case that $\mathcal{L}_P = \{\mathcal{L}_f \mid f \in \mathcal{F}\}$ where $\mathcal{F}$ is the set of all Boolean combinations of satisfaction sets of strong-fairness constraints on $V(P)$, and for $f \in \mathcal{F}$, $\mathcal{L}_f = \{\mathbf{x} \in S(L)^\omega \mid \mathbf{x}$ has a run in $f\}$, although it is true that $\{\mathcal{L}_f \mid f \in \mathcal{F}\}$ is closed under complementation. The reason is that strong fairness alone cannot capture sequentiality. For example, let P be the $\mathbb{B}$-process with $V(P) = I(P) = \{0, 1\}$, $R(P) = Z(P) = \phi$ and $P(i, j) = j$ for $i, j \in \{0, 1\}$. Then $\mathcal{L} = (0 + 1)^+ (01)^\omega \in \mathcal{L}_P$ but $\mathcal{L} \neq \mathcal{L}_f$ for any $f \in \mathcal{F}$.

Let $P, Q_1, \ldots, Q_k$ be arbitrary L-processes. Set $G_0 = \{V(P)\}$ and for $i \geq 1$ set

$$G_i = \left\{ \Pi_{V(P)} C \mid C \in \mathcal{B}((P \otimes Q_i)|_{D \times V(Q_i)}), \ D \in G_{i-1} \right\}$$

(where $(P \otimes Q_i)|_{D \times V(Q_i)}$ is the restriction (2.9) of $P \otimes Q_i$ to $D \times V(Q_i) \subset V(P \otimes Q_i)$, and $\Pi_{V(P)} C$ is the projection of C to $V(P)$).

The following theorem shows that (3.4) may be tested in time linear in k provided each Q_i is P-adic. The algorithm consists of consecutively testing for emptiness the k sets G_i. This test has complexity $O(km)$ where $m = \max_i |E(Q_i)|$, which, incidentally, is the same complexity as testing emptiness for a deterministic Streett automaton P with k fairness constraints [Saf88].

3.14 Theorem *For P, Q_i and G_i as above,*

a) $G_k = \phi \Rightarrow \mathcal{L}(P \otimes Q_1 \otimes \cdots \otimes Q_k) = \phi;$
b) $\mathcal{L}(P \otimes Q_1 \otimes \cdots \otimes Q_k) = \phi \Rightarrow G_k = \phi,$ *provided each Q_i is P-adic.*

Proof.
a) Suppose $\mathbf{x} \in \mathcal{L}(P \otimes Q_1 \otimes \cdots \otimes Q_k)$ has an accepting run $\mathbf{v}$. Then $\mathbf{v}$ has the form

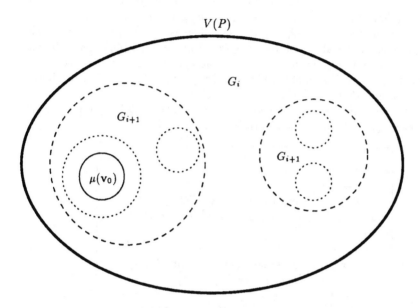

Fig. 1. Situation in the proof of (3.14)

$\mathbf{v} = (\mathbf{v}_0, \mathbf{v}_1, \ldots, \mathbf{v}_k)$ for $\mathbf{v}_0$ an accepting run of $\mathbf{x}$ in P and $\mathbf{v}_i$ an accepting run of $\mathbf{x}$ in Q_i, for $1 \le i \le k$. Thus, for some $K \in \mathcal{B}(P \otimes Q_1 \otimes \cdots \otimes Q_k)$, $\mu(\mathbf{v}) \subset K$ and so, in particular, for some $C_1 \in \mathcal{B}(P \otimes Q_1)$, $\phi \ne \mu(\mathbf{v}_0, \mathbf{v}_1) \subset \Pi_{V(P \otimes Q_1)} K \subset C_1$ and thus

$$\mu(\mathbf{v}_0) \subset \Pi_{V(P)} K \subset \Pi_{V(P)} C_1 \in G_1 .$$

Moreover, if $\mu(\mathbf{v}_0) \subset D_i \subset D_{i-1} \subset \cdots \subset D_1$ with $D_j \in G_j$ for $1 \le j \le i$, $\Pi_{V(P)} K \subset D_i$ and $i < k$, then, since $\Pi_{V(P \otimes Q_{i+1})} K$ is strongly connected and contained in $V(P \otimes Q_{i+1})^0$, for some $C_{i+1} \in \mathcal{B}((P \otimes Q_{i+1})|_{D_i \times V(Q_{i+1})})$,

$$\mu(\mathbf{v}_0, \mathbf{v}_{i+1}) \subset \Pi_{V(P \otimes Q_{i+1})} K \subset C_{i+1}$$

and thus $D_{i+1} \equiv \Pi_{V(P)} C_{i+1} \in G_{i+1}$ and $\mu(\mathbf{v}_0) \subset D_{i+1} \subset D_i$. Hence, by induction on k, $G_k \ne \phi$.
(b) Suppose $D_k \in G_k$. Then there is some $\mathbf{x} \in \mathcal{L}(P \otimes Q_k)$ with a run $(\mathbf{v}_0, \mathbf{v}_k)$ such that $\mu(\mathbf{v}_0, \mathbf{v}_k) \subset C \in \mathcal{B}((P \otimes Q_k)|_{D_{k-1} \times V(Q_k)})$ for some C with $\Pi_{V(P)} C = D_k$ and some $D_{k-1} \in G_{k-1}$. Thus, $\mu(\mathbf{v}_0) \subset D_k \subset D_{k-1}$. Now, suppose $\mu(\mathbf{v}_0) \subset$

$D_k \subset \cdots \subset D_i$ with $D_j \in G_j$, for $i \leq j \leq k$, and $(\mathbf{v}_0, \mathbf{v}_{i+1}, \ldots, \mathbf{v}_k)$ is an accepting run of $\mathbf{x}$ in $P \otimes Q_{i+1} \otimes \cdots \otimes Q_k$, for some i, $1 < i < k$. Since $D_i \in G_i$, there exists some accepting run $(\mathbf{w}_0, \mathbf{w}_i)$ in $P \otimes Q_i$ of (say) $\mathbf{y} \in \mathcal{L}(P \otimes Q_i)$, with $\mu(\mathbf{w}_0) \subset D_i$. Since $D_i = \Pi_{V(P)} C$ for some $C \in \mathcal{B}((P \otimes Q_i)|_{D_{i-1} \times V(Q_i)})$ where $D_{i-1} \in G_{i-1}$, it follows that $D_i \subset D_{i-1}$. Since $D_k \subset D_i$ and D_i is strongly connected, we may suppose that in fact, for some n, $w_{0j} = v_{0j}$ for $j \geq n$ (redefining w_{0j} as necessary). Thus, for $j \geq n$, $x_j \leq P(v_{0j}, v_{0j+1}) \leq s_P(v_{0j})$, while $y_j \leq P(v_{0j}, v_{0j+1}) * Q_i(w_{ij}, w_{ij+1})$. In particular, $y_j \leq s_P(v_{0j})$ and $y_j \leq Q_i(w_{ij}, w_{ij+1})$, so $s_P(v_{0j}) * Q_i(w_{ij}, w_{ij+1}) \neq 0$, for all $j \geq n$. Since Q_i is P-faithful, $s_P(v_{0j}) \leq Q_i(w_{ij}, w_{ij+1})$, whereas $x_j \leq s_P(v_{0j})$, and thus $x_j \leq Q_i(w_{ij}, w_{ij+1})$ for all $j \geq n$. Since Q_i is infinitary, $\mathbf{x} \in \mathcal{L}(Q_i)$. Let $\mathbf{v}_i$ be an accepting run of $\mathbf{x}$ in Q_i. Then $(\mathbf{v}_0, \mathbf{v}_i)$ is an accepting run of $\mathbf{x}$ in $P \otimes Q_i$ and so $(\mathbf{v}_0, \mathbf{v}_i, \mathbf{v}_{i+1}, \ldots, \mathbf{v}_k)$ is an accepting run of $\mathbf{x}$ in $P \otimes Q_i \otimes \cdots \otimes Q_k$. It follows by induction on k that $\mathbf{x} \in \mathcal{L}(P \otimes Q_1 \otimes \cdots \otimes Q_k)$.

This theorem gives a way to check (3.4) for arbitrary Q_i's (irrespective of whether each Q_i is P-adic). The algorithm is as follows:

```
i = 0
while i < k :
        i → i + 1
        if Gᵢ = φ, report (3.4) holds;  EXIT
find   x ∈ L(P) with accepting run³ v, μ(v) ⊂ D ∈ Gₖ
i = 1
while i < k :
        if x ∉ L(Qᵢ), repeat⁴ algorithm with P ⊗ Qᵢ
                in place of P, for {Qⱼ|j ≠ i}
        i → i + 1
report (3.4) fails — x ∈ L(P ⊗ Q₁ ⊗ ··· ⊗ Qₖ)
```

The complexity of this algorithm is $O(m^k)$ for $m = \max_i |E(Q_i)|$, but reduces to $O(km)$ in case the Q_i's are P-adic. Moreover, even in the general case, the empirical complexity often may look like $O(km)$. The algorithm can be implemented either through explicit state enumeration, or in terms of a BDD fixed point routine, as in [TBK91].

4 Conclusion

We have described a general algorithm for testing that a model P defined in terms of L-processes satisfies its specification. This algorithm has complexity which is linear in the number of component L-processes, when most of these

---

[3] It always is possible to find $\mathbf{x}$ of the form $\mathbf{x} = \mathbf{y}' \cdot \mathbf{y}^\omega$ for $\mathbf{v}$ of the form $\mathbf{v} = \mathbf{w}' \cdot \mathbf{w}^\omega$ with $\mathbf{w}, \mathbf{w}' \in V(P)^*$. By Lemma (3.6), if the Q_i's are P-faithful, then the choice of $\mathbf{x}$ for a given $\mathbf{v}$ is immaterial.

[4] If the Q_i's are all P-adic, then this recursive call is unreachable.

L-processes are P-adic, a class which generalizes strong fairness with sequential constraints. Currently, this algorithm is being implemented into the verification tool COSPAN [HK90]; however, as this implementation is not complete, there are no concrete results to report. Nonetheless, the linear bound speaks for itself.

Acknowledgement The author thanks Ken McMillan for his careful reading and helpful comments.

References

[GJ79]   M. R. Garey and D. S. Johnson. *Computers and Intractability.* Freeman, 1979.

[GL91]   O. Grumberg and D. E. Long. Model Checking and Modular Verification. In *Proc. CONCUR'91*, volume 527 of *Lec. Notes Comput. Sci. (LNCS)*. Springer-Verlag, 1991.

[GS91]   S. Graf and B. Steffen. Compositional Minimization of Finite State Systems. *Lec. Notes Comput. Sci. (LNCS)* **531**, pages 186–196, (1991).

[Hal74]   P. Halmos. *Lectures on Boolean Algebras.* Springer-Verlag, 1974.

[HK90]   Z. Har'El and R. P. Kurshan. Software for Analytical Development of Communications Protocol. *AT&T Tech. J.* **69**, pages 45–59, (1990).

[HSB94]   R. Hojati, V. Singhal, and R. K. Brayton. Edge-Street/Edge-Rabin Automata Environment for Formal Verification Using Language Containment. LICS (to appear), 1994.

[Koz77]   D. Kozen. Lower Bounds for Natural Proof Systems. *Proc 18th Symp. Found. Comput. Sci. (FOCS)*, pages 254–266, (1977).

[Kur90]   R. P. Kurshan. Analysis of Discrete Event Coordination. *Lec. Notes in Comput. Sci. (LNCS)* **430**, pages 414–453, (1990).

[Lon93]   D. E. Long. *Model Checking, Abstraction, and Compositional Verification.* PhD thesis, CMU, 1993.

[Saf88]   S. Safra. On the complexity of ω-automata. In *Proc. 29th Found. Comput. Sci. (FOCS)*, pages 319–327, 1988.

[Sik69]   R. Sikorski. *Boolean Algebras.* Springer-Verlag, 1969.

[TBK91]   H. Touati, R. Brayton, and R. P. Kurshan. Testing Language Containment for ω-Automata Using BDD's. *Lec. Notes in Comput. Sci. (LNCS)*, 1991.

On the Automatic Computation of Network Invariants

Felice Balarin* and Alberto L. Sangiovanni-Vincentelli

Department of Electrical Engineering and Computer Science
University of California, Berkeley, CA, USA 94720

Abstract. We study *network invariants*, abstractions of systems consisting of arbitrary many identical components. In particular, we study a case when an instance of some fixed size serves as an invariant. We study the decidability of the existence of such an invariant, present a procedure that will find it, if one exists, and finally give conditions under which such an invariant does not exist. These conditions can be checked in finite time, and if satisfied, they can be used in further searches for an invariant.

1 Introduction

The ability to create abstractions has been key in formal verification of complex digital systems (for example, see [3]). Usually, an abstraction is generated manually, at the considerable expense of time by the expert with the deep understanding of both the verification tool, and the system being designed.

One specific class of abstractions applies to systems with many identical components (also referred to as networks or iterative systems). Ideally, an abstraction of such a system should not depend on the actual number of components. Such an abstraction is called a *network invariant*. Once an invariant of manageable size is found it allows:

- a verification of a large system with a fixed number of components; and at the same time also
- a verification of the entire class of systems with the same structure but with different number of components.

This is of particular interest for distributed systems where algorithms (e.g. mutual exclusion) are usually designed to be correct for systems of any number of concurrent processes.

Although iterative systems have been studied for a long time [4], only recently there has been a significant interest in the formal verification of such systems. Browne, Clarke and Grumberg [2], and Shtadler and Grumberg [8] have studied conditions under which the satisfaction of formulae of certain temporal logics is independent of the size of the system. In [8] the conditions seem to be quite restrictive, while in [2] the conditions cannot in general be checked automatically.

* Supported by SRC under grant # 94-DC-008.

Wolper and Lovinfosse [9] have studied formal verification of iterative systems generated by interconnecting identical processes in certain regular fashion. They also present some decidability results for related problems. Kurshan and McMillan [5] present slightly more general results which can be applied both to process algebra and automata-based approaches. In both cases, automatic tools are used only to verify that a finite state system suggested by the user is indeed an invariant. Kurshan and McMillan have hinted that it might be a good idea to check whether a system of some fixed size serves as an invariant. This idea was further developed by Rho and Somenzi [6, 7], who have studied different network topologies and presented several sufficient conditions for the existence of such an invariant.

In this paper we address a problem of finding an invariant automatically. More precisely, we introduce a notion of a *tight* invariant, and give some results that can help a search for it. Intuitively, an invariant of a class of systems is a finite-state system that can exhibit any behavior that some system in the class can, and possibly some additional behaviors. Thus, in pre-order based formal verification paradigms (where a system is verified if it does not exhibit any undesirable behavior), an invariant is a conservative abstraction: if an abstraction is verified, so is every system in the class, but not vice versa. A tight invariant is an exact abstraction: if an abstraction is verified, so is every system in the class, and if an abstraction is not verified than there exists a system in the class which exhibits undesirable behavior. Thus, a tight invariant must exhibit *exactly* those behaviors that are exhibited by systems in the class.

Finding a tight invariant is easy if a *finite invariant* exists, i.e. if there exists a *finite* subclass such that any behavior exhibited by any system in the class is also exhibited by some system in the subclass. The main contribution of this paper is the test that can show that a finite invariant does not exist. The test is constructive in a sense that if successful, it identifies a set of behaviors that cannot be "covered" by any finite subclass, but must be exhibited by a tight invariant. Once identified, such a behavior can be added to the behaviors of some finite subclass to possibly generate a tight (but not finite) invariant. Unfortunately, we can not hope for a general algorithm that identifies all such sets of behaviors, because the existence of a finite invariant is undecidable (see Theorem 1). To the best of our knowledge no other algorithmic tests for the non-existence of a finite invariant are available.

In this paper we consider only networks of chain structure, i.e. every component can communicate only with its left and right neighbors. Also, we consider only automata on finite tapes. Thus, using our approach only safety properties can be verified.

The rest of the paper is organized as follows. In Sect. 2 we formally define the class of automata we consider, as well as rules by which these automata can be combined to form iterative systems. In Sect. 3 we illustrate these concepts on typical examples. The focus of our paper is the computation of the finite invariant presented in Sect. 4, where results on the decidability and the existence as well as the test for the non-existence of a finite invariant are given.

2 Basic Definitions

In this section we formalize the notion of an iterative system consisting of many identical automata. We assume that a state of the basic cell is fully observable and that transitions of the basic cell can depend on the present and next states of its left and right neighbors (see Fig. 1). These restrictions still enable us to model many regular hardware arrays, such as stacks, FIFO buffers and counters. Other examples that fit into our framework are a token passing mutual exclusion protocol [9] and the ever-so-popular Dining Philosophers Problem (e.g. [5]).

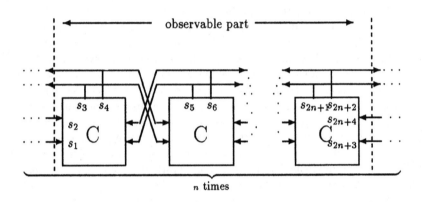

Fig. 1. An open network N_n.

In this paper, we adopt a standard definition of an automaton. More precisely, an *automaton* A over some finite alphabet Σ is a quadruple (S, I, T, F), where S is some finite *set of states*, $I \subseteq S$ is a *set of initial states*, $T \subseteq S \times \Sigma \times S$ is a *transition relation*, and $F \subseteq S$ is a *set of final states*. The language of A (denoted by $\mathcal{L}(A)$) is a set of all strings $x_1 x_2 \ldots x_k \in \Sigma^*$ for which there exists a sequence of states $s_0, s_1, \ldots, s_k$ such that $s_0 \in I$, $s_k \in F$ and $(s_{i-1}, x_i, s_i) \in T$ for all $i = 1, \ldots, k$.

We say that an automaton $A = (S, I, T, F)$ is a *cell* if it is defined over alphabet S^6 and the following condition holds:

$$(s, (s_1, s_2, s_3, s_4, s_5, s_6), t) \in T \text{ if and only if } s = s_3 \text{ and } t = s_4 \ .$$

Intuitively, the alphabet of a cell consists of three parts: s_1 and s_2 are the present and the next state of the left neighbor, s_3 and s_4 are the present and the next state of the cell itself, and finally s_5 and s_6 are the present and the next state of the right neighbor, as shown in Fig. 2.

To formalize connecting cells into larger units in [1] we define *concatenation* ".". Given two automata A and B the automaton the automaton $A \cdot B$ is well defined only if the alphabet of A is S^n and the alphabet of B is S^{k-n+4}, where

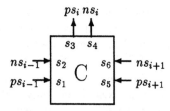

Fig. 2. A basic cell; *ps* and *ns* denote present and next states.

S is the state space of the basic cell and $k \geq n \geq 4$. The alphabet of $A \cdot B$ if S^k, and $\mathcal{L}(A \cdot B)$ satisfies the following:

$$\left(\begin{bmatrix} s_{1,1} \\ \vdots \\ s_{1,n-3} \\ \vdots \\ s_{1,n} \\ \vdots \\ s_{1,k} \end{bmatrix} \cdots \begin{bmatrix} s_{|s|,1} \\ \vdots \\ s_{|s|,n-3} \\ \vdots \\ s_{|s|,n} \\ \vdots \\ s_{|s|,k} \end{bmatrix} \right) \in \mathcal{L}(A \cdot B) \text{ if and only if}$$

$$\left(\begin{bmatrix} s_{1,1} \\ \vdots \\ s_{1,n} \end{bmatrix} \cdots \begin{bmatrix} s_{|s|,1} \\ \vdots \\ s_{|s|,n} \end{bmatrix} \right) \in \mathcal{L}(A) \text{ and } \left(\begin{bmatrix} s_{1,n-3} \\ \vdots \\ s_{1,k} \end{bmatrix} \cdots \begin{bmatrix} s_{|s|,n-3} \\ \vdots \\ s_{|s|,k} \end{bmatrix} \right) \in \mathcal{L}(B) .$$

Given a cell $C = (S, I, T, F)$ and an automaton E over alphabet S^4 (called an *environment*) a *network* of length n is defined by:

$$N_n = E \cdot \underbrace{C \cdot C \cdot \ldots \cdot C}_{n \text{ times}} .$$

If the behavior of the environment is unrestricted, i.e. if $\mathcal{L}(E) = (S^4)^*$ we say that the network is open. Otherwise, we say that the network is left-closed. In this paper, we will consider only open and left-closed networks, but the results are easily dualized for right-closed networks.

To compose networks into larger ones, only "peripheral" components of their languages need to be considered (see Fig. 1). Therefore, we introduce a notion of an *observable part*, first for elements of S^{2n+4}:

$$O((s_1, s_2, \ldots, s_{2n+4})) = (s_1, s_2, s_3, s_4, s_{2n+1}, s_{2n+2}, s_{2n+3}, s_{2n+4}) ,$$

and then, we extend the notion naturally to strings and languages (for technical details see [1]). For clarity, we write $\mathcal{L}_n$ instead of $O(\mathcal{L}(N_n))$.

An *iterative system* is the class of all networks of different length generated by the same cell and environment. If $\{N_n | n \geq 1\}$ is an iterative system and $\mathcal{L}_\infty = \bigcup_{n=1}^\infty \mathcal{L}_n$, then an *invariant* is any finite-state automaton A over alphabet S^8 satisfying $\mathcal{L}(A) \supseteq \mathcal{L}_\infty$. If $\mathcal{L}(A) = \mathcal{L}_\infty$, we say that A is a *tight invariant*. If in addition $\mathcal{L}(A) = \bigcup_{n=1}^{n^*} \mathcal{L}_n$ for some $n^* < \infty$, we say that A is a *finite invariant*.

Obviously, a tight invariant exists if and only if $\mathcal{L}_\infty$ is regular. One may try to find a "tightest regular invariant", i.e. an invariant that is not tight, but that has a language contained in the languages of all other invariants. Unfortunately, if $\mathcal{L}_\infty$ is not regular, such an invariant does not exist. To see this assume that A is such an invariant. Let x be some string in $\mathcal{L}(A) - \mathcal{L}_\infty$ (since $\mathcal{L}(A)$ is regular and $\mathcal{L}_\infty$ is not, such a string always exists). Since a language containing just x is also regular, one can construct an automaton A' such that $\mathcal{L}(A') = \mathcal{L}(A) - \{x\}$. Now, A' is an invariant and $\mathcal{L}(A) \not\subseteq \mathcal{L}(A')$, contradicting the assumption that A is the tightest regular invariant.

3 Examples

The following three examples illustrate three possible cases: when a finite invariant exists (Example 1), when a tight invariant exists, but a finite one does not (Example 2), and finally when a tight invariant does not exists (Example 3). All three examples are abstractions of buffers with different discipline of passing a token. In all three cases cells are initially in *idle* state. The state *token* indicates that a particular cell holds a token. In Examples 1 and 2, a cell moves to a special *dead* state once it has delivered a token. In the description that follows variables ps_{n-1}, ns_{n-1}, ps_{n+1} and ns_{n+1} take value of the present and the next state of immediate neighbors of the cell under consideration. In all examples, all states are final and all systems are open.

Example 1. In this example a cell can hold a token for any (possibly infinite) number of steps before delivering it to its neighbor. A cell can deliver only one token. More precisely the transition relation of the cell is defined by:

$$
\begin{array}{lll}
idle \longrightarrow idle : & \text{if } ps_{n-1} \neq token \text{ or } ns_{n-1} \neq dead \\
idle \longrightarrow token : & \text{if } ps_{n-1} = token \text{ and } ns_{n-1} = dead \\
token \longrightarrow token : & \text{always} \\
token \longrightarrow dead : & \text{always} \\
dead \longrightarrow dead : & \text{always}
\end{array}
$$

In this case a finite invariant exists. In fact, it is achieved for $n^* = 3$. For $n \geq 3$, a language $\mathcal{L}_n$ can be described by languages of the leftmost and the rightmost cell and the following additional constraint:

"If the rightmost cell ever moves from *idle* to *token* it will happen *at least $n - 2$ steps* after the leftmost cell leaves the *token* state."

Clearly, $\mathcal{L}_3$ (strictly) contains all $\mathcal{L}_n$'s, $n > 3$.

Example 2. In this example a cell holds a token for exactly one step. Again, once it delivers a single token, a cell will move to the *dead* state. The transition relation of the basic cell is:

$$idle \longrightarrow idle : \quad \text{if } ps_{n-1} = idle$$
$$idle \longrightarrow token : \quad \text{if } ps_{n-1} = token$$
$$token \longrightarrow dead : \text{if } ps_{n-1} = dead$$
$$dead \longrightarrow dead : \quad \text{if } ps_{n-1} = dead$$

In this case a finite invariant does not exist. All strings in $\mathcal{L}_n$ ($n \geq 2$) that are long enough must satisfy the following constraint:

"The rightmost cell moves *idle* to *token* *exactly* $n - 2$ step after the leftmost cell leaves the *token* state."

Obviously, for all $n \geq 3$ there are some strings in $\mathcal{L}_{n+1}$ which are not in $\mathcal{L}_n$. However, a tight invariant exists, and it is similar to the one in the previous example, except that the peripheral cells are restricted to remain in the *token* state for one step only.

Example 3. This example is similar to the previous one, except that once a cell delivers a token it will move back to the *idle* state and become ready to accept a new token.

$$idle \longrightarrow idle : \quad \text{if } ps_{n-1} \neq token$$
$$idle \longrightarrow token : \text{if } ps_{n-1} = token$$
$$token \longrightarrow idle : \text{always}$$

In this case $\mathcal{L}_\infty$ is not regular, so a tight invariant cannot exist. Indeed, $\mathcal{L}_\infty$ must include $\mathcal{L}_1$ and all strings for which there exists $k \geq 0$ such that the rightmost cell moves from *idle* to *token* exactly k steps after the leftmost cell leaves the *token* state. Notice that for any given string k must be constant. It is straightforward to show that such a language is not regular.

However, if we include in the description of the system an environment which allows at most one token in the system, a tight invariant exists and is similar to the one in Example 2.

4 Computing a Finite Invariant

4.1 Decidability and Existence

In this section we give some results on decidability as well as a simple semi-decision procedure that will find a finite invariant if one exists. Due to the space limitations we omit the proofs. They can be found in the extended version of this paper [1].

Theorem 1. *The existence of finite invariant for a left-closed iterative system is undecidable.*

The proof is by reduction of the finite memory problem for Turing machines. At present, it is not clear whether that proof can be extended to open systems. In fact, this result is similar to Theorem 4.3 in [9] and Theorem 4 in [4]. In all cases, the proofs substantially rely on the ability to distinguish one cell in the network: in our case, it is the environment, in [9] the first cell is explicitly distinguished, and in [4] one cell is distinguished by different boundary condition. Therefore, the decidability of the existence of a finite invariant for open systems is still an open problem.

Theorem 2. *Let $\{N_n | n \geq 1\}$ be an iterative system, and let A be some automaton. If $\mathcal{L}_1 \subseteq \mathcal{L}(A)$ and $O(\mathcal{L}(A \cdot C)) \subseteq \mathcal{L}(A)$, then A is an invariant of $\{N_n | n \geq 1\}$.*

The proof is by induction. Both Kurshan and McMillan [5] and Wolper and Lovinfosse [9] require by definition that an invariant satisfy the conditions similar to those in Theorem 2. We have adopted a broader definition, motivated by the application to language containment. Still, Theorem 2 provides the only finite procedure known to us, for verifying that a given automaton with non-trivial language is indeed an invariant.

Theorem 3. *A finite invariant of an iterative system $\{N_n | n \geq 1\}$ exists if and only if there exists $n^* < \infty$ such that:*

$$\mathcal{L}_{n^*+1} \subseteq \bigcup_{n=1}^{n^*} \mathcal{L}_n . \tag{1}$$

In fact, a stronger claim follows from the proof of Theorem 3: if (1) holds, then an automaton with the language $\bigcup_{n=1}^{n^*} \mathcal{L}_n$ is an invariant. This immediately gives us the following semi-decision procedure:

for every integer n^*: **if** (1) holds **then** HALT.

If the procedure terminates, it will produce n^* which can be used to construct a finite invariant with the language $\bigcup_{n=1}^{n^*} \mathcal{L}_n$. However, if a finite invariant does not exist, the procedure will not terminate.

4.2 Proving Non-existence of Finite Invariant

In this section, we show a sufficient condition for non-existence of finite invariant. The condition can be checked algorithmically, and, if satisfied, it provides useful information on sets of strings that every invariant must include in its language.

In this section we consider a generic *open* iterative system induced by a cell (S, I, T, F). Unless stated otherwise, we assume that s is some string in $(S^{2n})^*$. We use $| \cdot |$ to denote the length of a string. To refer to parts of the string we use the naming scheme detailed in Fig. 3. We call a pair $s_{x,y} = (s_{x,y}^1, s_{x,y}^2)$ a

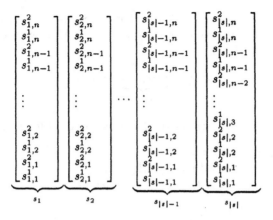

Fig. 3. A naming scheme for parts of the string.

transition, for all $x = 1, \ldots, |s|$ and all $y = 1, \ldots, n$. If s, t and u are transitions, for simplicity we write $(s, t, u) \in T$ instead of:

$$(t^1, (s^1, s^2, t^1, t^2, u^1, u^2), t^2) \in T .$$

To prove the non-existence of a finite invariant, we search for a sequence of strings: $x_1, x_2, \ldots$ satisfying $O(x_i) \in \mathcal{L}_i$, but $O(x_i) \notin \mathcal{L}_j$ for any $j < i$. We will show that in certain cases these relations can be established in a finite number of steps. We consider only a special case when x_{i+1} is obtained by extending x_i in a certain regular fashion. If that is the case we write $x_{i+1} = \alpha(x_i)$. Next, we define precisely the extension operator α.

Given strings $s \in (S^{2n})^*$ and $t \in (S^{2n+2})^*$ such that $|t| = |s| + 1$ we say that $t = \alpha_{ik}(s)$ if the following holds:

1. t obtained from s by adding one row and one column of transitions, i.e.:

$$t_{x,y} = s_{x,y} \text{ for all } x = 1, \ldots, |s|, \ y = 1, \ldots, n , \tag{2}$$

2. the observable part of t is the same as the observable part of s, except that the i'th column is repeated twice, i.e.:

$$O(t_x) = \begin{cases} O(s_x) & \text{for all } x = 1, \ldots, i , \\ O(s_{x-1})) & \text{for all } x = i+1, \ldots, |t| , \end{cases} \tag{3}$$

3. the last column of t is the same as the last column of s, except that the k'th row is repeated twice, i.e.:

$$t_{|t|,x} = \begin{cases} s_{|s|,x} & \text{for all } x = 1, \ldots, k , \\ s_{|s|,x-1} & \text{for all } x = k+1, \ldots, n+1 . \end{cases} \tag{4}$$

In the rest of this paper we assume that all extensions have common i and k, so without ambiguity we write $\alpha(s)$ for $\alpha_{ik}(s)$. Also, we use the following abbreviation for any $j \geq 0$:

$$\alpha^j(s) = \underbrace{\alpha(\alpha(\ldots\alpha(s)\ldots))}_{j \text{ times}} .$$

Proposition 4. *A string $\alpha(s)$ exists if and only if all of the following hold:*

C1: $s_{x,n} = s_{x,n-1}$, *for all* $x = 1, \ldots, i$,
C2: $s_{x+1,n} = s_{x,n-1}$, *for all* $x = i, \ldots, |s| - 1$,
C3: $s_{x,1} = s_{i,1}$, $s_{x,2} = s_{i,2}$, *for all* $x = i, \ldots, |s|$.

If $\alpha(s)$ exists, then it is unique.

Lemma 5. *If $s \in \mathcal{L}(N_{n-2})$, and s satisfies:*

C4: $s_{i,n-1} = s_{i+1,n-1}$,
C5: $s_{|s|,k} = s_{|s|,k+1} = s_{|s|,k+2}$,

then $t = \alpha(s)$ satisfies:

$$t^2_{x,y} = t^1_{x+1,y} \text{ for all } x = 1, \ldots, |s| - 1, \ y = 2, \ldots, n , \quad (5)$$

$$t^1_{1,x} \in I \text{ for all } x = 2, \ldots, n , \quad (6)$$

$$t^1_{|t|,x} \in F \text{ for all } x = 2, \ldots, n , \quad (7)$$

$$(t_{x,y-1}, t_{x,y}, t_{x,y+1}) \in T \text{ for all } x = 1, \ldots, |t|, y = 2, \ldots, n - 1 . \quad (8)$$

Proof. By definition, the assumption $s \in \mathcal{L}(N_{n-2})$ is equivalent to:

$$s^2_{x,y} = s^1_{x+1,y} \text{ for all } x = 1, \ldots, |s| - 1, \ y = 2, \ldots, n - 1 , (9)$$

$$s^1_{1,x} \in I \text{ for all } x = 2, \ldots, n - 1 , \quad (10)$$

$$s^1_{|s|,x} \in F \text{ for all } x = 2, \ldots, n - 1 , \quad (11)$$

$$(s_{x,y-1}, s_{x,y}, s_{x,y+1}) \in T \text{ for all } x = 1, \ldots, |s|, y = 2, \ldots, n - 1 . \quad (12)$$

Now, (5) follows from (2), (4), **C4**, and (9); (6) follows from (2), (4), and (10); (7) follows from (3), and (11); and finally (8) follows from (2), (3), **C5**, and (12). $\square$

From this point on, we do assume properties **C4** and **C5**. We make this assumption without loss of generality because they are always satisfied by $\alpha^2(s)$, which also satisfies all other restriction on s mentioned in this section.

The part of our strategy is to find a sequence of strings $x_1, x_2, \ldots$ satisfying $x_i \in \mathcal{L}_i$. The following lemma describes a case when all of these relations are satisfied if one of them is.

Lemma 6. *Let s satisfy **C1–C5**, and let $t = \alpha(s)$. If:*

C6: $(t_{x,n-1}, t_{x,n}, t_{x,n+1}) \in T$ *for all* $x = 1, \ldots, |t|$,

C7: $t^2_{|t|-1,x} = t^1_{|t|,x}$ *for all* $x = 2, \ldots, n$,

then:

$$s \in \mathcal{L}(N_{n-2}) \Longrightarrow t \in \mathcal{L}(N_{n-1}) , \tag{13}$$

$$s \in \mathcal{L}(N_{n-2}) \Longrightarrow \alpha^j(s) \in \mathcal{L}(N_{n-2+j}) \text{ for all } j \geq 0 . \tag{14}$$

Proof. Conditions (5)–(8), **C6** and **C7** are exactly the conditions for $t \in \mathcal{L}(N_{n-1})$ to be satisfied. Thus, (13) holds. If s satisfies **C6** and **C7** so does $\alpha(s)$. Therefore, we can repeatedly apply (13) to get (14). $\qquad\square$

The second part of our strategy is to find a sequence of strings $x_1, x_2, \ldots$ that $x_j \notin \mathcal{L}_i$ for any $j < i$. In Lemma 7 we establish a condition which enables us to prove this relation in a finite number of steps.

Lemma 7. *Let s satisfy* **C1–C5**, *and let* $t = \alpha(s)$. *If:*

C8: $s_{x,n-1}$ *is the unique element of the set* $\{u = (u^1, u^2) | \exists v, w, z = (u^2, z^2) : (v, t_{x,n}, u) \in T \wedge (w, t_{x+1,n}, z) \in T\}$ *for all* $x = 1, \ldots, |s|$, *and*
C9: *there exists a sequence of transition* $u_n, \ldots, u_4, u_3$ *such that* $u_n = t_{|t|,n}$, *and* u_x *(for all* $x = n-1, \ldots, 3)$ *is the unique element of the set:* $\{v | \exists w : (w, u_{x+1}, v) \in T\}$,

then:

$$O(t) \in \mathcal{L}_{n-1} \Longrightarrow O(s) \in \mathcal{L}_{n-2} , \tag{15}$$

$$O(s) \notin \mathcal{L}_{n-2} \Longrightarrow O(\alpha^j(s)) \notin \mathcal{L}_{n-2+j} \text{ for all } j \geq 0 . \tag{16}$$

Proof. Assume $O(t) \in \mathcal{L}_{n-1}$, let string t' be such that $t' \in \mathcal{L}(N_{n-1})$ and $O(t') = O(t)$. Also, let s' be the string obtained by removing from t' the $(n+1)$'st row and the last column. It follows from **C8** that the $(n-1)$'st row of s' is exactly equal to the $(n-1)$'st row of s. Since **C1–C3** also hold, we have that $O(s') = O(s)$. We claim that $s' \in \mathcal{L}(N_{n-2})$ and thus $O(s) \in \mathcal{L}_{n-2}$.

That s' satisfies conditions analog to (9), (10), and (12) follows directly from $t' \in \mathcal{L}(N_{n-1})$. It follows from **C9** that $t'_{|t'|,x} = u_x$ for all $x = n, \ldots, 3$, so $t' \in \mathcal{L}(N_{n-1})$ also implies $u^1_x \in F$, for all $x = 3, \ldots, n$. It follows from **C2** that $s'_{|s'|,n-1} = t'_{|t'|,n} = u_n$, so we can again apply **C9** to obtain $(s'_{|s'|,x})^1 = u^1_{x+1} \in F$, for all $x = n-1, \ldots, 2$.

It is easy to check that $\alpha(s)$ satisfies **C8** and **C9** if s does. Therefore, we can repeatedly apply (15) to get (16). $\qquad\square$

A reader will notice that the condition **C9** is used only to establish termination. If all the states of the basic cell are final, Lemma 7 holds even if **C9** is not satisfied.

We are now ready to postulate sufficient conditions for the non-existence of a finite invariant.

Theorem 8. *If s is such that it satisfies* **C1–C9** *and:*

$$s \in \mathcal{L}(N_{n-2}) , \tag{17}$$

$$O(\alpha^j(s)) \notin \mathcal{L}_{n-2}, \text{ for all } j > 0 , \tag{18}$$

$$O(\alpha^j(s)) \notin \mathcal{L}_m, \text{ for all } j \geq 0, \ m = 1, \dots, n-3 , \tag{19}$$

then:

a) *a finite invariant does not exist,*
b) $\mathcal{L}_\infty \supseteq \{O(\alpha^j(s)) | j \geq 0\}$.

Proof. From (17) and Lemma 6 we have:

$$O(\alpha^j(s)) \in \mathcal{L}_{n-2+j} \text{ for all } j \geq 0 . \tag{20}$$

We can rewrite (18) as $O(\alpha^{j-m}(s)) \notin \mathcal{L}_{n-2}$ for all $j > 0$, $0 \leq m < j$, and combine it with Lemma 7 to get:

$$O(\alpha^j(s)) = O(\alpha^m(\alpha^{j-m}(s))) \notin \mathcal{L}_{n-2+m} \text{ for all } j > 0, \ 0 \leq m \leq j . \tag{21}$$

Thus, for every $j \geq 0$ there exists a string (specifically $O(\alpha^j(s))$) in $\mathcal{L}_{n-2+j}$ (by (20)), which is not in $\mathcal{L}_m$ for any $m < n-2+j$ (by (19) and (21)), so a finite invariant cannot exist. Also, part **b)** follows from (20). □

Consider Example 2 in section 3 and the following string[2] in $\mathcal{L}_3$:

$$s = \begin{bmatrix} idle \\ idle \\ idle \\ idle \\ token \end{bmatrix} \begin{bmatrix} idle \\ idle \\ idle \\ token \\ dead \end{bmatrix} \begin{bmatrix} idle \\ idle \\ token \\ dead \\ dead \end{bmatrix} \begin{bmatrix} idle \rightarrow token \\ token \rightarrow dead \\ dead \rightarrow dead \\ dead \rightarrow dead \\ dead \rightarrow dead \end{bmatrix} .$$

It is straightforward to check that conditions **C1–C9** are satisfied for $\alpha_{2,1}$. It is also straightforward to define an automaton accepting $\{O(\alpha^j_{2,1}(s)) | j \geq 0\}$, thus (17)–(19) can be easily checked by a language containment checking tool. Since **C1–C9**, and (17)–(19) are all satisfied we conclude that a finite invariant does not exist and that $\mathcal{L}_\infty \supseteq (\mathcal{L}_1 \cup \mathcal{L}_2 \cup \{O(\alpha^j_{2,1}(s)) | j \geq 0\})$.

The following procedure shows how Theorem 8 can be used to search for an invariant:

1. construct A s.t. $\mathcal{L}(A) = \mathcal{L}_1$,
2. choose a string s and let n denote its "width" (i.e. $s \in (S^{2n})^*$),
3. construct B s.t. $\mathcal{L}(B) = \mathcal{L}(A) \cup \bigcup_{i=1}^{n-2} \mathcal{L}_i$; let $A := B$,
4. if $O(\mathcal{L}(A \cdot C)) \subseteq \mathcal{L}(A)$ then *HALT*,
5. if s satisfies **C1–C9**, (17)–(19) for some α_{ik}, where $1 \leq i \leq |s|$, $1 \leq k \leq n$
   then construct B s.t. $\mathcal{L}(B) = \mathcal{L}(A) \cup \{O(\alpha^j_{ik}(s)) | j \geq 0\}$; let $A := B$,
6. if $O(\mathcal{L}(A \cdot C)) \subseteq \mathcal{L}(A)$ then *HALT* else go to step 2.

---

[2] We omit writing $s^2_{x,y}$ for $x < 4$, and assume that $s^2_{x,y} = s^1_{x+1,y}$.

If the procedure terminates it will generate a tight (but possibly not finite) invariant A. Unfortunately, we can not claim that the procedure will terminate even if a tight invariant exists and all strings are systematically enumerated in step 2. It might be more efficient to apply this procedure interactively, i.e. to let the user choose a string and then execute other steps automatically.

5 Conclusions

We have studied the existence of the finite invariant of an iterative system consisting of many identical automata. We have shown that if constraints on the environment are allowed, the existence is undecidable, but we have also pointed out that the proof does not exist presently for the case of an unconstrained environment. We have presented a semi-decision procedure that will generate a finite invariant, if one exists. We have also provided sufficient conditions for the non-existence of a finite invariant. Those conditions can be checked in finite time so a semi-decision procedure can be defined that will recognize a pattern satisfying those conditions, if such a pattern exists. These results can then be used in a search for an invariant that is possibly tight but not finite. It is possible that neither of these procedures terminate. This is consistent with the decidability result (at least for closed systems).

This work can be naturally extended in several ways. From the theoretical point of view, the decidability of existence of a finite invariant for open iterative systems needs to be studied. From the practical point of view, it would be useful to generalize the conditions for non-existence. This can be done by analyzing sequences of strings where not only a single element, but a whole substring is repeated many times. It is also possible to construct cases where the non-existence can be proved by analyzing a sequence of sets of string, rather then just a sequence of strings. Finally, in order to verify liveness properties, these results need to be extended to the automata on infinite tapes.

References

1. Felice Balarin and Alberto L. Sangiovanni-Vincentelli. On the automatic computation of network invariants, 1994. UCB/ERL M94/18.
2. M.C. Browne, E.M. Clarke, and O. Grumberg. Reasoning about networks with many identical finite state processes. *Information and Computation*, 81(1):13–31, 1989.
3. E. M. Clarke, O. Grumberg, H. Hiraisi, S. Jha, D. E. Long, K. L. McMillan, and L. A. Ness. Verification of the Futurebus+ cache coherence protocol. In *Proc. 11th Intl. Symp. on Comput. Hardware Description Lang. and their Applications*, 1993.
4. Frederick C. Hennie. *Iterative Arrays of Logical Circuits*. MIT Press and John Eiley Sons, Inc., 1961.
5. R. P. Kurshan and K. L. McMillan. A structural induction theorem for processes. In *Proceedings of the 8th ACM Symp. PODC*, 1989.
6. J.K. Rho and F. Somenzi. Inductive verification of iterative systems. In *Proceedings of the 29th ACM/IEEE Design Automation Conference*, pages 628–33, June 1992.

7. J.K. Rho and F. Somenzi. Automatic generation of network invariants for the verification of iterative sequential systems. In Costas Courcoubetis, editor, *Computer Aided Verification: 5th International Conference, CAV'93, Elounda, Greece, June/July 1993, Proceedings*, pages 123–137. Springer-Verlag, 1993. LNCS vol. 697.
8. Z. Shtadler and O. Grumberg. Network grammars, communication behaviors and automatic verification. In J. Sifakis, editor, *Automatic Verification Methods for Finite State Systems, International Workshop Proceedings, Grenoble, France, 12-14 June 1989*, pages 151–65. Springer-Verlag, 1990. LNCS vol. 407.
9. P. Wolper and V. Lovinfosse. Verifying properties of large sets of processes with network invariants. In J. Sifakis, editor, *Automatic Verification Methods for Finite State Systems, International Workshop Proceedings, Grenoble, France, 12-14 June 1989*, pages 68–80. Springer-Verlag, 1990. LNCS vol. 407.

Ground Temporal Logic:
A Logic for Hardware Verification

David Cyrluk[1]* and Paliath Narendran[2] **

[1] Dept. of Computer Science, Stanford University, Stanford CA 94305 and
Computer Science Laboratory, SRI International, Menlo Park, CA 94025
cyrluk@cs.stanford.edu
[2] SUNY-ALBANY, Albany NY
dran@cs.albany.edu

Abstract. We present a new temporal logic, GTL, appropriate for specifying properties of hardware at the register transfer level. We argue that this logic represents an improvement over model checking for some natural hardware verification problems. We show that the validity problem for this logic is Π_1^1 complete. We then identify a fragment of the logic that is decidable. We show that in this fragment we are still able to encode many interesting problems, including the correctness of pipelined microprocessors.

1 Introduction

Temporal logic is a natural logic for hardware verification. Specifically model checking for various propositional temporal logics has proven to be a very practical tool for the fully automatic verification of many hardware circuits and finite state protocols. However these approaches suffer from various drawbacks.

One such drawback is the requirement that hardware implementations be carried out to the bit-level. This can lead to the state explosion problem as the number of states can increase exponentially with the number of bits in the implementation. It also necessitates a bit-level description of alus and adders. To deal with this problem current research relies on tools such as BDD's to encode a large number of states into a small representation [2, 5, 4]. [7] makes use of abstractions to significantly reduce the state space that needs to be explored.

However, the correctness argument for many of these circuits does not depend on a bit-level description of the circuit but only on a RTL description of the circuit. In such cases the correct abstraction is to abstract away from the bit-level using uninterpreted function symbols. Thus, perhaps, a first order temporal logic might be more appropriate for this type of hardware verification. The main drawback with using a full first-order temporal logic is that the validity problem now becomes incomplete, thus making automatic verification impossible.

* This research was partially supported by SRI International, DARPA contract NAG2-703, and NSF grants CCR-8917606, CCR-8915663.
** Much of this research was done while a visiting scientist at SRI International.

We thus propose a Ground Temporal Logic (*GTL*) that falls in between first-order and propositional temporal logic. To make this logic more expressive than propositional temporal logic we need to add a new *Next* temporal operator that operates on terms instead of on formulas. This allows us to relate successive states using uninterpreted functions without having to use quantified variables, and thus a first-order logic. For example, a typical statement one would want to encode is: *In the next state the value of C becomes $f(C)$*. In a first order temporal logic this would be encoded as $\forall x : C = x \supset \bigcirc(C = f(x))$, but in *GTL*, this could be more naturally stated without quantifiers: $\circ(C) = f(C)$.

Unfortunately this ground temporal logic is as undecidable as the full first-order logic. We identify a fragment of this logic that is straightforwardly decidable and yet still suitable for hardware verification. This fragment is expressive enough to express the correctness of the RSRE counter [8] verified using the interactive theorem prover, HOL. This example first motivated the definition of our language. On the one hand, we believed that the model-checking techniques associated with propositional temporal logic were more appropriate in verifying the correctness of the counter than theorem proving. On the other hand, by using a theorem prover we were able to abstract away from a bit-level description of the counter and thus obtain a more concise proof of correctness, that is independent of the size of the counter.

Our fragment is also expressive enough to express the correctness of the pipelined ALU circuit that has become a benchmark in the model checking community [7, 5, 4]. A goal of the model checking community is to find techniques that allow them to effectively verify the pipelined ALU with increasingly larger datapaths and register file. Our logic lets us verify the pipelined ALU once and for all for arbitrarily large datapaths and register file and for an arbitrary number of alu instructions. The cost we incur is that our fragment is much less temporally expressive than the decidable propositional temporal logics.

Using theorem proving techniques we have in the past verified several microprocessors such as Saxe's pipeline [20, 10]. We are currently verifying a more realistic microprocessor—a Verilog model of a much simplified MIPS R3000 processor. The correctness of these circuits is also expressible in our decidable fragment. In the future we can make use of this fragment by either implementing the fragment independently or by integrating it into a theorem proving environment.

The paper is organized as follows. Section 2 presents *GTL*. Section 3 gives a proof of the incompleteness of the logic. In section 4 we present a decidable fragment. In section 5 we extend this fragment so that various interesting verification problems can be described in it. Section 6 shows how to enode aspects of microprocessor verification in our logic. The final section presents conclusions and future work.

2 Ground Temporal Logic

In this section we give the syntax and semantics of the first-order temporal language we will be using. We follow the presentation of Kröger [14].

2.1 Syntax

Given a first order language, *FOL*, consisting of function symbols, constants (0-ary function symbols), predicate symbols, equality, but no variables, we define the language *GTL* as follows.

The alphabet of *GTL* is the alphabet of *FOL* along with $\bigcirc$, $\circ$, $\square$, and a set of state variables, V_s, whose values can change with time. There are no global variables.

The *terms* of *GTL* are defined inductively. Every state variable is a term. If f is an n-ary function symbol and $t_1, \ldots, t_n$ are terms then $f(t_1, \ldots, t_n)$ is a term. If t is a term then so is $\circ t$.

The *atomic formulas* of *GTL* are defined as follows:

If p is an n-ary predicate symbol and $t_1, \ldots, t_n$ are terms then $p(t_1, \ldots, t_n)$ is an atomic formula.

Formulas of *GTL* are defined inductively. Every atomic formula is a formula. If A and B are formulas then so are $\neg A, A \wedge B, \bigcirc A$, and $\square A$. The language of *GTL* is the smallest language generated by these rules.

Note that this language is the quantifier free, ground version of the language of First Order Temporal Logic as presented in [14] with the addition of our $\circ$ operator.

2.2 Semantics

Closely following the presentation in [14] we define the semantics of *GTL*.

We define a *model* $\mathbf{K} = (\mathbf{S}, \mathbf{W})$ for *GTL* as follows.

$\mathbf{S}$ is a model for the first-order language *FOL*. $\mathbf{S}$ consists of a non-empty universe $|\mathbf{S}|$, an n-ary function $\mathbf{S}(f) : |\mathbf{S}|^n \to |\mathbf{S}|$ for every n-ary function symbol f, and an n-ary relation $\mathbf{S}(p) \subset |\mathbf{S}|^n$ for every n-ary predicate symbol p other than $=$.

$\mathbf{W} = \{\eta_0, \eta_1, \ldots\}$ is an infinite sequence of state variable valuations (*states*): $\eta_i : V_s \to |\mathbf{S}|$.

We define two evaluation functions:

$\mathbf{S}_t^{(\eta_i)} : terms \to |\mathbf{S}|$, and

$\mathbf{S}_a^{(\eta_i)} : atomic\ formulas \to \{\mathbf{f}, \mathbf{t}\}$.

- $\mathbf{S}_t^{(\eta_i)}(a) = \eta_i(a)$ for state variable a.
- $\mathbf{S}_t^{(\eta_i)}(f(t_1, \ldots, t_n)) = \mathbf{S}(f)(\mathbf{S}_t^{(\eta_i)}(t_1), \ldots, \mathbf{S}_t^{(\eta_i)}(t_n))$, for f other than $\circ$.
- $\mathbf{S}_t^{(\eta_i)}(\circ t) = \mathbf{S}_t^{(\eta_{i+1})}(t)$
- $\mathbf{S}_a^{(\eta_i)}(p(t_1, \ldots, t_n)) = \mathbf{t}$ iff $(\mathbf{S}_t^{(\eta_i)}(t_1), \ldots, \mathbf{S}_t^{(\eta_i)}(t_n)) \in \mathbf{S}(p)$ for p other than $=$.
- $\mathbf{S}_a^{(\eta_i)}(t_1 = t_2) = \mathbf{t}$ iff $\mathbf{S}_t^{(\eta_i)}(t_1) \stackrel{=}{\scriptscriptstyle |\mathbf{S}|} \mathbf{S}_t^{(\eta_i)}(t_2)$.

We now define the truth value function $\mathbf{K}_i$: formulas $\to \{\mathbf{f}, \mathbf{t}\}$ for every $i \geq 0$.

- $\mathbf{K}_i(A) = \mathbf{S}_a^{(\eta_i)}(A)$ for atomic formula A.

- $\mathbf{K}_i(\neg A) = \mathbf{t}$ iff $\mathbf{K}_i(A) = \mathbf{f}$.
- $\mathbf{K}_i(A \wedge B) = \mathbf{t}$ iff $\mathbf{K}_i(A) = \mathbf{t}$ and $\mathbf{K}_i(B) = \mathbf{t}$.
- $\mathbf{K}_i(\bigcirc A) = \mathbf{t}$ iff $\mathbf{K}_{i+1}(A) = \mathbf{t}$.
- $\mathbf{K}_i(\Box A) = \mathbf{t}$ iff $\mathbf{K}_j(A) = \mathbf{t}$ for every $j \geq i$.

Definition. A formula A of GTL is valid in the model $\mathbf{K}$ if $\mathbf{K}_i(A) = \mathbf{t}$ for every $i \geq 0$. A is *valid* iff A is valid in every model $\mathbf{K}$. A is *satisfiable* iff A is valid in some model $\mathbf{K}$.

An alternate method for giving the semantics of the language is to divide the set of function symbols and constants into rigid and flexible sets. The approach we have taken uses V_s as the set of flexible constants. Allowing a set of flexible non-constant function symbols does not make the language more expressive.

3 Undecidability of Ground Temporal Logic

The logic presented in Section 2 is the simplest and in some way the smallest extension to propositional temporal logic that makes the problem of determining validity undecidable.

This is captured in the following theorem:

Theorem. *The Validity Problem of* GTL *is* Π_1^1*-complete.*

Proof. The validity of a formula, A in GTL, can be stated as: $\forall \mathbf{K} \forall i (\mathbf{K_i}(A) = \mathbf{t})$, which is a Π_1^1 sentence. Thus the validity is in Π_1^1.

That the validity problem is Π_1^1-hard is proven by using a reduction from the recurrence problem for a two-counter machine [9]. □

Although this is a discouraging result, it provides useful guidance in developing logics that are both expressive and can be automated; we have an upper bound on how expressive such a logic can be. The rest of this paper presents a fragment of GTL that can both express interesting properties of hardware and be decided.

4 A Decidable Fragment

We now present a fragment of GTL that is decidable. While this fragment is very simple, we illustrate its usefulness by encoding the correctness of a simple counter in it. In later sections we will extend this fragment to make it more useful until we eventually are able to express the correctness of a simple microprocessor.

GTL was motivated by a desire to combine the best features of model-checking and theorem proving into one logic. The algorithms from model-checking yields relatively efficient decision procedures that allow the automatic verification of hardware circuits. Theorem proving allows the verifier to abstract away from some of the bit-level details that are irrelevant for the proof of correctness, but is in general undecidable and requires a large amount of human effort.

By adding uninterpreted function symbols to a propositional temporal logic, we are trying to make use of the most natural type of abstraction, but still do something similar to model-checking.

The model checking problem is, given a model $\mathcal{M}$ and a specification S, to determine whether:

$$\mathcal{M} \models S$$

is true.

In propositional temporal logic $\mathcal{M}$ is a finite Kripke structure, and S is a propositional temporal formula.

In GTL $\mathcal{M}$ may or may not be a finite Kripke structure. It would be if the interpretations of the uninterpreted function symbols were functions with a finite domain and range. For example, in Figure 1, a possible interpretation for the *inc* function would be *plus one mod four*, yielding a 4-bit counter. In this case the model-checking problem for GTL would reduce to that of a propositional temporal logic.

However, in some cases there are no finite models or, more importantly, we want to verify that all models satisfy the specification S. Thus, instead of model checking we want to determine the validity of formulas of the form

$$I \supset S \tag{1}$$

where I is a GTL formula defining an implementation, and S is as before, a GTL formula.

GTL has been designed so as to be able to encode transition systems. The use of the $\circ$ operator allows the encoding of a next-state relation. If we restrict I to only those formulas that encode transition systems, and S to properties of the system that can be checked by exploring fixed length paths, then the validity problem for the formulas of the form in equation 1 is decidable. Note that I is not capable of stating an initial state for the transition system. This is crucial for our decidability argument. Note also that the restrictions that S refer only to fixed length paths is similar to the restrictions explored in a propositional temporal logic in [3].

Before giving details we illustrate this with an example. The transition from the Inc1 state to the Fetch state in Figure 1 would be encoded in GTL as:

$$\Box[\text{state} = \text{Inc1} \wedge \text{double} = \text{false} \supset \circ \text{state} = \text{Fetch} \wedge \circ \text{Cnt} = \text{inc}(\text{Cnt})] \tag{2}$$

Similar formulas express the remaining transitions in Figure 1. The conjunction of these formulas make up the implementation I. A finite model of I, perhaps by instantiating Inc with addition by 1 mod 4, corresponds to the traditional Kripke models of a propositional temporal logic.

The correctness statement for this counter states that, depending on the input, the value of Cnt is eventually updated correctly when the counter reaches the Fetch state again. We cannot state this most naturally in GTL, but rather have to give the exact number of transitions required for the counter to reach the fetch state again.

Thus, for example, the statement of correctness for the case when the input is Inc2 is:

$$\Box[\text{state} = \text{Fetch} \wedge \text{input} = \text{Inc2} \supset \circ \circ \circ \text{state} = \text{Fetch} \wedge \circ \circ \circ \text{Cnt} = \text{inc}(\text{inc}(\text{Cnt}))] \tag{3}$$

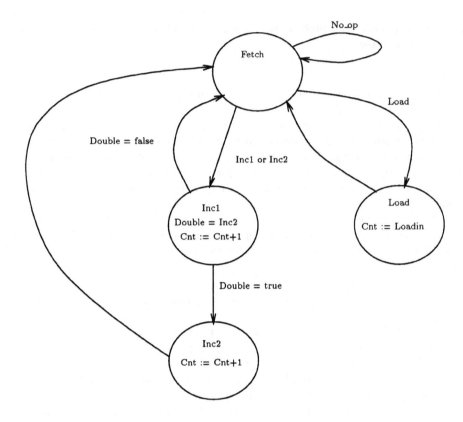

Fig. 1. A Counter

The statement of correctness for the counter as a whole would consist of the conjunction of similar formulas for the case of each of the different inputs.

This example provides the prototype for our first decidable fragment. I is of the form:

$$\bigwedge_i \Box[\text{condition}_i \supset \bigwedge_j \circ \text{state\_var}_j = T_j] \tag{4}$$

where condition$_i$ is a *GTL* formula that contains no temporal operators and T_j is a *GTL* term that contains no instance of $\circ$. Furthermore, the conditions are mutually exclusive, i.e., $i \neq j \supset \neg(\text{condition}_i \wedge \text{condition}_j)$

Mutual exclusivity of the conditions can either be implicitly assumed or guaranteed explicitly by allowing I to include constraints of the form:

$$\bigwedge_i \Box[T_i \neq T_i'] \tag{5}$$

where T_i and T_i' are *GTL* terms that contain no instance of the $\circ$ operator.

S is of the form:

$$\bigwedge_i \Box[\text{condition}_i \supset \bigwedge_k T_k = T_k'] \tag{6}$$

where condition$_i$ is a GTL formula that can contain arbitrary number of os, but no other temporal operators and T_k and T_k' are GTL terms that can contain arbitrary number of os.

We define the logic $GTL1$ to be the fragment of GTL that consists only of formulas of the form $I \supset S$ where I is restricted to be of the form described by equations 4 and 5 and S is restricted to be of the form described by equation 6.

The validity problem for $GTL1$ is obviously decidable. We can eliminate o from the terms T_k and T_k' in 6 by conditional rewriting using the formulae from 4. This reduces the problem to an instance of the validity problem for ground conditional equational logic which is decidable.

Obviously this logic is not very expressive. There are many useful statements even about this simple counter that $GTL1$ cannot express. Without detailing these deficiencies we incrementally expand the expressiveness of $GTL1$ and show that even with very few additions we can achieve a logic that is useful and still decidable.

5 Useful Extensions

We now describe some extensions that make $GTL1$ more useful for hardware verification.

The motivation for the first extension is simply notational convenience. In describing hardware we want to be able to give names to wires in the circuit. To do this we allow I to additionally include formulas of the form:

$$\Box[T = T'] \tag{7}$$

where neither T nor T' contain any temporal operators.

However, to make the logic truly useful for hardware verification we also need to have a representation for some sort of memory or register file. Thus, we add to our logic the special interpreted symbols, *read* and *write*. These symbols are related by the following axioms:

$$\forall \text{regfile}, \text{addr1}, \text{addr2}, \text{data} : \\ \text{addr1} = \text{addr2} \supset \text{read}(\text{write}(\text{regfile}, \text{addr1}, \text{data}), \text{addr2}) = \text{data} \tag{8}$$

$$\forall \text{regfile}, \text{addr1}, \text{addr2}, \text{data} : \\ \text{addr1} \neq \text{addr2} \supset \text{read}(\text{write}(\text{regfile}, \text{addr1}, \text{data}), \text{addr2}) = \text{read}(\text{regfile}, \text{addr2}) \tag{9}$$

We call this new logic $GTL2$. It is still decidable as methods such as those found in [18] can be used to decide combinations of ground equalities with other decidable theories. We can also add to $GTL2$ further interpreted functions that have decidable quantifier free theories [18].

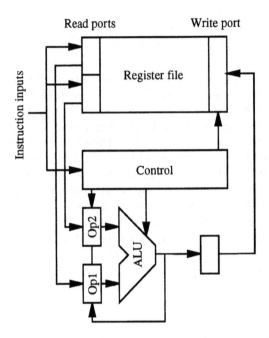

Fig. 2. A pipelined ALU

Just these simple extensions let us state the implementation and correctness for the pipelined ALU in Figure 5. In *GTL* the correctness of the pipelined ALU becomes:

$$\Box[\neg\text{stall} \supset \text{read}(\circ\circ\circ\text{file}, \text{dstn}) = \text{alufun}(\text{op}, \text{read}(\circ\circ\text{file}, \text{src1}), \text{read}(\circ\circ\text{file}, \text{src2}))$$
(10)

In comparing this to the work done in [7] we note that we can abstract away the size of the data paths and the specific alu function without resorting to logarithms or any other clever means, and yet we do not lose decidability.

This is still a toy example. We are, however, also able to state the correctness of microprocessors in this simple fragment.

6 Microprocessor Correctness

In [20] microprocessor correctness is stated in a form similar to equation 1, where I and S are conditional equations with a universally quantified time variable. In [21, 10] the microprocessor correctness is stated in a form that does not mention time, but rather uses explicit next-state relationships. We now summarize the approach to microprocessor correctness in [21, 10] and show that it can be encoded in a decidable fragment of *GTL*.

Microprocessors can be described as state transition systems. The state of the microprocessor consists of the state of the memory, register file, and internal registers of the processor (these would generally include the program counter, memory address register, and pipeline registers if the processor is pipelined, etc.). The approach taken in verifying microprocessors is to use a simple transition system as the specification of the microprocessor, and a more complex transition system as the implementation. In [20] the specification is actually a non-pipelined microprocessor. In [21] the specification is a transition system corresponding to the instruction set architecture. In both [21, 20] the implementation is a pipelined microprocessor. Early work [12, 13] used the instruction set architecture as the specification and a non-pipelined machine as the implementation.

The microprocessor verification problem is to show that the traces induced by the implementation transition system are a *subset* of the traces induced by the specification transition system, where *subset* has to be carefully defined by use of an abstraction mapping. The verification problem is illustrated in Figure 3(a), where I represents the implementation next-state function and A represents the specification next-state function. The details and complications of this approach are beyond the scope of this paper, and the reader is referred to [20, 21, 10, 1] for them.

Following the approach in [1, 21, 20, 16][3] the proof of correctness makes use of an *abstraction* function that maps an implementation state into a *corresponding* specification state. Correctness can then be reduced to showing that for any execution trace of the implementation machine there exists a *corresponding* execution trace of the specification machine. This is captured in Figure 3(a). The trace equivalence expressed in Figure 3 can be reduced to the commutativity of the diagram in Figure 3(b).

As discussed in [10, 21] the implementation machine may run at a different rate than the specification machine. For example, in the microprocessor described in [20] the specification machine takes one state transition to execute each instruction, but the implementation machine might take five cycles to execute branch instructions, but only one cycle for non-branch instructions. In the following we assume that the specification machine always takes one cycle to execute an instruction. We also assume that the number of cycles that the implementation machine takes to execute an instruction can be given as a function of the current state and current input. (This restriction can be slightly relaxed to deal with interrupts which might arrive a bounded number of cycles into the future.)

Let us denote the function that determines the number of cycles that the implementation machine takes to complete an instruction as num_cycles. We assume that it is provided by the hardware designer or verifier.

In [10, 20] this function is given by the following equation:

```
num_cycles =
  IF zero?(alu(IRD4, getOp(program(pc))))
     THEN 5
     ELSE 1
  ENDIF
```

---

[3] The precise details followed in these papers are somewhat different.

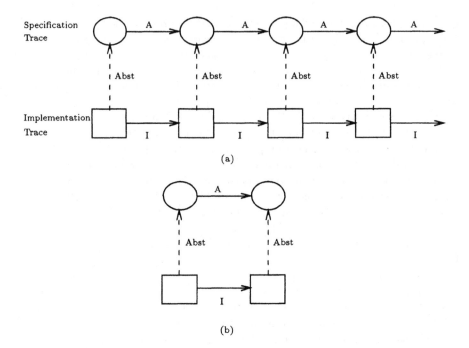

Fig. 3. Commutes

where **IRD4** and **pc** are registers in the implementation machine, i.e. state variables in *GTL*.

The first step in verifying the correctness of the microprocessor is to split the proof into cases based on the definition of **num_cycles**. Thus for each case we have a precise number of cycles that we have to cycle the implementation machine through. In the following we call this number **nc**. We denote the conditions under which **num_cycles** = **nc** by **cond$_{nc}$**.

In the microprocessor verifications we have looked at, the state variables of the specification state are simply a subset of the state variables of the implementation state. For example, in [20] the specification machine is characterized by the contents of the program counter and register file. The implementation machine is also characterized by the contents of the program counter and register file as well as additional pipeline registers. The abstraction mapping maps each specification register to the corresponding implementation register, but not necessarily from the exactly corresponding state. For example, in [20] the abstraction mapping is given by the following equations:

$$pc_A = pc_I \tag{11}$$

$$reg\_file_A = \circ \circ \circ \, reg\_file_I. \tag{12}$$

In other words the specification reg_file is the implementation reg_file, but three cycles into the future.

In the following we denote the ith state variable of the specification machine as V_A^i, and the ith state variable of the implementation machine as V_I^i. The abstraction mapping from the specification machine to the implementation machine can then be given by equations of the form:

$$\bigwedge_i (V_A^i = \circ^{a_i} V_I^i) \tag{13}$$

where a_i gives the number of lookahead cycles for state variable V^i.

Now, given our assumption that the specification machine takes only one cycle per instruction, we can *symbolically* execute the specification machine to obtain expressions capturing the state of the specification machine state variables after executing one instruction. For each specification machine state variable, V_A^i, we denote its symbolic next state expression as: $N\_V_A^i$.

Now, for each distinct value of nc, we can express the correctness of the microprocessor as:

$$I \supset S \tag{14}$$

where I encodes the transition system of the implementation machine and S is of the form:

$$\Box[\bigwedge_{nc} \mathbf{cond_{nc}} \wedge A \supset \bigwedge_i N\_V_A^i = \circ^{nc} V_I^i]. \tag{15}$$

where A encodes the abstraction mapping.

It is an easy exercise to show that equations 14 and 15 are in $GTL2$.

Pipeline Invariants The verification of some pipelined microprocessors requires the use of pipeline invariants. See [21] for example. Such microprocessors cannot be directly verified in $GTL2$ without being first provided the pipeline invariant. However, once the pipeline invariant is provided, the invariance of the pipeline invariant and the correctness of the microprocessor, assuming the pipeline invariant, can be stated in $GTL2$.

Theorem Proving We have carried out our experiments in processor verification in the context of a higher-order theorem prover [19]. In the theorem prover we state the correctness of the microprocessors in a more natural manner than indicated by equations 14 and 15. In the process of the verification we generate as intermediate goals statements that are instances of $GTL2$ formulas. We thus envision making use of logics such as $GTL2$, not just on their own, but as a way to integrate a user directed verification effort based on theorem proving with more automated verification tools such as model checkers or validity checkers for $GTL2$. Investigations concerning GTL can be viewed as a principled attempt to determine how much of the verification task can be automated.

7   Conclusions, Related Work, and Future Work

We have presented *GTL*, a new temporal logic that extends propositional temporal logic. We do this by providing an additional temporal operator, ○ that operates on terms. Other temporal logics have also provided notation equivalent to our ○ *t* [17, 15]. To our knowledge we are the first to analyze the complexity of ground temporal logics making use of this operator.

We showed that the full *GTL* is undecidable. We then identified fragments of *GTL* that are both decidable and useful for real hardware verification, including microprocessor verification. These fragments were in part motivated by our experience in using PVS for hardware verification. Using PVS, its ground decision procedures, and a BDD package we have verified the ALU pipeline in 90 seconds. Much, if not most, of this time is spent dealing with the overhead of a general-purpose higher-order theorem prover. There is current work in PVS to build better decision procedures for combining decidable theories. [6] reports on independent work that efficiently decides a fragment of *GTL2*. Efficient decision procedures for *GTL2* should be able to build upon this work.

In addition to using *GTL* as is, we envision it as a means to integrate decision procedures for a temporal logic into a theorem proving approach. By identifying fragments of *GTL* that are decidable we can identify instances of goals while doing theorem proving that can be directly dispatched. GTL *provides a logical/principled framework for exploring what fragments of hardware verification can be automated.*

In addition to identifying larger decidable fragments of *GTL*, we are currently exploring ways to extend *GTL* to be more expressive. One idea is to define new temporal operators that operate on terms much the same way the ○ does. One possible operator is `atnext(p) t`, which would denote the value of term t at the next time instance that formula p was true. While not identical, this line of research is similar in spirit to that in [11]. One difference is that in [11] all terms are considered rigid in order to obtain decidability.

To summarize, in *GTL* and its fragments we have identified a practically useful temporal logic, that allows us to overcome some of the disadvantages inherent in both the model-checking and theorem proving approaches to hardware verification.

References

1. Martín Abadi and Leslie Lamport. The existence of refinement mappings. In *Third Annual Symposium on Logic in Computer Science*, pages 165–175. IEEE, Computer Society Press, July 1988.
2. R. E. Bryant. Graph-based algorithms for boolean function manipulation. *IEEE Trans. Comput.*, C-35(8), 1986.
3. R.E. Bryant, D. L. Beatty, and C.-J. Seger. Formal hardware verification by symbolic trajectory evaluation. In *28th ACM/IEEE Design Automation Conference*, 1991.

4. J. R. Burch, E. M. Clarke, K. L. McMillan, and D. L. Dill. Sequential circuit verification using symbolic model checking. In *27th ACM/IEEE Design Automation Conference*, 1990.

5. J. R. Burch, E.M. Clarke, and D.E. Long. Representing circuits more efficiently in symbolic model checking. In *28th ACM/IEEE Design Automation Conference*, 1991.

6. J. R. Burch and D. L. Dill. Automated verification of pipelined microprocessor control. In *CAV '94*, 1994. Submitted.

7. E. M. Clarke, O. Grumberg, and D. E. Long. Model checking and abstraction. In *Nineteenth Annual ACM Symposium on Principles of Programming Languages*, pages 343–354, 1992.

8. W Cullyer and C Pygott. Hardware proofs using LCF-LSM and ELLA. Memorandum 3832, RSRE, September 1985.

9. D. Cyrluk and P. Narendran. Decision problems for ground temporal logics. Unpublished Manuscript.

10. David Cyrluk. Microprocessor verification in PVS: A methodology and simple example. Technical Report SRI-CSL-93-12, SRI Computer Science Laboratory, December 1993.

11. T. Henzinger. Half-order modal logic: How to prove real-time properties. In *Proceedings of the Ninth Annual Symposium on Principles of Distributed Computing*, pages 281–296. ACM Press, 1990.

12. W.A. Hunt. The mechanical verification of a microprocessor design. In *Proc. of IFIP Working Conference on From H.D.L Descriptions to Guaranteed Correct Circuit Designs*, 1986.

13. J. Joyce, G. Birtwistle, and M. Gordon. Proving a computer correct in higher order logic. Technical Report 100, Computer Lab., University of Cambridge, 1986.

14. Fred Kröger. *Temporal Logic of Programs*, volume 8 of *EATCS Monographs on Theoretical Computer Science*. Springer Verlag, 1987.

15. Leslie Lamport. The temporal logic of actions. Technical Report 79, Digital Systems Research Center, Palo Alto, California 94301, December 1991.

16. Paul Loewenstein and David Dill. Verification of multiprocessor cache protocol using simulation relations and higher-order logic. In *Computer-Aided Verification '90*, pages 187–205. DIMACS, American Mathematical Society, 1991.

17. Z. Manna and A. Pnueli. Verification of concurrent programs: A temporal proof system. In J. W. de Bakker annd J. van Leeuwen, editor, *Foundations of Computer Science IV, Distributed Systems: Part 2*, Mathematical Centre Tracts 159, pages 163–255. Center for Mathematics and Computer Science, Amsterdam, 1983.

18. G. Nelson and D. C. Oppen. Simplification by cooperating decision procedures. *ACM Transactions on Programming Languages and Systems*, 1(2):245–257, October 1979.

19. Sam Owre, John M. Rushby, and Natarajan Shankar. PVS: A prototype verification system. In Deepak Kapur, editor, *Automated Deduction - CADE-11, 11th International Conference on Automated Deduction, Lecture Notes in Artifical Intelligence*, pages 748–752. Springer Verlag, June 1992.

20. James B. Saxe, Stephen J. Garland, John V. Guttag, and James J. Horning. Using transformations and verification in circuit design. Technical Report 78, Digital Systems Research Center, Palo Alto, California 94301, September 1991.

21. Mandayam Srivas and Mark Bickford. Formal verification of a pipelined microprocessor. *IEEE Software*, 7(5):52–64, September 1990.

A Hybrid Model for Reasoning about Composed Hardware Systems

E. Thomas Schubert

Department of Computer Science
Portland State University

Abstract. To formally specify and reason about composed systems, a process algebra is developed that integrates an extended interpreter model. This approach utilizes the interpreter model for device decomposition, while also being able to reason about larger systems that require interdevice communication. By combining these approaches, convenient notations are available to specify and verify both device independent properties (e.g., instruction sets) and device interdependent properties (e.g., communication protocols).

1 Introduction

Previous approaches to system verification have decomposed systems into several hardware and software layers that may be independently verified [3, 9, 11, 17]. Each layer consists of an implementation description and a more abstract specification. The layers are joined, or "stacked", with each implementation description serving as the specification for the next lower layer. The hardware layers of these systems have been modeled as a microprocessor with memory. This simple specification of the hardware has been a convenient abstraction for verifying the correctness of software executing on the hardware base, but by inspection, is not representative of modern hardware implementations. A realistic hardware implementation consists of many different interacting components.

Our approach has developed a framework to formally specify and verify the correctness of the communication between hardware devices. The methodology (formalized within the HOL theorem proving system) allows a hardware system to be decomposed into a set of independently verified devices and provides a logic to specify and reason about the composed, aggregate system behavior. To demonstrate the technique, we present a system with concurrently executing devices that is verified to correctly pass information and coordinate activities. The system consists of a CPU, a memory subsystem with a memory management unit, a direct memory access device, and a bus controller. The remainder of this section will describe the overall approach. Subsequent sections describe the technique's formalization, a system verification example, and a brief conclusion.

1.1 Formal System Model of Communicating Interpreters

Formal reasoning about composed hardware systems requires that a device or system be represented as a well-defined mathematical object that denotes the important aspects of a system's behavior. To an outside observer, a system appears as a black box. We are interested in a system's behavior over time with respect to a state and environment. An input environment stream is provided and an output environment stream is observed. The observed behavior is typically a function of the inputs. Further, some systems will demonstrate sequential behavior where the outputs are a function of the input history, rather than just the most recent inputs. In this case, it is clear that the system maintains some sort of internal state.

The interpreter model provides an effective framework to describe and reason about device behavior, but the model does not provide sufficient structure to describe the concurrency inherent in composed systems. Several researchers have developed models for describing concurrent systems (e.g., temporal logic of actions, reactive systems). However, much of this work is too abstract to relate to concrete descriptions of programs or hardware. The literature on reactive systems and process algebras also ignores the internal or transformational aspects of a system while focusing on the system/environment interactions. Our model combines transition and reactive models to describe concurrent systems. This approach requires extensions to the interpreter model so that devices are modeled with a state and environment.

The interpreter model views a system as a single entity defined by a set of attributes that includes: a state, an environment, an operation list, and an operation choice function. A hardware system consisting of several devices (each modeled as an interpreter) will also be identified by these attributes. Due to the concurrent behavior of the composed system, the nature of these attributes is considerably more complex than for an individual interpreter. The interpreter model denotes state and environment as a function of time and the operation list as fixed over time. Concurrency will require a formal system model where the system operation list is also a function of time.

The role of the environment is elevated from its secondary status in the interpreter model. The environment is modeled as active and its behavior cannot be explicitly restricted. In a sense, the interplay between a system and its environment is like a game [1]. The environment is an opponent and the system must adapt a strategy that is guaranteed to "win" regardless of what the environment chooses to do. A system verification proof may wish to show that the strategy always succeeds. As in Lamport's model, we assume that the environment may take some number of actions (possibly zero) followed by at most, one action by the system. However, the environment player is not allowed to win by simply not giving the system a chance to play.

There are three general forms in which devices can be combined: shared input and output environments, shared state, and interlinked output environments (see Figure 1). Each model form requires that individual device (top level) interpreters are verified to perform their respective set of operations correctly. To describe a composed system, a combination of these forms may be required. The "glue logic" that links interpreters is provided by a process algebra.

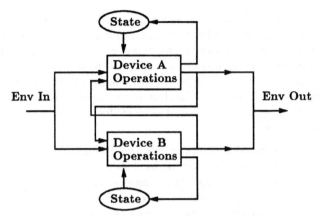

Fig. 1. Linked Environments Black Box Model

1.2 Process Algebras

Process algebras, such as CCS, have emerged as a framework to study the interaction of communicating, concurrent processes [2, 7, 15]. While processes are generally thought of as executing programs, hardware systems can also be studied under this framework. Such systems consist of many devices acting independently of each other, but communicating with one another to achieve a mutual goal. The behavior of a system is, in some sense, its entire capability to communicate, or, "what is observable." The behavior of each device can be described as a process and the interaction of devices studied in terms of a process algebra.

All elements of a system, including the medium used to transmit information between devices, are modeled as *agents*. For example, a hardware system model would not only describe the CPU and memory components as *agents*, but the bus that connects the two components (the medium) would also be represented by an *agent*. An agent's potential behavior is described by *actions*, which may be communications with other agents or independent concurrent actions. Two *agents* can only communicate only if they perform complementary send and receive actions (e.g., . $\overline{a}$ and a, respectively). Independent actions are represented by the reserved symbol τ and have no complement. Process algebras also provide *agent* constants, the most common being **0**, the "inactive" agent. A summary of the basic agent constructors appears in Table 1.

Operator	Example	Meaning
Prefix	$a.A$	Action a taken, followed by behavior of agent A
Summation	$A + B$	Behavior of either agent A or agent B
Composition	$A \mid B$	Agent A and agent B operate concurrently
Restriction	$A \backslash \ell$	ℓ labeled actions are not visibly performed
Relabelling	$A[f]$	Function f substitutes labels in agent A

Table 1. CCS Basic Operators

The basic axioms of CCS are presented in Figure 2. Most variations of CCS incorporate these laws and differ in which equational laws regarding the internal action are included. A brief summary of six such process algebra variations is presented in [6]. Agents with infinite behavior can also be defined with recursive definitions.

Summation (monoid) Laws

$$P + Q = Q + P$$
$$P + (Q + R) = (P + Q) + R$$
$$P + P = P$$
$$P + 0 = P$$

Restriction Laws

$$(\alpha.Q)\backslash L = \begin{cases} 0 & \text{if } \alpha \in L \\ \alpha.Q\backslash L & \text{otherwise} \end{cases}$$
$$(Q + R)\backslash L = Q\backslash L + R\backslash L$$
$$0\backslash L = 0$$

Composition Laws

$$P|Q = Q|P$$
$$P|(Q|R) = (P|Q)|R$$

$$P|0 = P$$

Relabelling Laws

$$(\alpha.Q)[f] = f(\alpha).Q[f]$$

$$(Q + R)[f] = Q[f] + R[f]$$
$$0[f] = 0$$

The Expansion Law

$$\alpha.P \,|\, \beta.Q = \alpha(P \,|\, \beta.Q) + \beta(\alpha.P \,|\, Q) + \begin{cases} \tau(P \,|\, Q) & \text{if } (\alpha = \overline{\beta}) \\ 0 & \text{otherwise} \end{cases}$$

Fig. 2. Basic Axioms of CCS

1.3 The Hybrid Approach

The relative benefit (or even presence) of various abstraction mechanisms differs depending on what level of the interpreter hierarchy is being verified. At the lowest levels of the hierarchy, structural and behavioral abstraction mechanisms relate a behavioral description to the structural definition of a device. Proofs between intermediate levels rely primarily on the use of temporal and data abstraction mechanisms. At all levels, the state based approach adopted by the interpreter model seems natural for this problem domain. As the problem domain shifts to verifying a system with multiple devices, we believe a process based approach is more appropriate. Lamport's transition axiom method proposes a state based approach to composing specifications [1, 10]. However, he points out that a process based approach is equivalent in expressive power. A process based approach seems more natural and frequently, abstract system properties such as security, are more easily expressed using a process based notation [8].

Our approach describes a system as a process algebra term that incorporates a set of individual device interpreters. This approach merges the interpreter (transition) model with a process algebra (reactive) model. Interpreters serve to describe device-private state changes while the process algebra calculus describes the interaction of devices. The system specification must also describe the set of all possible execution interleavings. The static operation selection mechanism provided by the interpreter model assumed that once an action is chosen (e.g., CPU add instruction), the operation runs until completion. In the composed model, operations cannot assume uninterrupted execution; the final result may be different depending on the actions of other devices. To satisfy the arbitrary interleaving condition requires that only one critical event occurs per action. This requires that the specification capture intermediate snapshots of interpreter executions when interference is possible.

The process algebra (reactive) model can describe the dynamic nature of device interactions. A process algebra term denotes a function that chooses which interpreter executes next. This in a sense, extends the selection function of the original interpreter model. Greater recognition is also given to the role the environment plays in systems. Process algebra terms capture the specification of environment interactions between devices.

The system is defined as an interpreter function, which given a state and environment, returns a new function, a new state and an output environment. At a given point in time, several devices (interpreters) may be able to act. To verify that the devices can work concurrently, we show that system requirements are satisfied regardless of the interpreter action interleaving.

2 Mechanization of the Process Algebra

There has been significant recent interest in mechanizing process algebras in HOL [4, 5, 14]. The process algebra developed here is based on CCS, with extensions to reason about interpreters. Using the HOL system type definition mechanisms, we define an initial algebra for *agents*, constructed with sequences of *actions*. HOL's recursive type definition facility [12] automates the process of defining new data types in terms of already existing types. Both new type constants and type constructors (operators) can be defined. Additional (recursive) functions can be defined to operate on the concrete data representation of the type. The properties of new types must be derived by formal proof. This guarantees that the type does not introduce inconsistency into the logic.

Actions: External actions represent enabled (boolean) signals that are part of an interpreter's environment. An external action requires a *label* that consists of a name and boolean value that denotes whether the action is a send or receive synchronization operation. The private communication action (τ action) is replaced by the **INTERNAL** action. Whereas CCS did not consider what activity was taken by an internal action, internal actions are now tagged with an interpreter that defines changes to private state.

```
define_type 'action' 'action = INTERNAL *interpreter | EXTERNAL label';;
```

Agents: The *agent* data type is constructed from sequences of *actions*. A preliminary data type consisting of the five CCS type constructors, described in Section 1.2, was developed. This type definition did cover the syntax of the concrete data type. However, we found that the semantics of some of the constructor operators can be defined in terms of others. Additionally, for our purposes, the relabelling operator has not been necessary. Its addition would be simple to add, if warranted by further application of the calculus. By minimizing the number of type operators used for the representation of any type, the complexity of inductive definitions is reduced.

Sequential Behavior: A term algebra for *agents* is created through a type definition with a single type constant (`INACTIVE`) and two type constructors (`PREFIX` and `SUMM`). These elements correspond to their CCS counterparts. An `INACTIVE` agent (denoted by "0" in CCS) cannot communicate with any other entity, and thus, appears to do nothing. The `PREFIX` operator creates a new agent from an action and an agent. The new agent will first communicate through the action operand and then behave as the agent operand. The `SUMM` operator creates a new agent that can behave as either of the agent operands. For determining the equivalence of two agents, we would like to define all agents in a normal form with agent terms consisting of only the prefix and summation operators.

```
define_type 'agent' 'agent = INACTIVE | PREFIX  action agent
                                     | SUMM    agent agent;;
```

Parallel/Concurrent Behavior: To describe processes that execute in parallel, we adapt a method described in [2]. This technique allows composed, concurrently executing agent expressions to be converted to an equivalent agent expressed only with the prefix and summation operators (as described by the expansion law in Figure 2). We define three mutually recursive functions to replace a composition type constructor (COMM, LMERGE, and COMPOSE). The communication operator, COMM, declares that two agents will communicate if they have complementary, enabled actions (e.g., a and $\overline{a}$). The left-merge operator, LMERGE creates a new agent from two agents, such that the new agent must first behave as though only the left agent (first argument) were active. Subsequent behavior is determined by an agent constructed by the composition of the resulting left agent derivative and the original right agent. The compose operator, COMPOSE, creates an arbitrary interleaving of the two agents through use of COMM, LMERGE, and the summation type constructor.

Transition Semantics: Operational semantics for *agent* expressions can be provided through a labeled transition relation indicating what *actions* an *agent* may perform. This relation can be defined using the inductive relation definition package. The inductive relation definition package provides a set of theorem-proving tools based on a derived principle of definition in HOL for defining relations inductively by a set of rules [13]. The relation is inductively defined by a collection of such rules to be the least relation closed under all the rules. The defined TRANS predicate, states that for an agent to evolve to another agent, there must be an immediate transition by a prefixed action (e.g., $\alpha.A \xrightarrow{\alpha} A$). Symmetric versions of a summation law are provided to indicate that the transition is valid if either SUMM operand agent can make the transition.

2.1 Agent Equivalence

Several notions of equivalence between agents can be defined. The kinds of models that satisfy the same equations as the initial algebra will be determined by which notion of equivalence is used. Trace semantic equivalence can be established by defining the meaning of an agent as a set of traces where a trace is a sequence of actions. Trace semantics permit fairly broad equivalence classes to be constructed. We are generally interested in a narrower definition of equivalence where two agents are equivalent if an external agent cannot distinguish between the visible behavior (traces) of the two agents. When using the notion of strong equivalence, traces consist of both external and internal actions. For our application, a weaker notion of observation equivalence is sufficient. Observation equivalence states that two equivalent descriptions implement the same external behavior; that is, internal actions cannot be detected by the external agent.

A device specification is an abstraction of the implementation, and thus, will not describe all of the implementation's actions. To show that an implementation satisfies a specification, *one-way* observation equivalence, is appropriate. Formally, **I** implements **S**'s behavior if for every action α of **S**, every α-derivative of **S** is one-way observation equivalent to some α-descendant of **I**. For the remainder of the paper we will use the term observation equivalence to mean one-way observation equivalence. Observation equivalence can be defined using an inductive relation definition that includes the following rules:

1. Observation equivalence is reflexive: $A \overset{oe}{=} A$.
2. Symmetric SUMM terms are observation equivalent: $A + B \overset{oe}{=} B + A$.
3. Two agents prefixed with the same action are observation equivalent if the derivative agents (without the prefixed action) are observation equivalent: $A \overset{oe}{=} B \Rightarrow \alpha.A \overset{oe}{=} \alpha.B$.
4. If an agent A is observation equivalence to an agent C **and** there is no action for which a transition is possible for an agent B and C, then $(A + B)$ and C are observation equivalent: $A \overset{oe}{=} C \wedge (NOTRAN\ B\ C) \Rightarrow (A + B) \overset{oe}{=} C$.
$$A \overset{oe}{=} C \wedge (NOTRAN\ B\ C) \Rightarrow (B + A) \overset{oe}{=} C.$$
5. If both of the summation agents (A and B) satisfy observation equivalence with the right-hand side agent (C) then the summation is observation equivalent to C: $A \overset{oe}{=} C \wedge B \overset{oe}{=} C \Rightarrow (A + B) \overset{oe}{=} C$.
6. If the right-hand side agent is a summation agent, then the left-hand side agent must satisfy observation equivalence for both of the right-hand side summation agents. This is the symmetric case of the previous rule for a left-hand summation: $A \overset{oe}{=} B \wedge A \overset{oe}{=} C \Rightarrow A \overset{oe}{=} (B + C)$.
7. The final rule requires a side condition that uses an auxiliary reach predicate definition. The term "REACHES A C", is true if C can be derived from A through only internal actions. This rule states that agents A and B are one-way observation equivalent if the internal action derivative of A is one-way observation equivalent to B: $A \overset{\tau^*}{\to} C \wedge C \overset{oe}{=} B \Rightarrow A \overset{oe}{=} B$.

3 System Verification using the Interpreter Calculus

To demonstrate how system integration can be achieved, we show how several devices can be composed to form a system. The system specification abstracts away details that suggest the system is implemented by several devices. As a first example, we will examine the interaction between a modified AVM-1 microprocessor and the memory subsystem. The first example is rather simple system as there is no possibility for device interference. The second example integrates a larger number of devices and integrates interrupt processing among multiple interpreters. With the addition of a direct memory access device (DMA), both the CPU and the DMA may wish to access memory simultaneously. This contention requires that we add a bus controller to the system and redefine the CPU as a set of interpreters. The addition of a DMA also requires the CPU respond to interrupts.

Previous work has constructed examples of how the process algebra can be used to specify and verify the interaction between devices [16]. Communication between devices may require only a single message or a series of messages to be passed between devices. At a low level, the information is passed over a bus using a hardware protocol (e.g., 4-phase handshaking). The types of interactions can be loosely described as belonging to one of the following categories:

Message Passing: Devices pass information asynchronously to one another. Devices may not wait for a sent message to be received and devices may or may not wait (block) for message arrivals.

Process Creation: In the context of hardware, new devices are not "created", but devices may be idle. The process of initiating a new task on an idle processor is analogous to process creation.

Rendezvous: Concurrently-active devices may eventually need to reach a common synchronization point before either can continue execution. Devices will be delayed until all devices reach the rendezvous point.

Remote Procedure Call: This is perhaps the most common form of interaction. A device (e.g., the CPU) requests a service of another device (e.g., the memory) and must wait until the device satisfies the request.

3.1 CPU/MMU Interaction

```
let MEM = new_definition
('MEM', "! (rep:^rep_ty) write read exec addr data superV done ack.
 MEM rep write read exec addr data superV done ack = !t:time. ?t':time.
   Next done (t, t+t') /\
   write t => store rep (address rep (addr t), (data t), (superV t))
          | (read t \/ exec t ==>
              (data (t+t') =
                    (fetch rep (address rep (addr t), superV t))))
            /\ (ack (t+t') =
                    memMgt rep (address rep (addr t),data t,
                          superV t, write t, read t, exec t))" );;
```

The AVM-1 memory interface specification (above) defines the effect of read and write operations on the state variables **memory** and **data**. If a write request is made at time t, the memory will reflect this request at some future time $t + t'$, otherwise, the memory will remain unmodified. If a read request is made at time t, then the data value returned is a function of the memory contents at time t.

Since the actual fetch and store operations are performed by the memory and not the CPU, these functions are abstract and part of the generic representation **rep**. When combining a CPU and memory, the functions would be instantiated by the concrete memory state manipulation functions. The protection mechanism provided by the MMU makes a distinction between data and instruction memory fetch requests. The **exec** variable indicates when a memory request is for an instruction code value.

From the perspective of the CPU, all of the parameters to **MEM** in the specification, are part of the environment. For ease in presentation, we will allow the PREFIX operator to construct an agent from two agents. Thus, $(a + b).c$ abbreviates $a.c + b.c$.

> done $= ack + nack$
>
> cpu_write_request $= \overline{(userMode + \tau.superMode)}.\overline{write}.\overline{address}.$done
>
> cpu_read_request $= \overline{(userMode + \tau.superMode)}.\overline{read}.\overline{address}.$done.$data$
>
> cpu_exec_request $= \overline{(userMode + \tau.superMode)}.\overline{exec}.\overline{address}.$done.$data$
>
> CPUtoMEM $=$ cpu_write_request $+ \tau.$cpu_read_request $+ \tau.$cpu_exec_request
>
> CPU $=$ CPUtoMEM.CPU

The terms express the actions that the CPU may perform with the use of the internal action (τ) being particularly significant. The internal action is used to indicate that the CPU will behave in one of several possible ways—with the choice being made internally by the CPU. The process algebra term CPUtoMEM states that the CPU may request memory either write data, read data, or fetch an instruction. Without the use of τ, the term would suggest that any of these events could occur, depending on what an external agent might choose.

Figure 3 pictorially presents the semantic difference when using the τ action to prefix actions for the case where the CPU chooses to write. The CPU must indicate that process executing is either a user or supervisor process through the $\overline{userMode}$ or $superMode$ action, respectively. This choice is dependent on the internal state of the CPU. The left-hand side figure shows the case where the CPU presents one output synchronization action to its environment. The CPU will either make a userMode request or a superMode request. The right-hand side figure shows the case where both output synchronization events are possible. In this (unrealistic) figure, the CPU indicates that the choice of which synchronization event is up to an external agent.

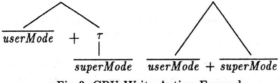

Fig 3. CPU Write Action Example

The MMU interpreter provides seven instructions: *superRead, superWrite, superExec, userRead, userWrite, userExec,* and *updateTblPtr*. Unlike the CPU (or the DMA described in Section 3.2), operation selection is dependent entirely on the environment. To express the notion that the MMU performs internal actions before responding, the τ action is inserted before the response is returned. Part of this action may be to update the segment table pointer value or cache. While the MMU is able to accept either a *userMode* or a *superMode* communication, the MMU chooses to respond with only $\overline{ack}$ or $\overline{nack}$ communication. The MMU specification also states when the MMU segment table pointer is updated. This action is not relevant to the CPU-MMU interface, so it is expressed as an internal (τ) action. The specification yields the process algebra terms:

done $= (\tau.\overline{ack} + \tau.\overline{nack})$

mmu_process_write $= (userMode + superMode).write.address.data.\tau.$done

mmu_process_read $= (userMode + superMode).read.address.\tau.$done$.\overline{data}$

mmu_process_exec $= (userMode + superMode).exec.address.\tau.$done$.\overline{data}$

MMUtoCPU $=$ mmu_process_write $+$ mmu_process_read $+$ mmu_process_exec

MMU $=$ MMUtoCPU.MMU

Proof of Correct Composition: The agents CPU and MMU are recursively defined and exhibit an infinite behavior. To show that the composed CPU and MMU communicate correctly, we must show that the system is always able to return to its initial state. It is also necessary to show that progress is made when the CPU initiates a dialogue with the MMU. The proof goal then can be stated as:

1. If either of the CPU actions are enabled, the memory subsystem will engage in communication.
2. The protocol will complete and the system returns to its initial state.

To reason about the finite protocol communication sequences, the recursive behavior of the agents is removed. The recursive agent reference is replaced with an (undefined) agent constant SUCCESS.

To express the goal in a general form, several auxiliary definitions are defined. (ENABLED) is defined to construct a list of all output actions that an agent can perform. A goal predicate definition BECOMES is defined. The predicate states that for all possible enabled output actions, a complement action exists such that a success agent is reached immediately or reached in a descendent.

```
let ENABLED = new_recursive_definition false AGENT 'ENABLED'
   "(ENABLED(INACTIVE)    = []) /\
    (ENABLED(PREFIX a A)  = ( (INTERNAL=a) => ENABLED A |
                            (( (TYP a) = F) => [a] | [] ))) /\
    (ENABLED(SUMM A B)    = APPEND (ENABLED A) (ENABLED B))

let BECOMES = new_definition('BECOMES',
   "!(system success :agent). BECOMES system success =
       (EVERY (\x. (TRANS system x success)) (ENABLED system) )");;
```

The success agent for the composed MMU-CPU is $SUCCESS \mid SUCCESS$. By unwinding the agent definitions and using the TRANS laws, all possible paths are found to reach the success agent through internal transitions. The MMU-CPU composed communication proof shows that:

$$\vdash BECOMES\ (CPU\ |\ MMU)\ (SUCCESS\ |\ SUCCESS)$$

3.2 Multiple Device System Specification

The inclusion of a DMA controller greatly increases the overall system complexity. A DMA adds interrupt behavior to the system as well as directly interacting with the other system devices. The DMA is programmed to supervise the transfer of a sequence of words between a peripheral device and memory. Upon completion of the transfer, the DMA will generate an interrupt to indicate the transfer has completed. The system described below, places the DMA before the MMU. Thus, DMA memory access requests are validated by the MMU. If the DMA acts as a user process, the MMU may also translate virtual address into real addresses.

Bus Controller Specification: We model the bus controller in a simplified manner, without a fairness property. The controller grants control to either the CPU or DMA and then waits for the controlling device to release the bus. This behavior may be described by the process algebra term:

$$BUS = (CPUreqBus.\overline{grantCPU}.CPUrelease.BUS)\ + \\ (DMAreqBus.\overline{grantDMA}.DMArelease.BUS)$$

The addition of the bus controller requires a change to the external behavior of the CPU as follows.

$$CPU = \overline{CPUreqBus}.grantCPU.CPUtoMEM.\overline{CPUrelease}.CPU$$

DMA Device Specification: The DMA provides four channels between memory and I/O devices. The behavior of each channel is determined by a set of registers, which are accessed as memory locations. We will include a DMA with a single write-to-memory channel. The register set for a channel consists of a memory address register, a channel counter register, and a control/status register. The DMA interpreter provides three operations: read DMA register, write DMA register, and service I/O device. The service I/O device operation is only available when the channel is enabled. A valid alternative interpreter specification might distinguish between the registers and define unique read and write operations for each of the registers.

To avoid a possible deadlock situation, the DMA agent is defined as two concurrent agents. One agent provides the interpreter operations that provide access to the DMA registers while the other agent describes the service I/O device operation. Deadlock might occur if the shared bus resource was being used by the CPU to access a DMA register while the DMA was attempting to obtain the bus in order to write a value to memory. The CPU would not free the bus until the DMA register operation completed and the DMA register operation would not begin until the DMA I/O service operation completed—which is blocked by the CPU use of the bus.

$$\text{DMAreg} = (write + read).DMAaddress.data.\overline{done}.\text{DMAreg}$$

$$\text{writeTransfer} = \overline{superMode}.\overline{write}.\overline{address}.\overline{data}.done$$

$$\text{IOwrite} = \overline{DMAreqBus}.grantDMA.\text{writeTransfer}.\overline{DMArelease}$$

$$\text{DMAio} = ioInt.\text{IOwrite}.(\text{DMAio} + \tau.\overline{cpuInt}.\text{DMAio})$$

$$\text{DMA} = \text{DMAreg} \mid \tau.\text{DMAio}$$

CPU Specification: The potential interference between the CPU and DMA requires a more realistic specification of the CPU. Interference may arise due to the shared use of memory. To avoid multiple critical events during the execution of an instruction, we divide the CPU interpreter specification into the three interpreters: FETCH, DECODE, and EXECUTE. We assume the CPU has an orthogonal instruction set so that each of these interpreters requires at most, one access to memory. The FETCH interpreter either detects an interrupt is pending (from the DMA) or obtains the next instruction from memory. For the present model, we assume that the CPU may not disable interrupt recognition. The DECODE interpreter may obtain an instruction operand from memory. The EXEC interpreter may write one word to memory.

Below we present the process algebra terms to describe the CPU behavior. Rather than assuming that the CPU is driven by an external clock, we assume that the CPU will continuously attempt to execute instructions. Note the use of the τ action in the agent CPU to indicate that a stage may not access memory.

$$\text{FETCH} = \overline{CPUreqBus}.grantCPU.cpu\_exec\_request.CPUrelease$$

$$\text{DECODE} = \overline{CPUreqBus}.grantCPU.cpu\_read\_request.CPUrelease$$

$$\text{EXEC} = \overline{CPUreqBus}.grantCPU.cpu\_write\_request.CPUrelease$$

$$\text{CPU} = (cpuInt+\text{FETCH}).(\tau + \text{DECODE}).(\tau + \text{EXEC}).\text{CPU}$$

Proof of Correct Composition: We may describe the composed system by the following term, with the system state being the union of all the device states.

$$\text{System} = \text{CPU} \mid \text{MMU} \mid \text{DMA} \mid \text{BUS}$$

The correctness proof shows that the composed system is observationally equivalent to a system that abstracts away all the communication between devices. One such proof we show is that the BUS can be abstracted away from the system:

$$\text{System} \stackrel{oe}{=} \text{CPU} \mid \text{MMU} \mid \text{DMA}$$

4 Conclusion

We have presented a framework to formally verify the correctness of communication between composed devices. Previous system verification research has developed *vertically verified systems*. However, the hardware bases for these systems have been simplistic. Our research is developing a framework to verify a more realistic *horizontally verified system*. This work demonstrates that CCS is a good choice for describing interdevice implementation-level connections within a computer system. Additional research will expand the calculus and address automating the derivation of process algebra expressions from interpreter specifications. Several improvements are being investigated, including an additional type constructor for recursive agents, greater proof support, and automation of the tedious aspects of the proofs.

References

1. Martin Abadi and Leslie Lamport. Composing specifications. Technical Report 66, Digital Systems Research Center, October 1990.
2. J. C. M. Baeten and W. P. Weijland. *Process Algebra.* Cambridge University Press, 1990.
3. William R. Bevier, Warren A. Hunt, and William D. Young. Toward verified execution environments. *IEEE Symposium on Security and Privacy*, 1987.
4. Albert Camilleri, Paola Inverardi, and Monica Nesi. *Combining Interaction and Automation in Process Algebra Verification.* Lecture Notes in Computer Science No. 494. Springer Verlag, 1991.
5. Albert John Camilleri. Mechanizing CSP trace theory in Higher Order Logic. *IEEE Transactions on Software Engineering*, 16(9):993–1004, September 1990.
6. R. DeNicola, P. Inverardi, and M. Nesi. Using the axiomatic presentation of behavioral equivalences for manipulating ccs specifications. In *International Worshop on Automatic Verification Methods for Finite State Systems*, Lecture Notes in Computer Science No. 407, pages 54–67. Springer Verlag, 1989.
7. C.A.R. Hoare. *Communicating Sequential Processes.* Prentice Hall, 1985.
8. J. L. Jacob. Specifying security properties. In C.A.R. Hoare, editor, *Development in Concurrency and Communication.* Addison-Wesley, 1990.
9. Jeffrey J. Joyce. Totally verified systems: Linking verified software to verified hardware. In M. Leeser and G. Brown, editors, *Hardware Specification, Verification and Synthesis: Mathematical Aspects*, Lecture Notes in Computer Science No. 408. Springer Verlag, July 1989.
10. Leslie Lamport. A simple approach to specifying concurrent systems. *Communications of the ACM*, 32(1), January 1989.
11. David May and David Shepherd. Towards totally verified systems. In *Conference on Mathematics of Program Construction*, Lecture Notes in Computer Science No. 375. Springer-Verlag, June 1989.
12. Tom Melham. Automating recursive type definitions in higher order logic. In G. Birtwhistle and P.A Subrahmanyam, editors, *Current Trends in Hardware Verification and Automated Theorem Proving*, pages 341–386. Springer-Verlag, 1989.
13. Tom Melham. A package for inductive relation definitions in HOL, 1991.
14. Tom Melham. A mechanized theory of the π-calculus in HOL. Technical Report 244, Computer Lab, University of Cambridge, 1992.
15. Robin Milner. *Communication and Concurrency.* Prentice Hall, 1989.
16. E. Thomas Schubert. *A Methodology for the Formal Verification of Composed Hardware Systems.* PhD thesis, University of California, Davis, 1992.
17. Phillip J. Windley. *The Formal Verification of Generic Interpreters.* PhD thesis, University of California, Davis, 1990.

Composing Symbolic Trajectory Evaluation Results*

Scott Hazelhurst and Carl-Johan H. Seger

Department of Computer Science, University of British Columbia, Vancouver,
Canada V6T 1Z4
E-Mail: {shaze, seger}@cs.ubc.ca

Abstract. Symbolic trajectory evaluation shows much promise as a
method for verifying large scale VLSI designs with a high degree of
automation. However, to verify today's designs, a method for composing partial verification results is needed. Consequently, we have proven
a number of inference rules for the composition of symbolic trajectory
evaluation results and developed a specialised theorem prover designed
specifically for combining verification results based on trajectory evaluation. In the paper we discuss the underlying inference rules of the prover
as well as more practical issues regarding the user interface. We conclude with an example in which we verify a design that could not have
been verified directly. In particular, the complete verification of a 64 bit
multiplier takes under 15 minutes on a Sparc 10/51 machine.

1 Introduction

The verification of computer systems has become more important as computer
systems grow in complexity and range of use. Verification — the proving under
some mathematical theory of the existence (or non-existence) of properties in the
system — comes at a cost: it is a difficult and computationally intensive task.
Although significant advances have been made, all verification methods suffer
from this problem in some way. There are a number of different approaches to
hardware verification [10]. This paper presents a new hybrid technique based on
theorem-proving and symbolic trajectory evaluation (STE).

1.1 Symbolic Trajectory Evaluation – STE

If the state space of a system (for example a circuit) can be embedded in a lattice,
the behaviour of the system can be expressed as a *trajectory*, a sequence of points
in the lattice determined by the initial state and the system functionality. We
define a partial order between sequences of states by extending the partial order
on the state space in a natural way.

---

* This research was supported by operating grant OGPO 109688 from the Natural Sciences Research Council of Canada, a fellowship from the Advanced Systems Institute,
and by research contract 92-DJ-295 from the Semiconductor Research Corporation.

The model we use of a system is simple and general. A *model structure* is a tuple $\mathcal{M} = [\langle \mathcal{S}, \sqsubseteq \rangle, Y]$, where $\langle \mathcal{S}, \sqsubseteq \rangle$ is a complete lattice ($\mathcal{S}$ being the state space and $\sqsubseteq$ a partial order on $\mathcal{S}$) and Y is a monotone successor function $Y : \mathcal{S} \to \mathcal{S}$. A sequence is a *trajectory* if and only if $Y(\sigma^i) \sqsubseteq \sigma^{i+1}$ for $i \geq 0$.

The key to the efficiency of trajectory evaluation is the restricted language that can be used to phrase questions about the model structure. The basic specification language we use is very simple, but expressive enough to capture many of the properties we need to check for.

A *predicate* over $\mathcal{S}$ is a mapping from $\mathcal{S}$ to the lattice $\{false, true\}$ (where $false \sqsubseteq true$). Informally, a predicate describes a potential state of the system: e.g., a predicate might be (A **is** x) which says that node A has the value x. A predicate is *simple* if it is monotone and there is a unique weakest $s \in \mathcal{S}$ for which $p(s) = true$. A *trajectory formula* is defined recursively as:

1. **Simple predicates**: Every simple predicate over $\mathcal{S}$ is a trajectory formula.
2. **Conjunction**: $(F_1 \wedge F_2)$ is a trajectory formula if F_1 and F_2 are trajectory formulas.
3. **Domain restriction**: $(e \to F)$ is a trajectory formula if F is a trajectory formula and e is a Boolean expression.
4. **Next time**: $(\mathbf{N}F)$ is a trajectory formula if F is a trajectory formula.

The truth semantics of a trajectory formula is defined relative to a model structure and a trajectory. Whether a trajectory $\tilde{\sigma}$ satisfies a formula F (written as $\tilde{\sigma} \models_{\mathcal{M}} F$) is given by the following rules.

1. $\sigma^0 \tilde{\sigma} \models_{\mathcal{M}} p$ iff $p(\sigma^0) = true$.
2. $\sigma \models_{\mathcal{M}} (F_1 \wedge F_2)$ iff $\sigma \models_{\mathcal{M}} F_1$ and $\sigma \models_{\mathcal{M}} F_2$
3. $\sigma \models_{\mathcal{M}} (e \to F)$ iff $\phi(e) \Rightarrow \sigma \models_{\mathcal{M}} F$, for all ϕ mapping Boolean expressions to $\{true, false\}$.
4. $\sigma^0 \tilde{\sigma} \models_{\mathcal{M}} \mathbf{N}F$ iff $\tilde{\sigma} \models_{\mathcal{M}} F$.

Given a formula F there is a unique *defining sequence*, δ_F, which is the weakest sequence which satisfies the formula[2]. The defining sequence can usually be computed very efficiently. From δ_F a unique *defining trajectory*, τ_F, can be computed (often efficiently). This is the weakest trajectory which satisfies the formula — all trajectories which satisfy the formula must be greater than it in terms of the partial order.

If the main verification task can be phrased in terms of "for every trajectory σ that satisfies the trajectory formula A, verify that the trajectory also satisfies the formula C", verification can be carried out by computing the defining trajectory for the formula A and checking that the formula C holds for this trajectory. We write such a result as $\models_{\mathcal{M}} [A \Longrightarrow C]$. The fundamental result of STE is given below.

Theorem 1. *Assume A and C are two trajectory formulas. Let τ_A be the defining trajectory for formula A and let δ_C be the defining sequence for formula C. Then $\models_{\mathcal{M}} [A \Longrightarrow C]$ iff $\delta_C \sqsubseteq \tau_A$.* $\square$

---

[2] "Weakest" is defined in terms of the partial order.

A key reason why STE is an efficient verification method is that the cost of performing STE is more dependent on the size of the formula being checked than the size of the system model.

STE uses ordered binary-decision diagrams (OBDDs or BDDs) [2] for efficient manipulation of boolean expressions. Using OBDDs, boolean expressions have canonical forms making comparison of expressions very efficient. Though BDDs have practical limitations, the use of BDD-based method has extended by orders of magnitude the size of systems that can be tackled by model-checkers.

The Voss system, a formal hardware verification system developed at the University of British Columbia, consists of three major components: a highly efficient implementation of OBDDs; an event driven symbolic simulator with very comprehensive delay and race analysis capabilities; and a general purpose, purely functional language. The language, called FL, is a strongly typed, polymorphic, and fully lazy language. Every object of type boolean in the system is internally represented as an OBDD. Consequently, FL is a very convenient language for developing prototype verification methodologies that require OBDD manipulations. Voss has been used to perform STE efficiently on large, sophisticated circuits. A full description of trajectory evaluation is given in [12]. Although many circuits can be verified very efficiently, there are limitations — these have have motivated this work.

1.2 Motivation

Our approach improves on STE by

1. Raising the level of abstraction, making verification more convenient for the user, and, more importantly, allowing us to overcome the limitations of BDDs.
2. Developing inference rules for the combination of STE results.
3. Using a specialised theorem-prover implementing these rules, allowing us to exploit the strengths of theorem-proving while still keeping the advantages of STE.

2 Related Work

There are two important influences on this work, compositionality and use of hybrid methods.

Compositionality is the divide-and-conquer approach of verification, promoting usability and efficiency. It allows re-use of results and, most importantly, it greatly reduces the state space of systems that need to be explored. Compositionality has been used in a number of different verification techniques, for example process algebras [11], model checking [3, 9], and theorem-proving.

By hybrid methods, we mean the combined use of different verification methods in a rigorous and sound way. One of the first systems which combined different approaches rigorously was the HOL-Voss system [13, 7], linking the HOL theorem-prover and the Voss STE system. This enables the proof of something within one system to be carried over into the other system. Although HOL-Voss is a much more powerful and general system, it is not clear that all the extra power is useful, and it is much more difficult to learn to use. Further, our system has the rules for composing symbolic trajectory evaluation results built in.

Kurshan and Lamport have combined the COSPAN model-checker with the TLP theorem prover [8]. The model checker proves properties of components of the system, which are then translated into a form suitable for the theorem-prover. In order to prove the overall result, a number of sub-results need to be proved (including a hand-checked step).

Hungar also links model checking and theorem-proving [6]. The model is given by a Kripke structure representing the semantics of an Occam program, and the properties are expressed in a variant of CTL. The results generated by model-checking are combined using the LAMBDA theorem-prover. Given an Occam program consisting of a number of processes, properties can be proven of each process using the model-checker. A number of rules — some analogous to the ones we propose in Section 3 — can be used by the theorem-prover to combine these sub-properties to prove properties of the entire program.

3 Inference Rules

The key result of STE is Theorem 1: from $\delta_C \sqsubseteq \tau_A$, we can infer $\models_{\mathcal{M}} [A \Longrightarrow C]$. This section develops other rules for inferring these results, enabling re-use and composition of old results, and incorporation of domain knowledge. To verify a circuit, we use STE to prove the low level properties of the circuit, and then use these rules to combine the low-level results into higher-level results. Throughout the section we shall use the circuit shown in Fig. 1 as a simple example. We omit all proofs for space reasons — the proofs can be found in [5].

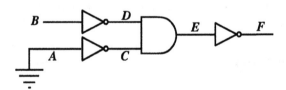

Fig. 1. Circuit C_2.

3.1 Identity Inference Rule

Although this theorem is very simple, it is useful.

Theorem 2. *If A is any trajectory formula then* $\models_{\mathcal{M}} [A \Longrightarrow A]$. □

3.2 Time Shift Inference Rule

This result allows us to move the "base time" of a result: if we can prove an assertion with all times relative to time zero, then the same result holds with all the times relative to some time t. This allows us to abstract details of time in certain places and promotes re-use of STE results.

Theorem 3. *For any trajectory formulas A and C, if $\models_{\mathcal{M}} [A \Longrightarrow C]$ then $\models_{\mathcal{M}} [N^t A \Longrightarrow N^t C]$ for any $t \geq 0$.* □

As an example, consider the circuit shown in Fig. 1. It is easy to see that trajectory evaluation can be used to prove that $\models_{\mathcal{M}} [B \text{ is } 1 \Longrightarrow N(D \text{ is } 0)]$. Applying the above theorem allows us to deduce $\forall t \geq 0$ that if B has the value 1 at time t then D has the value 0 at time $t+1$, i.e. $\models_{\mathcal{M}} [N^t(B \text{ is } 1) \Longrightarrow N^{(t+1)}(D \text{ is } 0)]$.

3.3 Post-condition Weakening and Pre-condition Strengthening

The following theorem is equivalent to the classical post-condition weakening and pre-condition strengthening. Importantly, semantic as well as syntactic information is used when deciding the ordering of two trajectories (shown later).

Theorem 4. *Let A, C, A_1 and C_1 be trajectory formulas. Suppose $\models_{\mathcal{M}} [A \Longrightarrow C]$. If $\delta_A \sqsubseteq \delta_{A_1}$, then $\models_{\mathcal{M}} [A_1 \Longrightarrow C]$. If $\delta_{C_1} \sqsubseteq \delta_C$, then $\models_{\mathcal{M}} [A \Longrightarrow C_1]$.* □

Example
In the example circuit, STE shows that $\models_{\mathcal{M}} [(B \text{ is } 0) \wedge (N(B \text{ is } 0)) \Longrightarrow (N^2(E \text{ is } 1) \wedge N^3(E \text{ is } 1))]$. Using post-condition weakening we can prove $\models_{\mathcal{M}} [(B \text{ is } 0) \wedge N(B \text{ is } 0)) \Longrightarrow N^2(E \text{ is } 1)]$.

3.4 Conjunction Rule

The following theorem is useful when properties about separate parts of the system have been proven and we now want to combine these results to reason about the combination. It is also useful when we have different results about the same part of the circuit and wish to combine them.

Theorem 5. *Let A_1, C_1, A_2, and C_2 be trajectory formulas. If $\models_{\mathcal{M}} [A_1 \Longrightarrow C_1]$ and $\models_{\mathcal{M}} [A_2 \Longrightarrow C_2]$, then $\models_{\mathcal{M}} [(A_1 \wedge A_2) \Longrightarrow (C_1 \wedge C_2)]$.* □

Example
Using the example circuit, we can prove $\models_{\mathcal{M}} [A \text{ is } 0 \Longrightarrow N(C \text{ is } 1)]$ and $\models_{\mathcal{M}} [B \text{ is } 0 \Longrightarrow N(D \text{ is } 1)]$. Combining these two results using conjunction gives $\models_{\mathcal{M}} [(A \text{ is } 0) \wedge (B \text{ is } 0) \Longrightarrow (N(C \text{ is } 1)) \wedge (N(D \text{ is } 1))]$.

3.5 Transitivity Inference Rule

This rule is analogous to the transitivity rule in logic: if $A \Rightarrow B$ and $B \Rightarrow C$ then $A \Rightarrow C$.

Theorem 6. *Let $A_1, C_1, A_2,$ and C_2 be trajectory formulas.*
Suppose $\models_{\mathcal{M}} [A_1 \Longrightarrow C_1]$ and $\models_{\mathcal{M}} [A_2 \Longrightarrow C_2]$. If $\delta_{A_2} \sqsubseteq \mathrm{lub}(\delta_{A_1}, \delta_{C_1})$ then $\models_{\mathcal{M}} [A_1 \Longrightarrow C_2]$. □

Example

We can use STE to show of the example circuit that $\models_{\mathcal{M}} [B \text{ is } 0 \Longrightarrow N^2(E \text{ is } 1)]$ and that $\models_{\mathcal{M}} [N^2(E \text{ is } 1) \Longrightarrow N^3(F \text{ is } 0)]$. Using transitivity, we can conclude, without having to compute anything else that $\models_{\mathcal{M}} [B \text{ is } 0 \Longrightarrow N^3(F \text{ is } 0)]$.

3.6 Specialisation Inference Rule

Specialisation is a rule which allows the generation from a general form of a trajectory assertion more specialised versions. For example from **[M is a $\Longrightarrow$ O is 2*a]** we may wish to generate **[M is 1 $\Longrightarrow$ O is 2*1]** or **[M is a+b $\Longrightarrow$ O is 2*(a+b)]**.

Definition 3.1 *A simple substitution σ is a function from a set of variables, $\tilde{I}$, to expressions over constants and these variables, $\mathcal{E}_{\tilde{I}}$.* □

Thus, if σ is a simple substitution and F is a trajectory formula, $\sigma(F)$ is F rewritten by replacing variables in F with the appropriate expressions given by the mapping.

Theorem 7 Simple Substitution Theorem. *Suppose $\models_{\mathcal{M}} [A \Longrightarrow C]$, and σ is a simple substitution. Then $\models_{\mathcal{M}} [\sigma(A) \Longrightarrow \sigma(C)]$.* □

A substitution defines what should be substituted for which variables throughout an assertion, and the simple substitution theorem shows that this is a valid result. A more sophisticated way of transforming an STE result is *specialisation*. Suppose we have two substitutions σ_1 and σ_2. Using the Simple Substitution Theorem we can derive the two results $\models_{\mathcal{M}} [\sigma_1(A) \Longrightarrow \sigma_1(C)]$ and $\models_{\mathcal{M}} [\sigma_2(A) \Longrightarrow \sigma_2(C)]$. Now using Theorem 5, we can derive the combined result $\models_{\mathcal{M}} [\sigma_1(A) \wedge \sigma_2(A) \Longrightarrow \sigma_1(C) \wedge \sigma_2(C)]$. We can generalise this in the following way.

Let $\tilde{\sigma} = [(c_1, \sigma_1), \ldots, (c_n, \sigma_n)]$, where each c_i is a boolean expression, and each σ_i a simple substitution. Then $\tilde{\sigma}$ is a *specialisation*, and if F is a trajectory formula, then $\tilde{\sigma}(F) = \wedge_i(c_i \rightarrow \sigma_i(F))$. The following result holds.

Theorem 8 Specialisation Theorem. *Let $\hat{\sigma}$ be specialisation. If $\models_{\mathcal{M}} [A \Longrightarrow C]$ then $\models_{\mathcal{M}} [\hat{\sigma}(A) \Longrightarrow \hat{\sigma}(C)]$*

Specialisation helps gluing together the different building blocks of a proof — often before using transitivity it is necessary to perform specialisation. As a simple example, suppose we have a circuit which takes in two numbers x and y and outputs $2*\max(x,y)$. The circuit consists of two parts: one part compares x and y and outputs the greater; the second part doubles the output of the first part. Using STE, natural results we might prove are:

$$T_1 = \models_{\mathcal{M}} [(A \text{ is } x) \wedge (B \text{ is } y) \Longrightarrow$$
$$\mathbf{N}(((x > y) \rightarrow (T \text{ is } x)) \wedge (\neg(x > y) \rightarrow (T \text{ is } y)))]$$

and

$$T_2 = \models_{\mathcal{M}} [\mathbf{N}(T \text{ is } a) \Longrightarrow \mathbf{N}^2(C \text{ is } (a + a))].$$

Now using specialisation, from the latter result we can obtain:

$$T_2' = \models_{\mathcal{M}} [\mathbf{N}(((x > y) \rightarrow (T \text{ is } x)) \wedge (\neg(x > y) \rightarrow (T \text{ is } y)))$$
$$\Longrightarrow \mathbf{N}^2((x > y) \rightarrow (C \text{ is } (x + x)) \wedge$$
$$((\neg(x > y)) \rightarrow (C \text{ is } (y + y))))].$$

Transitivity can now be applied between T_1 and T_2' to obtain the overall result which we want.

Specialisation also improves efficiency. In this example, it was faster to prove $T2$ than $T2'$ since the OBDDs needed in the verification of $T2$ are smaller than the ones in $T2'$. Although this example is so small that there is little real difference here, in larger examples, there can be an enormous practical difference. Thus, while it may be more indirect to prove a result using STE for a more general case (with simpler OBDDs) and then to specialise, it is far more efficient than directly using STE to obtain the desired result.

4 Implementing a useful tool

4.1 Implementation of Inference Rules

The theory described so far has been implemented, and integrated with Voss into a new system. This tool is a theorem prover with which a user can prove properties of circuits specified either in VHDL or netlist form. The verification proof is an FL program which uses the theorem prover at each step. The ability to use FL as a script language gives great versatility and power.

The system has defined abstract data types for representing and manipulating trajectories and the appropriate values (like integers). The inference rules return objects of type Theorem. Only the inference rules can manipulate these objects, guaranteeing soundness of results. By restricting the operations allowed on this type of object, we can ensure that the system is safe, while still having the power and flexibility of a fully-programmable script language.

All the inference rules described in the previous section have been implemented. In addition more complex rules which package the basic rules have been implemented so as to ease the job of the user. A very important part of the packaging is the provision of heuristics which help the user in finding specialisations and time-shifts. For example, before being able to use transitivity between two results T_1 and T_2, it may be necessary to specialise the latter or time-shift one of them. Of course, the user can provide these transformations, but it would be better if these transformations could be automatically found. Although the heuristics we have provided are fairly simple, they are very useful and in the example verifications we have done have proven quite adequate. What should be emphasised is that the heuristics are integrated into the system in a secure way. If the heuristics are inadequate (or error-prone) it may not be possible to prove correct results; it will not be possible to prove incorrect results. For space reasons a discussion of the heuristics is beyond the scope of this paper.

4.2 Raising the Level of Abstraction

Besides helping the verification process by reducing the semantic gap between the user's understanding and the circuit being verified, abstraction has key efficiency advantages.

All verification results are represented by our tool as an abstract data type. Being able to translate from this abstract data type to OBDDs is essential, but using a more abstract representation means we no longer have to rely on OBDDs alone. Although there are disadvantages in this, we can exploit the strengths of OBDDs wherever possible, yet overcome their limitations where necessary.

There are two ways in which this is done. First, using a more abstract representation increases the usefulness of specialisation. General forms of a result can be proven using STE where the general form has an efficient OBDD representation. Specialisation can then yield the result we want, even if the specialised version of the result has no efficient OBDD representation.

Second, we no longer rely on OBDDs as the sole means of arguing about the equality of two expressions. In the HOL-Voss system it was shown how to prove in an integrated system that two expressions have the same OBDD (and hence are semantically identical) without having to construct the OBDDs [7]. We adopt a similar approach by supplying a simple integer arithmetic theorem prover. With the domain knowledge this prover provides and the inference rules described earlier, we are able to reason about circuits that we would not be able to reason about otherwise.

As a simple example, we may have proven that a certain node obtains some value e_1 where e_1 is a complicated arithmetic expression. Using domain knowledge and trivial postcondition weakening, we may be able to show that the node obtains the value e_2 where e_2 is a much simpler expression denoting the same value as e_1.

Additional abstraction is provided by allowing the user to describe a mapping between the names of the nodes in the physical circuit being verified and logical names used in the proof. This is useful in referring to a group of physical nodes by a group name, rather than by referring to all their individual names. Instead of specifying the values of all 64 bit-valued nodes representing some integer value, the proof can refer to one integer-valued node. This syntactic sugar makes proofs much pleasanter, and library routines are provided to aid the creation of the mapping. The user needs to bear in mind that the results obtained actually are about the low-level circuit and not some "abstract" machine.

5 Example

For space reasons, we only give one example; however, we have applied this method successfully to a number of other examples. The problem we concentrate on as an example here is the verification of a simple combinational multiplier circuit. (We have verified a Wallace-tree multiplier too — although the proof is not much more complicated than the example we show here, it is easier to describe this proof.)

Ideally, the verification would be single STE result relating the output to the inputs. However, properties concerning multiplication cannot be verified using BDD based tools alone, since the representation of multiplication by BDDs needs exponentially sized BDDs [1]. In this example we show how a multiplication circuit can be verified using our tool by breaking the proof into parts and composing the results, thereby achieving a result equivalent to the straightforward STE verification.

5.1 The Multiplication Circuit

The multiplication circuit consists of a series of adders with some additional circuitry. If the bit-width of the of the circuit is b, there will be b stages. The figure below shows an overview of the circuit, which is implemented as a VHDL program and consists of approximately 25 000 gates (in our case, $b = 64$). The algorithm used is the standard long multiplication algorithm $xy = \Sigma_{i=1}^{b}(2^{i-1}x_iy)$ where x and y are b bit numbers and x_i is the i-th least significant bit of x. The function of the i-th stage is to compute $(2^{i-1}x_iy)$ and add this to the result obtained so far. All arithmetic is done modulo 2^b. An implication of this is that the i-th stage does not use the i higher order bits of y. Therefore the i-th stage has as input b bits which will give the partial sum so far, one bit of x, and $b - i$ bits from y. The input of x and y is on nodes A and B respectively.

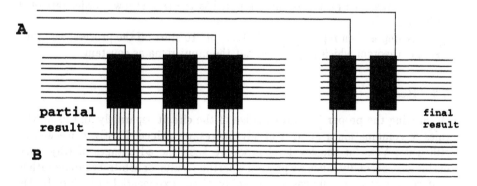

Fig. 2. Overview of Multiplier

5.2 Verification Strategy

The verification of the multiplier entails:

1. Verifying the individual adders;
2. Showing the individual adders are connected correctly;
3. The overall effect of the circuit is to perform multiplication.

The theory described earlier shows that the proof is equivalent to performing STE on the entire circuit. We gain significant performance improvements by using domain-knowledge (to show that two BDDs are the same without having to construct them), and composing a number of smaller STE results. The proof itself is a two page FL program which applies the proof rules described earlier: STE is used to prove local properties, and the composition rules show that that the parts are connected correctly and the overall result is multiplication. An outline of the proof follows, and an edited (for length and clarity – declarations are omitted) version of the proof appears in the appendix.

The key to the proof is to recognise that after the i-th stage, the result computed by the circuit is y multiplied by the i lower order bits of x. Suppose that at the end of the i-th stage we have proven that given the initial input for the circuit, the output of the i-th stage is indeed the i lower order bits of x multiplied by y. At the $(i + 1)$-st step we do three things.

Firstly, the local property of the stage is checked — that it actually does the addition and so on. This proof is done using Voss, and, crucially for the efficiency of the checking, is done for arbitrary input values rather than the actual values the circuit will use when executing. Secondly, once this check has been done, we compose this result with the result from the i-th stage. Specialising and time-shifting the new result[3] allows us to compose the transformed new result with the i-th stage result, yielding the $(i + 1)$-st stage's output in terms of the whole circuit's input. Finally, the consequent of this theorem is not quite in the form we want, and we use post-condition weakening to obtain that at the end of the $(i + 1)$-st stage the partial result computed is the $(i + 1)$ lower order bits of x multiplied by y.

This step is then repeated for $i = 1, \ldots, n$. To start off the process, we prove a simple theorem which just states that the input remains constant.

5.3 Results

By proving the properties of each stage of the circuit separately and using the rules of combination a number of advantages are gained. Firstly, since we are still using Voss for the low-level proof we keep the advantages of trajectory evaluation. Secondly, since we prove the partial results for arbitrary inputs rather that complicated specific cases, we can avoid the exponential growth of BDDs and thereby make the method tractable. All time-shifting and specialisations are derived automatically, simplifying the task of the user.

---

[3] These transformations are found by the heuristics.

This process verifies the entire circuit, including ensuring that different parts of the circuit are correctly connected. The verification of a 64-bit multiplier took just less than 800 CPU seconds on a Sun Sparc 10/51 processor. The performance is roughly quadratic in the bit-width, so this problem can easily be dealt with.

It is difficult to estimate the amount of human effort involved in the proof, since the proof went hand-in-hand with system development (and circuit debugging!). It turned out that one of the most difficult parts of the verification was getting the timing constraints correct. Our estimate is that the proof itself took two days to get right.

6 Conclusion

6.1 Summary

We have proposed a theorem-prover based on STE. This hybrid method is a powerful theory and tool for low-level hardware verification. In our largest example, we verified the correctness of a 64-bit multiplier (a circuit consisting on the order of twenty-five thousand gates) using approximately fifteen minutes of CPU time.

We believe that we have shown that the use of hybrid methods give us flexibility and power without losing rigour. An important contribution is that domain knowledge can be incorporated — a very important factor in improving efficiency. From a usability and safety point of view it is important that a single, integrated system implement the hybrid method so that the user does not have to switch between systems or perform any translation.

Composition of results is important: whether we have proved results of different parts of the circuit, or want to combine smaller results of the same part of the circuit, or re-use results, composition is a very powerful technique. Verification is made easier to understand, and, for reasons discussed earlier, much more efficient. In particular, symbolic trajectory evaluation results can be composed. This enables the derivation of symbolic trajectory evaluation results without having to explicitly perform the trajectory evaluation.

6.2 Problems and Extensions

This work is a prototype for future exploration. Besides issues of dealing with the rough edges, there are a number of problems and areas for future research.

As the use of domain knowledge is important, we need to see how this can be integrated in a cleaner fashion, and also extend the range and type of arguments we can make. We are also looking at enriching the logic in which we can express circuit properties. The importance of both of these points was illustrated in the verification of the Wallace-tree multiplier, where the proof would be made simpler and cleaner by such improvements.

One problem with theorem-provers is the "false implies everything problem". In our system, an inconsistent antecedent can be used to "prove" almost anything, leading to theorems which are mathematically valid but have no basis in reality thereby lulling human users into a false sense of security. It would be good to provide automatic detection of such inconsistencies.

Systems which have little identifiable structure, or in which the interactions of the sub-parts of the system are very complex and fine-grained spatially and temporally will not be appropriate targets for this method.

Though the heuristics developed for the tool performed well, providing different and more powerful ones would improve the the system's usability.

STE applies to any type of system, but in practice it has only been applied to hardware systems. The use of abstraction and composition increases the range of systems for which STE will be tractable, and we are investigating how STE can be extended to more general systems.

References

1. R. E. Bryant. On the Complexity of VLSI Implementations and Graph Representations of Boolean Functions with Application to Integer Multiplication. *IEEE Transactions on Computers*, 20(2):205–213, Feb. 1991.

2. R. E. Bryant. Symbolic Boolean Manipulation with Ordered Binary-Decision Diagrams. *ACM Computing Surveys*, 24(3):293–318, Sept. 1992.

3. E. Clarke, D. Long, and K. McMillan. Compositional Model Checking. In *IEEE Fourth Annual Symposium on Logic in Computer Science*, Washington, D.C., 1989. IEEE Computer Society.

4. C. Courcoubetis, editor. *Proceedings of the 5th International Conference on Computer-Aided Verification*, Lecture Notes in Computer Science 697, Berlin, July 1993. Springer-Verlag.

5. S. Hazelhurst and C.-J. H. Seger. A Simple Theorem Prover Based on Symbolic Trajectory Evaluation and OBDDs. Technical Report 93-41, Department of Computer Science, University of British Columbia, Nov. 1993. Available by anonymous ftp as ftp.cs.ubc.ca:/pub/local/techreports/1993/TR-93-41.ps.gz.

6. H. Hungar. Combining Model Checking and Theorem Proving to Verify Parallel Processes. In Courcoubetis [4], pages 154–165.

7. J. J. Joyce and C.-J. H. Seger. Linking BDD-based Symbolic Evaluation to Interactive Theorem-Proving. In *Proceedings of the 30th Design Automation Conference*. IEEE Computer Society Press, June 1993.

8. R. Kurshan and L. Lamport. Verification of a Multiplier: 64 Bits and Beyond. In Courcoubetis [4], pages 166–179.

9. D. E. Long. *Model Checking, Abstraction, and Compositional Verification*. PhD thesis, Carnegie-Mellon University, School of Computer Science, July 1993. Technical report CMU-CS-93-178.

10. M. C. McFarland. Formal Verification of Sequential Hardware: A Tutorial. *IEEE Transactions on Computer-Aided Design of Integrated Circuits and Systems*, 12(5):633–654, May 1993.

11. R. Milner. *Communication and Concurrency*. Prentice-Hall International, London, 1989.

12. C.-J. H. Seger and R. E. Bryant. Formal Verification by Symbolic Evaluation of Partially-Ordered Trajectories. Technical Report 93-8, Department of Computer Science, University of British Columbia, Apr. 1993.

13. C.-J. H. Seger and J. J. Joyce. A Mathematically Precise Two-Level Hardware Verification Methodology. Technical Report 92-34, Department of Computer Science, University of British Columbia, Dec. 1992.

A Proof of Multiplier

```
let Ainp="A";
let Bout= "B";
let Ground = "TC0";

// variable maps
....omitted for clarity
// Mathematical results
let theory = ...

// Preamble theorem-------------------------------------------
let signal_length = 100000;

let preambleThm=IDENTITY(("A" ISINT x)_&_("B" ISINT y)_&_(Ground ISINT zero)
                    FROM 0 TO signal_length);

// GENERAL STEP-----------------------------------------------
// This is the proof that the n-th stage in the multiplier works

// Timing considerations -- ....
let answer_delay  = CPA_DELAY*bit_width;
let start_stage n = n*answer_delay;
let signal_len  n = signal_length-n*answer_delay;

let stage n  =
      let Cinp= "TC"^(num2str n) in
      let Cout= "TC"^(num2str (n+1)) in
      let jpart   = BWID ('(bit_width - n)) j in
      (((Ainp ISINT i) _&_ (Binp ISINT j) _&_  (Cinp ISINT k))
         FROM 0  TO signal_len n)
       ==>>
        (Cout ISINT (k '+ (jpart '* ((BIT2  ('(n+1)) i) '* (POW2 ('n) )))
                    FROM    answer_delay TO   signal_len n);

let multhm n = VOSS (varmap n)  (stage n);

// Show partial result computed from the composition of stages 1..n is correct

let induc  n =
      let n1 = n+1 in
      let Cout= "TC"^(int2str n1) in
         Cout ISINT ((BWID ('n1) x) '* y)
                    FROM start_stage n+answer_delay TO signal_length;

// Each step in the proof:
//    1. prove the n-th stage works
//    2. use ALIGNSUB to compose (1) with the proof that the
//       previous stage computed what it should be
//    3. Use POSTWEAK to show the "induction" step

let  inferencestep n start =
   let newthm2 = ALIGNSUB theory start (multhm n) in
      POSTWEAK theory  newthm2 (induc n);

// Postamble:  Use POSTWEAK to get the result in the form we want
let postamble=
      let prop_delay = start_stage bit_width in
      let gate = "TC" ^ (num2str c_size) in
        (gate ISINT ( x '* y)) FROM prop_delay TO signal_length;

// Proof
let  do_proof i sofar  =
            i=bit_width =>  POSTWEAK theory sofar postamble
                         | do_proof (i+1) (inferencestep i sofar);

let  MultTheorem = do_proof 0 preambleThm;
```

The Completeness of a Hardware Inference System*

Zheng Zhu, C-J Seger

The Integrated System Design Laboratory
Department of Computer Science
The University of British Columbia
Vancouver, B.C. Canada V6T 1Z4

Abstract. Symbolic trajectory evaluation (STE) is a method for efficient circuit verification [1]. In [2] a set of inference rules was introduced for combining STE results. These inference rules were also proven sound. In this paper we show that, with one additional inference rule, the inference system is complete. Here, complete means that any formula $A \Rightarrow C$, that is valid in every model satisfying some collection Φ of STE assertions, can be derived from Φ by a finite applications of the inference rules. The completeness proof is based on the method of model construction—given Φ, a most general circuit model (in which every assertion in Φ holds) can be generated.

1   Introduction

In [1], Seger and Bryant introduced the underlying theory for symbolic trajectory evaluation. In general, symbolic trajectory evaluation—as implemented in the Voss system [3] for example—is an efficient and highly automated circuit verification method that has been applied to the verification of quite complex and large VLSI circuits. However, in order to successfully verify modern large and complex circuits, a method of breaking down the verification task into smaller, more manageable pieces, is needed. One step towards this goal is developing inference rules that allow the user to combine "smaller" verification results in a safe and sound manner. In [2], such a set of inference rules was introduced and a special purpose theorem prover aimed specifically at manipulating STE results was discussed. Although the inference rules were shown to be sound, they were not shown to be complete. In this paper we remedy this by first adding one more inference rule and then prove that the obtained inference system is complete. Intuitively, complete in this context means that the set of rules is powerful enough to derive all the logical consequences of a given set of trajectory assertions.

* This research was supported, in part, by operating grants OGPO 109688 and OGPO 046196 from the Natural Sciences and Engineering Research Council of Canada, fellowships from the Province of British Columbia Advanced Systems Institute, and by research contract 92-DJ-295 from the Semiconductor Research Corporation.

1.1 Trajectory Evaluation

If the state space of a system (for example a circuit) can be embedded in a complete lattice, the behavior of the system can be expressed as a *trajectory* $\sigma = \sigma^0 \cdots \sigma^n \cdots$, a sequence of values in the lattice determined by the initial state and the system functionality. We define a partial order between sequences of states by extending the partial order on the state space in a natural way.

The model we use of a system is simple and general. A *model structure* is a tuple $\mathcal{M} = [\langle \mathcal{S}, \sqsubseteq \rangle, Y]$, where $\langle \mathcal{S}, \sqsubseteq \rangle$ is a complete lattice ($\mathcal{S}$ being the state space and $\sqsubseteq$ a partial order on $\mathcal{S}$) and Y is a monotone successor function $Y: \mathcal{S} \to \mathcal{S}$. Intuitively, the ordering relation orders the state space according to "information content", and the monotonicity requirement guarantees that we cannot loose information about the future by adding information about the present. A sequence is a *trajectory* if and only if $Y(\sigma^i) \sqsubseteq \sigma^{i+1}$ for $i \geq 0$.

The key to the efficiency of trajectory evaluation is the restricted language that can be used to phrase questions about the model structure. The basic specification language we use is very simple, but expressive enough to capture many of the properties we would like to check.

A *predicate* over $\mathcal{S}$ is a mapping from $\mathcal{S}$ to the lattice $\{0, 1\}$. Informally, a predicate describes a potential state of the system: e.g., a predicate might be $(A \text{ is } x)$ which says that node A has the value x. A predicate is *simple* if it is monotonic and there is a unique weakest $s \in \mathcal{S}$ (measured by $\sqsubseteq$), called the *defining value*, for which $p(s) = 1$. A special simple predicate is the unc, or "unconstrained", that holds for every state in $\mathcal{S}$ and thus has $\perp$ as defining value. A *trajectory formula* is defined recursively as:

1. **Simple predicates**: Every simple predicate over $\mathcal{S}$ is a trajectory formula.
2. **Conjunction**: $(F_1 \wedge F_2)$ is a trajectory formula if F_1 and F_2 are trajectory formulas.
3. **Domain restriction**: $(e \to F)$ is a trajectory formula if F is a trajectory formula and e is either 1 or 0.
4. **Next time**: $(\mathbf{N}F)$ is a trajectory formula if F is a trajectory formula.

Note that, in general, symbolic trajectory evaluation, as described in [1], gains its power from extending trajectory formulas to a symbolic domain—and thus concisely encode a very large collection of formulas. However, it should be emphasized that the extension to a symbolic domain only increases the computational efficiency and not the expressiveness of the logic. Consequently, we will only deal with non-symbolic assertions in this paper.

The *depth* of a formula F, written $d(F)$, is defined recursively as:

1. $d(p) = 1$ if p is a simple predicate.
2. $d(F_1 \wedge F_2) = \max(d(F_1), d(F_2))$.
3. $d(e \to F) = d(F)$.

4. $d(\mathbf{N}F) = 1 + d(F)$.

The depth of a formula is simply the maximum number of nested next time operators plus one. A trajectory formula is said to be *instantaneous* if it does not contain any $\mathbf{N}$ operators, i.e., if the depth of the formula is 1. It is straightforward to show that any trajectory formula, A, can be written in the form:

$$A^0 \wedge \mathbf{N}A^1 \wedge \mathbf{N}^2 A^2 \wedge \cdots \mathbf{N}^m A^m$$

where $N^2 A^2$ is a shorthand for $N(NA^2))$, etc., all A^i's are instantaneous formulas, and m is a natural number.

The truth semantics of a trajectory formula is defined relative to a model structure and a trajectory. Whether a trajectory $\sigma = \sigma^0 \tilde{\sigma}$ satisfies a formula F, written $\sigma \models_{\mathcal{M}} F$, is defined recursively as:

1. $\sigma^0 \tilde{\sigma} \models_{\mathcal{M}} p$ iff $p(\sigma^0) = 1$.
2. $\sigma \models_{\mathcal{M}} (F_1 \wedge F_2)$ iff $\sigma \models_{\mathcal{M}} F_1$ and $\sigma \models_{\mathcal{M}} F_2$
3. (a) $\sigma \models_{\mathcal{M}} (0 \rightarrow F)$ always holds
   (b) $\sigma \models_{\mathcal{M}} (1 \rightarrow F)$ iff $\sigma \models_{\mathcal{M}} F$
4. $\sigma^0 \tilde{\sigma} \models_{\mathcal{M}} \mathbf{N}F$ iff $\tilde{\sigma} \models_{\mathcal{M}} F$.

Given a model structure $\mathcal{M} = [\langle \mathcal{S}, \sqsubseteq \rangle, Y]$, let $\mathcal{S}^\omega$ denote the set of all infinite sequences of elements from $\mathcal{S}$. Before introducing the concept of defining sequences, it is convenient to introduce an infix "choice" function mapping $\{0, 1\} \times \mathcal{S}^\omega$ to $\mathcal{S}^\omega$ and which is defined as:

$$e?\sigma = \begin{cases} \sigma & \text{if } e = 1 \\ \bot\bot \ldots & \text{otherwise} \end{cases}$$

Given a formula F, we can define the *defining sequence* of F, denoted by $\delta(F)$ as follows:

1. $\delta(p) = \overline{p} \bot\bot \ldots$ if p is a simple predicate with defining value $\overline{p}$.
2. $\delta(F_1 \wedge F_2) = \delta(F_1) \sqcup \delta(F_2)$.
3. $\delta(e \rightarrow F) = e?\delta(F)$.
4. $\delta(\mathbf{N}F) = \bot \delta(F)$.

The *defining trajectory* of F in the model structure $\mathcal{M} = [\langle \mathcal{S}, \sqsubseteq \rangle, Y]$, denoted by $\tau(F)$, is a sequence defined as [1]:

1. $\tau^0(F) = \delta^0(F)$
2. $\tau^{i+1}(F) = Y(\tau^i(F)) \sqcup \delta^{i+1}(F)$ for $i \geq 0$.

When it becomes necessary to indicate explicitly that it is the defining trajectory of the model structure $\mathcal{M} = [\langle \mathcal{S}, \sqsubseteq \rangle, Y]$, we use $\tau_{\mathcal{M}}(F)$ to denote the sequence.

The fundamental result of STE is the following theorem [1]:

Theorem 1. *Assume A and C are two trajectory formulas. Let $\tau(A)$ be the defining trajectory for formula A and $\delta(C)$ be the defining sequence for formula C. Then $\models A \Rightarrow C$ iff $\delta(C) \sqsubseteq \tau(A)$.*

Finally, for somewhat technical reasons, define an assertion $A \Rightarrow C$ to be *prime* if for every $i \geq 0$, $\delta(A^i) \sqcap \delta(C^i) = \perp$. Intuitively, by requiring the assertions to be prime, we avoid the trivial cases when we are both assuming and checking that a node has a particular value at the same time. Practically speaking, requiring assertions to be prime is no real restriction since non-prime assertions usually are indications of some error(s) in the assertions.

1.2 Circuit Model Structure

In circuit verification, we restrict ourselves to a model structure of the form

$$(\langle I \times V, \sqsubseteq \rangle, Y)$$

where I and V are both complete lattices under partial orderings $\sqsubseteq_I$ and $\sqsubseteq_V$ respectively. Intuitively, I is the internal state space, and V is the space of visible nodes of a circuit, which is a space of cartesian products of $\{0, 1, \perp, \top\}$ and $\sqsubseteq_V$ is the cartesian (pair-wise) extension of the partial ordering $\preceq: \perp \preceq 0, 1 \preceq \top$. When it is clear by the context, we will use $\sqsubseteq$ to replace $\sqsubseteq_I$ and $\sqsubseteq_V$, and use $\perp$ to refer the bottom elements of V and I, as well as the bottom element of $I \times V$. Given an element $(i, v) \in I \times V$, we use $(i, v) \downarrow_I$ and $(i, v) \downarrow_V$ to denote i and v respectively.

Finally, we use the convention that the simple predicates used in circuit verification can only refer to the visible state. Consequently, for any circuit trajectory formula A, $\delta(A) = (\perp, v)$ for some $v \in V$. We call the trajectory formulas which specify properties of the values in V *the formulas defined on V*.

1.3 Conventions

Throughout this paper, the following notational conventions are adopted: Φ denotes a set of trajectory assertions. $A, A_1, \cdots, C, C_1, \cdots$ denote trajectory formulas. Also, for every trajectory formula A, we define:

$$A = A^0 \wedge \mathbf{N} A^1 \wedge \mathbf{N}^2 A^2 \wedge \cdots \mathbf{N}^m A^m$$

where all A^i are instantaneous formulas and $m = d(A) - 1$. Also, for $0 \leq i \leq m$, let

$$A^{\leq i} = A^0 \wedge \mathbf{N} A^1 \wedge \mathbf{N}^2 A^2 \wedge \cdots \mathbf{N}^i A^i$$

and

$$A^{\geq i} = A^i \wedge \mathbf{N} A^{i+1} \wedge \cdots \wedge \mathbf{N}^{m-i} A^m.$$

We use $\mathcal{M}, \mathcal{M}_\Phi$ to denote model structures. If an assertion φ holds in a model structure $\mathcal{M}$, we say $\mathcal{M}$ is *a model structure of* of φ. If every assertion in Φ holds in a model $\mathcal{M}$, we say that Φ is a model structure of Φ.

2 Constructing Circuit Models from Trajectory Assertions

In this section, we present a constructive method which generates a circuit model from a given set of trajectory assertions Φ such that every assertion of Φ holds true in the model [4].

2.1 Circuit Model Construction

Given an assertion $A \Rightarrow C$, *the set of all suffixes of* $A \Rightarrow C$, denoted by $P_{A,C}$, is:

$$P_{A,C} = \{(A^{\geq i}, C^{\geq i}) \mid 0 \leq i \leq depth(C) - 1\}.$$

Let $A \Rightarrow C$ be a prime assertion. *The model structure constructed from* $A \Rightarrow C$, denoted by $\mathcal{M}_{A \Rightarrow C}$, is

$$\mathcal{M}_{A \Rightarrow C} = ((Q_{A,C} \times V, \sqsubseteq), Y_{A,C})$$

where

- $V = \{0, 1, X, \top\}^n$ is the visible circuit state partially ordered as before,
- $Q_{A,C}$ is the largest subset of $2^{P_{A,C}}$ such that for every $q \in Q_{A,C}$, $(A, C) \in q$,
- $\sqsubseteq$ is a binary relation of $Q_{A,C} \times V$: for arbitrary $q_1, q_2 \in Q_{A,C}$ and $v_1, v_2 \in V$, $(q_1, v_1) \sqsubseteq (q_2, v_2)$ if and only if $q_1 \subseteq q_2$ and $v_1 \sqsubseteq v_2$, and
- $Y_{A,C} : Q_{A,C} \times V \to Q_{A,C} \times V$ is the next-state function of the model. For arbitrary $q \in Q_{A,C}$ and $v \in V$, $Y_{A,C}(q, v) = (q', v')$ where

$$q' = \{(a^{\geq 1}, c^{\geq 1}) \mid (a, c) \in q \text{ and } \delta(a^0) \downarrow_V \sqsubseteq_V v\} \cup \{(A, C)\}$$

and

$$v' = \bigsqcup_{(a,c) \in q'} \delta(c^0) \downarrow_V \tag{1}$$

Let $\Phi = \{A_j \Rightarrow C_j \mid 0 \leq j \leq n - 1\}$ be a set of n prime assertions. Define

$$P_\Phi = \bigcup_{A \Rightarrow C \in \Phi} P_{A,C}$$

and Q_Φ as the largest subset of P_Φ such that every $q \in Q_\Phi$ contains Φ. *The model structure constructed from* Φ is

$$\mathcal{M}_\Phi = ((Q_\Phi \times V, \sqsubseteq), Y_\Phi)$$

where:

- $\sqsubseteq$ is a binary relation of $Q_\Phi \times V$ which is the pair-wise extension of $\subseteq$ and $\sqsubseteq_V$ to $Q_\Phi \times V$.

- $Y_\Phi : Q_\Phi \times V \to Q_\Phi \times V$ is the next-state function of the model. For any $q \in Q_\Phi$ and $v \in V$, $Y_\Phi(q, v) = (q', v')$ where v' is as in (1) and

$$q' = \{ (a^{\geq 1}, c^{\geq 1}) \mid (a, c) \in q,\ \delta(a^0) \downarrow_V \sqsubseteq_V v \} \cup \{ (A_i, C_i) \mid i = 0, \cdots, n-1 \}$$

Given the above construction, the following properties can easily be shown:

Lemma 2. *Let Φ be a set of prime trajectory formulas and A, C be arbitrary trajectory formulas.*

1. *For every $\varphi \in \Phi$, $\mathcal{M}_\Phi \models \varphi$.*

2. *Let $\tau^i_{\mathcal{M}_\Phi}(A) = (q, d)$ and $A' \Rightarrow C' \in \Phi$, if for some $k :\ 0 \leq k \leq i$, $(A'^{\geq k}, C'^{\geq k}) \in q$, then $\delta(\mathbf{N}^{i-k} A'^{\leq k}) \sqsubseteq \delta(A)$.*

2.2 An Example of Model Construction

To illustrate the above construction, we will apply the method to the assertion $A \Rightarrow C$, where:

$$A : (a \text{ is } 0) \wedge (b \text{ is } 0) \wedge \mathbf{N}(a \text{ is } 0) \wedge \mathbf{N}(b \text{ is } 0)$$
$$C : \mathbf{N}^2(c \text{ is } 1).$$

We assume the visible nodes are a, b, and c, and consequently that the state of the visible components is drawn from $\{0, 1, X, \mathsf{T}\}^3$.

To construct the model structure $\mathcal{M}_{A,C}$ as defined earlier, we first need to compute the set of all suffixes of $A \Rightarrow C$. In our case we get:

$$\{(A, C),\ ((a \text{ is } 0) \wedge (b \text{ is } 0),\ \mathbf{N}^1(c \text{ is } 1)),\ (\text{unc},\ (c \text{ is } 1))\}$$

Therefore, $Q_{A,C} = \{ Q_1, Q_2, Q_3, Q_4 \}$, where

$$Q_1 = \{(A, C),\ ((a \text{ is } 0) \wedge (b \text{ is } 0),\ \mathbf{N}^1(c \text{ is } 1)),\ (\text{unc},\ (c \text{ is } 1))\}$$
$$Q_2 = \{(A, C),\ ((a \text{ is } 0) \wedge (b \text{ is } 0),\ \mathbf{N}^1(c \text{ is } 1))\}$$
$$Q_3 = \{(A, C),\ (\text{unc},\ (c \text{ is } 1))\}$$
$$Q_4 = \{(A, C)\}.$$

Note that $Q_4 \subseteq Q_i$ and $Q_i \subseteq Q_1$ for all $Q_i \in Q_{A,C}$ and that Q_2 is neither a subset of nor a superset of Q_3. Thus, Q_4 is the bottom element and Q_1 is the top element in the partial order $(Q_{A,C}, \subseteq)$. It follows trivially that $(Q_{A,C}, \subseteq)$ is a complete lattice. Consequently,

$$S_{A,C} = Q_{A,C} \times \{0, 1, X, \mathsf{T}\}^3$$

is also a complete lattice, where $(a, b) \sqsubseteq (c, d)$ iff $a \subseteq c$ and $b \sqsubseteq_T d$.

Finally, the function $Y_{A,C}$ is defined as follows: For any $q \in Q_{A,C}$ and $d \in \{0, 1, X, \top\}^3$, we have $Y_{A,C}(q, d) = (q'(q, d), d'(q, d))$, where

$$q'(q, d) = \begin{cases} Q_1 & \text{if } q \in \{Q_1, Q_2\} \text{ and } (0, 0, X) \sqsubseteq_T d \\ Q_2 & \text{if } q \in \{Q_3, Q_4\} \text{ and } (0, 0, X) \sqsubseteq_T d \\ Q_4 & \text{otherwise,} \end{cases}$$

and

$$d'(q, d) = \begin{cases} (X, X, 1) & \text{if } q \in \{Q_1, Q_2\} \text{ and } (0, 0, X) \sqsubseteq_T d \\ (X, X, X) & \text{otherwise.} \end{cases}$$

To illustrate the use of this derived model structure, consider computing the defining trajectory in this model structure for the formula:

$$A' = (a \text{ is } 0) \wedge (b \text{ is } 0) \wedge \mathbf{N}(a \text{ is } 0) \wedge \mathbf{N}(b \text{ is } 0) \wedge \mathbf{N}^2(a \text{ is } 0) \wedge \mathbf{N}^2(b \text{ is } 0).$$

First, by definition, $\delta(A')$ equals:

$$(Q_4, (0, 0, X))\ (Q_4, (0, 0, X))\ (Q_4, (0, 0, X))\ (Q_4, (X, X, X))\ (Q_4, (X, X, X))\ \cdots$$

By the definition of τ,

$$\begin{aligned} \tau^0(A') &= \delta^0(A') = (Q_4, (0, 0, \boxed{X})) \\ \tau^1(A') &= Y(Q_4, (0, 0, X)) \sqcup \delta^1(A') \\ &= (Q_2, (X, X, X)) \sqcup (Q_4, (0, 0, X)) \\ &= (Q_2, (0, 0, \boxed{X})) \\ \tau^2(A') &= Y(Q_2, (0, 0, X)) \sqcup \delta^2(A') \\ &= (Q_1, (X, X, 1)) \sqcup (Q_4, (0, 0, X)) \\ &= (Q_1, (0, 0, \boxed{1})) \\ \tau^3(A') &= Y(Q_1, (0, 0, 1)) \sqcup \delta^3_{A'} \\ &= (Q_1, (X, X, 1)) \sqcup (Q_4, (X, X, X)) \\ &= (Q_1, (X, X, \boxed{1})) \\ \tau^4(A') &= Y(Q_1, (X, X, 1)) \sqcup \delta^4_{A'} \\ &= (Q_4, (X, X, X)) \sqcup (Q_4, (X, X, X)) \\ &= (Q_4, (X, X, \boxed{X})) \\ \tau^5(A') &= Y(Q_4, (X, X, X)) \sqcup \delta^5_{A'} \\ &= (Q_4, (X, X, X)) \sqcup (Q_4, (X, X, X)) \\ &= (Q_4, (X, X, \boxed{X})) \\ &\ \ \vdots \end{aligned}$$

The values surrounded by boxes represent the first 6 values on node "c".

3 A Simple Inference System

There are 7 inference rules in the system. They are presented in the form:

$$\Phi \frac{cond_1, cond_2, \cdots, cond_n}{A \Rightarrow C} \tag{2}$$

where Φ is a set of assertions which are premises of inferences. A, C are trajectory formulas. (2) reads: if the conditions $cond_1, \cdots, cond_n$ are true, then $A \Rightarrow C$ can be derived from Φ.

Rule 1. (Identity)

$$\{\}\,\frac{}{A \Rightarrow A}$$

Rule 2. (TimeShift)

$$\{A \Rightarrow C\}\frac{}{\mathbf{N}^t A \Rightarrow \mathbf{N}^t C} \qquad t \geq 0$$

Rule 3. (AntecedentStrengthen)

$$\{A \Rightarrow C\}\frac{\delta(A) \sqsubseteq \delta(A_1)}{A_1 \Rightarrow C}$$

Rule 4. (ConsequentWeaken)

$$\{A \Rightarrow C\}\frac{\delta(C_1) \sqsubseteq \delta(C)}{A \Rightarrow C_1}$$

Rule 5. (Conjunct)

$$\{A \Rightarrow C_1, \quad A \Rightarrow C_2\}\frac{}{A \Rightarrow C_1 \wedge C_2}$$

Rule 6. (Transitivity)

$$\{A_1 \Rightarrow C_1, \ A_2 \Rightarrow C_2\}\frac{\delta(A_2) \sqsubseteq \delta(C_1)}{A_1 \Rightarrow C_2}$$

Rule 7. (AntecedentTruncate)

$$\{A \Rightarrow C\}\frac{\forall i \leq depth(C).\ \delta(A)^i \sqcap \delta(C)^i = \bot}{A^{\leq t} \Rightarrow C^{\leq t+1}} \qquad t \geq 0$$

For proofs of the soundness of Rules 1–6, see [2].

The soundness of AntecedentTruncate can be shown as follows: Let $t \geq 0$ and $\mathcal{M}$ be a model structure of $A \Rightarrow C$, i.e., $\delta(C) \sqsubseteq \tau_{\mathcal{M}}(A)$ (Theorem 1).

1. Since $\delta(C) \sqsubseteq \tau_{\mathcal{M}}(A)$, $\delta(C^{\leq t}) \sqsubseteq \tau_{\mathcal{M}}(A^{\leq t})$. Therefore, $\mathcal{M} \models A^{\leq t} \Rightarrow C^{\leq t}$.
2. If $t = 0$, then $A^{\leq 0} = C^{\leq 0} = $ unc. By the definition of τ, $\delta(C)^0 \sqsubseteq \delta(A)^0$. Because $\delta_C^0 \sqcap \delta(A)^0 = \bot$, $C^0 = $ unc. Therefore, AntecedentTruncate is reduced to

$$\{A \Rightarrow C\}\frac{\forall i \leq depth(C).\ \delta(A)^i \sqcap \delta(C)^i = \bot}{\text{unc} \Rightarrow \text{unc}} \qquad t \geq 0$$

which is trivially true.

Assume $t > 0$. Because $\tau_{\mathcal{M}}^t(A)$ depends only on $A^{\leq t}$, $\tau_{\mathcal{M}}^t(A) = \tau_{\mathcal{M}}^t(A^{\leq t})$. Therefore,

$$\delta_C^t \sqsubseteq \tau_{\mathcal{M}}^t(A^{\leq t}) = Y(\tau_{\mathcal{M}}^{t-1}(A^{\leq t})) \sqcup \delta_A^t$$

Since $\delta_A^{t+1} \sqcap \delta_C^{t+1} = \bot$, $\delta_C^t \sqsubseteq Y(\tau_{\mathcal{M}}^{t-1}(A^{\leq t}))$. In both cases,

$$\mathcal{M} \models A^{\leq t} \Rightarrow \mathbf{N}^t C^t$$

Combining the results from 1 and 2 by Conjunct:

$$\mathcal{M} \models A^{\leq t} \Rightarrow C^{\leq t+1}$$ □

4 The Completeness of the Inference System

In this section, we prove the main result of this paper: the inference system given in Section 3 is powerful enough to derive all the logical consequences of Φ. A logical consequence of Φ is an assertion which is true in every model structure of Φ. We proved this claim by showing that if an assertion (φ) is true in the model structure $\mathcal{M}_\Phi$, then it can be derived by the inference system from Φ.

Theorem 3. *Let Φ be a set of trajectory formulas, A be any trajectory formula, C be any instantaneous trajectory formula, and $m \geq 0$. If $\mathcal{M}_\Phi \models A \Rightarrow N^m C$ then $\Phi \vdash A \Rightarrow N^m C$.*

A corollary of the theorem is the major conclusion of this paper:

Corollary 4. *Let Φ be a set of trajectory assertions and $\mathcal{M}_\Phi$ be the model structure of Φ. For any assertion $A \Rightarrow C$, $\mathcal{M}_\Phi \models A \Rightarrow C$ if and only if $\Phi \vdash A \Rightarrow C$.*

Proof. The proof that $\Phi \vdash A \Rightarrow C$ implies $\mathcal{M}_\Phi \models A \Rightarrow C$ follows the soundness proofs of the inference rules.

Assume $\mathcal{M}_\Phi \models A \Rightarrow C$. For every $t \geq 0$, $\mathcal{M}_\Phi \models A \Rightarrow N^t C^t$. By Theorem 3, $\Phi \vdash A \Rightarrow N^t C^t$ for every $t \geq 0$. Then by Conjunct (inference rule),

$$\Phi \vdash A \Rightarrow \bigwedge_{t \geq 0} N^t C^t$$

which is equivalent as saying that $\Phi \vdash A \Rightarrow C$. □

We now prove Theorem 3. Although the theorem is equally applicable to general trajectory assertions, the following proof assumes that Φ contains only prime assertions.

Proof. Prove by induction on $m \geq 0$.

The base case is when $m = 0$. $\mathcal{M}_\Phi \models A \Rightarrow N^0 C$ implies that

$$\delta(C) \sqsubseteq \tau^0_{\mathcal{M}_\Phi}(A) = \delta^0_A \qquad (3)$$

Then $\begin{array}{ll} (a)\ \Phi \vdash A \Rightarrow A & \text{By the Identity Axiom} \\ (b)\ \Phi \vdash A \Rightarrow C & \text{By Consequent Weaken and } (a), (3) \end{array}$

Assume $m \geq 0$, and for every $i \leq m$, $\mathcal{M}_\Phi \models A \Rightarrow N^i C$ implies $\Phi \vdash A \Rightarrow N^i C$.

Assume $\mathcal{M}_\Phi \models A \Rightarrow C$ and let $\tau_{\mathcal{M}_\Phi}^{m+1}(A) = (q,d)$ where q is a set of suffixes of the assertions in Φ: without losing generality, assume there exists $l \leq n$, such that

$$q = \{\, (A_i^{\geq k_i}, C_i^{\geq k_i}) \mid 0 \leq i \leq l,\ 1 \leq k_i \leq m \,\}$$

By the definition of $\mathcal{M}_\Phi$,

$$\delta(C) \sqsubseteq d' = \delta(A)^{m+1} \sqcup \bigsqcup_{i \leq l} \delta(C_i^{k_i}) \tag{4}$$

By Lemma 2, $\tau_{\mathcal{M}_\Phi}^{m+1}(A) = (q,d)$ implies that for every $i : 0 \leq i \leq l$,

$$\delta(\mathbf{N}^{m-k_i+1} A_i^{\leq k_i}) \sqsubseteq \delta(A)$$

Therefore, $\mathcal{M}_\Phi \models A \Rightarrow \mathbf{N}^{m-k_i+1} A_i^{\leq k_i}$. This means that for every $j : 0 \leq j \leq k_i - 1$, $\mathcal{M}_\Phi \models A \Rightarrow \mathbf{N}^{m-k_i+1} \mathbf{N}^j A_i^j$. By the induction hypothesis,

$$\Phi \vdash A \Rightarrow \mathbf{N}^{m+j-k_i+1} A^j \qquad 0 \leq j \leq k_i - 1 \tag{5}$$

By Conjunct, (5) implies

$$\Phi \vdash A \Rightarrow \mathbf{N}^{m-k_i+1} A_i^{\leq k_i} \tag{6}$$

What follows derives that $\Phi \vdash A \Rightarrow \mathbf{N}^{m+1} C$.

Step 1. Because $A_i \Rightarrow C_i \in \Phi$ for every $i : 0 \leq i \leq l$, by TimeShift,

$$\Phi \vdash \mathbf{N}^{m-k_i+1} A_i \Rightarrow \mathbf{N}^{m-k_i+1} C_i \qquad 0 \leq i \leq l \tag{7}$$

Step 2. By ConsequentWeaken, (7) implies

$$\Phi \vdash \mathbf{N}^{m-k_i+1} A_i \Rightarrow \mathbf{N}^{m-k_i+1} \mathbf{N}^{k_i} C_i^{k_i} \qquad 0 \leq i \leq l$$

Therefore,

$$\Phi \vdash \mathbf{N}^{m-k_i+1} A_i \Rightarrow \mathbf{N}^{m+1} C_i^{k_i} \qquad 0 \leq i \leq l \tag{8}$$

Step 3. By AntecedentTruncate, (8) implies

$$\Phi \vdash \mathbf{N}^{m-k_i+1} A_i^{\leq k_i} \Rightarrow \mathbf{N}^{m+1} C_i^{k_i} \qquad 0 \leq i \leq l \tag{9}$$

Step 4. By Transitivity, (6) and (9) imply

$$\Phi \vdash A \Rightarrow \mathbf{N}^{m+1} C_i^{k_i} \qquad 0 \leq i \leq l \tag{10}$$

Step 5. By Conjunct, (10) implies

$$\Phi \vdash A \Rightarrow \bigwedge_{0 \leq i \leq l} \mathbf{N}^{m+1} C_i^{k_i} \tag{11}$$

Step 6. By Identity and ConsequentWeaken,

$$\Phi \vdash A \Rightarrow N^{m+1} A^{m+1} \tag{12}$$

By Conjunct, (11) and (12) imply

$$\Phi \vdash A \Rightarrow N^{m+1} A^{m+1} \wedge \bigwedge_{0 \leq i \leq l} N^{m+1} C_i^{k_i} \tag{13}$$

Step 7. Let

$$B = N^{m+1} A^{m+1} \wedge \bigwedge_{0 \leq i \leq l} N^{m+1} C_i^{k_i}$$

By the definition of δ,

$$\delta(B)^{m+1} = \delta(A^{m+1}) \sqcup \bigsqcup_{0 \leq i \leq l} C_i^{k_i}$$

Therefore, by (4) and ConsequentWeaken, (13) implies $\Phi \vdash A \Rightarrow N^{m+1} C$. $\quad\square$

5 Model Construction Revisited

When constructing a circuit model from a give set of assertions, we assumed that every assertion in the given set is prime and non-symbolic. However, the method can be extended to non-prime and symbolic assertions as well.

First, a symbolic trajectory assertion can be viewed as a compact representation of a set of assertions. For example, an inverter specification:

$$(a \text{ is } x) \Rightarrow N(b \text{ is } \neg x)$$

is equivalent to two assertions:

$$(a \text{ is } 0) \Rightarrow N(b \text{ is } 1) \quad \text{and} \quad (a \text{ is } 1) \Rightarrow N(b \text{ is } 0)$$

Therefore, a model of a symbolic assertion can be constructed by means of constructing the model of the corresponding set of assertions. The following theorem relates models of general trajectory assertions to models of prime assertions.

Theorem 5. *Let $A \Rightarrow C$ be an arbitrary trajectory assertion defined on V. There exists a unique trajectory formula C' such that*

1. *$\delta_A^i \sqcap \delta(C')^i = \bot$, and $\delta(C')^i \sqsubseteq \delta(C)^i$ for every $i \geq 0$.*

2. *for any trajectory assertion $A'' \Rightarrow C''$, $\mathcal{M}_{A \Rightarrow C} \models A'' \Rightarrow C''$ if and only if $\mathcal{M}_{A \Rightarrow C'} \models A'' \Rightarrow C''$.*

It can be shown that for any given formulas A and C defined on V, such a C' can be computed. Therefore, given a non-prime assertion $A \Rightarrow C$, its constructed model structure is defined as that of $A \Rightarrow C'$. Note that, in general, such C' may not exist in arbitrary lattices. Interested readers are referred to [4].

Finally, a consequence of Theorem 3 is that given a set of trajectory assertions Φ, $\mathcal{M}_\Phi$ is the most-abstract model structure of Φ.

Theorem 6. *Let $\mathcal{M}$ be an arbitrary model of Φ. If $A \Rightarrow C$ is an assertion of $\mathcal{M}_\Phi$ defined on the domain of Φ, then $A \Rightarrow C$ is also an assertion of $\mathcal{M}$.*

Proof. We prove the theorem by contradiction: assume there exists model structure $\mathcal{M}$ of Φ and an assertion $A \Rightarrow C$ of $\mathcal{M}_\Phi$ such that

$$\mathcal{M}_\Phi \models A \Rightarrow C \qquad \text{but} \qquad \mathcal{M} \not\models A \Rightarrow C$$

By Corollary 4, $\mathcal{M}_\Phi \models A \Rightarrow C$ implies $\Phi \vdash A \Rightarrow C$. By the soundness of the inference system, $A \Rightarrow C$ is an assertion of every model structure of Φ. In particular, it is an assertion of $\mathcal{M}$, i.e., $\mathcal{M} \models A \Rightarrow C$. This contradicts to our assumption that $\mathcal{M} \not\models A \Rightarrow C$. This proves that $\mathcal{M} \models A \Rightarrow C$. □

6 Conclusions

There are two main results in this paper: the construction of a most general model structure from a set of trajectory assertions, and the soundness and completeness proofs for the inference system. Although the completeness result is more of theoretical, than practical value, it is nevertheless useful in that no further "basic" inference rules are needed. Consequently, a very safe—but still very practical—theorem prover for composing trajectory evaluation results can easily be constructed by simply implementing these few inference rules as an abstract data type and exporting only these construction functions, as pioneered by LCF[5] and extensively used by offsprings of LCF.

The method of constructing the most-general circuit model from a given set of trajectory assertions Φ is interesting in its own right. For example, it makes it possible to "simulate" a specification. This is often very useful in avoiding simple errors in the specifications. Another interesting possibility is to use the construction to create a most general model structure for some part of the system that has not yet been designed. This model structure could then be composed with parts that has been designed to allow verification of the complete system. However, for either of these applications to be practical, the construction must be implemented to work over a symbolic domain efficiently. We have in fact done so and are currently in the process of building a small prototype system to determine the practicality of the approach.

Acknowledgments We are indebted to the reviewers who have made many helpful and constructive remarks on an earlier draft of this paper.

References

1. SEGER, C.-J., AND BRYANT, R. Formal verification of digital circuits by symbolic evaluation of partially-ordered trajectories. Tech. Rep. Technical Report 93-8, The Computer Science Department, The University of British Columbia, The Computer Science Department, The University of B.C. Vancouver B.C. V6T 1Z4, 1993.
2. HAZELHURST, S., AND SEGER, C.-J. A simple theorem prover based on symbolic trajectory evaluation and obdds. Tech. Rep. Technical Report 93-41, The Computer Science Department, The University of British Columbia, The Computer Science Department, The University of B.C. Vancouver B.C. V6T 1Z4, 1993. (An abridged version of this work appears in this proceedings).
3. SEGER, C.-J. Voss – a formal hardware verification system, user's guide. Tech. Rep. Technical Report 93-45, The Computer Science Department, The University of British Columbia, The Computer Science Department, The University of B.C. Vancouver B.C. V6T 1Z4, 1993.
4. ZHU, Z. Construction of circuit models from trajectory specifications. In progress, 1994.
5. GORDON, M., MILNER, A., AND C., W. *Edinburgh LCF*, vol. 78 of *Lecture Notes in Computer Science*. Springer-Verlag, 1979.

Efficient Model Checking by Automated Ordering of Transition Relation Partitions

Daniel Geist and Ilan Beer

IBM Science and Technology, Haifa, Israel

Abstract. In symbolic model checking, the behavior of a model to be verified is captured by the transition relation of the state space implied by the model. Unfortunately, the size of the transition relation grows rapidly with the number of states even for small models, rendering them impossible to verify. A recent work [5] described a method for partitioning the transition relation, thus reducing the overall space requirement. Using this method, actions that require the transition relation can be executed by using one partition at a time. This process, however, strongly depends on the order in which the partitions are processed during the action.

This paper describes a criterion for ordering partitions which is independent of the circuit details. Based on this criterion, a heuristic algorithm for ordering partitions is described. The algorithm, which may be used in preparation for each symbolic simulation step, has been successfully implemented and has resulted in significant speed-ups of symbolic model checking. Specifically, this algorithm has made it possible to verify blocks inside an example microprocessor. The run time results are given here.

1 Introduction

The feasibility of formal verification methods is usually limited by the "state explosion problem", where the representation size of the model to be verified grows exponentially with the number of model states. Therefore, although formal verification methods have existed for many years, they have become practical only when research on reducing the representation size became successful.

The work on Binary Decision Diagrams (BDDs), and in particular Ordered Binary Decision Diagrams (OBDDs) done by Bryant [3, 2, 4], has had a big impact on formal methods [11, 7]. As a result, formal methods have become practical for moderate size models, successful systems such as SMV [11] were developed, and formal verification of real circuit designs has become feasible [10, 6].

However, even with the use of BDDs, the size of a verifiable model is still limited, and research has continued in order to find other methods of reducing the size of model representation. In systems that search the state transition graph, the model is represented as a BDD of its transition relation. The size of the BDD of the transition relation is usually the obstacle to verifying bigger circuit designs.

Recent research has shown that it is possible to significantly reduce the amount of space required by the transition relation, if it is not kept as a whole [5, 7, 12]. The space required by the transition relation can be much smaller if it is partitioned into many small relations whose combination produces the transition relation. This partition usually increases computation time but allows the verification of bigger circuit designs.

Specifically, in a recent paper by Burch, Clarke and Long [5], it was shown that the transition relation of a model can be partitioned into smaller relations, each handling the next value of a single state variable. They showed that it is possible to do the same operations as with the transition relation without the same space requirements. Operations with the partitioned transition relation can be executed iteratively, one partition at a time. This partitioning has been successfully used in the verification of models not previously possible [5].

The above process, however, strongly depends on the order in which the partitions are introduced to the operation. In [5] it was stated that a good order can be derived by examining the model to be verified and its semantics. This process, however, requires intimate knowledge of the circuit. Furthermore, this process requires manual intervention in order to derive and input the order. Thus, in order to fully automate the use of a partitioned transition relation, it is necessary to find a method of ordering the partitions that is independent of knowledge of the semantics of the circuit.

This paper describes a criterion for ordering partitions which is independent of the model details. Based on this criterion, a heuristic algorithm for ordering partitions is described, which can be run prior to symbolic model checking. This algorithm has been successfully implemented and has resulted in significant speed ups of symbolic model checking. Specifically, this algorithm has made it possible to verify blocks inside an example microprocessor. The run time results are given here.

The rest of the paper is partitioned as follows: Section 2 gives a brief background of basic computation operations performed in symbolic model checking. Section 3 reviews the work on partitioning done by Burch et al. [5] on partitioning the transition relation. Section 4 describes the criterion for ordering and Section 5 presents the algorithm which was devised, based on this criterion. Section 6 presents experimental results of the use of this work.

2 State Set Computation in Symbolic Model Checking

Throughout the paper it is assumed that the model to be verified is a design of a synchronous circuit, and the work in the paper is only relevant to this kind of a model.

The basic operations performed in the process of symbolic model checking are computations of the next or previous set of states of a given set of states. Symbolic Model Checking [11] does not operate on individual states (or paths). Rather, it performs its processing on sets of states that satisfy a certain predicate (condition). Finding the next set of states of a given set is called a *forward simulation step*. Similarly, finding the previous set of states of a given set is called a *backward simulation step*. Thus, a forward (or backward) simulation step is done from one set of states to another set of states. Additionally, the simulation is exhaustive, i.e., **all** possible states of the next step are generated.

The model checker that this work relates to is SMV [9, 11]. SMV uses OBDDs to represent sets of states. Assume that the model to be verified has n binary state variables. Let $M \subseteq \{0, 1\}^n$ be its state space, and let $V = \{v_1, \ldots, v_n\}$ be its set of state variables. A state $q \in M$ of the model is an assignment of binary values to v_i. Given a set $S \subseteq \{0,1\}^n$, we associate with S the boolean function $S(V)$, where $S(q) = 1$ iff $q \in S$.

In order to perform a forward (or backward) simulation step, a transition relation of the model to be verified must first be constructed. Let $V' = \{v'_1, \ldots, v'_n\}$ also denote the model set of state variables but designating some possible "next state" of V. The transition relation $N(V, V')$ of a model is a boolean function of $2n$ variables, such that $N(q, q') = 1$ iff $q' \in M$ is a possible "next state" of $q \in M$.

The transition relation captures the model's behavior inside a boolean function. We can then efficiently store it as a OBDD. For example, Figure 1 depicts the transition relation of a 3 bit counter. b_0, b_1 and b_2 are the current state variables where b_0 is the least significant bit. Respectively, b'_0, b'_1 and b'_2 are the next state variables.

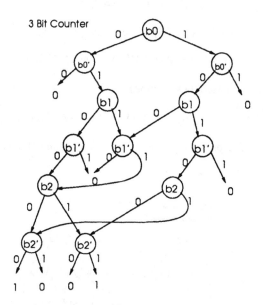

Fig. 1. The OBDD of a transition relation for a 3 bit counter

Let $S(V)$ be the characteristic function (or its OBDD) representing a set of states S. A forward simulation step to compute the next set $S'(V')$ is done as follows [11]:

$$S'(V') = \exists V\, [S(V) \wedge N(V, V')] \tag{1}$$

where $\exists V$ denotes $\exists v_1 \exists v_2 \ldots \exists v_n$. Similarly a backward simulation step is computed

$$S(V) = \exists V'\, [S'(V') \wedge N(V, V')]. \tag{2}$$

The methods to compute the operations in 1 and 2 are well known when $S(V)$ and $N(V, V')$ are represented as OBDDs (see [2]). Notice that existential quantification eliminates the dependency of the result OBDD upon the quantified variables.

3   Partitioning the Transition Relation

The main obstacle for checking bigger designs in SMV is the size of the transition relation. Burch, Clarke and Long [5] showed that it is possible to preserve the transition relation in parts whose sum of sizes is orders of magnitude smaller than the size of a full transition relation. This was based on the observation that the transition relation, is in fact, a conjunction of the set of transition relations for each state variable.

Specifically, the next value of each state variable is a boolean function of the current state.

$$v_i' = f_i(V) \quad i = 1 \ldots n.$$

A transition relation partition, $N_i(V, V')$, is defined by the following equation:

$$N_i(V, V') = v_i' \Leftrightarrow f_i(V).$$

The full transition relation is a conjunction of all partitions:

$$N(V, V') = N_1(V, V') \wedge \cdots \wedge N_n(V, V')$$

and a forward simulation step thus becomes

$$S'(V') = \exists V \left[S(V) \wedge N_1(V, V') \wedge \cdots \wedge N_n(V, V') \right]. \tag{3}$$

For example, Figure 2 depicts the partitions of the 3 bit counter transition relation from Figure 1.

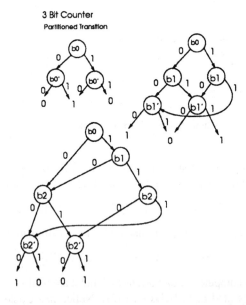

Fig. 2. OBDD representation of the partitioned transition relation for a 3 bit counter

Using 3 we can compute a simulation step. However, care should be taken how the operation is performed. We can, for example, begin by conjuncting all the partitions, leaving $S(V)$ as last, but then we will eventually create the full transition relation which we are trying to avoid. In order to truly exploit partitioning, the simulation step must be performed iteratively without creating a full transition relation in the process.

The technique suggested by Burch et al. [5] is to iteratively conjunct in the partitions, and to quantify out variables as soon as further steps do not depend on them: More explicitly, the user must choose an ordering ρ of $\{1, \ldots, n\}$ which determines the order in which the $N_i(V, V')$ are conjuncted.

Let D_i be the set of variables $N_i(V, V')$ depend on. Also let

$$E_i = D_{\rho(i)} - \bigcup_{k=i+1}^{n} D_{\rho(k)}.$$

A simulation step is done iteratively as follows:

$$S_1(V, V') = \exists E_1 [S(V) \land N_{\rho(1)}(V, V')]$$
$$S_2(V, V') = \exists E_2 [S_1(V, V') \land N_{\rho(2)}(V, V')]$$
$$\vdots \tag{4}$$
$$S'(V') = \exists E_n [S_{n-1}(V, V') \land N_{\rho(n)}(V, V')]$$

where the expression $\exists E_1$ means that we quantify out all variables in E_1. Therefore, S_1 depends only on $V - E_1$, S_2 depends only on $V - E_1 - E_2$, etc. For example, consider the consider the counter in Figure 2. Assume the partitions are ordered $(3, 2, 1)$ (from b'_2 to b'_0). In this case $D_1 = \{b_0, b_1, b_2\}$, $D_2 = \{b_0, b_1\}$, $D_3 = \{b_0\}$, and $E_1 = \{b_2\}$, $E_2 = \{b_1\}$, $E_3 = \{b_0\}$. Thus in the first iteration we quantify out b_2, then b_1, and finally b_0. This happens to be a good order. The method in which to evaluate it is given in the next section.

4 The Conjunction Ordering Criterion

The success of using conjunctive partitioning depends on finding a good order (ρ). Without it, the process of 4 will usually be much slower and consume more space than using the full transition relation as in 1. The difficulties in choosing an order manually were already pointed out in Section 1.

As mentioned before, in symbolic model checking sets of states are stored as OBDDs, and processing is conducted via OBDD operations. The space and time used in the computation depend primarily on the size of the OBDDs generated. Our goal is to minimize the maximal OBDD generated in the process of performing 4. However, predicting the size of the maximal OBDD is difficult, especially since there are going to be many $S(V)$ sets generated during the verification and it will mean considering each one of them. Instead, the following criterion is much simpler to compute.

Criterion: Minimize the maximal number of state variables that participate in any OBDD generated in the process of performing the computation of 4.

The criterion has the following advantages:

1. This criterion does not depend on understanding the semantics of the verified model and the role each variable has in the original circuit. It can be derived from examining the OBDDs of the transition partitions and deriving the set of variables (D_i), each partition N_i depends on.
2. This criterion is independent of any other optimization criteria used in symbolic model checking. Specifically, it is independent of variable ordering which is done to reduce OBDD sizes.
3. This criterion is independent of the set of states $S(V)$, that we wish to perform a simulation step on. The reasons are explained in the next section. Thus we only have to perform ordering once for all forward simulation steps that will be computed during verification. The same applies to backward simulation.

5 The Heuristic Algorithm

An optimal ordering for the criterion discussed in the previous section may be difficult to compute. The ordering problem may very well be intractable. However, it is possible to generate a heuristic ordering that will give good results in practice.

We assume that the set $S(V)$ depends on all n state variables. Our experience has shown that when $S(V)$ depends on only a few state variables then the simulation step will be fast even without ordering, and that space problems only arise when sets on which a simulation step is performed, depend on the full set of variables.

The heuristic algorithm is based on the following observations:

1. Each simulation step begins with an OBDD ($S(V)$ or $S'(V')$) that depends on n state variables and generates an OBDD that depends on n state variables (V' or V). In the process, an OBDD that depends at most on $2n$ variables may be generated (if the ordering is poor).
2. Each partition depends on exactly one variable of V'. In particular N_i depends on v_i'.
3. In a backward simulation step, a varying number of state variables is added in each iteration and exactly one variable of V' is quantified out, since each v_i' appears in exactly one partition.
4. In a forward simulation step a varying number of variables is quantified out in each iteration, and exactly one more variable (the v_i' variable) is added.

The above observations, assist in finding an order for the N_i partitions such that the maximal number of variables over all steps is minimized. From Observations 3 and 4 it follows that the way the number of state variables change in a forward simulation step is different than the way it changes in a backward simulation step. Let us first concentrate on a forward simulation step.

A forward simulation step begins with an OBDD that depends on all the current state variables and iteratively the next state variables (v_i') are added in, while current state variables (v_j) are removed. Thus the order should be such that many v_j as possible will be eliminated in the earlier iterations, before many v_i' are added in.

For example, consider the partition of the counter in Figure 2. The partition of the first bit depends only on b_0 and b_0'. While the partition of the last bit depends on b_2' and the whole set of b_0, b_1 and b_2. If the partitions are ordered from b_0' to b_2' it is not possible to quantify out any state variables until the last iteration and an OBDD that depends on $2n$ variables is generated. However, the partitions are ordered b_2' to b_0', it is possible to eliminate one state variable at each iteration and the maximal number of state variables will be $n + 1$. This example shows how crucial the ordering can be since the OBDD size is in the worst case an exponent of the number of variables. The latter ordering is in fact optimal according to the criterion.

The algorithm we implemented was a greedy algorithm that searched at each step for the partition which would result in the maximum number of variables quantified out. This partition chosen had the most number of *unique* variables (not appearing in any other partition that was not previously chosen). The minimization is therefore locally optimal at each iteration but is not necessarily globally optimal. If more than one partition is a possible candidate then a second criterion is applied: From the candidate partitions the algorithm looks for the one that shares most variables with other partitions that have not been conjuncted yet. The rational for this criterion is that it will set the stage for quickly quantifying out many variables in the next steps since all of the partitions that depend on them have been conjuncted.

Algorithm - Obtaining the order ρ for a forward simulation step

1. **Initialize remaining partition set to the whole partition set.**
2. **Loop**
   (a) **From the remaining partitions find the partition(s) that has the maximal number of unique variables. If there is only one, chose it and go to step (d).**
   (b) **From the selected partitions in step (a), find the partition(s) that has the maximal number of variables occurring in other remaining partitions. If there is only one, chose it and go to step (d).**
   (c) **If there is more than one partition satisfying (b), then choose from those one arbitrarily.**
   (d) **Make the chosen partition the next in the ρ order. Remove it from the remaining partition set.**
   Until the remaining partition set is empty

A backward step begins with an OBDD that depends on all the next state variables, and iteratively the current state variables (v_i) are added in, while the next state variables (v_j' are removed. Thus, the order should be such that the number of current state variables remains small until many of the v_i' are eliminated.

For example, consider the partition of the counter in Figure 2. The partition of the first bit depends only on b_0 and b_0'. While the partition of the last bit depends on b_2' and the whole set of b_0, b_1 and b_2. Ordering the partitions from b_2' to b_0' will generate, in the first iteration, an OBDD that depends on $2n$ variables. On the other hand, ordering from b_0' to b_2' will generate OBDDs that depend on at most $n + 1$ variables.

The ordering algorithm for a backward simulation step is similar to the forward one. It too, minimizes locally. A partition is preferred if it does not increase the number of variables participating in the conjunction, or if it contains the minimum number of variables not introduced before. It also uses a secondary criterion which is the same as for the forward simulation step ordering but for a different reason: From the candidate partitions, the algorithm looks for the one that shares most variables with other partitions that have not been conjuncted yet. The rational here, is that it will set the stage for choosing a partition in the next step, that will introduce only a small number of new variables.

Algorithm - Obtaining the order ρ for a backward simulation step

1. Initialize remaining partition set to the whole partition set.
2. Loop
   (a) From remaining partitions find the partition(s) that has the smallest number of variables not introduced before. If there is only one, chose it and go to step (d).
   (b) From the selected partitions in step (a), find the partition(s) that has the maximal number of variables occurring in other remaining partitions. If there is only one, chose it and go to step (d).
   (c) If there is more than one partition satisfying (b), then choose from those one arbitrarily.
   (d) Make the chosen partition the next in the ρ order. Remove it from the remaining partition set.
   Until the remaining partition set is empty

Notice that the ordering does not depend on $S(V)$ if we assume that $S(V)$ (or $S'(V')$) depends on the full set of n state variables.

A final remark is that Burch et al. [5] state that it is usually not beneficial to completely partition the transition relation. Rather, it is more useful to work with conjunctions of a small number of partitions. In this case, the algorithms need to be slightly modified to accommodate the fact that the number of v_i' in each partition is not exactly one. The details are left to the reader.

6 Results

The algorithms described in the previous section were implemented as an addition to the SMV system running on an IBM RISC System 6000. They are completely automatic and no extra human intervention is necessary for their operation. They were tested in the verification of various example circuits and have proven to be very beneficial. The extra cost of performing ordering is insignificant both in time and space as compared to other operations that are performed while running SMV.

Our experience has shown that partitioning becomes significantly beneficial only when the verified model becomes relatively big (above 20 state variables). For small models partitioning may unnecessarily result in slower execution times.

In general, the results show that performing a simulation step with partitions is slower. However, computing the transition relation partitioned is much faster and of course requires much less space. Therefore, excluding the case where a full transition relation cannot fit in memory, the payoff of using partitions should be measured by the time it requires to construct the transition relation versus the time to perform a simulation step, and also by the number of times a simulation step is performed. As the number of simulation steps increases, the advantage of using partitions becomes smaller.

The transition relation can be downsized considerably if the reachable state space is computed and the transition relation is computed assuming only reachable states (this is an SMV option). The process of computing the reachable space is done incrementally with computation of a partial transition relation at each increment. This transition relation is computed to be confined to the "thus far" discovered reachable set. It is used exactly once in a forward simulation step to compute a next set of states for the "thus far" discovered reachable set, and then it is discarded. Using partitions is by far a superior method for performing this incremental process, since only one simulation step is performed for each transition relation construction. The advantage in the transition relation construction is many times greater than the loss in one simulation step.

Therefore, the most effective way to perform model checking, as the results indicate, is to find the reachable state space using a partitioned transition relation, construct a full transition relation, and then do the model checking and counter examples non-partitioned. An exception is the cases where the full transition relation is too big and there is no alternative, but to use partitions.

In summary, partitioning is used simply because it is faster, and for some cases also because without partitioning the execution would reach core memory sizes (about 130MB) where it eventually exits without completing verification. We used partitioning in backward steps only when we could not fit a full transition relation in memory. This happened only in our bigger examples (around 90 state variables). There were also some examples where even with ordering the space required became too large.

Table 1 contains results of some examples we ran. Six runs were executed for each example: Two runs without using partitioning, two runs using partitioning with some arbitrary order and two runs where the partitions were ordered. The runs were done twice for the following reason: Once without proving any SPECs (temporal propositions [9]) - measuring only the time to create the transition relation and another run with some SPECs evaluated. The difference in the run times is the time it took to evaluate SPECs. The transition relation was built incrementally in each run.

The second column in the table is the number of model state variables. The next there numbers are the CPU times for the runs without SPEC evaluation. The next three numbers are the runs with SPEC evaluation.

The next two numbers are the maximum number of variables that participate in any conjunction of a forward step. One number for partitioning without ordering and another for partitioning with ordering. Since we used the incremental option of building the transition relation, we performed partition construction and ordering many times. Thus the two numbers are actually an average of all transition relation constructions. The numbers are an increment to the minimal number of n state variables. For example if there were 86 variables in the model and the average maximum number is 4 then it means that the BDDs that participated in the conjunction depended on at most 90 state variables (and not $2 \times 86 = 172$). Presenting the increment gives a clearer picture of how the algorithm performs since an increment of 1 is always the smallest possible value we can hope to reach.

The last two columns describe the maximum number of variables that participate in a backward step. The transition relation for backward steps was constructed only once. As with forward chaining the number given is the increment over the number of state variables of the model. Here, 0 is the smallest possible we can hope to reach.

Table 1. Results

Unit Name	Num. of States	Transition CPU (sec)			Transition+SPEC CPU (sec)			Max Frwrd (avg.)		Max Bckwrd	
		Full	Unord.	Ord.	Full	Unord.	Ord.	Unord.	Ord.	Unord.	Ord.
Bus Interface	86	∞	∞	1684	∞	∞	1902	31	4	- N/A -	
Request Que.	41	85	72	48	108	465	169	22	10	28	18
Bus Slave	35	44	20	18	45	27	20	25	6	15	2
Bus Master	53	1062	503	345	1091	1148	436	40	5	28	0

The first example in table 1 is a bus interface unit of a microprocessor. This unit accepts one request for a read or write from the cpu and executes it by initiating and controlling bus cycles. In addition the unit has many variants. It can pipeline two requests. It can also execute a burst cache line read or write, etc. The data and address information are reduced [1] before the process of verification takes place. This example has 86 state variables and it was not possible to verify it without partitioning or without ordering. The maximum number of variables for backward steps is not given since we could not produce it for the unordered mode.

The second example is a unit that interfaces the BIU with the cache. It queues cache requests from the cache and sends them one at a time to the BIU. It can send the requests out of order according to certain control priorities. Again the data and addresses are reduced. This is a typical example of a circuit that can be verified both ways. Notice that although we can generate the transition relation twice as fast when using partitions, the SPECs can be checked much faster if the full transition relation is used.

The third example is a bus slave. The unit verified transfers data to and from an internal FIFO. This is a small example. It runs quickly in all modes but ordered partitions are the fastest. Again we can see that checking SPECs with partitioning is slower. Notice that the maximum number in backward steps was 0. This may happen if one of the state variables remains constant throughout the reachable state space. The partition for this variable will be ordered first in backward steps since it reduces the number of variables that the conjunction depends on without adding new variables. Such a partition is ordered last in forward steps (in fact, an optimization to the algorithm which we do is to treat such constant partitions separately).

The fourth example is a bus master. It is a medium sized example. The transition relation is built significantly faster if the partitions are ordered. Notice how dramatic the difference is in computation of SPECs. If we use partitions without ordering we lose all the advantage we gained by building the transition relation partitioned. Notice that, again the fastest way to compute SPECs is with the full transition relation.

We can see from the results that running in ordered partitions mode is always faster for creating transition relations. In small examples the difference is insignificant, but in the big examples it is not possible to verify the model in the other modes. Verifying SPECs is faster with a full transition relation (if possible). Ordering for backward steps gave a 2.5 times speedup in one case and a 6.5 times speedup in another. This depends on the number and kind of SPECs we are verifying.

We have also noticed that checking a model that has bugs tends to consume much more memory and CPU time. Once the bugs are fixed the CPU time required for verification decreases dramatically. Our intuitive explanation for this is that incorrect models diverge into areas in the state space that the designer never intended. Thus, the state space tends to grow much larger when bugs are present in the model.

7 Conclusion

Partitioning the transition relation is a method which allows us to increase the size of verified circuits when doing model checking. The success of this method depends on ordering the partitions.

This paper has shown that ordering can be automated by minimization of the number of state variables as an ordering criterion. An algorithm was devised and implemented to order partitions according to this criterion and a few circuits were tested to demonstrate its advantage.

The results have shown that ordered partitions have a clear advantage over working without ordered partitions. In some cases, the improvement resulted in twice as fast execution times, and in others it was not possible to verify without ordering. In all cases that we have encountered, it was always faster to use ordered partitions mode.

Partitioned backward simulation steps were only useful in the cases where a full transition relation could not fit in memory. For those cases ordering has resulted in significant speedups (see Table 1).

Finally, partitioning does not solve the state explosion problem, but rather postpones it. Formal verification is still far from being able to swallow, in one gulp, the designs that are out in the real world. Once the storage space of the transition relation is reduced then other OBDD bottlenecks appear. A recent paper by Hu and Dill has dealt with reducing the OBDD size of invariants [8].

8 Acknowledgements

We would like to thank, Michael Yoeli, Raanan Gewirtzman and Yaron Wolfsthal for the time they spent in reviewing the paper and the invaluable suggestions they had in improving it.

References

1. Ilan Beer, Michael Yoeli, Shoham Ben-David, and Daniel Geist. Methodology and System for Practical Formal Verification of Reactive Hardware. Accepted to CAV 94, 1994.

2. Karl. S. Brace, Richard L. Rudell, and Randal E. Bryant. Efficient Implentation of a BDD Package. In *27th ACM/IEEE Design Automation Conference*, pages 40–45. ACM/IEEE, 1990.

3. Randal E. Bryant. Graph based algorithms for boolean function manipulation. *IEEE Transactions on Computers*, C-35, 1986.

4. Randal E. Bryant. Symbolic Boolean Manipulation with Ordered Binary-Decision Diagrams. *ACM Computing Surveys*, 24:298–318, September 1992.

5. Jerry R. Burch, Edmund M. Clarke, and David E. Long. Symbolic Model Checking with Partitioned Transition Relations. In *International Conference on Very Large Scale Integration*, Edinburg, Scotland, August 1991. IFIP.

6. Edmund M. Clarke, Orna Grumberg, Hiromi Hiraishi, Somesh Jha, David L. Long, Kenneth L. McMillan, and Linda A. Ness. Verification of the Futurebus+ Cache Coherence Protocol. In *Proceedings of the 11th International Conference on Computer Hardware Description Languages*, pages 15–30, 1993.

7. Olivier Coudert, Jean C. Madre, and Christian Berthet. Verifying Temporal Properties of Sequential Machines Without Building their State Diagrams. In R. Kurshan and E. M. Clarke, editors, *Workshop on Computer Aided Verification, DIMACS*, pages 75–84. American Mathematical Society, Providence, RI, 1990.

8. Alan J. Hu and David L. Dill. Efficient Verification with BDDs using Implicitly Conjoined Invariants. In *Proceedings of the Conference on Computer Aided Verification (CAV 93)*, 1993.

9. K. L. McMillan. *The SMV System DRAFT*. Carnegie Mellon University, Pittsburgh, PA, 1992.

10. K. L. McMillan and J. Schwalbe. Formal verification of the Encore Gigamax cache consistency protocol. In *Proceedings of the 1991 International Symposium on Shared Memory Multiprocessors*, April 1991.

11. Kenneth L. McMillan. *Symbolc Model Checking*. PhD thesis, Carnegie Mellon University, May 1992.

12. H. J. Touati, H. Savoj, B. Lin, R. K. Brayton, and A. Sangiovanni-Vincentelli. Implicit State Enumeration of Finite State Machines usin BDD's. In *IEEE International Conference on CAD*, pages 130–133, 1990.

The Verification Problem for Safe Replaceability

Vigyan Singhal [*]
Computer Science Division
University of California
Berkeley, CA 94720
vigyan@ic.eecs.berkeley.edu

Carl Pixley
Motorola Inc., MD OE321
6501 Wm Cannon Drive West
Austin, TX 78735
pixley@math.sps.mot.com

Abstract. This paper addresses the problem of verifying that a sequential digital design is a safe replacement for an existing design without making any assumptions about a known initial state of the design or about its environment. We formulate a safe replacement condition which guarantees that if an original design is replaced by a new design, the interacting environment cannot detect the change by observing the input-output behavior of the new design. Examples are given to show that safe replacement ($\preceq$) allows simplification of the state transition diagram of an original design. It is showed that if D_1 is a safe replacement for design D_0 then every closed strongly connected component of D_1 is contained in D_0. We present a decision procedure for determining whether a replacement design satisfies our safe replacement condition.

1 Introduction

This paper addresses the problem of implementation verification for synchronous sequential designs. Although we will address the problem for gate-level designs, our theory is equally applicable for designs at the state transition level. We want to guarantee that if we replace a sequential design then no surrounding environment will be able to detect the change based on the input-output behavior of the replacement (a *safe* replacement). The problem comes up because frequently designers work separately on separate pieces of a large design, and the objective is to modify one's design so that any of the interacting designs will not notice the modification (Figure 1). The designers do not want to make any assumptions about the surrounding designs outside their own, not even about any initializing sequence coming from outside when their design is powered up. Latches (or memory elements) in the designs may not even have reset lines. In Figure 1, the designer working on design $D0$ would like to replace it by design $D1$ without making any assumptions about the other interacting designs. In this paper, we will refer to the design outside of $D0$ as the environment of $D0$.

The problem of implementation verification for sequential designs is not a new one. Efficient methods exist for the verification of sequential designs [3, 8, 1].

---
[*] Research supported by NSF/DARPA Grant MIP-8719546 and a summer internship from Motorola, Inc.

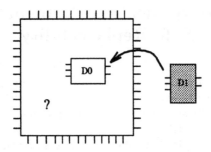

Fig. 1. Replacement of a sequential design

However, these methods only work for designs where all latches have a reset line which determines the designated initial state for the circuit. One key point where our work differs from previous work on sequential verification is that we do not assume any reset lines running to the latches. While it is well-known that data pipeline and memory designs frequently have latches with no reset lines, it is also true that many industry-level control designs have some latches or flip-flops without reset lines. An important reason for having latches without reset lines is the large saving in area by avoiding the routing of the reset lines all over the design. Also, latches without reset lines cost less (consume fewer transistors) than latches with reset lines. Since the latches can power-up in any state, we cannot assume a designated start state for the design. For such designs, the question that needs to be answered is "when can we replace a design with another, so that while the replacement design can power up in any state, there is no way the environment can detect the replacement"?

A notion of sequential hardware equivalence for designs which may not have a designated initial state is presented in [7]. Efficient BDD-based techniques are presented which verify this equivalence for two given designs. We will show that for sub-designs embedded in a large design (or the environment), this notion of equivalence is not always applicable.

In this paper, we will present our condition for safe design replacement. Although this condition is strong enough so that the interacting environment cannot detect the replacement, it does not require that every state in one design be equivalent to one in the other design (the classical notion of machine equivalence, as presented in [5]). Our condition also preserves possible interactions with the environment during initialization. We also explore the methods which can be used to verify the necessary and sufficient conditions for a new (replacement) design to be a safe replacement for an old (existing) design.

An orthogonal problem to the verification problem is the problem of using our replaceability notion to do sequential resynthesis on the existing design. We are working on this problem and, based on preliminary work, the notion of replaceability does seem to provide sufficient flexibility to achieve some optimizations.

2 Terminology and Background

Here we define some notation and a little background that we will need later in this paper.

Definition 1. A *deterministic Finite State Machine (DFSM)* M is a quintuple, $(Q, I, O, \lambda, \delta)$, where Q is the set of states, I is the set of input values, O is the set of output values, λ is the output function, and δ is the next state function. The output function λ is a completely-specified function from domain $(Q \times I)$ to range O. The next state function is a completely-specified function from domain $(Q \times I)$ to range Q. A hardware *design* D is a DFSM with n input wires, m output wires and t latches, Q has 2^t states, I has 2^n values and O has 2^m values.

We will also use λ and δ to denote the output and next state functions on sequences of inputs. So, if $\pi = a_1 \cdot a_2 \cdot a_3 \cdots a_p \in I^p$ is a sequence of p inputs, these functions are recursively defined as $\lambda(s, \pi) = \lambda(s, a_1) \cdot \lambda(\delta(s, a_1), \pi')$ and $\delta(s, \pi) = \delta(\delta(s, a_1), \pi')$, where $\pi' = a_2 \cdot a_3 \cdots a_p$. So, the range-domain relationships are $\lambda : Q \times I^p \rightarrow O^p$ and $\delta : Q \times I^p \rightarrow Q$.

Two designs are said to be *compatible* with each other if they have the same number of input and output wires. All notions of equivalence and replaceability developed in this paper are meaningful only for pairs of compatible designs. Henceforth, when talking about two different designs we will implicitly assume compatibility of the two. In this paper we will assume that designs D_0 and D_1 denote the quintuples $(Q_{D_0}, I, O, \lambda_{D_0}, \delta_{D_0})$ and $(Q_{D_1}, I, O, \lambda_{D_1}, \delta_{D_1})$, respectively.

Definition 2. A set of states, $S \subseteq Q$, is *closed (under all inputs)* if for each state $s \in S$, for each input $a \in I$, $\delta(s, a) \in S$.

Definition 3. A set of states, $S \subseteq Q$, is called a *closed strongly connected component (closed SCC)* if S is closed and for any two states $s_1, s_2 \in S$, there exists a finite input sequence $\pi \in I^*$ such that $\delta(s_1, \pi) = s_2$. It is easy to show that for any state s of design D, a closed SCC is reachable from s.

Definition 4. Given two states $s_0 \in Q_{D_0}$ and $s_1 \in Q_{D_1}$, state s_0 is *equivalent* to state s_1 ($s_0 \sim s_1$) if for any sequence of inputs $\pi \in I^*$, $\lambda_{D_0}(s_0, \pi) = \lambda_{D_1}(s_1, \pi)$. It can be easily shown that if $s_0 \sim s_1$, then for any input sequence $\pi \in I^*$, $\delta_{D_0}(s_0, \pi) \sim \delta_{D_1}(s_1, \pi)$. We say that π *distinguishes* s_0 and s_1 if $\lambda_{D_0}(s_0, \pi) \neq \lambda_{D_1}(s_1, \pi)$.

We now give a classical notion of equivalence between two DFSM's [5, page 23].

Definition 5. Two DFSM's M_1 and M_2 are *equivalent* to each other ($M_1 \equiv M_2$) if for each state s in M_1 there is a state t in M_2 such that $s \sim t$, and for each state t in M_2 there is a state s in M_1 such that $s \sim t$.

3   Sequential Hardware Equivalence (SHE)

In this section we will briefly review the work presented in [7] about the theory of
sequential hardware equivalence for equivalence between two gate-level hardware
designs without assuming any knowledge of initial state. When the design powers
up, the state it powers up in cannot be predicted. The desired input/output
behavior is achieved from the design by driving a fixed synchronizing sequence
of input vectors through the design after the power-up.

Definition 6. A sequence of inputs $\pi \in I^*$ is called an *essential reset sequence*
(or a *synchronizing sequence*) if for any pair of states $s_0, s_1 \in Q_{D_0}$, $\delta_{D_0}(s_0, \pi) \sim
\delta_{D_0}(s_1, \pi)$. A state $s \in Q_{D_0}$ is called an *essential reset state* if there exists a
state $s_0 \in Q_{D_0}$ and a synchronizing sequence π such that $\delta(s_0, \pi) \sim s$. A design
which has an essential reset state is called *essentially resettable*.

Definition 7. A state pair $(s_0, s_1) \in Q_{D_0} \times Q_{D_1}$ is *alignable* if there is a sequence
of inputs $\pi \in I^*$ such that $\delta_{D_0}(\pi, s_0) \sim \delta_{D_1}(\pi, s_1)$. The sequence π is called an
aligning sequence.

Definition 8. Designs D_0 and D_1 are *equivalent* ($D_0 \approx D_1$) if all state pairs are
alignable.

Definition 8 defines the notion of sequential hardware equivalence. The following results were shown in [7].

Theorem 9. $D_0 \approx D_1$ *if and only if there is a* <u>*single (but not necessarily unique)*</u>
aligning sequence that aligns <u>*all*</u> *state pairs in* $\overline{Q_{D_0}} \times Q_{D_1}$.

Theorem 10. *The relation* $\approx$ *is symmetric and transitive, but not reflexive.*

For the class of essentially resettable designs, the relation $\approx$ is an equivalence
relation. The non-reflexivity of SHE comes from the fact that a non-essentially-
resettable design does not have an aligning sequence with itself. Thus the design
is not equivalent to itself. An example of such a design is shown in Figure[2] 2. In
this design the state pair $(10, 11)$ is not alignable.

4   Sequential Replaceability

Here we start by justifying why the notion of sequential hardware equivalence,
discussed in Section 3, does not work for safe replacement of sequential designs.
Then we present our new notion of sequential replaceability, followed by some
properties of this new notion.

---

[2] We will frequently represent designs by state transition graphs (STG's). A label a/b
on an edge denotes that under input a, the source state outputs b. The destination
of the edge denotes the next state for that input. The t-bit binary-valued label on a
state denotes that in the design the state is implemented by that assignment of the t
latches. Notice that because a combinational function can be implemented in many
different ways, the design-to-STG transformation is a many-to-one mapping.

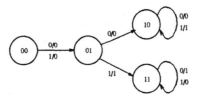

Fig. 2. Example of a design which does not have an essential reset sequence

4.1 Problems with Sequential Hardware Equivalence

From Theorem 9, two designs are considered equivalent if there exists a universal aligning sequence. This sequence is a synchronizing sequence for either design. However, in the design process, often the designer does not know the synchronizing sequence for his/her design $D0$. Even if they can determine such a sequence π for their design, it may not be possible for the environment to generate π. So, for a safe replacement of $D0$ we need to preserve <u>all</u> initializing sequences, and not just one.

The notion of SHE does not place any constraints on the outputs of the designs during the synchronization phase. However, we claim that this condition is too weak for a safe replacement[3]. *A priori*, we cannot assume that the external environment is not sensitive to the outputs during the synchronization phase. This is especially important because there may be another interacting design whose synchronizing sequence may be driven by an output of design $D0$, and affecting the outputs of $D0$ during synchronization may destroy that synchronizing sequence.

Finally, the notion of SHE does not work for designs which are not essentially resettable. As Theorem 10 states, such designs are not even equivalent to themselves. It will be presumptuous on our part to assume that such designs do not exist in real designs, and to present a theory which fails to replace such designs even by themselves. There can be two classes of real designs which are not essentially resettable. First, if the environment has some flexibility for the input/output behavior it can accept from the design, the design may have multiple non-equivalent closed SCC's (for example, Figure 2). In this example, the environment has a don't care condition so that the design is acceptable as long as it always toggles the input (state 11) or always outputs the input (state 10), after the synchronization phase. For the second class, consider the design in Figure 3. It can be seen that there is no synchronizing sequence for this design, and hence this design is not essentially resettable. However, once the design powers up, its state can be determined from its outputs, and based on the outputs the design can be driven to state 0. Thus, the behavior of this design can be controlled.

---

[3] We are indebted to Dr. Richard Rudell of Synopsys, Inc. for comments about sequential replacement.

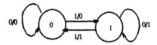

Fig. 3. Example of a design which does not have an synchronizing sequence

4.2 Conditions for Valid Replacement

Here we present our new condition for safe replacement of a sequential design. We assume that no latches have any reset lines[4]. Since it cannot be predicted which state the design powers up in, we can safely assume that no matter which state the original design powers up in, the subsequent input/output behavior of the design is acceptable to the environment. This motivates the following condition (the *safe replacement condition*):

Definition 11. Design D_1 is a *safe replacement* for design D_0 ($D_1 \preceq D_0$) if given any state $s_1 \in Q_{D_1}$ and any finite input sequence $\pi \in I^*$, there exists some state $s_0 \in Q_{D_0}$ such that the output behavior $\lambda_{D_1}(s_1, \pi) = \lambda_{D_0}(s_0, \pi)$.

For example, consider the design D_0 in Figure 4 consisting of 1 input wire, 1 output wire and 3 latches. Design D_1 in Figure 5 satisfies the safe replacement condition ($D_1 \preceq D_0$). States 00, 11 and 11 in D_1 behave like states 000, 011 and 101, respectively, in D_0 for all input sequences. The remaining state 10 in D_1 behaves like state 010 for all input sequences starting with 0, and like state 101 for all input sequences starting with 1.

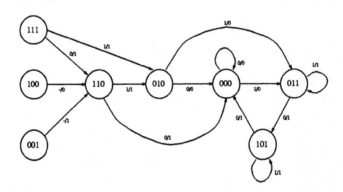

Fig. 4. Example design D_0

We would like to emphasize some interesting properties of the safe replacement condition:

---

[4] If some latches do have a reset line, they can be modeled by a latch without a reset line if we treat the reset line as another primary input.

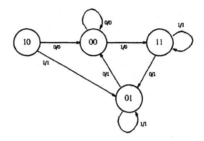

Fig. 5. Replacement design D_1

Remark 1: If $D_1 \preceq D_0$, then there may be states $s_0 \in Q_{D_0}$ and $t_0 \in Q_{D_1}$ such that for all states $s \in Q_{D_0}$ and $t \in Q_{D_1}$, $s_0 \not\sim t$ and $t_0 \not\sim s$. For example, state 111 in D_0 (Figure 4) is not equivalent to any state in D_1 (Figure 5); also, state 10 in D_1 is not equivalent to any state in D_0. Thus classical machine equivalence [5], that requires that every state in each design be equivalent to be some state in the other design, is not necessary for safe replacement, although it is sufficient.

Remark 2: As is obvious from Definition 11, the relation $\preceq$ is reflexive and transitive. However, the relation $\preceq$ is not symmetric. Although $D_1 \preceq D_0$, it is not true that $D_0 \preceq D_1$ because the state 110 in design D_0 produces an output sequence of $1 \cdot 0$ on the input sequence $1 \cdot 1$ and there is no state in D_1 which exhibits this behavior. However, an equivalence relation can easily be defined from a transitive and reflexive one.

Definition 12. Designs D_0 and D_1 are *replacement equivalent* if $D_0 \preceq D_1$ and $D_1 \preceq D_0$.

Consider the equivalence classes of designs modulo replacement equivalence—a design D belongs to an equivalence class $[D_0]$ if and only if it is replacement equivalent to D_0. We say that $[D_1] \preceq [D_0]$ if and only if $D_1 \preceq D_0$. Now, $\preceq$ is a partial ordering on these design equivalence classes (since it can be easily shown that it is reflexive, transitive and anti-symmetric).

Remark 3: Although the replacement design has to be compatible with the original design (same number of inputs and outputs), it does not have to the same number of latches (for example, D_0 in Figure 4 has 3 latches whereas the replacement design D_1 in Figure 5 has only 2 latches).

Remark 4: While the theory of sequential hardware equivalence Section 3 cannot be applied to designs which are not essentially resettable, our safe replacement conditions work for any designs. For essentially resettable designs, sequential replaceability is a stronger condition than SHE (sequential replaceability implies SHE, because as the following Theorem 13 shows, if $D_1 \preceq D_0$, any synchronizing sequence for D_0 can align all state pairs in $Q_{D_0} \times Q_{D_1}$.

Theorem 13. *If $D_1 \preceq D_0$ and ρ is a synchronizing sequence for D_0 then ρ is also a synchronizing sequence for D_1 and for any states $s_0 \in Q_{D_0}$ and $s_1 \in Q_{D_1}$, $\delta_{D_0}(s_0, \rho) \sim \delta_{D_1}(s_1, \rho)$.*

Proof. Pick a state $s' \in Q_{D_0}$. For the proof, we just need to show that for any state $t \in Q_{D_1}$, $\delta_{D_1}(t, \rho) \sim \delta_{D_0}(s', \rho)$. Suppose not. Then $\delta_{D_1}(t, \rho) \not\sim \delta_{D_0}(s', \rho)$. There exists a sequence π such that $\lambda_{D_1}(\delta_{D_1}(t, \rho), \pi) \neq \lambda_{D_0}(\delta_{D_0}(s', \rho), \pi)$. However, since $D_1 \preceq D_0$, there exists a state $s \in Q_{D_0}$ such that $\lambda_{D_1}(t, \rho \cdot \pi) = \lambda_{D_0}(s, \rho \cdot \pi)$. This also means that $\lambda_{D_1}(\delta_{D_1}(t, \rho), \pi) = \lambda_{D_0}(\delta_{D_0}(s, \rho), \pi)$. However, since ρ is a synchronizing sequence for D_0, $\delta_{D_0}(s, \rho) \sim \delta_{D_0}(s', \rho)$. Thus, $\lambda_{D_1}(\delta_{D_1}(t, \rho), \pi) = \lambda_{D_0}(\delta_{D_0}(s, \rho), \pi) = \lambda_{D_0}(\delta_{D_0}(s', \rho), \pi)$, which is a contradiction. ∎

Remark 5: The idea of safe replacement implicitly uses the fact that the original design D_0 can power up in any state. Power-up states are generally beyond the control of designers for physical reasons. It may be possible that, by design, D_0 cannot power up in some states or that the likelihood of powering up in some states is so remote that D_0 is never observed to do so. The notion of safe replacement still applies with replacing Q_{D_0} and Q_{D_1}, in Definition 11, by the power-up states of D_0 and D_1, respectively.

Necessary Conditions for a Safe Replacement Even though there is some flexibility for the implementation of the replacement design, it cannot have arbitrarily few states; in fact, as the following results show, each closed SCC in the replacement design must be equivalent (Definition 5) to some closed SCC in the original design.

Lemma 14 (Lemma 2 in [2]). *Suppose that DFSM's M_0 and M_1 have no equivalent states then there is an input sequence π such that for any states s_0 of M_0 and s_1 of M_1, $\lambda_{M_0}(s_0, \pi) \neq \lambda_{M_1}(s_1, \pi)$.*

Lemma 15. *If $D_1 \preceq D_0$, and $t \in Q_{D_1}$ lies in a closed SCC of D_1, then there exists state $s \in Q_{D_0}$ such that $s \sim t$.*

Proof. (by contradiction). Assume that $D_1 \preceq D_0$. Let M be a closed SCC of D_1. Then M is a DFSM. Suppose that no state of M is equivalent to any state of D_0. By Lemma 14, there is a sequence π that differentiates every state of M from every state of D_0. *A fortiori*, π differentiates a particular state, say s of M from every state of D_0. Therefore, the assumption that $D_1 \not\preceq D_0$ is false. ∎

Theorem 16. *If $D_1 \preceq D_0$, and M_1 is a closed SCC in design D_1, then there must be a closed SCC M_0 in design D_0 such that $M_0 \equiv M_1$.*

Proof. Consider a state t in M_1. From Lemma 15, we know that there exists a state $s \in Q_{D_0}$ such that $s \sim t$. We will show that if M_0 is any closed SCC reachable from s then M_0 is equivalent to M_1.

Consider any state s' in M_0. Let π be an input sequence such that $\delta_{D_0}(s, \pi) = s'$. Since M_1 is closed under all inputs, $t' = \delta_{D_1}(t, \pi)$ lies in M_1. Also, since $s \sim t$, we have $s' \sim t'$.

Similarly, consider any state t'' in M_1. Let ρ be an input sequence such that $\delta_{D_1}(t', \rho) = t''$. Again, $s'' = \delta_{D_0}(s', \rho)$ lies in M_0, and $s'' \sim t''$.

Thus, DFSM's M_0 and M_1 are equivalent. ∎

As remarked previously, the equivalence classes modulo replacement equivalence are partially ordered by safe replacement ($\preceq$). Theorem 16 shows that each closed SCC of design D_0 defines a minimal element that is $\preceq$ to the equivalence class $[D_0]$. Designs with unique minimal predecessors are therefore ones with unique closed SCC's, or ones where all SCC's are equivalent to each other.

Special Case - Known Initializing Sequence Set

Sometimes, the designer knows the initializing sequences for the design and does not need a replacement condition as strong as in Definition 11. The designer knows that whenever the design powers up, one of sequences from an initializing sequence set Π is applied and the design is reset to some desired behavior. The designer does not care about the outputs while an initializing sequence is applied[5], and knows that the design "works" as long as some sequence from Π is applied after power-up. As an example, consider a design with two input lines a and b. The designer knows that after power-up the input a will be set at 1 for the first 2 clock cycles to initialize the design. For this example, the initializing set $\Pi = \{(10 \cdot 10, 10 \cdot 11, 11 \cdot 10, 11 \cdot 11\}$, where a represents the first input and b the second input. For an initializing sequence set Π we can modify our safe replacement condition from Definition 11 to the following:

Definition 17. Design D_1 is a *safe replacement* for design D_0 under the initializing sequence set Π ($D_1 \overset{\Pi}{\preceq} D_0$) if given any state $s_1 \in Q_{D_1}$, an initializing sequence $\pi_1 \in \Pi$, and any finite input sequence $\rho \in I^*$, there exists some state $s_0 \in Q_{D_0}$ and an initializing sequence $\pi_0 \in \Pi$ such that the output behavior $\lambda_{D_1}(\delta_{D_1}(s_1, \pi_1), \rho) = \lambda_{D_0}(\delta_{D_0}(s_0, \pi_0), \rho)$.

We can also derive the following result (the proof is similar to that of Theorem 16 and is omitted for brevity).

Theorem 18. *If $D_1 \overset{\Pi}{\preceq} D_0$ and M_1 is a closed SCC in design D_1, then there must be a closed SCC M_0 in design D_0 such that $M_0 \equiv M_1$.*

It should be noted that for a single design a designer may have more than one set of initializing sequences. Each of these set might be used to initialize the design to a different behavior. For such a situation the designer would like to verify that the replacing design is a safe replacement under each initializing sequence set.

4.3 Verification for Sequential Replaceability

Although we have safe replacement conditions in Definitions 11 and 17, we still need a decision procedure to verify if a replacement design satisfies these conditions. In this section we develop two methods to answer this verification question.

---

[5] If the designer does care about the outputs during the initialization phase, it is easy to modify our safe replacement condition for such a case also.

Method I - Finding a Discriminating Sequence

The following procedure decides whether (new) design D_1 is a safe replacement for (original) design D_0, i.e, if $D_1 \preceq D_0$. If D_1 is not a safe replacement for D_0 then this algorithm finds a state s_1 of D_1 and an input sequence π that distinguishes s_1 from every state of D_0.

We construct a multiple rooted, acyclic directed graph whose nodes are labeled by pairs of the form (s, A) where s is a state of D_1 and A is a subset of states of D_0. Nodes are either marked or unmarked; the markings may be **FAIL**, **JUMP**(N) or **SUCCEED**, where N is another node. Edges of the graph are labeled by a single input $a \in I$. We presume that the equivalent state pairs of D_0 and D_1 have already been computed, see [7].

The roots of the digraph are pairs of the form (s_0, Q_{D_0}) where s_0 is a state of D_1, Q_{D_0} is the set of all states of D_0 and s_0 is not equivalent to any state in Q_{D_0}. If the set of roots is empty, then clearly $D_1 \preceq D_0$.

loop until some leaf is marked **FAIL** or all leaves are marked **JUMP** or **SUCCEED**:

Choose a node N labeled (s_1, A) of the existing tree that has no **JUMP** or **SUCCEED** mark and choose an input a such that no edge out of N has the label a.

Let $s_1' = \delta_{D_1}(s_1, a)$ and $A' = \{s' |$ for some s in A, $s' = \delta_{D_0}(s, a)$ and $\lambda_{D_1}(s_1, a) = \lambda_{D_0}(s, a)\}$. If a node labeled (s_1', A') already exists, say node N', create an edge labeled a from N to N', and goto the beginning of the loop. Else, create a new edge out of N labeled a and pointing to a new node N' labeled (s_1', A').

Mark the new node N' as follows:

1. If A' is empty, mark the new node N **FAIL** and exit the program.
2. If s_1' is state equivalent to any state in A', mark N **SUCCEED**, and go to the beginning of the loop.
3. If there exists a node N'' labeled (s_1', A'') such that $A'' \subset A'$, mark N' as **JUMP**(N''), and go to the beginning to the loop.
4. For each node N'' labeled (s_1', A'') such that $A'' \supset A'$, mark N'' as **JUMP**(N').

End **loop**

Proof. Termination: Each node must have a distinct label and there can only be finitely many labels. Furthermore each node has an upper bound on the number of edges emanating from it – the number of primary input combinations. Therefore the program must terminate.

If the program terminates because a **FAIL** node is created, any path from a root (s, Q_{D_0}) to the **FAIL** node gives a sequence that distinguishes s from any state of Q_{D_0}. If there is no **FAIL** node then all leaf nodes are marked **SUCCEED** or **JUMP**.

Claim: If all leaf nodes are marked **SUCCEED** or **JUMP** then no input sequence will distinguish a state of D_1 from all the states of D_0.

Observation: We first observe that there cannot be loop of **JUMP** nodes—
N_1 -**JUMP**$(N_2) \to N_2$-**JUMP**$(N_3) \to \cdots \to N_k$ -**JUMP**$(N_1) \to N_1$ because

each **JUMP** reduces the cardinality of the second coordinate of the label of a node.

The proof of the claim is by contradiction. Suppose there were a state s_1 of D_1 and an input sequence $\pi = a_1 \cdot a_2 \cdots a_k$ that distinguishes s_1 from all states of D_0. Let $N_0 = (s_1, Q_{D_0})$. Notice that node N_0 cannot be marked **SUCCEED**, because then s_1 would be equivalent to a state in D_0, a contradiction. Construct the sequence of nodes $N_0, N_1, \ldots, N_k$, none of which is marked **SUCCEED**, by applying the following procedure recursively. If node N_i is unmarked then node N_{i+1} is the node reached by traversing the edge labeled a_{k+1} from N_i. Otherwise, N_i is marked **JUMP**(N); jump to N and keep jumping nodes until a node N' is reached which is not marked **JUMP** (see the observation above). This node cannot be marked **SUCCEED**. [This is because nodes marked **SUCCEED** have their left hand component equivalent to some state in their right hand component. But then N_i would have the same property and would have been marked **SUCCEED** rather than **JUMP** which is a contradiction.] Now, N_{i+1} is the node reached by traversing the edge labeled a_{k+1} from N'.

Node N_{i+1} labeled (s_{i+1}, A_{i+1}) cannot be marked **SUCCEED**, because s_{i+1} cannot be equivalent to any state in A_{i+1}. [If it was so, by backtracking the edges traversed, we can find a state in Q_{D_0} that cannot be distinguished from s_1.] Since there are no nodes marked **FAIL**, the last node N_k labeled (s_k, A_k) must have a non-empty set A_k. But then by choosing a state in A_k, and backtracking the edges traversed in constructing the sequence of nodes, one can find an element of Q_{D_0} that is not distinguished from s_1 by π. ∎

Before we execute the above algorithm, we could check to see if each closed SCC of D_1 has a state equivalent to some state of D_0. If not, then by Theorem 16, $D_1 \npreceq D_0$. Otherwise, if there is no state outside the closed SCC's of D_1, then $D_1 \preceq D_0$ and we are done. If there is state outside the closed SCC's of D_1, we use the method outlined above knowing that each root (s_0, Q_{D_0}) of the digraph is such that s_0 is outside all closed SCC's of D_1. Thus, if the number of states outside closed SCC's is small compared to the the number of states which lie in closed SCC's, we can probably expect our algorithm to be efficient. Also, note that the algorithm would be correct if we did not mark any states **JUMP** (remove substeps 3 and 4 in the marking step). Marking states as **JUMP** is just a way to prune the search space, and make the algorithm more efficient.

Known Initializing Sequence Set First, based on Theorem 18, we check if each closed SCC of D_1 has a state equivalent to some state of D_0. If this check fails, we know that D_1 is not a safe replacement for D_0 under the initializing sequence set Π. Otherwise, we compute sets $Q_0 = \{s|$ there exists $s_0 \in Q_{D_0}$ and $\pi \in \Pi$ such that $\delta_{D_0}(s_0, \pi) = s\}$, and $Q_1 = \{s|$ there exists $s_1 \in Q_{D_1}$ and $\pi \in \Pi$ such that $\delta_{D_0}(s_1, \pi) = s\}$. Now, we can use the same digraph-construction method described above, except that the roots of the digraph are pairs of the form (s, Q_0) such that $s \in Q_1$ and s is not equivalent to any state in Q_0. The proof of correctness is similar to the one above, and is omitted for brevity.

Method II - Using Language Containment

Definition 19. A <u>non-deterministic finite automaton</u> (NFA) is a 5-tuple $A = (Q, \Sigma, \delta, I, F)$, where Q (the set of states) and Σ (the alphabet) are finite non-empty sets and $\delta : Q \times \Sigma \to 2^Q$ is the transition relation, $I \subseteq Q$ is the set of non-deterministic initial states, and $F \subseteq Q$ is the set of final states. The language of A, denoted by $\mathcal{L}(A)$, is a set of finite strings of the alphabet, and is defined as in [6].

Given the original design D_0 and the new design D_1, we will construct two NFA's $A_0 = (Q_{D_0}, \Sigma, \delta_{A_0}, I_{A_0}, F_{A_0})$ and $A_1 = (Q_{D_1}, \Sigma, \delta_{A_1}, I_{A_1}, F_{A_1})$ such that $D_1 \preceq D_0$ if and only if $\mathcal{L}(A_1) \subseteq \mathcal{L}(A_0)$. Here $\Sigma = I \times O$. For $i \in \{0, 1\}$, $\delta_{A_i}(s, (a, b)) = t$ if and only if $\delta_{D_i}(s, a) = t$ and $\lambda_{D_i}(s, a) = b$. Also, $F_{A_i} = Q_{D_i}$ and $I_{A_i} = Q_{D_i}$.

However, since the problem of language containment between two NFA's is PSPACE-complete [4, page 265], this approach is not likely to be more efficient that Method I. Although Method I may also be inefficient in the worst case, we can probably hope to terminate with **SUCCESS** or **FAIL** without exploring the entire search space, especially if D_1 has been derived from D_0 using some synthesis algorithms.

5 Conclusions

We have defined a notion of safe replaceability ($\preceq$) that is independent of initial states of a design, is independent of the intended environment of a design, and applies to all sequential designs, resettable or not. This notion accomplishes for sequential designs what the notion of boolean equivalence accomplishes for combinational designs. We have shown by example that this notion is strictly weaker than the property that every state of the replacement design is equivalent to some state of the original design. We observed that safe replaceability is a reflexive and transitive relation (i.e., a partial ordering of designs). Finally, we gave two algorithms for deciding whether one design is a safe replacement for another.

6 Acknowledgements

We would like to thank the CAD research groups at the University of Colorado at Boulder, Motorola Austin and the University of California at Berkeley for many useful comments on this work.

References

1. H. Cho, G. D. Hachtel, S.-W. Jeong, B. Plessier, E. Schwarz, and F. Somenzi. ATPG Aspects of FSM Verification. In *Proc. Intl. Conf. on Computer-Aided Design*, pages 134–137, 1990.

2. H. Cho, S.-W. Jeong, F. Somenzi, and C. Pixley. Synchronizing Sequences and Symbolic Traversal Techniques in Test Generation. *Journal of Electronic Testing: Theory and Applications*, 4(12):19–31, 1993.

3. O. Coudert, C. Berthet, and J. C. Madre. Verification of Sequential Machines Based on Symbolic Execution. In J. Sifakis, editor, *Proc. of the Workshop on Automatic Verification Methods for Finite State Systems*, volume 407 of *Lecture Notes in Computer Science*, pages 365–373, June 1989.

4. M. R. Garey and D. S. Johnson. *Computers and Intractability.* W. H. Freeman and Co., 1979.

5. J. Hartmanis and R. E. Stearns. *Algebraic Structure Theory of Sequential Machines.* Intl. Series in Applied Mathematics. Prentice-Hall, Englewood Cliffs, N.J., 1966.

6. J. E. Hopcroft and J. D. Ullman. *Introduction to Automata Theory, Languages and Computation.* Addison-Wesley, 1979.

7. C. Pixley. Introduction to a Computational Theory and Implementation of Sequential Hardware Equivalence. In E. M. Clarke and R. P. Kurshan, editors, *Proc. of the Conf. on Computer-Aided Verification*, volume 531 of *Lecture Notes in Computer Science*, pages 54–64, June 1990.

8. H. Touati, H. Savoj, B. Lin, R. K. Brayton, and A. L. Sangiovanni-Vincentelli. Implicit State Enumeration of Finite State Machines using BDD's. In *Proc. Intl. Conf. on Computer-Aided Design*, pages 130–133, November 1990.

Formula-Dependent Equivalence for Compositional CTL Model Checking

Adnan Aziz   Thomas R. Shiple   Vigyan Singhal
Alberto L. Sangiovanni-Vincentelli
Email: {adnan,shiple,vigyan,alberto}@ic.eecs.berkeley.edu

Department of EECS, University of California, Berkeley, CA 94720

Abstract. We present a state equivalence that is defined with respect to a given CTL formula. Since it does not attempt to preserve all CTL formulas, like bisimulation does, we can expect to compute coarser equivalences. We use this equivalence to manage the size of the transition relations encountered when model checking a system of interacting FSMs. Specifically, the equivalence is used to reduce the size of each component FSM, so that their product will be smaller. We show how to apply the method, whether an explicit representation is used for the FSMs, or BDDs are used. Also, we show that in some cases our approach can detect if a formula passes or fails, without composing all the component machines. The method is exact and fully automatic, and handles full CTL.

1 Introduction

Formal design verification is the process of verifying that a design has certain properties that the designer intended. A well known verification technique is computation tree logic (CTL) model checking. In this approach, a design is modeled as a finite state machine (FSM), properties are stated using CTL formulas, and a "model checker" is used to prove that the FSM satisfies the given CTL formulas [6]. The complexity of model checking a formula is linear in the number of states of the FSM.

Oftentimes, large designs are constructed by linking together a set of FSMs. The straightforward approach to model checking such a design is to first form the product of the component FSMs to yield a single FSM, and then proceed to model check this single FSM. However, the size of the product machine can be exponential in the number of component machines, and hence the model checker may take exponential time. This is known as the "state explosion problem" when using explicit representations, or the "representation explosion problem" when using implicit representations, like ordered binary decision diagrams (BDDs). As it turns out, we cannot hope to do better than this in the worst case, because the problem of model checking a system of interacting FSMs is PSPACE-complete [1].

Our goal is to develop an algorithm that alleviates the explosion problem by identifying equivalent states in each component machine. These equivalent states are then used to simplify the components before taking their product, thus leading to a smaller product machine. It is well known that *bisimulation equivalence* is the coarsest (or weakest) equivalence that preserves the truth of *all* CTL formulas [4]. However, in general we are interested in model checking a system with respect to just a few formulas, and hence preserving all CTL formulas is stronger than needed. Thus, we investigate a formula-dependent equivalence that preserves the truth of a particular formula of interest, but possibly not of other formulas. This leads to a coarser equivalence, and thus to a greater opportunity for simplification. If an explicit representation is used for the FSMs, then this equivalence is used to form the quotient machines of the components. If BDDs are used, then the equivalence relation is used to define a range of permissible transition relations, among which we want to use the one with the smallest BDD.

Consider for example the FSM M described in Figure 1. The CTL formula $\phi = \forall G(\text{REQ} \to \forall F \text{ACK})$ expresses the property that every request is eventually acknowledged. The behaviors from state 1 and 5 are different. However, since there are no behaviors from states 4 and 8 where REQ is produced, then ϕ is always true at these states. Hence, states 1 and 5 can actually be merged, with respect to ϕ. Consequently, M can be replaced by the 5-state machine M': verifying ϕ on a product machine containing the component M is equivalent to verifying ϕ on the product machine with M replaced by M'.

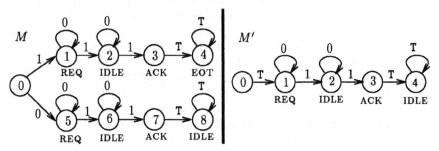

Fig. 1. Finite state machine M with inputs 0 and 1 and outputs REQ, ACK, IDLE and EOT. The symbol T means "true", the union of all input assignments.

The approach we have developed can be applied to *any* formula of CTL. Thus, we can handle formulas that refer to atomic propositions of any number of the component machines, and the formulas can be nested arbitrarily. The approach is fully automatic and it is exact, that is, it returns exactly the set of product states satisfying the formula of interest. Finally, in some cases the approach can detect if a formula passes or fails, without composing all the component machines.

Section 2 discusses related work, and Section 3 presents some preliminaries. In Section 4 we develop our formula-dependent equivalence, and in Section 5 we discuss how this equivalence can be used to simplify compositional model checking. Finally, Section 6 mentions future work and gives conclusions. Proofs for the propositions and theorems can be found in [2].

2 Related Work

Other researchers have addressed the problem of reducing the complexity of model checking. As mentioned in the introduction, bisimulation preserves the truth of all CTL formulas, and hence can be used to identify equivalent states to derive smaller component machines. This technique has been used by [3].

Clarke *et al.* presented the *interface rule*, which can be applied when a CTL formula refers to the atomic propositions of just one machine, the "main" machine [7]. In this case, the outputs of the other machines that cannot be sensed by the main machine, can be "hidden". After hiding such outputs, some states in the other machines may become equivalent, and hence the number of states can be reduced. This technique is orthogonal to our approach, and thus the two approaches could be combined. In general, any output not referred to by the formula, and not observable by other machines, can be hidden.

Grümberg *et al.* defined a subset of CTL, known as ACTL, which permits only universal path quantification, and not existential path quantification [11]. They go on to develop an approach to compositional model checking for ACTL. If an ACTL formula is true of one component in a system, then it is true of the entire system. Thus, in some cases the full product machine can be avoided. However, the formula may be true of the entire system, *without* being true of any one component in isolation, i.e. their approach is conservative, and not exact. In this case, some components must be composed, and the procedure repeated. The user has the option of manually forming abstractions for some of the machines. If the formula is false, then the product machine must always be formed. An asset of this approach is that it handles fairness constraints on the system.

Dams *et al.* have also devised an approach using ACTL [9]. Like our method, they compute an equivalence with respect to a *single* formula. Although they are limited to formulas of ACTL, it may turn out that coarser equivalences are possible by restricting to a subset of CTL. They do not address how their equivalence can be used in *compositional* model checking, where a formula may refer to the atomic propositions of several interacting machines.

Our experience indicates that existential path properties are useful for determining if a system *can* exhibit a certain behavior. This is especially true when ascertaining if the environment for a system has been correctly modeled so that it can produce the stimuli of interest. Hence, we are interested in techniques that can handle *full* CTL.

The work of Chiodo *et al.* [5] has similar aims as ours, and the current work can be seen as an outgrowth of that work. Both approaches are exact, fully automatic, and formula dependent. We have extended Chiodo's method (see Section 5.2), and have cast our extension as an equivalence on states.

3 Preliminaries

3.1 Finite States Machines

The systems that we want to verify are synchronous, interacting FSMs. Each component FSM receives a set of binary-valued inputs, and produces another set of binary-valued outputs. Formally, an FSM is a 5-tuple $M = (S, I, J, T, O)$, where S is a finite set of states, $I = \{\alpha_1, \ldots, \alpha_m\}$ is a set of m inputs supplied by the environment of the FSM, $J = \{\beta_1, \ldots, \beta_n\}$ is a set of n outputs, T is the transition relation, and O is the output function. T relates a starting state, an assignment to the inputs, and an ending state, i.e. $T \subseteq S \times \Sigma \times S$, where $\Sigma = 2^I$. We require the transition relation to be *complete*, so that for each $a \in \Sigma$ and $x \in S$, there exists at least one $y \in S$ such that $(x, a, y) \in T$. The output function takes a state in S and returns an assignment to the outputs, i.e. $O : S \to 2^J$. Our definition of FSM is equivalent to that of a Moore machine in [11].

Composition is defined in the usual way. In composing two interacting FSMs, some inputs of each machine may be equal to the outputs of the other machine, whereas other inputs may come from the environment of the composed FSM. Thus, the inputs of the composition are the inputs of the components that are not outputs of either component. The outputs of the composition are all the outputs of the components. Figure 2 shows an example, where M_1 has states $\{1, 2, 3\}$, and M_2 has states $\{1', 2'\}$. The sets of inputs and outputs for M_1 are $\{a, q\}$ and $\{p\}$ respectively; and for M_2 are $\{a, p\}$ and $\{q\}$ respectively. For the composition $M_1 \times M_2$, the sets of inputs and outputs are $\{a\}$ and $\{p, q\}$ respectively.

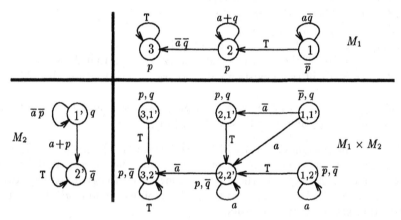

Fig. 2. Example of FSM composition: p is the output of M_1, q is the output of M_2, and a is an external input. $a\bar{q}$ is shorthand for the subset $\{\{a\}\} \subseteq 2^{\{a,q\}}$. The union of $a\bar{q}$, $\bar{a}q$ and aq is denoted by $a+q$.

3.2 Computation Tree Logic

Computation tree logic is a language used to describe properties of state transition systems. We are interested in checking CTL formulas that describe properties of the composition of a set of interacting FSMs. Since the composition of a set of FSMs is again an FSM, we give the syntax and semantics of CTL for a single FSM M. We allow two types of atomic propositions:

1. each output variable is an atomic proposition, and
2. each subset of states is an atomic proposition

The second type arises naturally when recursively checking formulas. With this, the set of CTL formulas is defined inductively as follows:

- p is a CTL formula, where p is an output variable or a subset of states, and
- if ψ_1 and ψ_2 are CTL formulas, then so are $\neg\psi_1$, $\psi_1 \vee \psi_2$, $\exists X\psi_1$, $\exists G\psi_1$, and $\exists[\psi_1 \ U \ \psi_2]$.

Note that inputs are *not* allowed as atomic propositions. However, by modeling an input by an FSM whose output describes the expected behavior of the input, one can implicitly use an input as an atomic proposition.

The semantics of CTL is usually defined on finite Kripke structures, which are directed graphs where each node is labeled by a set of atomic propositions [6]. To extend these semantics to FSMs, we just ignore the labels on the transitions of the FSMs, and we view the outputs as atomic propositions. Let $M = (S, I, J, T, O)$ be an FSM. A *path* from state x_0 is an infinite sequence of states $x_0x_1x_2\ldots$ such that for every i, there exists an $a \in \Sigma$ such that $(x_i, a, x_{i+1}) \in T$. The notation $M, x_0 \models \phi$ means that ϕ is true in state x_0 of FSM M. The semantics of CTL is defined inductively as follows:

- $M, x_0 \models p$, where $p \in J$, iff $p \in O(x_0)$.
- $M, x_0 \models p$, where $p \subseteq S$, iff $x_0 \in p$.
- $M, x_0 \models \neg\psi_1$ iff $M, x_0 \not\models \neg\psi_1$.
- $M, x_0 \models \psi_1 \vee \psi_2$ iff $M, x_0 \models \psi_1$ or $M, x_0 \models \psi_2$.
- $M, x_0 \models \exists X\psi_1$ iff there exists a path $x_0x_1x_2\ldots$ such that $M, x_1 \models \psi_1$.
- $M, x_0 \models \exists G\psi_1$ iff there exists a path $x_0x_1x_2\ldots$ such that for all i, $M, x_i \models \psi_1$.
- $M, x_0 \models \exists[\psi_1 \ U \ \psi_2]$ iff there exists a path $x_0x_1x_2\ldots$ and some $i \geq 0$ such that $M, x_i \models \psi_2$ and for all $j < i$, $M, x_j \models \psi_1$.

For example in machine $M_1 \times M_2$ of Figure 2, state $\langle 1, 2' \rangle$ satisfies the formula $\exists G(\neg p \wedge \neg q)$, whereas none of the other state do. The expression $\exists F\psi$ is an abbreviation for $\exists[true \ U \ \psi]$, where *true* is a logical tautology. Lastly, we define the *CTL model checking problem* as the problem of determining *all* states of the system that satisfy a given formula.[1]

4 Formula-Dependent Equivalence

Our goal is to define an equivalence on the states of each component machine that is as coarse as possible with respect to a given CTL formula ϕ, while being efficiently computable. Section 5 explains how we intend to apply this equivalence to model checking, but the main idea is to merge equivalent states to minimize the size of each component. The minimized machines are then composed. Optionally, the product can be computed incrementally by composing a

---

[1] If a set of initial states is known, then we can restrict our attention to the reachable state space. In this case, we can apply known techniques for exploiting the unreachable states, such as minimizing the transition relation with respect to unreachable states; these techniques are orthogonal to those discussed in this paper.

few of the minimized machines, and then computing a new equivalence for this sub-product. When the top level is reached and just a single machine remains, the usual CTL model checking algorithm is applied to determine the states that satisfy ϕ.

Our formula dependent equivalence can be best explained by comparing it to bisimulation. ("strong bisimulation" of Milner [12, p. 88]) Given an FSM $M = (S, I, J, T, O)$, the bisimulation equivalence relation, denoted by $\sim$, is the coarsest equivalence relation satisfying the following:

For all $x, y \in S$, $x \sim y$ implies

- $O(x) = O(y)$ and
- for all $a \in \Sigma$ (recall from Section 3 that $\Sigma = 2^I$)
    - whenever $x \xrightarrow{a} t$, then for some w, $y \xrightarrow{a} w$ and $t \sim w$, and
    - whenever $y \xrightarrow{a} w$, then for some t, $x \xrightarrow{a} t$ and $t \sim w$.

The soundness of this definition follows from the observation that the class of equivalence relations satisfying the above definition contains the identity, and is closed under union. Intuitively, two states are bisimilar if their corresponding infinite computation trees[2] "match". This means that the two states have the same outputs, and on each input, the two states have next states whose infinite computation trees again match.

We use the notion of $PASS$ and $FAIL$ states to ease the strict requirement of bisimulation that the infinite computation trees of two states match. Loosely, if a state is a $PASS^\phi$ state with respect to a CTL formula ϕ, then it satisfies ϕ in all environments; likewise, if a state is $FAIL^\phi$, then it does not satisfy ϕ in any environment. Given $PASS^\phi$ and $FAIL^\phi$ states, the first modification to bisimulation we make is that subtrees rooted at $FAIL^\phi$ states are ignored. This means that transitions to $FAIL^\phi$ states from one state need not be matched by the other state. This works because only potential witnesses to a formula need to be preserved. The second modification is that two states are equivalent if they are both $PASS^\phi$ states. A consequence of this is that whereas bisimulation requires the infinite computation trees of next states to match, now it is sufficient that the next states are both $PASS^\phi$ states. This is what we mean by two infinite trees matching $up\ to\ PASS^\phi$ states. Essentially then, we say that two states are equivalent with respect to ϕ if

1. they are equivalent with respect to the immediate subformulas of ϕ, and
2. either they are both $PASS^\phi$ states or both $FAIL^\phi$ states, or the infinite computation trees of the two states match up to $PASS^\phi$ states, ignoring all subtrees rooted at $FAIL^\phi$ states.

Before formally defining our equivalence relation, we define the $PASS^\phi$ and $FAIL^\phi$ sets. For a given formula ϕ, $PASS^\phi$ and $FAIL^\phi$ sets are defined for each component. In the following definition, we assume a system of just two components, M and M'. In defining the $PASS^\phi$ and $FAIL^\phi$ sets for M, M' is referenced because the atomic propositions in ϕ may refer to M'. The symbols p_o and p_i

---

[2] The infinite computation tree of a state is formed by "unrolling" the FSM starting from that state.

are used to distinguish those output atomic propositions produced by M and those produced by M'.

Definition 1. Let $M = (S, I, J, T, O)$ and $M' = (S', I', J', T', O')$ be FSMs, and let ϕ be a CTL formula. Let $p_o \in J$, $p_i \in J'$, and $p_s \subseteq S \times S'$. $PASS^\phi$ and $FAIL^\phi$ for M are subsets of S, as follows:

ϕ		
p_i	$PASS^\phi$	$\emptyset$
	$FAIL^\phi$	$\emptyset$
p_o	$PASS^\phi$	$\{x \in S \mid p_o \in O(x)\}$
	$FAIL^\phi$	$S \setminus PASS^\phi$
p_s	$PASS^\phi$	$\{x \in S \mid \forall s' \in S', (x, s') \in p_s\}$
	$FAIL^\phi$	$\{x \in S \mid \forall s' \in S', (x, s') \notin p_s\}$
$\neg\psi$	$PASS^\phi$	$FAIL^\psi$
	$FAIL^\phi$	$PASS^\psi$
$\psi_1 \vee \psi_2$	$PASS^\phi$	$PASS^{\psi_1} \cup PASS^{\psi_2}$
	$FAIL^\phi$	$FAIL^{\psi_1} \cap FAIL^{\psi_2}$
$\exists X \psi$	$PASS^\phi$	$\{x \in S \mid \forall a \in \Sigma, \exists t \text{ s.t. } x \xrightarrow{a} t \text{ and } t \in PASS^\psi\}$
	$FAIL^\phi$	$\{x_0 \in S \mid \text{for every path } x_0 x_1 x_2 \ldots, x_1 \in FAIL^\psi\}$
$\exists G \psi$	$PASS^\phi$	greatest fixed-point of: $R_0 = PASS^\psi$; $R_{i+1} = R_i \cap \{x \in S \mid \forall a \in \Sigma, \exists t \text{ s.t. } x \xrightarrow{a} t \text{ and } t \in R_i\}$
	$FAIL^\phi$	$\{x_0 \in S \mid \text{for every path } x_0 x_1 x_2 \ldots, \text{there exists } i \geq 0$ s.t. $x_i \in FAIL^\psi\}$
$\exists[\psi_1 U \psi_2]$	$PASS^\phi$	least fixed-point of: $R_0 = PASS^{\psi_2}$; $R_{i+1} = R_i \cup \{x \in S \mid x \in PASS^{\psi_1}, \text{ and } \forall a \in \Sigma, \exists t$ s.t. $x \xrightarrow{a} t \text{ and } t \in R_i\}$
	$FAIL^\phi$	$\{x_0 \in S \mid \text{for every path } x_0 x_1 x_2 \ldots, \text{either}$ 1) there exists $i \geq 0$ s.t. $x_i \in FAIL^{\psi_1}$ and $\forall j \leq i, x_j \in FAIL^{\psi_2}$, or 2) $\forall i \geq 0, x_i \in FAIL^{\psi_2}\}$

As an example of $PASS^\phi$ and $FAIL^\phi$, consider the FSM in Figure 3. For $\psi = p$, states 1, 2, 3, 5, 6 and 7 lie in $PASS^\psi$ and states 4 and 8 lie in $FAIL^\psi$. For $\phi = \exists G p$, states 3 and 7 lie in $PASS^\phi$, while states 4 and 8 lie in $FAIL^\phi$, and states 1, 2, 5 and 6 lie in neither. The following proposition says that, indeed, if x is in $PASS^\phi$, then any product state with x as a component satisfies ϕ.

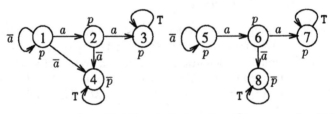

Fig. 3. Illustrating $PASS^\phi$ and $FAIL^\phi$, and the fact that $\mathcal{E}^\phi$ is coarser than bisimulation.

Proposition 2. Let ϕ be a CTL formula, let x be a state of M, and let t be a state of any FSM M'. If $x \in PASS^\phi$, then $M \times M', \langle x, t \rangle \models \phi$. Likewise, if $x \in FAIL^\phi$, then $M \times M', \langle x, t \rangle \not\models \phi$.

Note that the converse is not true. For example, consider a component M and the formula $\phi = q \wedge \neg q$, where q is an output of some other component. Then $FAIL^\phi$ for M is empty (because $FAIL^q$ and $PASS^q$ are empty by case p_i), even though ϕ is not satisfiable (i.e. for any component M', no state in $M \times M'$ satisfies ϕ). In fact, by generalizing this reasoning, we can show that if $FAIL^\phi$ were defined in such a way that the converse of Proposition 2 did hold, then $FAIL^\phi$ would be EXPTIME-hard to compute. The reduction is from CTL satisfiability, which is known to be EXPTIME-complete [10]. To check if a formula ϕ is satisfiable, compute $FAIL^\phi$ for the component M shown in Figure 4, where p is some atomic proposition *not* in ϕ. We can show that $x \in FAIL^\phi$ if and only if ϕ is not satisfiable, and thus satisfiability can be answered if we could compute $FAIL^\phi$ exactly. Similarly, since $x \in FAIL^\phi$ if and only if $x \in PASS^{\neg\phi}$, the same reduction shows that $PASS^\phi$ would also be EXPTIME-hard to compute.

Fig. 4. Component machine used to show that computing $FAIL^\phi$ exactly is EXP-TIME-hard.

Now we formally define our equivalence relation. Let $M = (S, I, J, T, O)$ and $M' = (S', I', J', T', O')$ be FSMs, and let ϕ be a CTL formula. Following Milner's development of bisimulation, we define the equivalence relation $\mathcal{E}^\phi$ on the states of FSM M as the coarsest equivalence relation satisfying the following:

For $x, y \in S$, $\mathcal{E}^\phi(x, y)$ iff:
Case $\phi = p_i$: $(x, y) \in S \times S$.
Case $\phi = p_o$: $x \in PASS^\phi$ and $y \in PASS^\phi$, or $x \in FAIL^\phi$ and $y \in FAIL^\phi$.
Case $\phi = p_s$: for all $s' \in S'$, $(x, s') \in p_s$ iff $(y, s') \in p_s$.
Case $\phi = \neg\psi$: $\mathcal{E}^\psi(x, y)$.
Case $\phi = \psi_1 \vee \psi_2$: $\mathcal{E}^{\psi_1}(x, y)$ and $\mathcal{E}^{\psi_2}(x, y)$.
Case $\phi = \exists X \psi$: $\mathcal{E}^\psi(x, y)$ and
    1. $x \in FAIL^\phi$ and $y \in FAIL^\phi$, or $x \in PASS^\phi$ and $y \in PASS^\phi$, or
    2. $O(x) = O(y)$, and for all $a \in \Sigma$
        • whenever $x \xrightarrow{a} t$ and $t \notin FAIL^\psi$, $\exists w$ s.t. $y \xrightarrow{a} w$ and $\mathcal{E}^\psi(t, w)$, and
        • whenever $y \xrightarrow{a} w$ and $w \notin FAIL^\psi$, $\exists t$ s.t. $x \xrightarrow{a} t$ and $\mathcal{E}^\psi(t, w)$.
Case $\phi = \exists G \psi$: $\mathcal{E}^\psi(x, y)$ and
    1. $x \in FAIL^\phi$ and $y \in FAIL^\phi$, or $x \in PASS^\phi$ and $y \in PASS^\phi$, or
    2. $O(x) = O(y)$, and for all $a \in \Sigma$
        • whenever $x \xrightarrow{a} t$ and $t \notin FAIL^\phi$, $\exists w$ s.t. $y \xrightarrow{a} w$ and $\mathcal{E}^\phi(t, w)$, and
        • whenever $y \xrightarrow{a} w$ and $w \notin FAIL^\phi$, $\exists t$ s.t. $x \xrightarrow{a} t$ and $\mathcal{E}^\phi(t, w)$.
Case $\phi = \exists[\psi_1 \ U \ \psi_2]$: $\mathcal{E}^{\psi_1}(x, y)$ and $\mathcal{E}^{\psi_2}(x, y)$ and

1. $x \in FAIL^{\phi}$ and $y \in FAIL^{\phi}$, or $x \in PASS^{\phi}$ and $y \in PASS^{\phi}$, or
2. $O(x) = O(y)$, and for all $a \in \Sigma$
   - whenever $x \xrightarrow{a} t$ and $t \notin FAIL^{\phi}$, $\exists w$ s.t. $y \xrightarrow{a} w$ and $\mathcal{E}^{\phi}(t, w)$, and
   - whenever $y \xrightarrow{a} w$ and $w \notin FAIL^{\phi}$, $\exists t$ s.t. $x \xrightarrow{a} t$ and $\mathcal{E}^{\phi}(t, w)$.

In a manner similar to Milner, we can show that $\mathcal{E}^{\phi}$ is the maximum fixed-point of a certain functional. Hence, using a standard fixed-point computation, $\mathcal{E}^{\phi}$ can be computed in polynomial time.

Notice that $\mathcal{E}^{\phi}$ requires equivalence on all subformulas. As the following example shows, this requirement is warranted. Consider M_1 in Figure 5. For $\phi = \exists F(p \wedge \exists F(\overline{p} \wedge q))$, states 2, 3 and 5 lie in $FAIL^{\phi}$ because p is false in these states. So with respect to ϕ, the infinite computation trees of 1 and 4 match when $FAIL^{\phi}$ states are ignored, and if we did not require equivalence on subformulas, they would be $\mathcal{E}^{\phi}$-equivalent. However, if we were to compose M_1 with M_2, ϕ holds in state $\langle 1, 1' \rangle$ but does not hold in state $\langle 4, 1' \rangle$. Thus, it would be wrong to have 1 and 4 be $\mathcal{E}^{\phi}$-equivalent. Requiring equivalence on all subformulas fixes this problem.

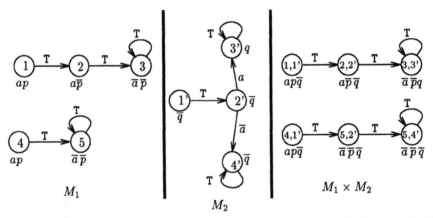

Fig. 5. Equivalence on subformulas is required. Only the states reachable from $\langle 1, 1' \rangle$ and $\langle 4, 1' \rangle$ are shown in $M_1 \times M_2$.

Since we define CTL so that formulas may refer directly to states via atomic propositions, then any formula-*independent* equivalence (e.g. bisimulation) will distinguish every pair of states, whereas $\mathcal{E}^{\phi}$ may make some states equivalent. However, even if we could not refer to states, $\mathcal{E}^{\phi}$ is still coarser than bisimulation. As stated earlier, one reason for this is that the subtrees rooted at $FAIL^{\phi}$ states are ignored. This is illustrated in Figure 3: if $\phi = \exists Gp$, then 4 is a $FAIL^{\phi}$ state, and thus 1 and 5 are $\mathcal{E}^{\phi}$-equivalent; however, they are not bisimilar.

On the other hand, there are cases where $\mathcal{E}^{\phi}$ distinguishes two states that can actually be merged. Consider the FSM M_1 in Figure 6 and the formula $\phi = \exists Gq$, where q is an output of some component not shown. Since q is an input to M_1, the sets $PASS^{\phi}$ and $FAIL^{\phi}$ are empty, and hence $\mathcal{E}^{\phi}$ reduces to bisimulation.

States 1 and 3 are not bisimilar because 2 and 4 have different outputs, and thus 1 and 3 are not $\mathcal{E}^\phi$-equivalent. However, q must be false to reach states 2 and 4, and thus the difference between states 1 and 3 does not affect the validity of ϕ. Hence, states 1 and 3 can be merged with respect to ϕ, but $\mathcal{E}^\phi$ will not merge them.

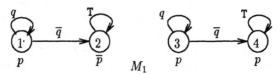

Fig. 6. $\mathcal{E}^\phi$ equivalence is incomplete. The input to M_1 is q, and the output is p. States 1 and 3 can be safely merged with respect to the formula $\phi = \exists G q$.

The following proposition says that $\mathcal{E}^\phi$-equivalent states cannot be distinguished, with respect to ϕ, by any environment. This is key in proving Theorem 4, the theorem of correctness.

Proposition 3. Let ϕ be a CTL formula, and let x and y be states of M such that $\mathcal{E}^\phi(x, y)$. Then for any state t of any FSM M', the following holds: $M \times M', \langle x, t \rangle \models \phi$ iff $M \times M', \langle y, t \rangle \models \phi$.

As an aside, note that the converse of Proposition 3 is not true. In fact, just because two states cannot be distinguished with respect to ϕ by any environment, this does not imply that they can be merged. For example, consider M_1 in Figure 5, and the formula $\phi = \exists F(p \wedge \exists F(\overline{p} \wedge q))$. As stated earlier, states 2 and 5 lie in $FAIL^\phi$, and thus for any state t of any FSM M', $M_1 \times M', \langle 2, t \rangle \models \phi$ iff $M_1 \times M', \langle 5, t \rangle \models \phi$ (i.e. by Proposition 2, neither $\langle 2, t \rangle$ nor $\langle 5, t \rangle$ satisfies ϕ). However, if we were to merge states 2 and 5 into a single state, states 1 and 4 would become equivalent. But, as discussed earlier, it would be wrong to have 1 and 4 be $\mathcal{E}^\phi$-equivalent.

Ultimately, the purpose of computing $\mathcal{E}^\phi$ is to be able to merge equivalent states, thus leading to smaller component machines. Given an equivalence relation on the states of an FSM, we define the *quotient machine* in the usual way. As we describe in the next section, if we are using an explicit representation for FSMs, then we use the quotient machine of each FSM in place of the original component. The following theorem asserts that doing this does not alter the result returned by the model checker. If we are using an implicit representation, then we use $\mathcal{E}^\phi$ to define a range of permissible transition relations for each component, among which we want to use the smallest.

Theorem 4. Let ϕ be a CTL formula, and let $M_1, \ldots, M_n$ be FSMs. Let $M_i/\mathcal{E}_i^\phi$ be the quotient of M_i with respect to $\mathcal{E}_i^\phi$, and let $[s_i]$ denote the equivalence class of $\mathcal{E}_i^\phi$ containing s_i. Then for all product states $\langle s_1, \ldots, s_n \rangle$,

$$M_1 \times \ldots \times M_n, \langle s_1, \ldots, s_n \rangle \models \phi \text{ iff } M_1/\mathcal{E}_1^\phi \times \ldots \times M_n/\mathcal{E}_n^\phi, \langle [s_1], \ldots, [s_n] \rangle \models \phi.$$

5 Compositional Model Checking

The equivalence relation $\mathcal{E}^\phi$ can be used to manage the size of the transition relations encountered in compositional model checking. The assumptions are that each component machine is relatively small and easy to manipulate, and that the full product machine is too large to build and manipulate. The general idea is to minimize each component machine, with respect to $\mathcal{E}^\phi$ for that machine, before composing it with other machines. We can incrementally build the product machine by composing machines into clusters, and again applying minimization to each cluster. When just one machine remains, we apply a standard CTL model checker. Figure 7 outlines a procedure for this approach.

```
function compositional_model_checker(φ, M₁, ..., Mₙ) {
    if (n = 1)
        return model_checker(φ, M₁);
    for (i = 1; i ≤ n; i⁺⁺)
        Mᵢ* = minimize(Mᵢ, φ);
    M₁', ..., Mₗ' = form_clusters(M₁*, ..., Mₙ*);
    compositional_model_checker(φ, M₁', ..., Mₗ');
}
```

Fig. 7. Outline of procedure for compositional model checking: minimize and form product incrementally.

The question of how to minimize a component with respect to $\mathcal{E}^\phi$ depends on what sort of data representation is used for the transition relations. If an explicit representation is used (e.g. adjacency lists), then minimization is simply a matter of forming the quotient machines $M_i/\mathcal{E}_i^\phi$. After the model checker is applied to the product of the quotient machines, Theorem 4 can be directly applied to recover the product states in the original state space that satisfy ϕ.

If an implicit representation is used, then minimization becomes more complicated. We focus on the case where BDDs are used. There is no correlation between the size of the BDD for a transition relation, and the number of transitions in the relation. Thus, the idea behind minimization in this case is to use $\mathcal{E}^\phi$ to define a range of transition relations, any of which can be used in place of the original transition relation, and then choose the relation in this range with the smallest BDD. It should be noted however, that smaller component BDDs do not *guarantee* a smaller product BDD—this is only a heuristic.

For a component M, we take the upper bound of the range to be T^{max}, which is the relation formed by adding to T any transition between two states for which there exists a transition between equivalent states (e.g. if $s \xrightarrow{a} s'$ is in T and $\mathcal{E}^\phi(x, s)$ and $\mathcal{E}^\phi(x', s')$, then $x \xrightarrow{a} x'$ is added). The lower bound is T itself. Given these bounds, a heuristic like *restrict* [8] is used to find a small BDD between T and T^{max}. It can be shown that any transition relation between T and T^{max} can be used without altering the result returned by the model checker. Alternatively, instead of looking for a small relation between T and T^{max}, we can just use T^{min}, which is the transition relation of the quotient machine, if it turns out that T^{min} is small.

5.1   Early Pass/Fail Detection

Sometimes the model checking problem is posed as: given a formula ϕ and a subset of product states Q, is Q contained in the set of states satisfying ϕ? For example, Q may be the set of initial states. Since our method returns all states satisfying ϕ, a simple containment check answers the question. However, in some cases, we may be able to answer the question without composing all the machines, yielding a further savings in time. This is known as early pass/fail detection.

Let $Q = \{q^1, q^2, \ldots, q^m\}$, where q^j is the product state $\langle s_1^j, s_2^j, \ldots, s_n^j \rangle$, and let $FAIL_i^\phi$ be the $FAIL^\phi$ states in component i. If $s_i^j \in FAIL_i^\phi$, then any product state $\langle t_1, \ldots, t_{i-1}, s_i^j, t_{i+1}, \ldots, t_n \rangle$ does not satisfy ϕ, so in particular, q^j does not satisfy ϕ. Hence, the answer to the above question is "no". So in summary, if for any i, $FAIL_i^\phi$ intersects the ith state component of the set Q, then the answer is "no".

On the other hand, to reach an early "yes" answer, we need each state in Q to be "covered" by at least one $PASS^\phi$ state. If $s_i^j \in PASS_i^\phi$, then every state in Q with s_i^j as its ith component is guaranteed to satisfy ϕ. So in summary, if for every state in Q, at least one of its component states is a $PASS^\phi$ state, then the answer is "yes".

5.2   Processing Subformulas

As the number of subformulas in ϕ increases, the equivalence $\mathcal{E}^\phi$ becomes finer because equivalence on all subformulas is required. However, if some of the subformulas of ϕ are first replaced by fresh atomic propositions representing the product states satisfying the subformulas, then this may lead to a coarser equivalence. This follows since knowing which product states satisfy a subformula adds information to what was originally known, information that can be used at the component level in computing $\mathcal{E}^\phi$ (for the new ϕ).

This is illustrated by the system in Figure 2, where $\phi = (\exists G(p \wedge q)) \wedge Q$, and Q is the set $\{\langle 1, 1' \rangle, \langle 2, 1' \rangle\}$ of product states. Lines 1 through 6 of Table 1 show the equivalence classes calculated for M_1 on the subformulas of ϕ. The end result (line 6) is that no states are equivalent; hence, we have gained nothing. Instead of processing all of ϕ, we could stop after computing the equivalence for $\exists G(p \wedge q)$. In this case, states 2 and 3 are equivalent (line 4), and thus a smaller machine can be built for M_1. When this quotient machine is composed with M_2 and the model checker is applied, we discover that no product states satisfy $\exists G(p \wedge q)$. At this point, we can create a fresh atomic proposition, Q', to represent this (empty) set of states. Then when we calculate the equivalence on M_1 for $Q' \wedge Q$ (which is the same as the original ϕ), we see that states 1 and 2 are now equivalent (line 8), so we can again construct a smaller machine for M_1.

Thus, we may want to follow a strategy where a nested formula is recursively decomposed into simpler subformulas, and the compositional model checker of Figure 7 is applied to each subformula. Note that whereas Chiodo *et al.* [5] recursively decompose a formula into its *immediate* subformulas, we can decompose

	ϕ	$PASS^\phi$	$FAIL^\phi$	equiv classes
1	q	$\emptyset$	$\emptyset$	$\{1,2,3\}$
2	p	$\{2,3\}$	$\{1\}$	$\{1\},\{2,3\}$
3	$p \wedge q$	$\emptyset$	$\{1\}$	$\{1\},\{2,3\}$
4	$\exists G(p \wedge q)$	$\emptyset$	$\{1\}$	$\{1\},\{2,3\}$
5	Q	$\emptyset$	$\{3\}$	$\{1,2\},\{3\}$
6	$(\exists G(p \wedge q)) \wedge Q$	$\emptyset$	$\{1,3\}$	$\{1\},\{2\},\{3\}$
7	Q'	$\emptyset$	$\{1,2,3\}$	$\{1,2,3\}$
8	$Q' \wedge Q$	$\emptyset$	$\{1,2,3\}$	$\{1,2\},\{3\}$

Table 1. Equivalence classes for M_1 of Figure 2 on $(\exists G(p \wedge q)) \wedge Q$.

a formula into *arbitrary* subformulas, since our equivalence works on nested formulas.

Of course, even though we may be able to compute coarser equivalences with this strategy, the drawback is that a reduced product machine must be reconstructed for each subformula. Experiments are required to determine how to decompose a formula to achieve a balance between these conflicting demands.

6   Future Work and Conclusions

We have presented a formula-dependent equivalence that can be used to manage the size of the transition relations encountered in compositional CTL model checking. We have yet to implement the method, and the ultimate effectiveness of the method can be confirmed only by experimentation. Given an arbitrary CTL formula ϕ, the method works by first computing an equivalence on the states of each component machine, which preserves ϕ. If an explicit representation for transition relations is used, then the quotient machine is constructed for each component, and the quotient machines are used to build a smaller product machine.

If BDDs are used, then the equivalence for each component is used to determine a range of permissible transition relations. More work remains to derive a procedure for efficiently choosing a relation from this range that will ultimately lead to a smaller product machine.

Our approach can be applied incrementally to build the product machine by clustering some minimized machines, forming their product, and repeating the equivalence computation. Research is needed to understand how best to cluster the components to achieve the smallest sub-products. Also, we outlined how our approach can be applied to the subformulas of a formula, to achieve a coarser equivalence. We need to devise a heuristic to intelligently decompose a formula into subformulas to take advantage of this.

An important part of a CTL model checker is the ability to generate counter-examples. Since we are altering the product machine, a counter-example in the altered product may not actually exist in the full product. A method needs to be developed to handle this. Finally, we plan to extend our method to fair-CTL model checking, and we would like to apply similar ideas to the language containment paradigm.

Acknowledgments

We wish to thank the reviewers for their helpful comments. This work was supported by SRC grant 94-DC-008, SRC contract 94-DC-324, and NSF/DARPA grant MIP-8719546. In addition, the second author was supported by an SRC Fellowship.

References

1. A. Aziz and R. K. Brayton. Verifying interacting finite state machines. Technical Report UCB/ERL M93/52, Electronics Research Laboratory, College of Engineering, University of California, Berkeley, July 1993.
2. A. Aziz, T. R. Shiple, V. Singhal, R. K. Brayton, and A. L. Sangiovanni-Vincentelli. Formula-dependent equivalence for compositional CTL model checking. Technical report, Electronics Research Laboratory, College of Engineering, University of California, Berkeley, 1994.
3. A. Bouajjani, J.-C. Fernandez, N. Halbwachs, P. Raymond, and C. Ratel. Minimal state graph generation. *Science of Computer Programming*, 18(3):247–271, 1992.
4. M. C. Browne, E. M. Clarke, and O. Grumberg. Characterizing Kripke structures in temporal logic. Technical Report CS 87-104, Department of Computer Science, Carnegie Mellon University, 1987.
5. M. Chiodo, T. R. Shiple, A. L. Sangiovanni-Vincentelli, and R. K. Brayton. Automatic compositional minimization in CTL model checking. In *Proc. Int'l Conf. on Computer-Aided Design*, pages 172–178, Nov. 1992.
6. E. M. Clarke, E. A. Emerson, and A. P. Sistla. Automatic verification of finite-state concurrent systems using temporal logic specifications. *ACM Trans. on Programming Languages and Systems*, 8(2):244–263, Apr. 1986.
7. E. M. Clarke, D. E. Long, and K. L. McMillan. Compositional model checking. In *4th Annual Symposium on Logic in Computer Science*, Asilomar, CA, June 1989.
8. O. Coudert, C. Berthet, and J. C. Madre. Verification of synchronous sequential machines based on symbolic execution. In J. Sifakis, editor, *Proceedings of the Workshop on Automatic Verification Methods for Finite State Systems*, volume 407 of *Lecture Notes in Computer Science*, pages 365–373. Springer-Verlag, June 1989.
9. D. Dams, O. Grumberg, and R. Gerth. Generation of reduced models for checking fragments of CTL. In C. Courcoubetis, editor, *Proceedings of the Conference on Computer-Aided Verification*, volume 697 of *Lecture Notes in Computer Science*, pages 479–490. Springer-Verlag, June 1993.
10. E. A. Emerson. Temporal and modal logic. In J. van Leeuwen, editor, *Handbook of Theoretical Computer Science*, pages 995–1072. Elsevier Science Publishers B.V., 1990.
11. O. Grumberg and D. E. Long. Model checking and modular verification. In J. C. M. Baeten and J. F. Groote, editors, *CONCUR '91, International Conference on Concurrency Theory*, volume 527 of *Lecture Notes in Computer Science*. Springer-Verlag, Aug. 1991.
12. R. Milner. *Communication and Concurrency*. Prentice Hall, New York, 1989.

An Improved Algorithm for the Evaluation of Fixpoint Expressions*

David E. Long[1], Anca Browne[2], Edmund M. Clarke[2],
Somesh Jha[2], Wilfredo R. Marrero[2]

[1] AT&T Bell Laboratories, Murray Hill, NJ 07974
[2] School of Computer Science, Carnegie Mellon University, Pittsburgh, PA 15213

Abstract. Many automated finite-state verification procedures can be viewed as fixpoint computations over a finite lattice (typically the powerset of the set of system states). Hence, fixpoint calculi such as the propositional μ-calculus have proven useful, both as ways to describe verification algorithms and as specification formalisms in their own right. We consider the problem of evaluating expressions in a fixpoint calculus over a given model. A naive algorithm for this task may require time n^q, where n is the maximum length of a chain in the lattice and q is the depth of fixpoint nesting. In 1986, Emerson and Lei presented a method requiring about n^d steps, where d is the number of alternations between least and greatest fixpoints. More recent algorithms have reduced the exponent by one or two, but the complexity has remained at about n^d. In this paper, we present a new algorithm that makes extensive use of monotonicity considerations to solve the problem in about $n^{d/2}$ steps. Thus, the time required by our method is only about the square root of the time required by the earlier algorithms.

1 Introduction

Many automated finite-state verification algorithms can be viewed as fixpoint computations over a finite lattice. Examples include: model checking procedures for logics such as CTL [6] and PDL [12], methods for computing strong and weak bisimulation equivalence in CCS [16], and language containment and emptiness algorithms for ω-automata [5]. Approaches based on fixpoint logics such as the

* This research was sponsored in part by the Wright Laboratory, Aeronautical Systems Center, Air Force Material Command, USAF, and the Advanced Research Projects Agency (ARPA) under grant number F33615-93-1-1330, and in part by the Semiconductor Research Corportation (SRC) under contract 92-DJ-294, and in part by the National Science Foundation under contract number CCR-9217549. The views and conclusions contained in this document are those of the authors and should not be interpreted as necessarily representing the official policies or endorsements, either expressed or implied, of Wright Laboratory, the U. S. Government, the Semiconductor Research Corporation, or the National Science Foundation. The U. S. Government is authorized to reproduce and distribute reprints for Government purposes notwithstanding any copyright notation thereon.

propositional μ-calculus [13] are tied even more directly to fixpoint computation. With the increasing use of binary decision diagrams (BDDs) [3] for finite-state verification [4, 10], algorithms based on set manipulations and fixpoints have become even more important, since methods that require the manipulation of individual states do not take advantage of the representation. In this paper, we consider the complexity of evaluating fixpoint expressions over finite lattices. Our main result is a new algorithm that makes extensive use of monotonicity considerations to reduce the complexity of evaluation. The number of steps required by our method is roughly the square root of the number of steps required by the best previously known algorithms.

Our ideas are independent of the particular fixpoint calculus used, but for concreteness, we will be using the propositional μ-calculus of Kozen [13]. This logic is designed for expressing properties of transition systems, and formulas in the logic (with no free propositional variables) evaluate to sets of states. There have been many algorithms proposed for evaluating a formula of the logic with respect to a given transition system. These mostly fall into two categories: local and global. Local procedures, like those developed by Cleaveland [7], Stirling and Walker [17], and Winskel [19], are designed for proving that a specific state of the transition system satisfies the given formula. Because of this, it is not always necessary to examine all the states in the transition system. However, the worst-case complexity of these approaches is generally larger than the complexity of the global methods, though recent work by Andersen [1], Larsen [14], and Mader [15] has improved the bounds. Global procedures generally work bottom-up through the formula, evaluating each subformula based on the value of its subformulas. Iteration is used to compute the fixpoints. Because of fixpoint nesting, a naive global algorithm may require about n^q steps to evaluate a formula, where n is the number of states in the transition system and q is the depth of nesting of the fixpoints. Emerson and Lei [11] improved on this by observing that the complexity of evaluating a formula really depends only on the number of alternations d of least and greatest fixpoints. Emerson and Lei gave an algorithm requiring only about n^d steps. Subsequent work by Cleaveland, Klein, Steffen, and Andersen [1, 8, 9] has reduced the overhead, but the overall number of steps has remained at about n^d. Our new algorithm is also a global method. By using extensive monotonicity considerations, we are able to show that only about $n^{d/2}$ steps are required to evaluate a formula with d alternations.

The remainder of this paper is organized as follows. Section 2 summarizes the syntax and semantics of the propositional μ-calculus and reviews Emerson and Lei's work. In Sect. 3 we present our new algorithm and discuss its complexity. We consider some open questions and directions for future research in Sect. 4.

2   The Propositional μ-Calculus

In the propositional μ-calculus, formulas are built up from:

1. atomic propositions p, p_1, p_2, ...;

2. atomic propositional variables $R, R_1, R_2, \ldots$;
3. logical connectives $\cdot \wedge \cdot$ and $\cdot \vee \cdot$;
4. modal operators $\langle a \rangle \cdot$ and $[a] \cdot$, where a is one of a set of program letters a, b, $a_1, a_2, \ldots$; and
5. fixpoint operators $\mu R_i. (\cdots)$ and $\nu R_i. (\cdots)$.

Formulas in this calculus are interpreted relative to a transition system that consists of:

1. a nonempty set of states T;
2. a mapping L that takes each atomic proposition to some subset of T (the states where the proposition is true); and
3. a mapping T that takes each program letter to a binary relation over T (the state changes that can result from executing the program).

The intuitive meaning of the formula $\langle a \rangle \phi$ is "it is possible to execute a and transition to a state where ϕ holds." $[\cdot]$ is the dual of $\langle \cdot \rangle$; for $[a]\phi$, the intended meaning is that "ϕ holds in all states reachable (in one step) by executing a." The μ and ν operators are used to express least and greatest fixpoints, respectively. We could also allow negation (with some restrictions); in this case, greatest fixpoints could be expressed using the duality $\nu R. \phi(R) = \neg \mu R. \neg \phi(\neg R)$. To emphasize this duality, we write the empty set of states as $\perp$.

Formally, a formula ϕ over the free propositional variables $R_1, R_2, \ldots, R_k$ is interpreted as a k-argument predicate transformer. (A predicate transformer is simply a mapping from sets of states to a set of states.) We denote this predicate transformer by ϕ^M. ϕ^M is defined inductively by giving its value for the arguments $S_1, \ldots, S_k$. We write this value as $\phi^M(\bar{S})$.

1. $p^M(\bar{S}) = L(p)$.
2. $R_i^M(\bar{S}) = S_i$.
3. $(\phi \wedge \psi)^M(\bar{S}) = \phi^M(\bar{S}) \cap \psi^M(\bar{S})$. Disjunction is similar.
4. $(\langle a \rangle \phi)^M(\bar{S}) = \{ s \mid \exists t \, [(s,t) \in T(a) \wedge t \in \phi^M(\bar{S})] \}$.
   $([a]\phi)^M(\bar{S}) = \{ s \mid \forall t \, [(s,t) \in T(a) \rightarrow t \in \phi^M(\bar{S})] \}$.
5. $(\mu R. \phi)^M(\bar{S})$ is the least fixpoint of the predicate transformer $\tau : 2^{\mathsf{T}} \rightarrow 2^{\mathsf{T}}$ defined by:

$$\tau(S) = \phi^M(S, S_1, \ldots, S_k) \ ,$$

where the first parameter of ϕ^M is the value for R. The interpretation of $\nu R. \phi$ is similar, except that we take the greatest fixpoint.

Within formulas, there is no negation, and so the fixpoints are guaranteed to be well-defined. Formally, each possible τ is monotonic ($S \subseteq S'$ implies $\tau(S) \subseteq \tau(S')$). This is enough to ensure the existence of the fixpoints [18]. For finite transition systems, the fixpoints can be computed by iterative evaluation. More precisely, for some $i \leq n = |\mathsf{T}|$, the fixpoint is equal to $\tau^i(\perp)$ (for a least fixpoint) or $\tau^i(\mathsf{T})$ (for a greatest fixpoint). In what follows, we will often abuse notation and identify the formula ϕ with its meaning ϕ^M.

Since we will be using the concept of alternation depth, we briefly summarize Emerson and Lei's observations [11]. Consider the expression

$$\mu R_1. (\langle a \rangle R_1) \vee (\mu R_2. R_1 \vee p \vee \langle b \rangle R_2) \ .$$

The subformula $\mu R_2. (\cdots)$ defines a monotonic predicate transformer τ taking one set (the value of R_1) to another (the value of $\mu R_2. (\cdots)$). When evaluating the outer fixpoint, we start with the approximation $\bot$ and then compute $\tau(\bot)$. Now R_1 is increased (say to S_1), and we want to compute the least fixpoint $\tau(S_1)$. Since $\bot \subseteq S_1$, we know that $\tau(\bot) \subseteq \tau(S_1)$. To compute a least fixpoint, it is enough to start iterating with any approximation known to be below the fixpoint. This implies that we can start iterating with $\tau(\bot)$ instead of $\bot$. At the next step, R_1 will be even larger, and so we will start the inner fixpoint computation with $\tau(S_1)$. We never restart the inner fixpoint computation, and so we can have at most n increases in the value of the inner fixpoint variable. Overall, we only need about n steps to evaluate this expression, instead of n^2. Emerson and Lei showed that this type of simplification makes it possible to evaluate a formula ϕ in about n^d steps, where d is the alternation depth of the formula. The alternation depth of a formula is intuitively equal to the number of alternating nestings of least and greatest fixpoints. For the formula above, the alternation depth is 1, so n^1 steps suffice. Note: throughout this paper, when we speak of the number of steps used by an algorithm, we mean the number of fixpoint approximations produced during the evaluation process. Thus, we avoid details of how sets and relations are represented and manipulated.

3 The Algorithm

We first illustrate the essential idea behind our new algorithm on a formula involving three fixpoints (with alternation depth three):

$$\mu R_1. \psi_1(R_1, \nu R_2. \psi_2(R_1, R_2, \mu R_3. \psi_3(R_1, R_2, R_3))) \ . \tag{1}$$

To compute the outer fixpoint, we start with $R_1 = \bot$, $R_2 = \top$ and $R_3 = \bot$. Call these values R_1^0, R_2^{00}, and R_3^{000} respectively. The superscript on R_i gives the iteration indices for the fixpoints involving $R_1, \ldots, R_i$. So R_3^{000} means that all three fixpoints are at their the initial approximations. We then iterate to compute the inner fixpoint; call the value of this fixpoint $R_3^{00\omega}$. (The ω stands for whatever number of steps were needed for the fixpoint iteration to converge.) We now compute the next approximation R_2^{01} for R_2 by evaluating $\psi_2(R_1^0, R_2^{00}, R_3^{00\omega})$, and then we go back to the inner fixpoint. Eventually, we reach the fixpoint for R_2, having computed R_2^{00}, $R_3^{00\omega}$, R_2^{01}, $R_3^{01\omega}$, $\ldots$, $R_2^{0\omega}$, $R_3^{0\omega\omega}$. Now we proceed to $R_1^1 = \psi_1(R_1^0, R_2^{0\omega}, R_3^{0\omega\omega})$. We know that $R_1^0 \subseteq R_1^1$, and we are now going to compute $R_2^{1\omega}$. Note that the values $R_2^{0\omega}$ and $R_2^{1\omega}$ are given by

$$R_2^{0\omega} = \nu R_2. \psi_2(R_1^0, R_2, \mu R_3. \psi_3(R_1^0, R_2, R_3))$$

and

$$R_2^{1\omega} = \nu R_2. \psi_2(R_1^1, R_2, \mu R_3. \psi_3(R_1^1, R_2, R_3)) \ .$$

By monotonicity, we know that $R_2^{1\omega}$ will be a superset of $R_2^{0\omega}$. However, since R_2 is computed by a greatest fixpoint, this information does not help; we still must start computing with $R_2^{10} = \top$. At this point, we begin to compute the inner fixpoint again. But now let us look at $R_3^{00\omega}$ and $R_3^{10\omega}$. We have

$$R_3^{00\omega} = \mu R_3.\, \psi_3(R_1^0, R_2^{00}, R_3)$$

and

$$R_3^{10\omega} = \mu R_3.\, \psi_3(R_1^1, R_2^{10}, R_3) \ .$$

Since $R_1^0 \subseteq R_1^1$ and $R_2^{00} \subseteq R_2^{10}$, monotonicity implies that $R_3^{00\omega} \subseteq R_3^{10\omega}$. Now R_3 is a least fixpoint, so starting the computation of $R_3^{10\omega}$ anywhere below the fixpoint value is acceptable. Thus, we can start the computation for $R_3^{10\omega}$ with $R_3^{00\omega}$ (i.e., we take $R_3^{100} = R_3^{00\omega}$). Since $R_3^{00\omega}$ is in general larger than $\bot$, we obtain faster convergence. Also note that

$$R_2^{01} = \psi_2(R_1^0, R_2^{00}, R_3^{00\omega})$$

and

$$R_2^{11} = \psi_2(R_1^1, R_2^{10}, R_3^{10\omega}) \ .$$

Since $R_1^0 \subseteq R_1^1$, $R_2^{00} \subseteq R_2^{10}$, and $R_3^{00\omega} \subseteq R_3^{10\omega}$, we will have $R_2^{01} \subseteq R_2^{11}$. This means that we can use the same trick when computing $R_3^{11\omega}$. Thus, we will use $R_3^{01\omega}$ for the approximation R_3^{110}. In general, we can start computing $R_3^{1j\omega}$ from $R_3^{1j0} = R_3^{0j\omega}$. Eventually we find another fixpoint for R_2. Then, once we compute R_1^2 (or in general, R_1^{k+1}), we can use the fixpoints $R_3^{1j\omega}$ ($R_3^{kj\omega}$) as the initial approximations R_3^{2j0} ($R_3^{(k+1)j0}$) to $R_3^{2j\omega}$ ($R_3^{(k+1)j\omega}$).

If we use this idea, how many steps does the computation take? The dominating term is the number of steps made when computing the inner fixpoint. With previously known algorithms, this inner computation starts from $\bot$ each time, and hence may involve about n^3 steps (one factor of n for each of the three fixpoints). In our case, if we fix a particular j, then we have

$$R_3^{0j0} \subseteq R_3^{0j\omega} = R_3^{1j0} \subseteq R_3^{1j\omega} = R_3^{2j0} \subseteq \cdots = R_3^{\omega j0} \subseteq R_3^{\omega j\omega} \ .$$

This implies that for each j, we can have at most n strict inclusions among the values of R_3^{ijm} that we compute, and so for each j we take only about n steps. Since there can be up to n different j values, we take only about n^2 steps while computing the inner fixpoint, thus saving a factor of n. (Again, we are using "number of steps" to mean the number of fixpoint approximations produced.)

The relationship between the different approximations to R_3 is shown in Fig. 1. The computation of least fixpoints proceeds from bottom to top, and the computation of greatest fixpoints proceeds from left to right. The chain mentioned above corresponds to one of the vertical columns in this figure. When computing with approximation R_1^j, we save the "frontier" values $R_3^{j0\omega}, \ldots, R_3^{j\omega\omega}$ and use them as the initial approximations $R_3^{(j+1)00}, \ldots, R_3^{(j+1)\omega0}$ when computing with R_1^{j+1}. We have at most n strict inclusions within each vertical chain in the figure.

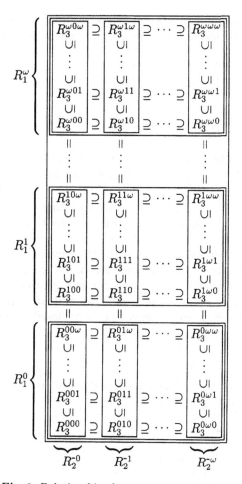

Fig. 1. Relationships between approximations for R_3

Note that we can build this type of table for arbitrarily nested fixpoints. Suppose, for example, that we were also computing an outer greatest fixpoint for a relation R_0. Figure 1 would correspond to a series of computations with R_0 at $\top$. If we then compute the next approximation for R_0, it will be smaller than the initial approximation. Then by monotonicity, when we go through the computations for R_1, R_2, and R_3 again, we will get at each stage something smaller than during the first set of computations. For R_2, this means that we can use the frontier fixpoint values produced during the first set of computations as initial approximations when doing the second set of computations. The effect is to build a second table like the one in the figure to the right of the previous table.

To argue in more detail that $n^{d/2}$ steps suffices to evaluate a formula with alternation depth d, we now present a special-case algorithm. This algorithm handles strictly alternating fixpoints and only saves frontier values for least fix-points. Assume that the formula that we wish to evaluate has the form:

$$F_1 \equiv \mu R_1.\, \psi_1(\nu R_1'.\, \psi_1'(F_2))$$
$$F_2 \equiv \mu R_2.\, \psi_2(\nu R_2'.\, \psi_1'(F_3))$$
$$\vdots$$
$$F_q \equiv \mu R_q.\, \psi_q(\nu R_q'.\, \psi_q'(R_1, R_1', \ldots, R_q, R_q')) \ ,$$

where $\equiv$ denotes syntactic equality. This formula has alternation depth $d = 2q$.

The special-case algorithm is given in Fig. 2. The algorithm uses an array A_i to store the frontier values for the fixpoint variable R_i. The array A_i is indexed by iteration indices for all the greatest fixpoints enclosing R_i. There are $i-1$ of such enclosing fixpoints. Each iteration index is between 0 and n (inclusive), and so A_i has $(n+1)^{i-1}$ entries. Initially, all array values are $\bot$. When evaluating R_i, we start with the array value indicated by the current iteration indices for the enclosing greatest fixpoints, and iterate until convergence. At the end of the iteration, the array holds the fixpoint value. For each greatest fixpoint variable R_i', we have an associated iteration index j_i. When evaluating R_i', we start with $\top$ and iterate $n+1$ times (even if convergence is achieved earlier). We update j_i after each iteration.

```
function eval(φ)
Handle atomic propositions, logical operations, etc.
if φ = μRi.ψi(···) then
      Ri := Ai[j1, . . . , ji−1]
      repeat
            Oi := Ri
            Ri := eval(ψi)
            Ai[j1, . . . , ji−1] := Ri
      until Ri = Oi
      return Ri
else if φ = νRi'.ψi'(···)
      Ri' := ⊤
      for ji from 0 to n
            Ri' := eval(ψi')
      endfor
      return Ri'
endif
```

Fig. 2. Pseudo-code for the special-case algorithm

Note that this algorithm implements the ideas described previously. For the three fixpoint example that we used earlier, the array for R_3 would have $n+1$ entries because R_3 is within one enclosing greatest fixpoint. Initially these are all $\bot$, corresponding to the values R_3^{0j0} (for $0 \leq j \leq n$). During the computation of the fixpoint for R_2, the entries are updated to hold the values $R_3^{0j\omega}$. When we compute the approximation R_1^1 and begin computing the inner fixpoints again, the entries are used as the initial approximations R_3^{1j0}.

In proving this algorithm correct, we would show that any given array entry increases monotonically. While space limitations prevent us from giving the proof, we can use this fact to derive a complexity bound. Let T_i denote the number of approximations computed for R_i, and let T_i' denote the number of approximations for R_i'. Clearly, $T_1 \leq n+1$. From the algorithm, we see that the fixpoint for R_i' is evaluated T_i times, and for each evaluation, we produce $n+1$ approximations. Thus, $T_i' \leq (n+1)T_i$. For R_i, each entry in A_i increases monotonically, so for any one entry, we can make at most n steps in which the value strictly increases. There are $(n+1)^{i-1}$ entries in A_i, so this gives at most $(n+1)^i$ steps. We evaluate the fixpoint for R_i at most T_{i-1}' times. Thus, we make at most T_{i-1}' extra steps to detect convergence. In total, we have $T_i \leq (n+1)^i + T_{i-1}'$. Expanding out the values, we get

$$T_1 \leq n+1$$
$$T_1' \leq (n+1)T_1 = (n+1)^2$$
$$T_2 \leq (n+1)^2 + T_1' = 2(n+1)^2$$
$$T_2' \leq (n+1)T_2 = 2(n+1)^3$$
$$\vdots$$
$$T_q \leq q(n+1)^q$$
$$T_q' \leq q(n+1)^{q+1}.$$

Summing over all fixpoints and expressing the result in terms of the alternation depth $d = 2q$, we get $O(d^2(n+1)^{d/2+1})$ steps. In contrast, previously known algorithms may require about n^d steps to evaluate this formula. Generalizing to formulas with odd alternation depth yields the bound $O(d^2(n+1)^{\lfloor d/2 \rfloor +1})$.

In the general algorithm, we handle arbitrary formulas, save information for both types of fixpoints and always stop computations on detecting convergence. This version of the algorithm does not use tables to store the frontier values, since just initializing the tables requires about $n^{d/2}$ steps. If all of the fixpoint computations converged immediately, this would represent mostly wasted effort. Instead, frontiers will be represented by queues. We will write queues using square brackets, with the head of the queue at the left. The last element in the queue corresponds to a fixpoint and is conceptually replicated as many times as required. As an example, consider (1) again. During the computations with R_1^0, we will be building up the frontier values $R_3^{00\omega}$, $R_3^{01\omega}$, etc. Within n steps, we will find the fixpoint value $R_2^{0\omega}$. Say this happens after three steps. Then the frontier for R_3 will be represented by the queue

$$[R_3^{00\omega}, R_3^{01\omega}, R_3^{02\omega}, R_3^{03\omega}].$$

If we were to continue iterating on R_2, the value of R_2 would not change, and so $R_3^{0j\omega}$ would be equal to $R_3^{03\omega}$ for all $j > 3$. Rather than actually computing these values, we just view $R_3^{03\omega}$ as being duplicated as many times as needed. When we start computing with R_1^1, we will pull elements from the above queue to start each computation of R_3. If R_2 now takes more than three steps to converge, we will use $R_3^{03\omega}$ more than once. Each time we pull it from the queue and find that the queue is empty, we put another copy back in the queue in case another iteration on R_2 is required.

More deeply nested fixpoints give rise to nested queues. Consider a formula with five fixpoints: $\mu R_1 . \nu R_2 . \mu R_3 . \nu R_4 . \mu R_5 . (\cdots)$. For R_5, a frontier will be represented by a queue, each element of which is a queue of state sets. The elements of the outer queue are subfrontiers corresponding to the values R_2^{i0}, R_2^{i1}, R_2^{i2}, etc. The subfrontier corresponding to R_2^{ij} holds fixpoints corresponding to R_4^{ijk0}, R_4^{ijk1}, R_4^{ijk2}, etc. In general, a frontier for a least fixpoint nested inside q greatest fixpoints is represented by queues nested to depth q.

As the computation proceeds, existing frontiers will be decomposed to get at the starting fixpoint approximations, and we will be constructing new frontiers from the fixpoint values. In order to keep track of this decomposition and construction process, we use two stacks for each fixpoint variable R. One stack, I_R, will be associated with the frontier being decomposed, and the other stack, F_R, will hold the frontier being constructed. Each stack element is either a set of states (representing an approximation or a fixpoint value), or a queue (representing a subfrontier). We will write stacks using angle brackets, with the top of a stack on the left. Initially, the I_R stack for a top-level fixpoint variable R holds either $\top$ or $\bot$, depending on whether R is a greatest or least fixpoint. The stack I_R for a least fixpoint variable R nested inside q greatest fixpoints holds $\bot$ nested inside q queues. The initial value for a stack corresponding to a greatest fixpoint variable nested inside a number of least fixpoints is similarly defined.

Pseudo-code for the algorithm is shown in Fig. 3. During each iteration for the fixpoint $\mu R . \psi(\cdots)$, we pull out the next sub-frontier for the inner ν variables (lines 8–12). These subfrontiers are pushed onto the initial stacks corresponding to the ν variables. For a top-level ν-subformula $\nu R' . (\cdots)$ of ψ, these subfrontiers will be state sets representing the initial approximation to use for R'. We then recursively evaluate the inner fixpoints. Afterwards, we build up sub-frontiers for subsequent evaluations of the inner greatest fixpoints (lines 15–19). If the computation of R has not yet converged, we discard the old frontiers for the inner μ fixpoints and replace them with the new frontiers that have been built up (lines 21–24). Note that with two successive μ fixpoints, this simply results in picking up the inner fixpoint from the previous stopping point. Hence the algorithm also makes use of Emerson and Lei's observation [11].

As an example of the algorithm's operation, we consider (1) again. Initially, $I_{R_1} = \langle \bot \rangle$, $I_{R_2} = \langle [\top] \rangle$, and $I_{R_3} = \langle [\bot] \rangle$. All the stacks F_{R_1}, F_{R_2}, and F_{R_3} are empty. The computation proceeds as shown in Fig. 4. In the figure, $\mu R_1 : \bot$ denotes a call to the evaluation routine for the formula $\mu R_1 . (\cdots)$ with $\bot$ on the top of the stack I_{R_1} (i.e., the evaluation of the fixpoint starting from $\bot$). The notations "start R_1" and "end R_1" denote the start of an iteration for computing R_1 and the end of an iteration, respectively. The notation "return $R_2^{0\omega}$" indicates returning a fixpoint value for R_2. Finally, $\mu R_3 : \bot \rightarrow R_3^{00\omega}$ denotes the evaluation

```
1   function eval(φ)

2   Handle base cases, logical operations, etc.
3   if φ = μR. ψ(R) then
4       set R to the top element of I_R
5       for each inner ν variable R'
6           push [] on F_{R'}
7       repeat
8           for each inner ν variable R'
9               let Q be the queue on top of I_{R'}
10              dequeue e from Q
11              if Q is now empty, enqueue e again
12              push e on I_{R'}
13          R_old := R
14          R := eval(ψ)
15          for each inner ν variable R'
16              pop e from F_{R'}
17              let Q be the queue on top of F_{R'}
18              enqueue e in Q
19              pop I_{R'}
20          if R ≠ R_old then
21              for each inner μ variable R'
22                  pop I_{R'}
23                  pop e from F_{R'}
24                  push e on I_{R'}
25      until R = R_old
26      push R on F_R
27      return R
28  if φ = νR. ψ(R) then
29      Analogous code to the above
```

Fig. 3. Pseudo-code for the general algorithm

of $\mu R_3.\,(\cdots)$ starting with $\bot$ and yielding $R_3^{00\omega}$ as the result. The figure shows how the stacks evolve during the computation.

Unfortunately, the general algorithm still has a worst case complexity of about $n^{d/2}$. We have constructed transition systems and classes of formulas where the algorithm takes at least $(n+1)^{d/2+1}$ steps to evaluate a formula in the class with alternation depth d (even). We do not have space to present the construction, but we will try to give an intuitive idea of the kind of behavior that leads to the high complexity. The formulas involve $d/2$ pairs (R_i, R_i') of fixpoint variables. R_i is a least fixpoint variable, and R_i' is a greatest fixpoint variable. Let the states of the system be $\{s_0, \ldots, s_{n-1}\}$. Suppose that each of the pairs $(R_1, R_1'), \ldots, (R_i, R_i')$ has one of the following values:

$$(\bot, \top), (\{s_0\}, \top - \{s_0\}), \ldots, (\top - \{s_{n-1}\}, \{s_{n-1}\}), (\top, \bot).$$

Call such a situation a "diagonal configuration." Obviously, the number of diagonal configurations for these variables is $(n+1)^i$. Any two diagonal configurations

	I_{R_1}	F_{R_1}	I_{R_2}	F_{R_2}	I_{R_3}	F_{R_3}
start	$\langle\bot\rangle$	$\langle\rangle$	$\langle[\top]\rangle$	$\langle\rangle$	$\langle[\bot]\rangle$	$\langle\rangle$
$\mu R_1\colon\bot$	$\langle\bot\rangle$	$\langle\rangle$	$\langle[\top]\rangle$	$\langle[\,]\rangle$	$\langle[\bot]\rangle$	$\langle\rangle$
start R_1	$\langle\bot\rangle$	$\langle\rangle$	$\langle\top,[\top]\rangle$	$\langle[\,]\rangle$	$\langle[\bot]\rangle$	$\langle\rangle$
$\nu R_2\colon\top$	$\langle\bot\rangle$	$\langle\rangle$	$\langle\top,[\top]\rangle$	$\langle[\,]\rangle$	$\langle[\bot]\rangle$	$\langle[\,]\rangle$
start R_2	$\langle\bot\rangle$	$\langle\rangle$	$\langle\top,[\top]\rangle$	$\langle[\,]\rangle$	$\langle\bot,[\bot]\rangle$	$\langle[\,]\rangle$
$\mu R_3\colon\bot\to R_3^{00\omega}$	$\langle\bot\rangle$	$\langle\rangle$	$\langle\top,[\top]\rangle$	$\langle[\,]\rangle$	$\langle\bot,[\bot]\rangle$	$\langle R_3^{00\omega},[\,]\rangle$
end R_2	$\langle\bot\rangle$	$\langle\rangle$	$\langle\top,[\top]\rangle$	$\langle[\,]\rangle$	$\langle[\bot]\rangle$	$\langle[R_3^{00\omega}]\rangle$
start R_2	$\langle\bot\rangle$	$\langle\rangle$	$\langle\top,[\top]\rangle$	$\langle[\,]\rangle$	$\langle\bot,[\bot]\rangle$	$\langle[R_3^{00\omega}]\rangle$
$\mu R_3\colon\bot\to R_3^{01\omega}$	$\langle\bot\rangle$	$\langle\rangle$	$\langle\top,[\top]\rangle$	$\langle[\,]\rangle$	$\langle\bot,[\bot]\rangle$	$\langle R_3^{01\omega},[R_3^{00\omega}]\rangle$
end R_2	$\langle\bot\rangle$	$\langle\rangle$	$\langle\top,[\top]\rangle$	$\langle[\,]\rangle$	$\langle[\bot]\rangle$	$\langle[R_3^{00\omega},R_3^{01\omega}]\rangle$
$\vdots$	$\vdots$	$\vdots$	$\vdots$	$\vdots$	$\vdots$	$\vdots$
$\mu R_3\colon\bot\to R_3^{0\omega\omega}$	$\langle\bot\rangle$	$\langle\rangle$	$\langle\top,[\top]\rangle$	$\langle[\,]\rangle$	$\langle\bot,[\bot]\rangle$	$\langle R_3^{0\omega\omega},[R_3^{00\omega},\ldots]\rangle$
end R_2	$\langle\bot\rangle$	$\langle\rangle$	$\langle\top,[\top]\rangle$	$\langle[\,]\rangle$	$\langle[\bot]\rangle$	$\langle[R_3^{00\omega},\ldots,R_3^{0\omega\omega}]\rangle$
return $R_2^{0\omega}$	$\langle\bot\rangle$	$\langle\rangle$	$\langle\top,[\top]\rangle$	$\langle R_2^{0\omega},[\,]\rangle$	$\langle[\bot]\rangle$	$\langle[R_3^{00\omega},\ldots,R_3^{0\omega\omega}]\rangle$
end R_1	$\langle\bot\rangle$	$\langle\rangle$	$\langle[\top]\rangle$	$\langle[R_2^{0\omega}]\rangle$	$\langle[R_3^{00\omega},\ldots,R_3^{0\omega\omega}]\rangle$	$\langle\rangle$
start R_1	$\langle\bot\rangle$	$\langle\rangle$	$\langle\top,[\top]\rangle$	$\langle[R_2^{0\omega}]\rangle$	$\langle[R_3^{00\omega},\ldots,R_3^{0\omega\omega}]\rangle$	$\langle\rangle$
$\nu R_2\colon\top$	$\langle\bot\rangle$	$\langle\rangle$	$\langle\top,[\top]\rangle$	$\langle[R_2^{0\omega}]\rangle$	$\langle[R_3^{00\omega},\ldots,R_3^{0\omega\omega}]\rangle$	$\langle[\,]\rangle$
start R_2	$\langle\bot\rangle$	$\langle\rangle$	$\langle\top,[\top]\rangle$	$\langle[R_2^{0\omega}]\rangle$	$\langle R_3^{00\omega},[R_3^{01\omega},\ldots,R_3^{0\omega\omega}]\rangle$	$\langle[\,]\rangle$
$\mu R_3\colon R_3^{00\omega}\to R_3^{10\omega}$	$\langle\bot\rangle$	$\langle\rangle$	$\langle\top,[\top]\rangle$	$\langle[R_2^{0\omega}]\rangle$	$\langle R_3^{00\omega},[R_3^{01\omega},\ldots,R_3^{0\omega\omega}]\rangle$	$\langle R_3^{10\omega},[\,]\rangle$
end R_2	$\langle\bot\rangle$	$\langle\rangle$	$\langle\top,[\top]\rangle$	$\langle[R_2^{0\omega}]\rangle$	$\langle[R_3^{01\omega},\ldots,R_3^{0\omega\omega}]\rangle$	$\langle[R_3^{10\omega}]\rangle$
start R_2	$\langle\bot\rangle$	$\langle\rangle$	$\langle\top,[\top]\rangle$	$\langle[R_2^{0\omega}]\rangle$	$\langle R_3^{01\omega},[R_3^{02\omega},\ldots,R_3^{0\omega\omega}]\rangle$	$\langle[R_3^{10\omega}]\rangle$
$\mu R_3\colon R_3^{01\omega}\to R_3^{11\omega}$	$\langle\bot\rangle$	$\langle\rangle$	$\langle\top,[\top]\rangle$	$\langle[R_2^{0\omega}]\rangle$	$\langle R_3^{01\omega},[R_3^{02\omega},\ldots,R_3^{0\omega\omega}]\rangle$	$\langle R_3^{11\omega},[R_3^{10\omega}]\rangle$
$\vdots$	$\vdots$	$\vdots$	$\vdots$	$\vdots$	$\vdots$	$\vdots$
end R_1	$\langle\bot\rangle$	$\langle\rangle$	$\langle[\top]\rangle$	$\langle[R_2^{0\omega},\ldots,R_2^{\omega\omega}]\rangle$	$\cdots$	$\langle[R_3^{\omega0\omega},\ldots,R_3^{\omega\omega\omega}]\rangle$
return R_1^{ω}	$\langle\bot\rangle$	$\langle R_1^{\omega}\rangle$	$\langle[\top]\rangle$	$\langle[R_2^{0\omega},\ldots,R_2^{\omega\omega}]\rangle$	$\cdots$	$\langle[R_3^{\omega0\omega},\ldots,R_3^{\omega\omega\omega}]\rangle$

Fig. 4. Example computation of the algorithm in Fig. 3

are incomparable with respect to pairwise set containment. Hence, computing the inner fixpoints for one diagonal configuration gives us no information about the inner fixpoints for a different diagonal configuration. We can also arrange for all inner fixpoints to be $\top$ above the diagonal (i.e., when $R_i \cup R'_i = \top$ and they have a nonempty intersection) and to be $\bot$ below the diagonal. Under these circumstances, we can show that all diagonal configurations for $(R_1, R'_1), \ldots, (R_{d/2}, R'_{d/2})$ will occur during the computation. This means that the number of steps must be at least $(n+1)^{d/2}$.

Our method uses more space than previous approaches, since frontier values must be stored. In the worst case, we may have to store about $n^{d/2}$ state sets. There does not seem to be any way to avoid this, since we cannot rearrange the order in which the fixpoint approximations are computed. Note though that the space complexity is generally much better than the time complexity (since we only store "slices" of the approximations that have been computed). If needed, we can trade time for space during long computations by simplifying or discarding some of the frontiers.

4 Conclusion

We have presented a new algorithm for evaluating a formula in the propositional μ-calculus with respect to a finite transition system. Our algorithm takes about $n^{d/2}$ steps, where d is the alternation depth of the formula. The best previously known algorithms required about n^d steps. A straightforward implementation of our algorithm would require an extra factor of n or so for bookkeeping and set manipulations, but we believe that methods such as those used by Cleaveland, Klein, Steffen, and Andersen [1, 8, 9] could be used to reduce this extra complexity. It is not as clear whether efficient local procedures can be developed that make use of our ideas, but this is an interesting question.

Another line of research involves trying to place lower bounds on the complexity of the evaluation process. It can be shown that the language recognition version of the problem is in NP intersect co-NP. This suggests that it would be very difficult to prove that there is no polynomial time algorithm for the problem. However, it might be possible to prove something about a restricted class of algorithms. A natural class to consider is "oblivious" algorithms. These are methods that only make use of the structure of the nesting of fixpoints, and perhaps the fixpoint values. Given a formula like $\mu R_1. \psi_1(R_1, \nu R_2. \psi_2(R_1, R_2))$, we would view ψ_1 and ψ_2 as being given by oracles. The complexity of an algorithm would be measured in the number of calls to the oracles. This is a natural class of methods. For example, both Emerson and Lei's original algorithm and our new one can be viewed as members of this class. A proof that no algorithm of this class can make do with just a polynomial number of oracle queries would imply that any polynomial time algorithm would have to do something clever based on the structure of the formula. Another way of exploring the complexity of the problem is to look for links with classical complexity theory. Jha has obtained some results connecting fixpoint alternation and alternating Turing machines (ATMs). For example, a fixpoint formula with k alternations can be used to simulate an ATM with k alternations.

References

1. H. R. Andersen. Model checking and boolean graphs. In B. Krieg-Bruckner, editor, *Proceedings of the Fourth European Symposium on Programming*, volume 582 of *Lecture Notes in Computer Science*. Springer-Verlag, February 1992.

2. G. V. Bochmann and D. K. Probst, editors. *Proceedings of the Fourth Workshop on Computer-Aided Verification*, volume 663 of *Lecture Notes in Computer Science*. Springer-Verlag, July 1992.

3. R. E. Bryant. Graph-based algorithms for boolean function manipulation. *IEEE Transactions on Computers*, C-35(8), 1986.

4. J. R. Burch, E. M. Clarke, K. L. McMillan, D. L. Dill, and J. Hwang. Symbolic model checking: 10^{20} states and beyond. In *Proceedings of the Fifth Annual Symposium on Logic in Computer Science*. IEEE Computer Society Press, June 1990.

5. E. M. Clarke, I. A. Draghicescu, and R. P. Kurshan. A unified approach for showing language containment and equivalence between various types of ω-automata. In A. Arnold and N. D. Jones, editors, *Proceedings of the 15th Colloquium on Trees in Algebra and Programming*, volume 407 of *Lecture Notes in Computer Science*. Springer-Verlag, May 1990.

6. E. M. Clarke and E. A. Emerson. Synthesis of synchronization skeletons for branching time temporal logic. In *Logic of Programs: Workshop, Yorktown Heights, NY, May 1981*, volume 131 of *Lecture Notes in Computer Science*. Springer-Verlag, 1981.

7. R. Cleaveland. Tableau-based model checking in the propositional mu-calculus. *Acta Informatica*, 27(8):725–747, 1990.

8. R. Cleaveland, M. Klein, and B. Steffen. Faster model checking for the modal mu-calculus. In Bochmann and Probst [2].

9. R. Cleaveland and B. Steffen. A linear-time model-checking algorithm for the alternation-free modal mu-calculus. *Formal Methods in System Design*, 2(2):121–147, April 1993.

10. O. Coudert, C. Berthet, and J. C. Madre. Verification of synchronous sequential machines based on symbolic execution. In J. Sifakis, editor, *Proceedings of the 1989 International Workshop on Automatic Verification Methods for Finite State Systems, Grenoble, France*, volume 407 of *Lecture Notes in Computer Science*. Springer-Verlag, June 1989.

11. E. A. Emerson and C.-L. Lei. Efficient model checking in fragments of the propositional mu-calculus. In *Proceedings of the First Annual Symposium on Logic in Computer Science*. IEEE Computer Society Press, June 1986.

12. M. J. Fischer and R. E. Ladner. Propositional dynamic logic of regular programs. *Journal of Computer and System Sciences*, 18:194–211, 1979.

13. D. Kozen. Results on the propositional mu-calculus. *Theoretical Computer Science*, 27:333–354, December 1983.

14. K. G. Larsen. Efficient local correctness checking. In Bochmann and Probst [2].

15. A. Mader. Tableau recycling. In Bochmann and Probst [2].

16. R. Milner. *A Calculus of Communicating Systems*, volume 92 of *Lecture Notes in Computer Science*. Springer-Verlag, 1980.

17. C. Stirling and D. J. Walker. Local model checking in the modal mu-calculus. *Theoretical Computer Science*, 89(1):161–177, October 1991.

18. A. Tarski. A lattice-theoretic fixpoint theorem and its applications. *Pacific Journal of Mathematics*, 5:285–309, 1955.

19. G. Winskel. Model checking in the modal ν-calculus. In *Proceedings of the Sixteenth International Colloquium on Automata, Languages, and Programming*, 1989.

Incremental Model Checking in the Modal Mu-Calculus*

Oleg V. Sokolsky   Scott A. Smolka

Department of Computer Science
SUNY at Stony Brook
Stony Brook, NY 11794-4400
{oleg,sas}@sbcs.sunysb.edu

Abstract. We present an incremental algorithm for model checking in the alternation-free fragment of the modal mu-calculus, the first incremental algorithm for model checking of which we are aware. The basis for our algorithm, which we call *MCI* (for Model Checking Incrementally), is a linear-time algorithm due to Cleaveland and Steffen that performs global (non-incremental) computation of fixed points. *MCI* takes as input a set Δ of *changes* to the labeled transition system under investigation, where a change constitutes an inserted or deleted transition; with virtually no additional cost, inserted and deleted states can also be accommodated. Like the Cleaveland-Steffen algorithm, *MCI* requires time linear in the size of the LTS in the worst case, but only time linear in Δ in the best case. We give several examples to illustrate *MCI* in action, and discuss its implementation in the Concurrency Factory, an interactive design environment for concurrent systems.

1 Introduction

The Concurrency Factory [CGL+94] is a joint project between the State University of New York at Stony Brook and North Carolina State University to develop an integrated toolset for the specification, verification, and implementation of concurrent and distributed systems. Like the Concurrency Workbench [CPS93], the Factory employs bisimulation, preorder, and model checking as its main avenues of analysis.

A major underlying goal of the project is that the Factory be suitable for industrial application. One manner in which we are striving to achieve such applicability is through the use of *incremental computation*, which is basically an attempt to avoid repeating lengthy analyses of a system specification after the specification has undergone some relatively minor change.

This current paper is concerned with the incrementalization of the model checking routine of the Concurrency Factory, or, more generally, *incremental model checking in the modal mu-calculus*. The modal mu-calculus [Koz83] is a highly expressive logic that can be used to specify safety and liveness properties of concurrent systems represented as labeled transition systems (LTSs). Our focus here is on the *alternation-free* fragment of the modal mu-calculus [EL86] which, intuitively, means that the "level" of mutually recursive greatest and least fixed-point operators is one.

Our main result is an incremental algorithm for model checking in the alternation-free modal mu-calculus, which we call *MCI* (for Model Checking Incrementally). To our knowledge, *MCI* is the first incremental algorithm for model checking, of any logic,

---

* Research supported in part by NSF Grants CCR-9120995 and CCR-9208585, and AFOSR Grant F49620-93-1-0250DEF.

to be proposed in the literature. The basis for *MCI* is a linear-time algorithm due to Cleaveland and Steffen (henceforth referred to as the *CS* algorithm) which performs global (non-incremental) computation of fixed points.

MCI takes as input a set Δ of *changes* to the LTS under investigation. An element of Δ corresponds to an inserted or deleted transition, although with virtually no additional cost, inserted and deleted states can also be accommodated. Its output is a *variable assignment*, representing the desired fixed-point solution.

The main technique utilized by *MCI* is to first compute the immediate effects of Δ on the results of the previous computation and then restart the fixed-point iteration. As part of the correctness proof of *MCI*, we show that it is safe to restart the iterations only after making certain adjustments to the current variable assignment — raising it sufficiently high in the lattice of all variable assignments when computing greatest fixed points, and, dually, lowering it sufficiently when computing least fixed points.

The required adjustments to the variable assignment are realized by making certain *assumptions* about the connectivity of nodes in the *product graph*, a data structure capturing all dependencies between pairs of the form (s, X_i), for LTS state s and logical variable X_i. We show that it is the presence of strongly connected components in the product graph that leads to the existence of distinct greatest and least fixed point solutions. *MCI* later checks that the assumptions it made were correct, and undoes the effects of any that turned out to be invalid.

In terms of its computational complexity, *MCI*'s worst-case behavior is asymptotically the same as that of *CS*. This is to be expected for it is easy to construct an example in which the value of every variable changes as the result of adding a transition to the LTS. Thus, every node of the product graph must be visited during the incremental run. In fact we prove, via a reduction from SS-REACHABILITY (see [Ram93]), that model checking is an *unbounded* problem, meaning that the running time of an incremental update cannot, in general, be expressed solely in terms of Δ.

In the best case, however, *MCI* requires time linear only in the size of Δ, which is typically constant with respect to the size of the LTS. We show that *MCI* exhibits this kind of performance on an incremental computation involving Milner's scheduler [Mil80], an oft-used benchmark for verification tools.

The closest related work we are aware of is that of Ryder et al. [RMP88] which treats incremental solutions to graph problems in a very general setting. However, they only give sufficient conditions to ensure it is safe to restart iterations of the original algorithm after an incremental update. In practice, these conditions are very restrictive and, in general, an additional computation is needed before iterations can be safely restarted. An informal discussion of the role of cycles in fixed-point incremental computation on graphs is presented in [PS89].

The structure of the rest of the paper is as follows. Section 2 defines the syntax and semantics of the modal mu-calculus and the corresponding model checking problem. Section 3 contains our description of the *CS* algorithm, while Section 4 presents our *MCI* algorithm. Section 5 proves the correctness of *MCI* and analyzes its complexity. Section 6 discusses our implementation of *MCI* in the Concurrency Factory and illustrates the algorithm in action through examples. Finally, Section 7 concludes and outlines directions for future work.

2 Syntax and Semantics of the Modal Mu-Calculus

A *Labeled Transition System* (LTS) is a 4-tuple $\langle S, Act, \rightarrow, s_0 \rangle$ where S is the set of *states*, Act is the set of *actions*, $\rightarrow \subseteq S \times Act \times S$ is the *transition relation*, and s_0 is the *start state*.

We next give the syntax and semantics of a version of the alternation-free modal mu-calculus defined in [CS93], which we refer to as *CS-logic*. Formulas in CS-logic are of two types: basic formulas and equational blocks. The syntax of *basic formulas* is given by the following grammar:

$$\Phi ::= A \mid X \mid \Phi \vee \Phi \mid \Phi \wedge \Phi \mid [a]\Phi \mid \langle a \rangle \Phi$$

where $A \in \mathcal{AP}$, a fixed set of atomic propositions, and $X \in Var$, a countably infinite set of variables.

Basic formulas are interpreted with respect to an LTS $\mathcal{L} = \langle \mathcal{S}, Act, \rightarrow, s_0 \rangle$, a *valuation mapping* $\mathcal{V} : \mathcal{AP} \rightarrow \mathcal{P}(\mathcal{S})$, relating every atomic proposition A to the set of states in which A holds, and an *environment* $e : Var \rightarrow \mathcal{P}(\mathcal{S})$, mapping each variable X to the set of states that satisfy X. For a fixed environment e, the meaning of basic formulas is given by the semantical function $[\![\cdot]\!] e : \Phi \rightarrow \mathcal{P}(\mathcal{S})$, defined in Figure 1.

An *equational block* B is formed by applying operator *min* or *max* to a set E of mutually recursive equations of the form

$$X_1 = \Phi_1$$
$$\vdots$$
$$X_n = \Phi_n,$$

where each Φ_i is a basic formula and the X_i are pairwise distinct. Operators *min* and *max* are understood respectively as the least and greatest fixed points of E. Following [CS93], we assume that the Φ_i are *simple*, i.e., an atomic proposition, or constructed by the application of exactly one operator to variables. Every formula can be made simple with at most a linear blow-up in size.

Semantically blocks are understood as functions from environments to environments. Let a block B contain a set of equations E with variables $X_1, \ldots, X_n$ defined as left-hand sides. Let $\overline{S} = \langle S_1, \ldots, S_n \rangle \in (2^\mathcal{S})^n$ and let $e_{\overline{S}} = e[X_1 \mapsto S_1, \ldots, X_n \mapsto S_n]$. Then the function

$$f_E^e(\overline{S}) = \langle [\![\Phi_1]\!] e_{\overline{S}}, \ldots, [\![\Phi_n]\!] e_{\overline{S}} \rangle,$$

defined on the lattice of tuples of sets of states ordered by point-wise set inclusion is monotonic. By the Tarski-Knaster fixed-point theorem, f_E^e has both least and greatest fixed points given by:

$$\nu f_E^e = \bigcup \{ \overline{S} \mid \overline{S} \subseteq f_E^e(\overline{S}) \}$$
$$\mu f_E^e = \bigcap \{ \overline{S} \mid f_E^e(\overline{S}) \subseteq \overline{S} \}$$

Blocks can now be interpreted in the following fashion:

$$[\![max E]\!] e = e_{\nu f_E^e}$$
$$[\![min E]\!] e = e_{\mu f_E^e}.$$

Finally, a *formula* $\mathcal{B} = \{ B_1, \ldots, B_m \}$ is a set of blocks, with the following syntactic restrictions: all variables appearing on the left-hand sides in the set of blocks are distinct, and the formula's block graph is acyclic. The *block graph* of $\mathcal{B}$ is the directed graph with nodes $B_1, \ldots, B_m$ and edges $\langle B_i, B_j \rangle$ whenever a variable appearing as a left-hand side of an equation in B_i is used in B_j (we say that B_j *depends* on B_i in this case).

$$[\![A]\!]e = \mathcal{V}(A)$$
$$[\![X]\!]e = e(X)$$
$$[\![\Phi_1 \wedge \Phi_2]\!]e = [\![\Phi_1]\!]e \cap [\![\Phi_2]\!]e$$
$$[\![\Phi_1 \vee \Phi_2]\!]e = [\![\Phi_1]\!]e \cup [\![\Phi_2]\!]e$$
$$[\![[a]\Phi]\!]e = \{s \mid \forall s'.s \xrightarrow{a} s' \Rightarrow s' \in [\![\Phi]\!]e\}$$
$$[\![\langle a \rangle \Phi]\!]e = \{s \mid \exists s'.s \xrightarrow{a} s' \wedge s' \in [\![\Phi]\!]e\}$$

Fig. 1. Semantics of basic formulas.

Restricting the block graph to be acyclic ensures that no alternating fixed points [EL86] can occur.

The meaning $[\![\mathcal{B}]\!]e$ of the formula $\mathcal{B}$ containing blocks $B_1, \ldots, B_m$, topologically sorted by the dependency relation, can be computed through a sequence of environments

$$e_1 = [\![B_1]\!]e$$
$$\vdots$$
$$e_m = [\![B_m]\!]e_{m-1}$$

with $[\![\mathcal{B}]\!]e = e_m$. Due to the acyclicity restriction on block graphs, we are ensured that $[\![\mathcal{B}]\!]e_m = e_m$.

If $\mathcal{B}$ is a closed formula, i.e., every variable mentioned in the right-hand side of some equation appears on the left-hand side of an equation in one of the blocks, then for every two environments e and e', we have $[\![\mathcal{B}]\!]e = [\![\mathcal{B}]\!]e'$. Now, for every variable X defined in the formula we can compute the set of states in which X holds as $[\![X]\!][\![\mathcal{B}]\!]$. When the LTS is finite-state, f_E^s is continuous and the fixed points also have iterative characterizations which are used by the *CS* and *MCI* algorithms to compute fixed points.

The problem of *model checking in CS-logic* can now be defined as follows: given an LTS $\mathcal{L} = \langle S, A, \rightarrow, s_0 \rangle$ and a CS-logic formula $\mathcal{B}$ with a designated variable X defined within it, determine whether $s_0 \in [\![X]\!][\![\mathcal{B}]\!]$.

3 The Cleaveland-Steffen Model Checking Algorithm

The *CS* algorithm performs (non-incremental) global computation of fixed-points, i.e., the value of every variable is computed in every state. Due to the acyclicity restriction on block graphs (see Section 2), computation can proceed block-by-block: once the fixed point of a block is computed, the variable assignments in that block can no longer change due to dependencies on other blocks. Blocks are processed in the order resulting from topologically sorting the block graph.

The *CS* algorithm, as well as our incremental algorithm, uses an elaborate set of data structures to achieve its linear running time. To simplify its presentation, we describe the *CS* algorithm in terms of an intuitive structure called the "product graph" (cf. boolean graphs in [And92]). For efficiency reasons, the product graph is not computed by the algorithm explicitly, although its construction would not affect the asymptotic complexity. The correspondence between the product-graph-based presentation and the original *CS* algorithm is straightforward.

The *product graph* of an LTS $\mathcal{L} = \langle S, Act, \rightarrow, s_0 \rangle$ and a mu-calculus formula $\mathcal{B}$ is a directed graph with set of vertices $\{\langle s, X_i \rangle \mid s \in S, X_i \in Var\}$ and set of edges given by the following rules:

- if $X_i = X_j \vee X_k$ or $X_i = X_j \wedge X_k$, then for every $s \in S$, $\langle s, X_j \rangle \rightarrow \langle s, X_i \rangle$ and $\langle s, X_k \rangle \rightarrow \langle s, X_i \rangle$
- if $s \xrightarrow{a} s'$ and $X_i = \langle a \rangle X_j$ or $X_i = [a]X_j$, then $\langle s', X_j \rangle \rightarrow \langle s, X_i \rangle$.

If operator $\vee$ or $\langle a \rangle$ is used to define X_i, the node $\langle s, X_i \rangle$ is called an *or*-node of the product graph; otherwise, it is called an *and*-node. Note that the direction of edges is reversed compared to the LTS. The intuition for this comes from the fact that the truth of a variable in a node of the product graph is determined by truth of its immediate predecessors in the product graph, which, according to the semantics of the modal operators, is dependent on the immediate successors of the current state in the LTS.

For each node $\langle s, X_i \rangle$ of the product graph, the *CS* algorithm maintains the following variables:

- A boolean variable indicating whether or not variable X_i is true of state s in the current stage of the analysis. We simply use the name of the product graph node, i.e, $\langle s, X_i \rangle$, as the name of this variable, and sometimes refer to it as the *value* of the node. $\langle s, X_i \rangle$ is initialized to true if X_i is defined in a *max*-block, and, dually, is initialized to false if X_i is defined in a *min* block (we refer to these initializations as *trivial*), with the following exceptions:
    - The right-hand side of the equation for X_i is an atomic proposition A. Then $\langle s, X_i \rangle$ = true if $s \in \mathcal{V}(A)$, and false otherwise.
    - If state s has no a-derivatives and X_i is defined by $\langle a \rangle X_j$, then $\langle s, X_i \rangle$ = false, and if X_i is defined by $[a]X_j$, then $\langle s, X_i \rangle$ = true.
- A counter $C_{\langle s, X_i \rangle}$ that keeps track of the immediate predecessors of $\langle s, X_i \rangle$ in the product graph: if X_i is defined in a *max*-block and $\langle s, X_i \rangle$ is an *or*-node, then $C_{\langle s, X_i \rangle}$ records how many immediate predecessors of $\langle s, X_i \rangle$ are currently true. $C_{\langle s, X_i \rangle}$ is used dually in the case that X_i is defined a *min*-block and $\langle s, X_i \rangle$ is an *and*-node; i.e., it records how many immediate predecessors are currently false. In either case, $C_{\langle s, X_i \rangle}$ is initialized to the number of immediate predecessors of $\langle s, X_i \rangle$ in the product graph. These are the only cases in which counters are used.

Also, for every block B_j of the formula, a list M_j of nodes of the product graph is maintained, such that $\langle s, X_i \rangle$ is in M_j if X_i is defined in B_j, $\langle s, X_i \rangle$ recently changed its value, and the effect of this change on other nodes has yet to be determined. Initially, M_j contains all nodes $\langle s, X_i \rangle$ that were initialized non-trivially (see above).

The *CS* algorithm is captured by the following procedure, where $\mathcal{L}$ is a finite-state LTS and $\mathcal{B}$ is a CS-logic formula.

procedure $CS(\mathcal{L}, \mathcal{B})$
   topologically sort the blocks of $\mathcal{B}$
   initialize the $\langle s, X_i \rangle$, $C_{\langle s, X_i \rangle}$, and M_j as described above
   for each $B_j \in \mathcal{B}$ in topological order **do**
     if B_j is of type *max* **then** $MAX(B_j)$
     else $MIN(B_j)$

Procedure MAX, invoked on a block B_j, proceeds as follows:

procedure $MAX(B_j)$
  while M_j not empty **do**
    delete some $\langle s, X_i \rangle$ from M_j
    $DOWN(\langle s, X_i \rangle)$
  for each $\langle s, X_i \rangle$ such that X_i is defined in B_j and $\langle s, X_i \rangle$ is true **do**
    $UP(\langle s, X_i \rangle)$

Procedure MIN is dual to it, with all occurrences of UP and $DOWN$, as well as true and false, interchanged. MAX first propagates the changes to the variable assignment for block B_j recorded in M_j by repeatedly calling the procedure $DOWN$ (given below). When M_j is finally empty, the greatest fixed point for B_j will have been computed. MAX then invokes UP on each variable $\langle s, X_i \rangle$ such that X_i is defined in B_j and $\langle s, X_i \rangle$ was *not* falsified during the preceding while loop. This is necessary because there may exist variables $\langle s', X_j \rangle$ in *min*-blocks, trivially initialized to false and dependent on $\langle s, X_i \rangle$, whose values should now be true. The calls to UP will produce the desired effect.

Procedure $DOWN$ takes as a parameter a product graph node that has just changed its value from true to false, and checks whether any of its successors are affected by the change.[2]

procedure $DOWN(\ \langle s, X_i \rangle\ )$:
  for each *or*-node $\langle s', X_j \rangle$ such that $\langle s, X_i \rangle \rightarrow \langle s', X_j \rangle$ and $\langle s', X_j \rangle$ is true **do**
    decrement $C_{\langle s', X_j \rangle}$ by 1
    if $C_{\langle s', X_j \rangle} = 0$, **then**
    $\langle s', X_j \rangle := $ false
    add $\langle s', X_j \rangle$ to M_k /* X_j is defined in B_k */
  for each *and*-node $\langle s', X_j \rangle$ such that $\langle s, X_i \rangle \rightarrow \langle s', X_j \rangle$ and $\langle s', X_j \rangle$ is true **do**
    $\langle s', X_j \rangle := $ false
    add $\langle s', X_j \rangle$ to M_k /* X_j is defined in B_k */

The name of the procedure stresses the fact that the function δ on the environment (variable assignment) computed by $DOWN$ satisfies $\delta(e) \leq e$; i.e., the resulting assignment moves *down* in the lattice of tuples of sets of states described in Section 2. Procedure UP is dual to $DOWN$ and can be obtained by syntactically interchanging all occurrences of *and* and *or*, and true and false.

4   The Incremental Model Checking Algorithm

In this section, we modify the CS algorithm to obtain our MCI algorithm. MCI works incrementally in the following sense: Let $\mathcal{L}$ and $\mathcal{B}$ constitute a given instance of the model checking problem, and assume that the desired variable assignment has been previously computed (say, by an application of CS). Given a set Δ of *changes* to the LTS, where a change may correspond to either an inserted or deleted transition, MCI computes the new variable assignment by judiciously using the previously computed one as the "starting point" of the computation.

---

[2] Note that the change to $\langle s, X_i \rangle$ will affect $\langle s', X_j \rangle$ only if $\langle s, X_i \rangle$ is an immediate predecessor of $\langle s', X_j \rangle$ in the product graph and $\langle s', X_j \rangle$ was true. Collectively, these two conditions, and the fact that the block graph is acyclic, ensure that B_k, the block in which X_j is defined, is a *max*-block, as desired. See [CS93] for further details.

The top-level structure of MCI is basically the same as in CS: an initialization phase, in which the immediate effects of Δ on the previously computed variable assignment are ascertained, is followed by a for-loop in which blocks are processed in topological order of the block graph by calling modified MAX and MIN procedures. In the incremental case, however, it is necessary to start off the fixed-point computation of a block by making certain adjustments to the current variable assignment to ensure that the proper fixed point is computed.

As discussed in greater detail below, the adjustments will raise the variable assignment in the case of a max-block (by calling a modified UP procedure), and lower it in the case of a min-block (by calling a modified $DOWN$ procedure). Like before, iterations of the fixed-point computation will then lower the assignment in the case of a max-block, and raise it in the case of a min-block.

Since, for either type of block, we may need to shift the variable assignment up or down the lattice, it is no longer sufficient to provide procedures MAX and MIN, invoked on a block B_j, a single "work list" M_j, as in the non-incremental case. Rather, two such lists are now required, for both types of blocks: $down_j$, which records variables that change their values from true to false, and up_j, containing variables that change their values from false to true. Moreover, every product graph and- and or-node now has an associated counter, regardless of the type of the block they are defined in.

The initialization phase uses Δ to update both the product graph and the variable assignment of the previous computation.[3] Changes to the product graph reflect the semantics of basic formulas, in particular, the modal operators (the insertion and deletion of LTS transitions has no immediate effect on basic formulas constructed out of logical operators). When a transition $s \xrightarrow{a} s'$ is inserted into the LTS, for every pair of variables X_i, X_j such that $X_j = [a]X_i$ or $X_j = \langle a \rangle X_i$, the edge $\langle s', X_i \rangle \rightarrow \langle s, X_j \rangle$ is inserted into the product graph. Conversely, when a transition is deleted from the LTS, the corresponding set of edges is deleted from the product graph.

In response to changes in the product graph, counters are updated as one would expect. If $\langle s', X_j \rangle$ is an or-node and $\langle s, X_i \rangle$ is true, then $C_{\langle s', X_j \rangle}$ is incremented by 1 if the edge $\langle s, X_i \rangle \rightarrow \langle s', X_j \rangle$ is inserted into the product graph, and is decremented by 1 if this edge is deleted. The situation is dual for and-nodes.

The following cases require changing the value of a variable as an immediate result of inserting or deleting a product graph edge, independent of the block type. In each case, assume that X_j is defined in block B_k.

- An edge $\langle s, X_i \rangle \rightarrow \langle s', X_j \rangle$ such that $\langle s', X_j \rangle$ is an and-node is added to the product graph. If $\langle s', X_j \rangle$ is true and $\langle s, X_i \rangle$ is false, then $\langle s', X_j \rangle$ is changed to false and $\langle s', X_j \rangle$ is added to $down_k$.
- An edge $\langle s, X_i \rangle \rightarrow \langle s', X_j \rangle$ such that $\langle s', X_j \rangle$ is an or-node is added to the product graph. If $\langle s', X_j \rangle$ is false and $\langle s, X_i \rangle$ is true, then $\langle s', X_j \rangle$ is changed to true and $\langle s', X_j \rangle$ is added to up_k.
- An edge $\langle s, X_i \rangle \rightarrow \langle s', X_j \rangle$ such that $\langle s', X_j \rangle$ is an and-node is deleted from the product graph. If $\langle s, X_i \rangle$ is false and the node $\langle s', X_j \rangle$ had only one false predecessor (this number is recorded in $C_{\langle s', X_j \rangle}$), then $\langle s', X_j \rangle$ is changed to true and $\langle s', X_j \rangle$ is added to up_k.
- An edge $\langle s, X_i \rangle \rightarrow \langle s', X_j \rangle$ such that $\langle s', X_j \rangle$ is an or-node is deleted from the product graph. If $\langle s, X_i \rangle$ is true and the node $\langle s', X_j \rangle$ had only one true predecessor,

---

[3] Because MCI accepts as input a set of changes to the LTS, updates to the variable assignment made during initialization can potentially "overlap." Care needs to be taken to avoid unnecessary computations.

then $\langle s', X_j \rangle$ is changed to false and $\langle s', X_j \rangle$ is added to $down_k$.

The initializations described above depend only on the semantics of basic formulas and are therefore independent of the type of fixed point being computed. Simply restarting the fixed-point iteration (e.g., calls to $DOWN$ in the case of a max-block) at this point would bring us to a fixed point, but not necessarily the required fixed point! Rather, we must conclude the initialization phase by making certain *assumptions* about the existence of strongly connected components (SCCs) in the product graph. Assumptions made during initialization and their subsequent propagation through the product graph, will serve to adjust the variable assignment to a level where fixed-point iteration can be safely restarted.

To motivate our use of assumptions, consider the following scenario. The insertion of a transition in the LTS has resulted in the formation of a new SCC, call it C, in the product graph. Assume that C is contained in the subgraph of the product graph pertaining to a max-block B_j, and according to the results of the previous fixed-point computation, all nodes in C are false. Further assume that C is free of "external interference," that is, there is no edge $\langle s, X_i \rangle \rightarrow \langle s', X_j \rangle$ entering C such that the value of $\langle s, X_i \rangle$ uniquely determines the value of $\langle s', X_j \rangle$. For example, if $\langle s', X_j \rangle$ is an *and*-node and $\langle s, X_i \rangle$ is false, then $\langle s, X_i \rangle$ would be a source of external interference.[4] Then it is not difficult to see that the variable assignment in which all nodes in C are uniformly set to true or false is a fixed point. The point is, however, that the required fixed point for C is the largest one, i.e., the one in which all nodes are assigned the value true.

We will therefore, in general, assume that when an edge $\langle s, X_i \rangle \rightarrow \langle s', X_j \rangle$ is added to the product graph such that $\langle s, X_i \rangle$ and $\langle s', X_j \rangle$ are defined in the same max-block B_k and both are false, that a new SCC, free of external interference, has been created. We record this assumption by setting $\langle s, X_i \rangle$ to true and adding it to up_k and a new list called $assumptions_k$.[5] Note that the counter $C_{\langle s, X_i \rangle}$ is not updated to reflect $\langle s, X_i \rangle$'s new value and thus an inconsistency is introduced. This is intentional and will be used later to determine whether the assumption was a valid one.

The case of a min-block is dual: when an edge is added between two true nodes corresponding to variables defined in the same min-block, we change one of them to false and update the block's *assumptions* list. The changed value is reflected in the *down* list.

Deleted edges can also cause us to make assumptions during initialization. Suppose that a max-block SCC had only one source of external interference, which was eliminated when an edge was deleted from the product graph. The desired variable assignment in this case has all nodes in the SCC uniformly set to true. We therefore assume that whenever an edge $\langle s, X_i \rangle \rightarrow \langle s', X_j \rangle$ is deleted from the product graph such that X_j is defined in a max-block B_k and both nodes are false, $\langle s, X_i \rangle$ constituted the only source of external interference in the SCC containing $\langle s', X_j \rangle$. We record this assumption by setting $\langle s', X_j \rangle$ to true and adding it to both up_k and $assumptions_k$.

Consider now the propagation of the changes made to the variable assignment during initialization. These are recorded in up_j and $down_j$, for each block B_j. As before, blocks are processed in topological order of the block graph, by calling a modified MAX or MIN

---

[4] Because C is strongly connected, each node in C has at least one incoming edge and is therefore either an *and*- or *or*-node.

[5] A number of small optimizations can be made here to prevent obviously false assumptions, e.g. do not make the assumption if either of the nodes has no incoming edges or no outgoing edges, and thus cannot be on a cycle. We leave these to careful implementors.

procedure depending on the type of the block. *MAX* and *MIN* now commence with an *adjustment phase*, during which the variable assignment is shifted up (in the case of a *max* block) or down (in the case of a *min* block) in the lattice of assignments to ensure that the fixed-point computation can proceed normally. The new *MAX* procedure is given by:

procedure $MAX(B_j)$
  while up_j not empty **do**   /* Adjustment Phase */
    delete some $\langle s, X_i \rangle$ from up_j
    $UP(\langle s, X_i \rangle)$
  while $assumptions_j$ not empty **do**   /* Check validity of assumptions */
    delete some $\langle s, X_i \rangle$ from $assumptions_j$
    if $\langle s, X_i \rangle$ is an *and*-node **then**
      if $C_{\langle s, X_i \rangle} \neq 0$ **then**
        $\langle s, X_i \rangle :=$ false
        add $\langle s, X_i \rangle$ to $down_j$
    if $\langle s, X_i \rangle$ is an *or*-node **then**
      if $C_{\langle s, X_i \rangle} = 0$ **then**
        $\langle s, X_i \rangle :=$ false
        add $\langle s, X_i \rangle$ to $down_j$
  while $down_j$ not empty **do**   /* Iteration Phase */
    delete some $\langle s, X_i \rangle$ from $down_j$
    $DOWN(\langle s, X_i \rangle)$

The adjustment phase makes as many variables true as possible by iteratively invoking *UP*. As shown in Section 5, the resulting variable assignment will be high enough in the lattice to contain every fixed point of the semantic function. At this point, the validity of any previously made assumptions is determined by checking whether the value of $\langle s, X_i \rangle$ is consistent with $C_{\langle s, X_i \rangle}$, for each node $\langle s, X_i \rangle$ on the list of assumptions. If an inconsistency is detected, i.e., according to $C_{\langle s, X_i \rangle}$, $\langle s, X_i \rangle$ should be false, we reset the variable and let *DOWN* undo the effects of the assumption. When finished, the fixed-point iteration (applications of *DOWN*) can be safely restarted. Procedure *DOWN* is modified as follows:

procedure $DOWN(\langle s, X_i \rangle)$:
  for each *or*-node $\langle s', X_j \rangle$ such that $\langle s, X_i \rangle \rightarrow \langle s', X_j \rangle$, X_j defined in B_k **do**
    decrement $C_{\langle s', X_j \rangle}$ by 1
    if $\langle s', X_j \rangle =$ true and $C_{\langle s', X_j \rangle} = 0$ **then**
      $\langle s', X_j \rangle :=$ false
      add $\langle s', X_j \rangle$ to $down_k$ /* X_j is defined in B_k */
    else if $C_{\langle s', X_j \rangle} \neq 0$, $\langle s', X_j \rangle =$ true and B_k is a *min*-block **then**
      $\langle s', X_j \rangle :=$ false
      add $\langle s', X_j \rangle$ to $down_k$ /* X_j is defined in B_k */
      add $\langle s', X_j \rangle$ to $assumptions_k$
  for each *and*-node $\langle s', X_j \rangle$ such that $\langle s, X_i \rangle \rightarrow \langle s', X_j \rangle$, X_j defined in B_k **do**
    increment $C_{\langle s', X_j \rangle}$ by 1
    if $\langle s', X_j \rangle =$ true **then**
      $\langle s', X_j \rangle :=$ false
      add $\langle s', X_j \rangle$ to $down_k$ /* X_j is defined in B_k */

The overall structure of *DOWN* is retained, except that counters are updated in both cases and an assumption is made when *DOWN* encounters a true variable in a *min*-block. The intuition for the assumption is somewhat different from the one for assumptions made at initialization. Here we are assuming that $\langle s', X_j \rangle$ is a part of an SCC and the change to $\langle s, X_i \rangle$ eliminated a source of external interference for the SCC. As before, procedures *MIN* and *UP* are dual to those given above and are obtained by interchanging all occurrences of *UP* and *DOWN*, *up* and *down*, and *and* and *or*.

So far we have assumed that the changes to the LTS only concern transitions. States, however, can be added and deleted with almost no extra effort. The basic idea is to assume that the variable assignment for an isolated state, i.e., one devoid of incident transitions, is known — it can be computed during the first, nonincremental run of the algorithm. During incremental runs, state additions are processed before any other changes by setting variables of the form $\langle s, X_i \rangle$, where s is a new state, in accordance to the variable assignment of an isolated state. The processing of inserted and deleted transitions can now proceed as before. For state deletions, we assume that any incident transitions are deleted as well.

5   Correctness and Complexity

The proof of correctness of the *MCI* algorithm is given by the following theorem.

Theorem 1. *Let $\mathcal{L}$ be an LTS, $\mathcal{B}$ a CS-logic formula, Δ a set of changes to $\mathcal{L}$ in the form of inserted and deleted transitions, and $\mathcal{L}'$ the LTS obtained by applying Δ to $\mathcal{L}$. Furthermore, let e be the variable assignment obtained by algorithm CS on input $\mathcal{L}$ and $\mathcal{B}$, and, similarly, let e' be the variable assignment obtained by CS on input $\mathcal{L}'$ and $\mathcal{B}$. Then MCI, using e as the initial variable assignment, terminates on input $\mathcal{L}$, $\mathcal{B}$, and Δ with variable assignment e'.*

Proof Sketch: Consider a block B in $\mathcal{B}$. Without loss of generality, assume that B is a *max*-block; the case where B is a *min*-block is completely dual and therefore omitted. For O an arbitrary topological order of B's block graph, the proof is by induction on the position of B in O and proceeds in two main steps.

We first show that when B's *up* list is empty, the current variable assignment is higher in the lattice of variable assignments than any fixed point of B's recursive equations. In particular, it contains B's greatest fixed point. For this purpose it is convenient to define B's *subgraph*, the subgraph of the product graph induced by the set of nodes $\{\langle s, X_i \rangle \mid X_i \text{ is defined in } B\}$.

The proof now proceeds by induction on the topological order of the strongly connected components of B's subgraph (this is well defined since the acyclicity of the block graph guarantees that every SCC appears within one block). That is, fix an SCC C and assume the result for any SCC having edges leading into C. There are two cases to consider depending on whether or not C is a *trivial* SCC (consisting of one node). For the case when C is non-trivial, we have to worry about cycles of false nodes in it. The details of the case analysis are omitted but the crucial point is showing that no such cycle can exist unless some of its nodes are uniquely determined by the values of nodes outside C.

Now that we have established that when the *up* list has been emptied the current variable assignment contains B's greatest fixed point, the second step of the proof basically coincides with the proof of correctness of the *CS* algorithm. That is, we show that the processing of entries in B's *down* list monotonically lowers the variable assignment, and, when the list is empty, the greatest fixed point will have been reached (see [CS93] for details). $\qquad\square$

Consider now the computational complexity of the *MCI* algorithm. In the worst case, its complexity is the same as that of the *CS* algorithm: linear in the product of the size of the LTS and the size of the formula, where the size of the LTS is taken to be the total number of states and transitions, and the size of the formula is the total number of equations over all blocks. The proof is similar to that of [CS93], and is based on the fact that a product-graph node $\langle s, X_i \rangle$ can appear at most once in each list up_j and $down_j$, for X_i defined in block B_j. We ensure this property by checking (in constant time) if a node is already present in a list, before attempting to add it to the list. Thus *UP* and *DOWN* can only be invoked on a node at most once each, and each such invocation traverses each outgoing edge once.

We have also shown that the problem of model checking in the alternation-free fragment of the modal mu-calculus falls into the category of *unbounded* problems, i.e., the running time of an incremental update cannot be expressed solely in terms of the size of the change to the input. The proof of the unboundedness of the model checking problem is via a reduction from the single-source reachability problem (SS-REACHABILITY): given a directed graph (V, E) and a fixed vertex $s \in V$, determine, for every vertex $v \in V$, whether v is reachable from s.

In [RR91], it is shown that SS-REACHABILITY is unbounded in the *locally persistent* model of computation [AHR+90], which, intuitively, comprises all incremental algorithms in which no global information is maintained between updates. It is straightforward to show that *MCI* is locally persistent, and it thus follows that the performance of the algorithm is the best one could hope for in an incremental setting.

6 Implementation and Examples

The *MCI* algorithm has been implemented as part of the Concurrency Factory project. We started with the implementation of *CS*, which we later modified to make use of incremental computation. Although the non-incremental version is still needed for the initial computation of fixed points, we were able to avoid unnecessary duplication of code. In particular, with only minor changes, the incremental versions of *UP* and *DOWN* produce correct results in the initial computation.

We now consider an example of *MCI* in action, which is intended to demonstrate the best-case behavior of the algorithm. The assumptions concerning the design process, however, seem to be realistic. The system in question is Milner's scheduler [Mil80], consisting of a circular chain of simple "cycler" processes $C_0, \ldots, C_{n-1}$. Milner's scheduler is often used as a benchmark for verification tools, partly because its state space grows exponentially with the size of the scheduler (number of cyclers).

Each C_i is initiated by the previous one in the chain by means of action g_i, after which it carries out the sequence of observable actions a_i, b_i "in parallel" with initiating C_{i+1}. The LTS for C_i is depicted in Figure 2. C_0 must be furnished with a transition labeled by action $\overline{start}$ (the dashed line in Figure 2) that allows it to be initiated by a separate starter process.

Imagine that the designer has completed the scheduler and checks it for the absence of deadlocks. The property "there is a reachable deadlocked state" is expressed by X_1 in the following *min*-block, where '$-$' stands for "any action:"

$$min\{ \begin{aligned} X_1 &= X_2 \vee X_3 \\ X_2 &= \langle - \rangle X_1 \\ X_3 &= [-]X_4 \\ X_4 &= ff \end{aligned} \quad \}.$$

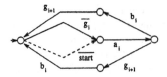

Fig. 2. The LTS for cycler C_i

The scheduler is correct and thus the formula is not satisfied. Imagine now that the designer, in an attempt to simplify the implementation, decides to omit the $\overline{start}$ transition from C_0 and then checks the scheduler again. The scheduler is now deadlocked in the start state, which disconnects the start state from the rest of the scheduler.

The second, incremental run of *MCI* finds the deadlock. Intuitively, the effects of the update to C_0 should not propagate very far, and Table 1 reveals this to indeed be the case. There, three rows of results are presented: (1) execution times of our implementation of *MCI* on the original, deadlock-free scheduler for increasing numbers of cyclers; (2) execution times for incremental runs of *MCI* on the updated, deadlocked scheduler; and (3) execution times of our implementation of *CS* on the original scheduler. The second row shows that the verification of the updated scheduler can be performed incrementally in *constant time*, independent of the number of cyclers. The third row allows us to compare how the *MCI* and *CS* algorithms perform on the first, necessarily non-incremental verification of the scheduler. As can be seen, the difference in execution times is negligible and, thus, the extra information we maintain in the incremental case does not significantly affect the constant factors of the algorithm.

We have also considered an example of the worst-case behavior of the incremental computation, involving a linear chain of transitions which we again test for deadlock. The update to the LTS is to extend the chain with yet another transition. Obviously, the LTS still has a deadlock, but the assumption made during the initialization phase of the incremental computation results in a wave of changes to the values of the variables that reaches the start state. At this point, it is determined that the assumption was wrong, so the wave of changes reverses direction and traverses the whole length of the chain again. The second application of *MCI* ends up taking about 75% more time than the first application.

7 Conclusions and Future Work

We have modified the algorithm of [CS93] to obtain *MCI*, an incremental algorithm for model checking in the alternation-free fragment of the modal mu-calculus. *MCI* can be easily modified to handle various kinds of incremental updates to the logical formula, such as the inversion of logical and modal connectives (e.g., changing $[a]$ to $\langle a \rangle$). Such updates affect the semantics of basic formulas but not the strongly connected components of the product graph.

We believe that our results on the non-uniqueness of fixed points of functions on graphs and their applications to incremental computation have wider applicability than just model checking CS-logic. For example, the *MCI* algorithm can be easily generalized to perform (global) incremental evaluation of *boolean equation systems* [Lar92]. Furthermore, if we do not confine ourselves to boolean variables in each node of the graph, then it appears that a variety of graph problems can be accommodated, including those pertaining to data-flow analysis [Hec77] (the relationship between data-flow

No. of cyclers		2	3	4	5	6	7	8	9	10
incr.	1st pass	0.01	0.02	0.05	0.19	0.48	1.36	3.06	7.52	18.90
	2nd pass	0.01	0.01	0.01	0.01	0.01	0.01	0.01	0.01	0.01
non-incr.		0.01	0.02	0.07	0.19	0.47	1.22	3.07	7.50	18.65

Table 1. Execution times, in seconds, for the scheduler example

analysis and model checking is investigated in [Ste91]).

Other directions for future work include the pursuit of incremental algorithms for model checking in the full modal mu-calculus [EL86] and for local model checking [SW91].

References

[AHR+90] B. Alpern, R. Hoover, B. K. Rosen, P. F. Sweeney, and F. K. Zadeck. "Incremental Evaluation of Computational Circuits". In *Proc. of the 1st Annual ACM-SIAM Symposium on Discrete Algorithms*, 1990.

[And92] H. R. Andersen. "Model Checking and Boolean Graphs". In *Proceedings of ESOP'92*. LNCS 582, 1992.

[CGL+94] R. Cleaveland, J. N. Gada, P. M. Lewis, S. A. Smolka, O. V. Sokolsky, and S. Zhang. "The Concurrency Factory – Practical Tools for Specification, Simulation, Verification and Implementation of Concurrent Systems". In *Proceedings of the DIMACS Workshop on Specification Techniques for Concurrent Systems, Princeton, NJ.*, May 1994.

[CPS93] R. Cleaveland, J. Parrow, and B. Steffen. "The Concurrency Workbench: A Semantics-Based Tool for the Verification of Concurrent Systems". *ACM TOPLAS*, 15(1), 1993.

[CS93] R. Cleaveland and B. Steffen. "A Linear-Time Model Checking Algorithm for the Alternation-Free Modal Mu-Calculus". *Formal Methods in System Design*, 2, 1993.

[EL86] E. A. Emerson and C.-L. Lei. "Efficient Model Checking in Fragments of the Propositional Mu-Calculus". In *Proc. LICS '86*. IEEE Computer Society Press, 1986.

[Hec77] S. M. Hecht. *Flow Analysis of Computer Programs*. Elsevier, North Holland, 1977.

[Koz83] D. Kozen. "Results on the Propositional Mu-Calculus". *Theoretical Computer Science*, 27:333–354, 1983.

[Lar92] K. G. Larsen. "Efficient Local Correctness Checking". In *Proceedings of CAV'92*, 1992.

[Mil80] R. Milner. *A Calculus of Communicating Systems*. LNCS 92, 1980.

[PS89] L. L. Pollock and M. L. Soffa. "An Incremental Version of Iterative Data Flow Analysis". *IEEE Trans. Software Engineering*, 15(12), 1989.

[Ram93] G. Ramalingam. *Bounded Incremental Computation*. PhD thesis, Computer Sciences Dept., University of Wisconsin-Madison, 1993.

[RMP88] B. G. Ryder, T. J. Marlowe, and M. C. Paull. "Conditions for Incremental Iteration: Examples and Counterexamples". *Sci. Program.*, 11(1), 1988.

[RR91] G. Ramalingam and T. Reps. "On the Computational Complexity of Incremental Algorithms". Technical Report TR-1033, Computer Sciences Dept., University of Wisconsin-Madison, 1991.

[Ste91] B. Steffen. "Data Flow Analysis as Model Checking". In *Proc. TACS'91*. LNCS 526, 1991.

[SW91] C. Stirling and D. Walker. "Local Model Checking in the Modal Mu-Calculus". *Theoretical Computer Science*, 89(1), 1991.

Performance Improvement of State Space Exploration by Regular & Differential Hashing Functions

Bernard Cousin
Jean-michel Hélary

IRISA - Université de Rennes-I
Campus universitaire de Beaulieu
35042 - Rennes cédex
FRANCE

phone : (33) 99.84.73.33
fax : (33) 99.38.38.32
E-mail : bcousin@irisa.fr

Abstract
This paper presents regular hashing functions. Used in conjunction with differential computation process, regular hashing functions enable searching time of global state to be optimized. After a formal definition of the regular property for hashing functions, we propose a characterization of this property. Then the formal definition of differential hashing function is given. Next, we show the performance acceleration produced by the precomputed differential computation process applied to three hashing functions commonly used. The observed accelerations can be significant because the complexity of proposed implementation is independent of key length or respectively of item difference contrary to the usual or respectively differential implementations. Last we study the performances of precomputed differential hashing computation process on reachability graph exploration of distributed systems specified by Petri net using the Bouster tool, and on state space exploration of protocols specified by Lotos using the Open/Caesar environment.

1. Introduction

Hashing method is a well suited method to achieve state-space exploration for verification of distributed systems. Hashing method is used as searching method to accelerate the retrieval process of a particular state among large state space under exploration. This method enables searching, insertion and suppression operations to be done on average at a constant cost in number of comparisons. But the usual computation process of the hashing value has the first following drawback : its complexity is at least proportional to the key length, and unfortunately, the state descriptor, from which the keys are based, is in general very large (several hundreds of bytes [Doldi 92, Holzmann 91, Wolper 93]), if accurate modelling is considered.

Moreover, some of the recent works on state-space exploration made an intensive use of hashing functions (bitstate method [Holzmann 88], multihash method [Wolper 93]) : two hashing function calls in Holzmann's Spin validation environment for each newly created state, Wolper recommends 20 calls for each newly created state to achieve large coverage of the state-space. These two methods reduce the amount of space needed to store the explored state-space but, as Wolper writes, due to the intensive function calls, they have the second following drawbacks : "computing 20 hash functions is quite expensive and will substantially slow down the search".

In previous work on improvement of state-space exploration we have introduced *differential* hashing functions [Cousin 93]. These hashing functions use differential computation process of the hashing value which replaces the usual computation process, and which optimizes their processing time. The proposed optimization is based on the following observation: key structure can be viewed as record of items. The computation process is called differential, if we can infer the hashing value of a key from the hashing value of another key which differs from the previous key in only few items. It can be more efficient to deduce the new hashing value knowing the value of some few new items rather than to apply the usual computation process on every item of the new key.

In opposition to the usual one, the differential computation process complexity is proportional to the difference number, which is the number of modified item between two successive keys. In general, studied systems have inherently successive states whose keys have few differing items (less than 10%) : the subsystems of a distributed system do not evolve simultaneously and at every moment. And our performance evaluation of differential computation process for hashing function has showed that it enables a substantial processing time improvement for typical distributed systems analysis.

However when the difference ratio is high the improvement can vanish because each difference in the key needs to be computed by a so-called mono-differential function. Even if the complexity of the mono-differential computation is far less than the complexity of the usual computation, the time spent into numerous function calls can exceed the gain in complexity.

This is why in this current work we propose an improvement of the differential computation process. This improvement is based on the knowledge of the mathematical operations which produce the new states. This knowledge enables the differential hashing value associated to each transition of the model to be computed during the initial loading phase of the model (or during the first firing of each transition). Similarly to the transition function associated to each transition which enables a new state to be produced from a previous state, the differential hashing value enables the hashing value of a new state to be computed from the hashing value of the previous state. The precomputed value is obtained from the same items as the original differential method : the items which differ. So we called it the differential hashing value.

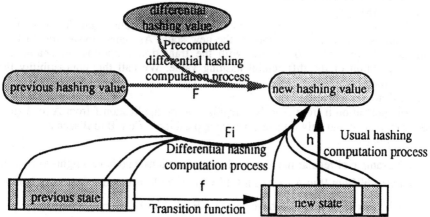

Figure 1 - Hashing value computation

From data point of view, there can be three different manners to implement the same hashing function (Figure 1): the usual hashing computation process which is computed from all the items of the new state; the differential hashing computation process which uses the previous hashing value and all the items of the previous state and of the new state which differ [Cousin 93]; and the precomputed differential hashing computation process which uses the previous hashing value and the differential hashing value associated with the transition which has produced the new state.

At storage and computing cost of one differential hashing value for each transition of the model, the precomputed and differential method enables the complexity of hashing value to be completely independent of the length of the key and the number of differences. Unfortunately, a precomputed differential computation process can not be associated with every existing hashing function : first, because the hashing function has to be differential and it has been shown in our previous work that all hashing functions are not differential; second, because the operations used by the hashing function have to be compatible with the operations used by the transition function - this property is called *regularity*.

In the second section, after a formal description of regular functions and of differential hashing functions, we produce a characterization of the regular function. This characterization can be useful to find a precomputed differential hashing computation process because all hashing functions as well as all differential hashing functions are not necessarily regular.

The third section addresses the following question: Are the precomputed differential computation processes always more efficient than the differential or usual computation processes? We show that this is always the case, and furthermore, the precomputed differential implementation, enabling very efficient coding, cuts down the processing time significantly.

The fourth section gives the performances achieved by the precomputed differential hashing function used in the Bouster verification tool [Bonafos 90], based on formal description of distributed systems by Petri net, and the Open/Caesar verification tool [Garavel 90] based on Lotos. These state-space exploration examples allow us to exhibit the advantages and the limits of our precomputed differential computation process.

2. Regular functions

2.1 Presentation

Regular functions are defined in relation to other applications which represent the transition functions used to produce the new state. As stated above, the hashing functions have to be compatible with the transition functions, to enable the precomputation of the differential hashing value. We call this compatibility the regular property.

Definition: regular property

An application h from A to B is regular for the application f from A to A (h is said f-regular) if and only if there exists an application F from B to B such as :

$$f \circ h = h \circ F. \tag{1}$$

An example of regular function is given by functions who have a right-inverse i.e. there exists h^r from B to A such as $h \circ h^r = Id$ where Id is the identity function on A.

In fact, $\forall f$, taking $F = h^r \circ f \circ h$ we have $h \circ F = h \circ h^r \circ f \circ h = f \circ h$ (q.e.d.).

2.2 Characterization

We characterize the applications h from A to B which are f-regular. First we restrict ourself to surjections, that is to say to applications h such as h(A) = B.

Theorem :

A surjection h from A to B is f-regular if, and only if, it satisfies the following property: $\forall x \in A, \forall y \in A, h(x) = h(y) \Rightarrow h(f(x)) = h(f(y))$. 　　(2)

Proof :

a) the condition (2) is necessary.

Suppose, by hypothesis, that h is f-regular ; let us assume that $\forall x \in A, \forall y \in A, h(x) = h(y)$.

According to the definition (1) the f-regular application h respects the following property : $\exists F, f \circ h = h \circ F$.

Therefore, since F is an application : $\forall x \in A, \forall y \in A, h(x) = h(y) \Rightarrow F(h(x)) = F(h(y))$. According to the definition (1) : $\forall x \in A, \forall y \in A, h(x) = h(y) \Rightarrow h(f(x)) = h(f(y))$, thus (2) is satisfied.

b) The condition is sufficient.

Let $\mathcal{R}$ be the equivalence relation on A : $x \mathcal{R} y \Leftrightarrow h(x) = h(y)$.

Let us denote the quotient set $A/\mathcal{R}$ by $\tilde{A}$, and, for any $x \in A$, let $\tilde{x}$ denote the class of x. With the surjection h can be associated the canonical injection $\tilde{h}$ from $\tilde{A}$ to B defined by $\forall \tilde{x} \in \tilde{A}, \tilde{h}(\tilde{x}) = h(x)$, where x is a particular element of $\tilde{x}$.

Clearly, $\tilde{h}$ is an application since if x' is another element of $\tilde{x}$, we have h(x') = h(x).

Similarly $\tilde{h}$ is an injection since $\tilde{h}(\tilde{x}) = \tilde{h}(\tilde{y}) \Rightarrow h(x) = h(y)$ where $x \in \tilde{x}$ and $y \in \tilde{y}$,

$$\Rightarrow x \mathcal{R} y$$

$$\Rightarrow \tilde{x} = \tilde{y} .$$

Finally, $\tilde{h}$ is an surjection since $\tilde{h}(\tilde{A}) = h(A) = B$. Thus $\tilde{h}$ is a bijection from $\tilde{A}$ onto B. With the application f from A onto A, we associate the relation $\tilde{f}$ from $\tilde{A}$ to $\mathcal{P}(\tilde{A})$, where $\mathcal{P}(\tilde{A})$ denotes the set of subsets of $\tilde{A}$, defined by : $\forall \tilde{x} \in \tilde{A}, \tilde{f}(\tilde{x}) = \{\widetilde{f(x)}$ such as $x \in \tilde{x}\}$. In others words : $\tilde{y} \in \tilde{f}(\tilde{x}) \Leftrightarrow \exists x \in \tilde{x}, \exists y \in \tilde{y}$, such as f(x)=y. 　　(3)

Let us show that $\tilde{f}$ is an application from $\tilde{A}$ on to $\tilde{A}$.

Suppose that $\tilde{f}$ has two images $\tilde{y}$ and $\tilde{z}$ for the same element $\tilde{x}$:

$$\tilde{y} \in \tilde{f}(\tilde{x}) \text{ and } \tilde{z} \in \tilde{f}(\tilde{x}).$$

By definition (3) :

$\exists x \in \tilde{x}, \exists y \in \tilde{y}$, such that f(x)=y and $\exists x' \in \tilde{x}, \exists z \in \tilde{z}$, such that f(x')=z. But $x \mathcal{R} x'$ and thus h(x) = h(x').

By the hypothesis (2) : $h(x) = h(x') \Rightarrow h(f(x)) = h(f(x'))$; hence $f(x) \mathcal{R} f(x')$ i.e. $y \mathcal{R} z$, and thus $\tilde{y} = \tilde{z}$ (q.e.d.).

Two images by $\tilde{f}$ of the same element are not distinct, so $\tilde{f}$ is an application. Consider the application F from B to B by composition of the previous applications via $\tilde{A}$: $F = \tilde{h}^{-1} \circ \tilde{f} \circ \tilde{h}$ (Figure 2). Finally we show that $h \circ F = f \circ h$.

Let φ from A to $\tilde{A}$ be the application defined by $\forall x \in A, \varphi(x) = \tilde{x}$. We have $\forall x \in A$,

$(f \circ \varphi)(x) = \varphi(f(x)) = \widetilde{f(x)} = \widetilde{f}(\widetilde{x}) = \widetilde{f}(\varphi(x)) = (\varphi \circ \widetilde{f})(x)$, and thus $f \circ \varphi = \varphi \circ \widetilde{f}$. (4)

Composing on the right with $\widetilde{h}$: $f \circ \varphi \circ \widetilde{h} = \varphi \circ \widetilde{f} \circ \widetilde{h}$,

which can be written : $f \circ \varphi \circ \widetilde{h} = \varphi \circ \widetilde{h} \circ \widetilde{h}^{-1} \circ \widetilde{f} \circ \widetilde{h}$. (5)

But , $\forall x \in A$, $(\varphi \circ \widetilde{h})(x) = \widetilde{h}(\varphi(x)) = \widetilde{h}(\widetilde{x}) = h(x)$, and thus : $\varphi \circ \widetilde{h} = h$. (6)

Putting this relation in (5) gives : $f \circ h = h \circ \widetilde{h}^{-1} \circ \widetilde{f} \circ \widetilde{h}$, that is to say

$f \circ h = h \circ F$ by definition of F. (q.e.d.)

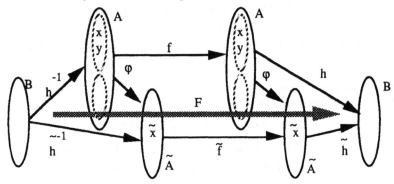

Figure 2 - F construction

The case where h is not a surjection is straightforward. Consider first F' from h(A) to h(A) such that $f \circ h = h \circ F'$, then extend arbitrarily F' to F from B to B.

2.3 Examples

For instance, let $\oplus$ be the application defined as the "exclusive or" between all key items. Each item belongs to the same set B. Formally :

$$\forall <x_1,...,x_i,...,x_N> \in \prod_{k=1}^{N} B, \oplus (<x_1,...,x_i,...,x_N>) = x_1 \oplus ... \oplus x_i \oplus ... \oplus x_N \text{ with } \oplus$$

the usual "exclusive Or" operator from B to B.

The application $\oplus$ is regular towards the usual "exclusive Or" operator ($\oplus$).

Proof :

As the composition of the operators $\oplus$ is commutative and associative, similarly the composition of the application $\oplus$ and the operator $\oplus$ is commutative and associative, and therefore, $\oplus$ is $\oplus$-regular: $\oplus \circ \oplus = \oplus \circ \oplus$ (q.e.d.).

So the precomputed differential hashing function of the hashing function $\oplus$ for the transition function $\oplus$ is $\oplus$ itself.

But the application $\oplus$ is not regular towards the usual addition (+).

Proof :
Counter-example of the characterization property (2):
Let be N $=3$, $x_1 = 1$, $x_2 = 2$, $x_3 = 3$.

$\bigoplus(<1,2,3>) = \bigoplus(<1,3,2>) = 0$ but $\bigoplus(<1,3,2+1>) \neq \bigoplus(<1,2,3+1>)$ because

$\bigoplus(<1,3,2+1>)=1\oplus3\oplus(2+1)=1\oplus3\oplus3=1$ and $\bigoplus(<1,2,3+1>)=1\oplus2\oplus(3+1)=1\oplus2\oplus4=7$
(q.e.d.).

This example explains the important of the choice of good hashing functions which are regular towards the transition function, all the more the hashing functions has to be differential.

2.4 Differential functions

Hashing function is an application with N variables from a product of sets

$\prod_{k=1}^{N} A_k$ to a set B. We denote $h(<x_1,...,x_i,...,x_N>)$ the hashing value of the key $<x_1,...,x_i,...,x_N>$.

Definition : mono-differential computation function

The hashing function h has a mono-differential computation function of the hashing value for its ith item if and only if it exists a function F_i from $B \times A_i \times A_i$ to B such that :

$$\forall <x_1,...,x_i,...,x_N> \in \prod_{k=1}^{N} A_k, \ \forall y_i \in A_i, \ F_i(h(<x_1,...,y_i,...,x_N>), \ y_i, \ x_i) =$$

$$h(<x_1,...,x_i,...,x_N>). \tag{7}$$

The mono-differential function (F_i) enables the computation of the hashing value of a key $(<x_1,...,x_i,...,x_N>$: called son_key) from the hashing value of another key $(<x_1,...,y_i,...,x_N>$: called father_key) having in common with the previous key every items but one (i: called modified item). Knowing the old value of the modified item of the father_key (y_i), the new value that this modified item must have in the son_key (x_i), and the hashing value computed from the father_key $(h(<x_1,...,y_i,...,x_N>))$, the mono-differential function associated with the modified item enables the computation of the hashing value associated with the son_key $(h(<x_1,...,x_i,...,x_N>))$.

Definition : differential hashing function

If for every item of the keys of the hashing function h studied there exists one mono-differential computation function, then the set of these mono-differential functions constitutes a complete differential computation function family. If a hashing function has a complete differential computation function family, we shall say that it is differential.

2.5 Characterization

We can characterize the hashing functions which are differential, that is, which admit a complete differential computation function family.

Not every hashing function admits differential computation functions. For instance, the hashing function "$\otimes$" built over a binary set product and defined as the binary operator "logical And" is not differential. In fact, no mono-differential computation function can be established: $F_1(0,0,1)$ can be equal either to $F_1(\otimes<0,1>,0,1)=\otimes(<1,1>)=1$, or to $F_1(\otimes<0,0>,0,1)=\otimes(<1,0>)=0$, which contradicts the image unicity property.

Theorem :

The hashing functions which admit a mono-differential computation function for their i^{th} item are characterized by the following property:

$$\forall<x_1,...,x_i,...,x_N>\in \prod_{k=1}^{N} A_k, \ \forall<y_1,...,y_i,...,y_N>\in \prod_{k=1}^{N} A_k, \ h(<x_1,...,x_i,...,x_N>) =$$

$$h(<y_1,...,x_i,...,y_N>) \Leftrightarrow h(<x_1,...,y_i,...,x_N>) = h(<y_1,...,y_i,...,y_N>). \tag{8}$$

The proof that the property (8) is a necessary and sufficient condition in order that the hashing functions admit a complete differential computation function family can be found in [Cousin 93b].

2.6 Example

Consider the same hashing function $\oplus$, defined previously as the "exclusive or" between all the key items, each item belonging to the same set B.

The mono-differential hashing computation functions F_i can be defined as

$$F_i(\oplus(<x_1,...,y_i,...,x_N>), y_i, x_i) = \oplus(<x_1,...,y_i,...,x_N>)\oplus y_i\oplus x_i$$

The proof that the hashing function $\oplus$ has a complete mono-differential computation functions ($\forall i\in [1,N], \exists F_i$) can be found in [Cousin 93].

3. Performance study

3.1 Presentation

In this section, we address the issue of performance improvement achieved by the precomputed differential computation process. For that purpose, we compare the performances obtained by hashing functions using precomputed differential computation process to usual or differential ones.

Due to the reduction of the image definition domain, hashing functions can associate several keys with one unique hashing value. Hashing methods associate specific functions with hashing functions to resolve collisions. As the precomputed and differential computation processes return exactly the same value than the usual computation process, the observed performance improvements are independent of the performance of the collision resolution function, and thus, are entirely preserved. So our study focuses on performances of hashing functions without studying the collision resolution functions. Nevertheless, previous studies have shown that the distribution of the three following hashing functions are good to very good. So, the preconceived idea according to which regular and differential properties induce inefficient hashing functions, is not well founded.

Three hashing functions have been chosen among those found in the literature. This sample does not cover all existing hashing functions, but these three functions have the advantage to be regular and differential. Besides this advantage, the fact that these functions are widely used and the good performance provided by their coding simplicity have also been taken in consideration as a criterion of choice. A simple algorithm with short code is often faster than a complex and long algorithm. Further studies are undertaken to search differential algorithms of other hashing functions, in order to increase the study spectrum.

The proposed hashing functions use the following basic operators: division/product, modulo, addition/subtraction, folding [Knott 75]. Other operators can been used: power/root, radix transformation, polynomial computation, etc [Lum 71] : since they have long computation time, we choose not to use them.

Let F_i^x denotes the differential hashing function for the ith item of the hashing function h^x, F^x denotes the precomputed differential function over the transition function f, and δ_f denotes the set of different items for the transition function f : $\delta_f = \{x_i \text{ such that } x_i \neq f(x_i)\}$.

The first function (h^1) uses the folding method based on the logical operator "exclusive or", as previously described :

$$h^1(<x_1,...,x_i,...,x_N> = \bigoplus_{i=1}^{N} x_i \ ,$$

$$F_i^1 \ (h^1(<x_1,...,x_i,...,x_N>), \ x_i, \ y_i) = h^1(<x_1,...,x_i,...,x_N>) \oplus y_i \oplus x_i \ ,$$

$$F^1(h^1<x_1,...,x_i,...,x_N>, \ f) = h^1(<x_1,...,x_i,...,x_N>) \oplus (\bigoplus_{x_i \in \delta_f} (f(x_i) \oplus x_i) \ ).$$

The second function (h^2) is a sum of the key items weighted by the power of the prime constant p:

$$h^2(<x_1,...,x_i,...,x_N> = \sum_{i=1}^{N} x_i.p^{(i-1)},$$

$$F_i^2 \ (h^2(<x_1,...,x_i,...,x_N>), \ x_i, \ y_i) = h^2(<x_1,...,x_i,...,x_N>) + (y_i - x_i).p^{(i-1)} \ ,$$

$$F^2(h^2<x_1,...,x_i,...,x_N>, \ f) = h^2(<x_1,...,x_i,...,x_N>) + \sum_{x_i \in \delta_f} (f(x_i) - x_i).p^{(i-1)} \ .$$

The last function (h^3), which can be found in [Knuth 73], combines multiple precision integer arithmetic and modulo operation. b is equal to the number of bits needed to code every item. K is a prime number. In short, the whole key is interpreted as multiple precision integer.

$$h^3(<x_1,...,x_i,...,x_N> = (\sum_{i=1}^{N} x_i.2^{(i-1).b} \ ) \bmod K \ ,$$

$$F_i^3 \ (h^3(<x_1,...,x_i,...,x_N>), \ x_i, \ y_i) = (h^3(<x_1,...,x_i,...,x_N>) + (y_i - x_i).2^{(i-1).b}) \bmod K,$$

$$F^3(h^3<x_1,...,x_i,...,x_N>, \ f) = (h^3(<x_1,...,x_i,...,x_N>) + \sum_{x_i \in \delta_f} (f(x_i) - x_i).2^{(i-1).b}) \bmod K.$$

The performances of the three hashing functions are displayed in two graphs. In both graphs, the time unit is $1/60.10^{-6}$ second.

The first graph shows the performance results of the usual algorithm and differential algorithm for various key lengths [Figure 3]. The codes of the mono-differential and precomputed processes are similar so their processing times are almost equal : to facilitate the graph reading, only the differential values are drawn. As the processing times can depend on the modified item, we use the following process to obtain meaningful results : we compute the mean of several processing time based on random drawings of the modified item.

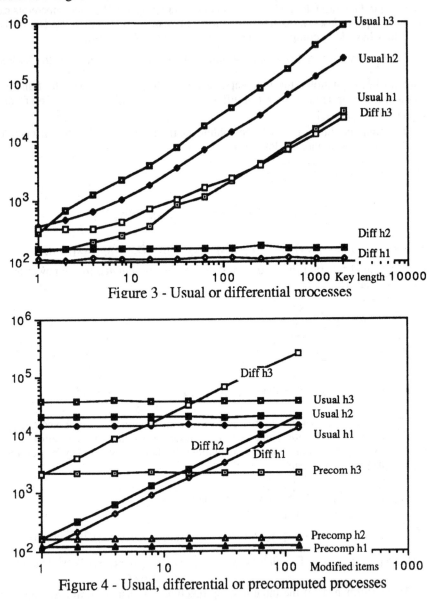

Figure 3 - Usual or differential processes

Figure 4 - Usual, differential or precomputed processes

The second graph shows the processing time obtained by the usual algorithm, the differential algorithm and the precomputed differential algorithm [Figure 4]. In this case the key length is fixed and equal to 128. We compute the processing time for various numbers of simultaneous modified items.

3.2 Preliminary discussion

Before going inside the comparison of hashing function computation processes (usual, differential or precomputed), it is worth to discuss some facts concerning the functions themselves. First, we observe that all the functions have an usual computation process which increases in time depending on the length of the key (Figure 3). In fact, to be efficient, hashing functions must use all the key items, if they are significant, to compute a hashing value [Knuth 73]. So the usual computation processes have similar behavior : their computation time increases according to the key length.

Recall that the displayed times do not show the overall performance times of hashing methods, because hashing methods are based on hashing/collision resolution function pair. We emphasize that an inadequate hashing function can generate numerous collisions which will degrade considerably the overall performance of the hashing method. However, we should not conclude that the obtained results on hashing computation time are not significant. In fact, the time due to resolution collision is independent of key length, so it becomes negligible for long keys compared to the computation time of the hashing functions [Deudon 92]. Furthermore experiments show that the three previous hashing functions are rather good scattering functions.

We have made these performance tests over numerous versions of the hashing functions: word length variation (1, 2, 4 bytes item), modification of the overflow treatment, table extraction or computation of the computational factor, etc. The previous and following described behaviors have been maintained, even if some local optimizations have been noted.

3.3 Results

If we compare the results obtained from the differential algorithm to those from the usual algorithm, we observe that the processing time of the differential algorithm is, on the one hand, shorter than the processing time of the usual algorithm, and on the other hand, constant with respect to key length, with the exception of the hashing function number 3 (Figure 3). In fact, the implementation of the differential algorithm of the third hashing function uses multiple precision integer arithmetic. So the processing time increases according to key length.

The influence of the modified item location can be observed over all the hashing functions and their differential algorithms, but this influence is very low over all the functions except the number 3 hashing function, as established in the previous paragraph. The other two functions, although their absolute duration is short, have some erratic variations. In fact, theses variations are generated by either the coincidence or not coincidence of the modified item location with some constants used during differential computation process.

We recall that the previous results have been established by algorithms based on mono-differential functions. In case of multiple differences (case where the son_key differs from the father_key by more than one item) the processing time can be deduced from the mono-differential result : it is equal to the product of this value and the number of differences. So the duration is proportional to the number of different items between the two keys.

The Figure 4 exhibits the following behaviors: the processing times of the usual or precomputed algorithms are constant : they are independent of the number of simultaneous modified items; even though the processing time of differential algorithm is proportional to the number of simultaneous modified items. As stated in our previous work, the processing time of differential algorithm is shorter than the one of usual algorithm, only if the number of modified item is low (>10%). Previous experiments have pointed out that this condition is verified in practice, and accordingly differential algorithm improves the processing time of state space exploration process.

Nonetheless the current measurements show real improvement of the processing time using the precomputed differential algorithm compared with the two other algorithms. Shorter are the processing time for usual or differential algorithm, shorter are the precomputed algorithm, for a ratio from 20/1 to more than 100/1.

4. Application to state space exploration

Usual state spaces can be very large. In fact, the graph size and the number of states depend on parallelism and accuracy of the modelized system. Previous studies lead to establish that real distributed system or protocol models have states whose size is significant: 1916 bytes for P-channel protocol [Doldi 92]; hundreds of bytes for Holzmann [Holzmann 91]; or from our own experiments several hundred of bytes (Transport class 4 protocol).

Large state space exploration raises two main problems : very huge storage and very long computation time. Precomputed differential computation process only address the second problem, but many methods, trying to reduce the storage requirement, increase their processing time (it is the usual time/space trade-off). Several of these methods make intensive use of hashing functions [Holzmann 88, Wolper 93]. In this case our proposition can be very effective if it can be combined with these methods : under those circumstances, state space exploration requires less storage space and shorter processing time.

An intensive performance test campaign has been carried out using either the Bouster validation tool [Campergue 91] or the Open/Caesar verification environment [Garavel 90] on a set of several distributed system models. These models had various number of places (2 to 50000), various numbers of transitions (1 to 50000), various numbers of arcs (10 to 100000) and they produce several hundred thousand states. These models either have been obtained from an automatic tool which generates models with specific or symmetrical topologies, either have been found in literature, or have been provided by the package of the tool itself.

Using the Unix profiler tool, our study established that the hashing function, the collision resolution and access function, the transition selection function and the compare function are the most time consumer functions: each of these functions consume on average 15 to 40 % of the processor time in user mode depending on the distributed system studied, and in various order. The order and the utilization time of these functions are variable because they depend on the collision rate which itself depends on the hashing function, the size of the hashing table and the model characteristics. All the other functions without exception used less than 10% of the processor time (most of them significantly less). Some specific methods, like bitstate method, can raise the processor utilization rate of hashing function to more than 50%, if the collision rate is kept low (i.e. hashing table size is close to explored state space size).

We have built three versions for both verification tools using the h^2 function which is faster than the h^3 function and has better distribution than the h^1 function. The first version uses the usual computation process, the second version the differential computation process, the third one the precomputed differential computation process. Let us recall that the first tool (Bouster) uses Petri net as description language, while the second one (Caesar) uses Lotos. We have selected some significant results (Table 5).

First, all the measurements show an hashing function speed-up produced by the precomputed differential computation process compared with the usual computation process from 100/1 to 70/1, so our technic is rather efficient. Second, in contrast with the differential speed-up which varies with the studied system, the precomputed speed-up is quite constant and independent of the context. Third, the results show that the acceleration is independent of the description language and of the verification method used.

	Total time	usual comp.	diff. comp.	precomp. proc.
TP4 protocol (P. net)	2064.6	621.5	147.8	18.9
alternate bit protocol (P. net)	880.3	367.1	67.5	7.3
alternate bit protocol (Lotos)	2357.0	1043.4	214.4	40.5
symmetrical ring (P. net)	5040.6	2405.3	1070.1	23.6
symmetrical ring (Lotos)	14867.1	8655.4	4367.2	96.3

Table 5 - processing time

These results exhibit at the same time the importance and the limits of the gain which can likely be achieved with the precomputed differential process. In fact, on average the processor spends about 40% in the average of its user mode time in the code of the hashing function and collision resolution function. A fast hashing function, judicious and balanced, should enable this processing time to be reduced, decreasing the collision rate and hashing value computation time. Nevertheless, the performance increase due to a differential technique can not magically reduce the inherent complexity of the system studied, in particular the huge number of states that we sometimes need to generate. Other methods like data densification, partial exploration, on the fly verification, etc can be combined to advantage with our method.

These promising results must not hide an important phenomena already raised by numerous performance researchers: the influence of the virtual memory mechanism on execution time. In fact, a considerable slow down is noticed as soon as the data application can not longer be kept in core memory. Nevertheless the direct access technique offered by the hashing method as long as the collision rate is kept low, favors this method against all other proposed methods because it reduces the inputs and outputs between secondary and main memory.

5. Conclusion

The results show that differential hashing speed-up increases with key length. In fact, typical hashing functions use all key items (this process is recommended to enable the hashing value distribution to be balanced); hence the usual algorithm complexity is proportional to the key length. In the context of many applications (large graph, very numerous states) long key lengths are generated, as corroborated by several examples. If the differential algorithm complexity is proportional to the modified item number between the original key and the new key, on the contrary, precomputed differential algorithm complexity is fixed, and its processing time is comparable to the processing time of one function call of the mono-differential algorithm, which is very short.

The differential hashing functions are not a general answer to all the computation time problems: they do not always exist, and when they exist, they are not always the most efficient. But our previous work establishes that, first, numerous hashing functions can be associated with a complete mono-differential function set (i.e. they are differential); second, for the majority of differential computation processes processing time is shorter than for the usual ones; third, differential techniques require applications where the modified items can be obtained at low cost (no need of modified item searching). The proposed application (state space exploration) has all these prerequisite characteristics, and consequently enables a substantial improvement of performances.

The current work shows that, if the hashing function is regular for the transition functions and if the hashing function is differential then its coding produces drastic improvements in processing time. In conjunction with state space compression methods the precomputed differential implementation of hashing functions have shown their great efficiency : reducing of both storage and processing time requirements to achieve the verification of distributed systems.

References

[Algayres 91] B.Algayres, & all, "Vesar: a Pragmatic Approach to Formal Specification and Verification", Computer Network and ISDN Systems, vol 25 n°7, February 1991.

[Bonafos 90] B.Bonafos, E.Domingo, "Leda : Structured Language for Automata Description and Verification", rapp. de recherche, Bordeaux-France, june 1990.

[Campergue 92] C.Campergue, C.Nouaille, "Bouster : génération parallèle du graphe des marquages accessibles", rapport ENSERB, Bordeaux-France, Juin 1992.

[Cousin 93] B.Cousin, "Differential Hashing Functions : Application to Reachability Graph Generation". ICCI'93. Sudbury - Canada, 26-29 May 1993.

[Cousin 93b] B.Cousin, "Les fonctions de hachage différentielles". CFIP'93. Montréal - Canada, 7-9 septembre 1993. Hermès, p525-541.

[Deudon 92] G.Deudon, C.Houillon, "Techniques de hachage", rapport interne ENSERB, Bordeaux-France, Juin 1992.

[Dimitrijevic 89] D.D.Dimitrijevic, M.S. Chen, "Dynamic State Explosion in Quantitative Protocol Analysis",PSTV-IX, Twente-Netherland, 6-9June 1989.

[Doldi 92] L.Doldi, P.Gauthier, "Veda-2: Power to the Protocol Designers", FORTE'92, Lannion-France, 13-16 Octobre 1992.

[Garavel 90] H.Garavel, J.Sifakis, "Compilation and Verification of Lotos Specifications", PSTV-X, Ottawa, june 1990.

[Holzmann 88] G.J.Holzmann, "An Improved Protocol Reachability Analysis Technique", Sofware, Practice and Experience, vol 18 n°2, Feb. 1988.

[Holzmann 91] G.J.Holzmann, "Design and Validation of Computer Protocols", Prentice-Hall, 1991.

[Knuth 73] D.E.Knuth, "The Art of Computer Programming: Sorting and Searching", vol 3, Addison-Wesley, 1973.

[Knott 75] G.D.Knott, "Hashing Functions", The Computer Journal, vol 18, n°3, August 1975, p265-278.

[Lum 71] V.Y.Lum, P.S.T.Yuen, M.Dodd, "Key-to-Adress Transform Techniques : a Fundamental Performance Study on Large Existing Formatted Files", Communications of the ACM vol14 n°4, April 1971.

[West 86] C.H.West, "Protocol Validation by Random State Exploration", PSTV-VI, Montréal - Canada, June 1986.

[Wolper 93] P.Wolper, D.Leroy, "Reliable Hashing without Collision Detection", CAV, Elounda - Greece, June 1993.

[Zhao 86] J.Zhao, G.Bochmann, "Reduced Reachability Analysis of Communication Protocols: a New Approach", PSTV-VI , Montréal - Canada, June 1986.

Combining Partial Order Reductions with On-the-fly Model-Checking

Doron Peled

AT&T Bell Laboratories
600 Mountain Avenue
Murray Hill, NJ 07974, USA

Abstract

Partial order model-checking is an approach to reduce time and memory in model-checking concurrent programs. On-the-fly model-checking is a technique to eliminate part of the search by intersecting the (negation of the) checked property with the state space during its generation. We prove conditions under which these two methods can be combined in order to gain from both reductions. An extension of the model-checker SPIN, which implements this combination, is studied, showing substantial reduction over traditional search, not only in the number of reachable states, but directly in the amount of memory and time used.

1 Introduction

Partial order model-checking is an approach to reduce time and memory when checking that concurrent programs satisfy their linear temporal specification. The main idea behind partial order methods is the observation that when modeling the executions of a program as interleaved sequences of atomic actions (and indeed most assertion languages used are based on such modeling), concurrent activities are interleaved in many possible orders. The checked properties are in many cases insensitive to the interleaving order. This allows fixing some arbitrary order among them, which allows reducing the size of the checked state space. In the kernel of such algorithms for generating a reduced state space are routines for exploring from each generated state a *subset* of the successor states rather than *all of them*.

Partial order methods were at first restricted to checking a constrained family of properties: the verification method of Katz and Peled [9] and the model-checking methods of Valmari [17] and Godefroid [4] were limited to dealing with safety properties, termination, local and stable properties. Later, Valmari [18] developed a way to check arbitrary nexttime-free temporal properties. Peled [16] generalized these ideas and showed how to gain more reduction by rewriting the checked formula, and how to do the model-checking under fairness assumptions.

We suggest here an algorithm that combines on-the-fly model-checking [11, 3, 2] with partial order reduction. That is, intersect the reduced state space during its generation with an automaton that represents the negation of the checked property. Then, besides the benefit of generating a reduced state space, the generation does not necessarily need to be completed: it might be that an error will be found before the end of the construction, or that parts of the (reduced) state space need not be present in the intersection. The method allows checking the class of stuttering-closed Büchi automata properties, which includes the nexttime-free temporal properties.

Godefroid and Wolper [5] proposed a method based on combining automata, one for each of the program's processes, with one for the checked formula. With this method, if the program can execute from some point two totally independent (i.e., concurrent) infinite tasks, one of them might be ignored by the reduction algorithm. Valmari [19] presented a different method for on-the-fly partial order reduction: in order to solve the above ignoring problem, the choice of a subset of successors taken from each node is based on an algorithm that prefers to include all the operations that can make a change to the current state of the property automaton.

Both of these methods handle specifications represented as automata over (illegal) sequences of the program *operations*. The edges of the specification automaton B are labeled with operations. The specification automaton in both of these methods synchronizes each of its transitions, labeled with some program operation, with the execution of that program operation in one or more processes. When a pair of program operations that can be executed concurrently in the checked program appear both on transitions of B, their relative order can no longer be fixed arbitrarily. Thus, the checked property restricts the partial order reduction of the state space.

Our method allows combining a Büchi automaton that defines illegal sequences of program *states*. This is the kind of automaton obtained from a temporal specification, e.g., by using Wolper's translation algorithm [20] and is also the kind of automaton used to check properties with the model-checker SPIN [6]. In this setting, the transitions of the specification automaton B are labeled with sets of propositions rather than with program operations. Each transition of B is synchronized with global program states. Again, the partial order reduction must be restricted by the checked property in order to guarantee that the truth of the checked property is preserved between the full and the reduced state space. This is done here by restricting the reduction, not allowing to arbitrarily fix the order between any pair of operations that can both change at least one of the checked predicates. Unlike [5, 19], in our approach, we are not always deprived of the ability to fix an arbitrary order between two concurrent operations when both can cause a change to the state of the specification automaton B when synchronized with it.

Special care is taken w.r.t. the above ignoring problem. A way to avoid this problem is to disallow a reduced set of successors from some state, when one of them closes a cycle. This was first suggested in [16]. The applicability of it to the on-the-fly case is not trivial and is carefully proven here. It should be mentioned that a requirement such as the above is very subtle: an earlier condition for the ignoring problem appeared in [7] and required only that at least one of the selected operations does not close a cycle. This turns out to be insufficient for preserving temporal properties.

Our strategy is in essence the reverse of [19]. Namely, we prefer selecting operations that *do not* change the predicates or propositional values in the checked formula. The advantage is the use of a simpler algorithm than [19] for calculating subsets of successors. The reduction method presented here also introduces partial order reductions done on-the-fly under various fairness assumptions.

The reduction method suggested here is presented as a collection of constraints on the selection of an appropriate subset of the enabled operation from each given program state. The constraints are simple and easy to implement within state space model-checkers. The cost of the additional calculations in time and space required to find appropriate subsets of operations is very small, hence making the additional calculations pay-off very quickly. The algorithm described here was implemented as an extension to the model-checker SPIN [6]. This is the first implementation of a partial order model-checker with the full power of stuttering closed Büchi automata [1]. Experiments with various known algorithms and protocols show substantial reductions in space and time.

2 Preliminaries

A *finite-state program* P is a triple $\langle T, Q, \iota \rangle$ where T is a finite set of operations, Q is a *finite* set of *states*, and $\iota \in Q$ is the *initial state*. The enabling condition $en_\alpha \subseteq Q$ of an operation $\alpha \in T$ is the set of states from which α can be executed. Each operation $\alpha \in T$ is a partial transformation $\alpha : Q \mapsto Q$ which needs to be defined at least for each $q \in en_\alpha$.

An *interleaving sequence* of a program is an infinite[1] sequence of operations $v = \alpha_1 \alpha_2 \ldots$ that *generates* the sequence of states $\xi = q_0 q_1 q_2 \ldots$ from Q, such that (1) $q_0 = \iota$, (2) for each $0 \leq i < n$, $q_i \in en_{\alpha_{i+1}}$ holds, and $q_{i+1} = \alpha_{i+1}(q_i)$. The interleaving semantics of a program sometimes involves a restricting condition on interleaving sequences called *fairness*. Then, only sequences satisfying the assumed fairness conditions are considered to be executions of the program.

An *admissible* sequence is an interleaving sequence or any segment, i.e., a suffix of a prefix, of such a sequence. We represent an admissible sequence either as a set of states from Q (denoted using ξ, $\xi' \ldots$), or a sequence of executed operations $\alpha_1 \alpha_2 \alpha_3 \ldots$ (denoted using v, u, w, v', v_i, $\ldots$). The fact that ξ is the sequence of states obtained by executing the sequence of operations v from initial state ι is denoted by $states(v, \xi)$. The last state of a finite admissible sequence obtained by executing the sequence of operations v from ι is denoted by fin_v.

Definition 2.1 *An* independence relation *is a binary reflexive and symmetric relation* $I \subseteq T \times T$ *such that for each pair of operations* $(\alpha, \beta) \in I$ *(called* independent *operations) it must hold that for each* $q \in Q$,

- *If* $q \in en_\alpha \cap en_\beta$ *(i.e.,* α *and* β *enabled from* q*), then* $\beta(q) \in en_\alpha$ *(due to symmetry, also* $\alpha(q) \in en_\beta$*).*
- *If* $q \in en_\alpha \cap en_\beta$ *then* α *and* β *are commutative as state transformers of* Q. *That is,* $\alpha(\beta(q)) = \beta(\alpha(q))$.

A dependency relation D is the complement of an independence relation, i.e., $D = (T \times T) \setminus I$.

Two strings v, $w \in T^*$ are considered equivalent [15], denoted $v \equiv_D w$, iff there exists a sequence of strings u_0, u_1, $\ldots$, u_n, where $u_0 = v$, $u_n = w$, and for each $0 \leq i < n$, $u_i = \bar{u}\alpha\beta\hat{u}$ and $u_{i+1} = \bar{u}\beta\alpha\hat{u}$ are admissible[2] for some $\bar{u}$, $\hat{u} \in T^*$, α, $\beta \in T$, $(\alpha, \beta) \notin D$. That is, w is equivalent to v iff it can be obtained from it by repeatedly commuting adjacent independent operations.

The definition of equivalence between finite strings is now extended to interleaving sequences [12]. Denote by $Pref(w)$ the set of finite prefixes of the (finite or infinite) string w. A relation '$\preceq_D$' is defined between pairs of finite or infinite strings over T as follows: $v \preceq_D v'$ iff $\forall u \in Pref(v) \exists w \in Pref(v') \exists z \in T^*(w \equiv_D z \wedge u \in Pref(z))$. That is, each finite prefix of v is a prefix of a permutation (obtained by commuting adjacent independent operations) of some prefix of v'. Extend now '$\equiv_D$' to infinite strings by defining $v \equiv_D v'$ for v, $v' \in T^\omega$ iff $v \preceq_D v'$ and $v' \preceq_D v$. It is easy to see that '$\equiv_D$' is an equivalence relation [12]. A *trace* is an equivalence class of admissible

---

[1]Finite executions can be avoided, for simplifying the representation, by adding a special operation that is enabled exactly when no other operation is enabled and does not change the state.

[2]The requirement of admissibility is essential here since the definition of independence does not rule out the case that $(\alpha, \beta) \in I$, β is disabled from some state q, and becomes enabled after α is executed from q.

finite [15] or infinite [12] sequences. Denote a trace σ by $[v]_D$, where v is any member of σ. The index D is omitted when clear from the context. It can be easily shown that for finite v, if $v \equiv_D v'$, then $fin_v = fin_{v'}$.

Concatenation of two traces $\sigma = [\alpha_0 \alpha_1 \ldots \alpha_n]$ and $\sigma' = [\beta_0 \beta_1 \ldots \beta_m \ldots]$, where σ is finite and σ' is either finite or infinite, is defined as $\sigma \sigma' = [\alpha_0 \alpha_1 \ldots \alpha_n \beta_0 \beta_1 \ldots \beta_m \ldots]$, provided that $\alpha_0 \alpha_1 \ldots \alpha_n \beta_0 \beta_1 \ldots \beta_m \ldots$ is admissible. Denote $\sigma \sqsubseteq_D \rho$ if $\sigma = [v]$, $\rho = [w]$ and $v \preceq_D w$. For finite traces, it means that there exists σ' such that $\rho = \sigma \sigma'$. A *run* π of a program P, defined with respect to some dependency relation D, is an infinite trace that contains interleaving sequences of P.

A nexttime-free LTL [14, 13] formula φ is constructed from propositional variables $p_0, p_1, p_2 \ldots$, the boolean connectives ('$\wedge$', '$\vee$', '$\neg$') and the modals '$\square$' (always), '$\diamond$' (eventually) and '$\mathcal{U}$' (until), but not the modal '$\bigcirc$' (next-time). Denote the propositions appearing in the checked formula φ by $\mathcal{P}$. An *interpretation mapping* is a function $\mathcal{F} : Q \mapsto 2^{\mathcal{P}}$, i.e., $\mathcal{F}(q)$ are the variables that are assigned a truth value T in q (the others are assigned the value F). We assume the existence of an interpretation mapping $\mathcal{F}_{(P, \mathcal{P})}$ for each pair of a program P and a set of propositional variables $\mathcal{P}$ (these indexes will be omitted when clear from the context). An interpretation mapping can be extended to sequences, mapping an execution sequence ξ of P into a *propositional sequence* $\mathcal{F}(\xi)$. The fact that a sequence ξ (more precisely, $\mathcal{F}(\xi)$) satisfies a temporal formula φ is denoted by $\xi \models \varphi$, and the fact that all the sequences of a program P satisfy φ is denoted by $P \models \varphi$.

A *state graph* of a program P, which can be used to represent the state space of P, is a graph $G = \langle \hat{s}, V, E \rangle$, where V is a finite set of nodes, $\hat{s} \in V$ is the *starting node*, and E is a finite set of edges, labeled with operations from T. For each node $s \in V$, $val(s)$ is a state of Q, and in particular, $val(\hat{s}) = \iota$. If $s \xrightarrow{\alpha} t \in E$, then $val(s) \in en_\alpha$ (we also say that α is *enabled from* s), and $val(t) = \alpha(val(s))$. A state graph *generates* a sequence of operations $\alpha_1 \alpha_2 \ldots$ (or their corresponding sequence of states), if there exists a (finite or infinite) path starting with $\hat{s}$ whose edges are labeled with $\alpha_1 \alpha_2 \ldots$ in this order.

An algorithm to generate the *full* state graph of a program can be obtained from the one in Figure 1, by replacing the underlined procedure call $ample(val(s))$ at line 3 by the set of *all* the operations enabled at the state $val(s)$ of the node s, denoted by $en(val(s))$. The flag $open(s)$ is a boolean flag that holds when s is not fully expanded, i.e., is still active, and thus is currently on the search stack.

3 Spawning Reduced State Graphs

We start the presentation with an algorithm **A1** that constructs reduced state spaces. It is related to the works of [9, 17, 18, 7, 16]. It uses a depth first search to expand the state space. We initially employ the following fairness assumption:

F if an operation α is enabled from some state of an interleaving sequence, then some operation that is dependent on α (possibly α itself) must appear later (or immediately) in this sequence.

Thus, we limit the following discussion to runs π that contain interleaving sequences satisfying **F**. Adding this fairness assumption slightly changes the algorithm and greatly simplifies its proof. Later, in Section 3.2, the fairness assumption will be removed and all the runs of the program P will be considered. Thus, two versions of the algorithm, one under the fairness requirement **F** or any stronger requirement, and one under no fairness assumption, will be presented.

3.1 Subsets Construction under Fairness

When algorithm **A1**, depicted in Figure 1, expands a node s, only a subset $ample(val(s))$ of the enabled operations $en(val(s))$ is used to generate successors for s (line 3 in Figure 1). Various algorithms to calculate the selected subset of successors from a given node s with value $val(s) = x$ can be found e.g., in the works of Katz and Peled [9], Valmari [17, 18], and Godefroid et al [4, 5, 7]. Such a subset $\mathcal{E}(x)$ must satisfy the following condition:

C1 No operation $\alpha \in T \setminus \mathcal{E}(x)$ that is dependent on some operation in $\mathcal{E}(x)$ can be executed in P after reaching the state $x = val(s)$ and before some operation in $\mathcal{E}(x)$ is executed.

It follows trivially that under the fairness assumption **F**, the condition **C1** guarantees that:

P1 For every run $\pi = [v][w]$, such that $v \in T^*$, $w \in T^{\omega}$, $fin_v = val(s) = x$, there exists $\alpha \in \mathcal{E}(x)$ s.t. $[\alpha] \sqsubseteq_D [w]$.

An *occurrence* of a state x in a run π is a string v with $fin_v = x$ such that $[v] \sqsubseteq_D \pi$. According to property **P1**, $\mathcal{E}(x)$ returns at least one immediate successor operation for each occurrence of x, in each partial order execution π of P (with respect to the dependency relation D). The algorithm **R1** to calculate $ample(x)$ appears in Figure 2. It uses the routine $check\_succ(x, i)$, which returns the operations enabled from the state x if they satisfy the property **C1** (at line 3 in Figure 2), or the empty set, otherwise. More details on how our implementation checks this condition are described in Section 5 and in [8]. A second condition [16] is enforced at lines 5–8 in Figure 2:

C2 If $\mathcal{E}(x)$ does not include *all* the operations enabled from $x = val(s)$, then no operation $\alpha \in \mathcal{E}(x)$, when applied to x, closes a cycle on the search stack (i.e., we do not allow that an open node with value $\alpha(x)$).

The correctness of our algorithm depends also on a simple property of the DFS algorithm:

P2 During the execution of a DFS algorithm, if the immediate successors of a node s are not open (i.e., not currently on the search stack), then when closing the node s, its immediate successors are already closed.

Notice that this does not hold if there is at least one successor of the currently expanded node s on the search stack: this successor is still open when the execution backtracks to s. The requirements **C1** and **C2** guarantee the following property, during and after the execution of the expansion algorithm **A1+R1**:

P3 Let s be a closed node and let $\pi = [v][\alpha w]$, with $\alpha \in T$, be a run of P, such that $fin_v = val(s)$. Then there exists a path $s_0 \xrightarrow{\beta_1} s_1 \xrightarrow{\beta_1} \ldots \xrightarrow{\beta_n} s_n \xrightarrow{\alpha} t$, with $s_0 = s$, such that $[\beta_1 \beta_2 \ldots \beta_n \alpha] = [\alpha \beta_1 \beta_2 \ldots \beta_n] \sqsubseteq_D [\alpha w]$.

In other words, even if α is enabled from $x = val(s)$, the reduced state space G' may not contain an edge labeled by α exiting s. However, for each run π in which that state x occurs, there exists a path in G' that starts with s, and labeled with a sequence of operations $\beta_1 \beta_2 \ldots \beta_n \alpha$, (where the operations $\beta_1 \beta_2 \ldots \beta_n$ are independent of α) such that $\beta_1 \beta_2 \ldots \beta_n \alpha$ is a possible continuation of π after the occurrence of x (i.e., $[v \beta_1 \beta_2 \ldots \beta_n \alpha] \sqsubseteq_D \pi$). The proof of this property is by induction on the *order of closing nodes* by the expansion algorithm (at line 14 in Figure 1). When closing a node s, there are two possibilities:

1. All the operations enabled from $x = val(s)$ are expanded from s. This includes in particular the case where by applying one of them to x we reach an open node. In this case, the appropriate path, with a length of 1, exists trivially.

```
1   create_node(s, ι); set open(s); expand_node(s); /* initialization */
2   proc expand_node(s);
3      working_set(s) := ample(val(s));
4      while working_set(s) ≠ φ do
5         α := some operation of working_set(s);
6         working_set(s) := working_set(s) \ {α};
7         succ_state := α(val(s)); /* the α-successor of val(s) */
8         if new(succ_state) then
9            create_node(s', succ_state); /* node s' has value succ_state */
10           set open(s'); /* set s' to open, i.e., on the search stack */
11           expand_node(s') fi; /* expand the successors of s' */
12        create_edge(s, α, s');
13     end while;
14     unset open(s); /* close s, i.e., remove it from search stack */
15  end expand_node.
```

Figure 1: Algorithm **A1**: an off-line reduced state space expansion algorithm

```
1   proc ample(x):set of T;
2      for i := 1 to num_proc /* repeat for every process */
3         Ε := check_succ(x, i); /* enabled Pᵢ operations, if satisfy C1 */
4         if Ε ≠ φ then /* ... otherwise, check_succ returns φ */
5            foreach α in Ε /* check cycle closing */
6               │if exists s' with val(s') = α(x) and open(s')│ /* α closes a cycle */
7                  then goto next_proc;
8            end foreach;
9            return(Ε) fi /* A subset satisfying C1 and C2 found */
10        next_proc: next i;
11        return(en(x)); /* cannot find a good subset */
12  end ample;
```

Figure 2: Routine **R1**: finding a subset of successors under the fairness **F**

2. A proper subset of operations enabled from x is calculated by $ample(x)$. By **C1** and hence **P1**, there exists an edge $s \xrightarrow{\gamma} s'$ such that γ is an immediate successor of this occurrence of x in the run π. By **C2**, none of the operations in $ample(x)$ applied to the state $x = val(s)$ closes a cycle. Thus, s' is not open. By **P2**, once completing the expansion of s, the node s' is already closed. Thus, the inductive hypothesis can be applied to s' to obtain a path that ends with α. Since both α and γ are immediately executed from the state $val(s)$ in π, they are interindependent and thus the edge $s \xrightarrow{\gamma} s'$ can be appended to the beginning of that path to form a longer path with the appropriate property.

The ability to use the reduced state graph is based on the following theorem [16]:

Theorem 3.1 *The reduced state graph G' generated by algorithm* **A1+R1** *satisfies the following properties: (1) all the sequences generated by G' are interleaving sequences of P, and (2) for each run π of the program P, there exists at least one sequence that corresponds to some path of G', starting from its initial node.*

The first property is trivial. The second is proved by constructing for an arbitrary interleaving sequence of P, inductively on the length of its prefixes, an equivalent sequence which is generated by G', using **P3**. This allows applying LTL model-checking algorithms to G', instead of to the full state space G for properties φ that satisfy:

P4 If $\xi \equiv_D \xi'$ (i.e., ξ and ξ' belong to the same run), then $\xi \models \varphi$ iff $\xi' \models \varphi$.

This is of course unsatisfactory, as (1) the checked property φ may not satisfy **P4**, and (2) it can be difficult to check if φ satisfies **P4**. To allow checking arbitrary nexttime-free temporal properties, we employ the following definition [18]:

Definition 3.2 *An operation $\alpha \in T$ is visible in φ if it can change the truth value of some predicate that appears in the checked property φ. Denote the set of operations that are visible in φ by $a(\varphi)$.*

Theorem 3.3 *Let φ be a nexttime-free LTL property and the dependency relation D' used satisfies $D' \supseteq (a(\varphi) \times a(\varphi))$. If $v \equiv_{D'} v'$, with $states(v, \xi)$, $states(v', \xi')$, then $\xi \models \varphi$ iff $\xi' \models \varphi$.*

The proof of this theorem [16] is based on showing that the two propositional sequences $\mathcal{F}(\xi)$ and $\mathcal{F}(\xi')$, are equivalent up to stuttering (see also Theorem 3.6). Since φ is next-time free, it cannot distinguish between these two propositional sequences [13]. Now, in order to force the premise of the theorem, instead of using a dependency relation D obtained by analyzing the commutativity between the operations of the program P, we can use $D' = D \cup (a(\varphi) \times a(\varphi))$. Notice that extending the dependency relation while maintaining its symmetry and reflexivity preserves the conditions in Definition 2.1. In [16] it is shown how to avoid adding all the pairs $a(\varphi) \times a(\varphi)$ to the dependency relation. This is based on the fact that if φ is written as a boolean combination of smaller temporal formulas φ_i, then we need to add to D the dependencies $\bigcup_i (a(\varphi_i) \times a(\varphi_i))$, which can be fewer than $a(\varphi) \times a(\varphi)$. This can be formalized as an additional requirement:

C3 The dependence relation D used satisfies besides the conditions of Definition 2.1 also that $D \supseteq \bigcup_i (a(\varphi_i) \times a(\varphi_i))$, where φ can be equivalently written as some boolean combination of the temporal properties φ_i.

Notice that there may be various ways to rewrite φ as a boolean combination (the most trivial of which is to use φ itself), some of which may give bigger dependencies relations than others. For example, if $\varphi = \square(p_1 \wedge p_2)$, then $a(\varphi)$ includes all the operations whose execution may change the value of the predicates p_1 or p_2. But φ is logically equivalent to $\square p_1 \wedge \square p_2$ with $a(\varphi_1) = a(\square p_1)$ includes dependencies between the operations that can change p_1, and similarly, for $a(\varphi_2)$. Thus, a pair of operations such that the first changes only p_1 but not p_2, and the second changes p_2 but not p_1 is in $a(\varphi) \times a(\varphi)$; however, these operations are not necessarily dependent according to **C3**.

When model-checking under a fairness assumption such as **F**, or any stronger assumption, e.g., process justice or process fairness, the algorithm **A1** with the routine **R1** are appropriate. One just has to apply a model-checking algorithm that is tuned to the fairness assumption used, as shown in [14], to the reduced state space. In case one is using a fairness assumption ψ that is strictly stronger than **F**, one needs also to add dependencies between operations, so that if $\xi \equiv_D \xi'$, then $\xi \models \psi$ iff $\xi' \models \psi$. To achieve this, one can apply requirement **C3** also to ψ, in the same way it is applied to φ [16].

3.2 Subsets Construction without Fairness Assumption

We remove now the fairness assumption **F**. Thus, we consider now all the runs π of P. In this case, the condition **C1** does not guarantee anymore the property **P1**: there may be a run π that has no immediate successors for an occurrence of $x = val(s)$ among the subset of successors $\mathcal{E}(x)$ chosen by the algorithm **R1**. Condition **C3** is replaced now with the following:

C3' If $\mathcal{E}(x)$ does not include all the operations enabled from x, then none of the operations in the selected subset of operations $\mathcal{E}(x)$ is visible.

As will be evident from the following theorems, requirement **C3'** enforces that visible operations cannot be commuted. This means effectively that the commutativity is restricted by the dependency relation $D \cup (a(\varphi) \times a(\varphi))$. Thus, this is a stronger requirement than **C3**, which allows adding dependencies to D after decomposing the property φ to subformulas, each one of which contributing a smaller number of dependencies. We modify algorithm **R1** into algorithm **R2**, which is the same as the one in Figure 2, except for replacing the frame at line 6 with the following:

| if $visible(\alpha)$ or (exists s' with $val(s') = \alpha(x)$ and $open(s')$) |

Instead of **P1**, the following weaker property now holds:

P1' For every run $\pi = [v][w]$, such that $x = fin_v = val(s)$, $\mathcal{E}(x)$ either

    a. there exists some $\alpha \in \mathcal{E}(x)$ such that $[\alpha] \sqsubseteq_D [w]$ or

    b. the operations in $\mathcal{E}(x)$ are invisible and independent of all the operations that appear in w.

Consider now a string w to be a (finite or infinite) vector of operations. Denote by $w(i)$ the $i + 1st$ operation in w (the first operation is $w(0)$), by $w(i..j)$ the operations in the places i through j, and by $w(n..)$ the operations of w except the first n. Let v be a finite string over T of length n. A *selection function* for v is a function $r : \{0 \ldots n-1\} \mapsto \{\text{T}, \text{F}\}$. Denote by v_r the string remaining after deleting all the symbols $v(i)$ where $r(i) = \text{F}$. Denote by v_{f} the string remaining after deleting the symbols $v(i)$ where $r(i) = \text{T}$. Selection functions are also extended to infinite strings in a natural way. Let $r \angle m$ be the selection function r shifted to the left m places, i.e., $r \angle m(i) = r(i + m)$.

The property **P3'** replaces now **P3**, where the underlined consequence is replaced by:

P3' $\ldots$ for $u = \beta_1 \beta_2 \ldots \beta_n$, there exists a selection function r such that (1) $[u\,\alpha] = [\alpha\,u_r,\,u_{\text{f}}]$, (2) there exists w' such that $[u_r\,w'] = [w]$ (and hence $[u_r] \sqsubseteq_D [w]$), (3) the operations $\beta_1 \beta_2 \ldots \beta_n$ are invisible, and (4) the operations in u_{f} are independent of the operations in w'.

The proof of **P3'** is similar in details to the proof of **P3**.

Definition 3.4 *Let $A \subset T$ be a set of letters. Let $v, w \in T^\omega$. Define $v \preceq_D^A w$ if there exists a selection function r for w such that $v \equiv_D w_r$, the operations of w_{f} are among A, and if $r(m) = \text{F}$, then the operations in $w(m+1..)_{r\angle m+1}$ are independent of $w(m)$.*

That is, $v \preceq_D^A w$ iff it is possible to remove from w some operations from A, which are independent of all the non-removed operations of w that come after them, and obtain a string which is equivalent to v. The fact that the reduced state space G' can be used instead of the full state space G to check that $P \models \varphi$ (although without the fairness assumption **F**, it does not satisfy consequence (2) of Theorem 3.1), is based on the following theorems:

Theorem 3.5 *For every interleaving sequence v of the program P, there exists an interleaving sequence v' of P generated by the reduced state graph G' obtained by* **A1+R2** *such that $v \preceq_{D'}^A v'$, with $D' = D \cup (a(\varphi) \times a(\varphi))$ and $A \cap a(\varphi) = \phi$.*

The proof is by constructing inductively (over the length of prefixes of v) a sequence v' from v, using **P3'**.

Theorem 3.6 *Let φ be a nexttime-free LTL formula with a set of visible operations $a(\varphi)$. Let $A \subseteq T$, $A \cap a(\varphi) = \phi$ and $D' = D \cup a(\varphi) \times a(\varphi)$. Let v, $v' \in T^\omega$ such that $v \preceq_{D'}^A v'$ and let $states(v, \xi)$, $states(v', \xi')$. Then $\xi \models \varphi$ iff $\xi' \models \varphi$.*

Sketch of proof. It can be shown that $\mathcal{F}(\xi)$ and $\mathcal{F}(\xi')$ are equivalent up to stuttering: for every prefix v' of v with n occurrences of operations from $a(\varphi)$ of v, there exists a prefix w' of w with n occurrences of visible operations, and vice versa. Because the visible operations are all interdependent, these occurrences are the same and appear in both prefixes in the same order. This relies on properties of traces [15]. Then it follows that the interpretation mapping $\mathcal{F}$ assigns the same subset of propositions to the states $fin_{v'}$ and $fin_{w'}$.

To summary, theorems 3.5 and 3.6 (or Theorems 3.1 and 3.3, under fairness, respectively) assert that the algorithm **A1** with **R2** (or with **R1**, resp.) constructs a state graph G' that generates for each propositional sequence (fair propositional sequence, resp.) of a program P a propositional sequence which is equivalent to it up to stuttering. To guarantee it, the dependency relation used does not allow to commute the visible operations, i.e., those that can change the value of the propositional variables (when fairness is assumed, certain commutativity between visible operations is allowed by **C3**). Since the checked property φ is restricted to be nexttime-free, it cannot distinguish between any two stuttering equivalent sequences [13], and thus φ holds in all the sequences generated by G' iff it holds in all the interleaving sequences of P.

4 Combining the Reduction with On-the-fly Intersection

One may view $P \models \varphi$ as language containment: let L_P be the language of the propositional sequences of P, obtained from the interleaving sequences of P using the interpretation mapping $\mathcal{F}$. Let L_φ be the language of the propositional sequences satisfying φ. Then $P \models \varphi$ is the same as $L_P \subseteq L_\varphi$ or equivalently $L_P \cap \overline{L_\varphi} = \phi$ [11] (where $\overline{L_\varphi}$ is the compliment of the language L_φ w.r.t. the alphabet $2^\mathcal{P}$). This view is very important to the correctness of our on-the-fly algorithm.

To implement a checker for the latter condition, one can transfer the property $\neg\varphi$ into a finite Büchi automaton $\mathcal{B}$ that generates exactly the sequences of the language $L_{\neg\varphi}$ (which is the same as $\overline{L_\varphi}$), as shown in [20]. Such an automaton $\mathcal{B}$ is a quintuple $\langle S, \Delta, \Sigma, \delta, F \rangle$, where S is the set of automaton states, $\Delta \subset S$ is the set of *initial states*, $\Sigma = 2^\mathcal{P}$ is the alphabet (the labels on the transitions), $\delta \subseteq S \times \Sigma \times S$ is the transition relation, and F is the set of *accepting* states. The automaton $\mathcal{B}$ *accepts* an infinite sequence ξ iff there exists an infinite path in $\mathcal{B}$, starting with some state in Δ, whose *edges* agree upon the propositional variables with the *states* of ξ, such that some state in F appears in this path infinitely many times. The specification may directly be given as an automaton that accepts exactly the sequences of the negation of the checked property φ over a set of propositional variables $\mathcal{P}$, instead of as an LTL formula [11, 6, 2].

A state graph G of P can be treated as an automaton that generates the propositional sequences of P. The state graph G can be constructed "on-the-fly", i.e., while intersecting it with the automaton $\mathcal{B}$. This allows sometimes to find a counter example for the checked property before the entire graph for G is generated, or to eliminate generating some subgraphs that do not synchronize with $\mathcal{B}$, gaining in memory and time.

Let $\mathcal{A}$ be the Büchi automaton that generates the intersection of the language L_P of G and the language $L_{\neg\varphi}$ of $\mathcal{B}$. A transition $\langle x, y \rangle \xrightarrow{\langle \alpha, \beta \rangle} \langle x', y' \rangle$ of $\mathcal{A}$ corresponds to a transition $x \xrightarrow{\alpha} x'$ of G and a transition $y \xrightarrow{\beta} y'$ of B, such that x and β agree on the atomic propositions; denote by $same\_label(x, \beta)$ the fact that the node x and the transition β agree, i.e., are labeled by the same subset of $\mathcal{P}$. The initial states of $\mathcal{A}$ are $\{\iota\} \times \Delta$, i.e., the initial state of the program, paired with any initial state of B. A combined state $\langle x, y \rangle$ of $\mathcal{A}$ is accepting iff y is accepting in B. Checking for emptiness of the language accepted by $\mathcal{A}$ is done by checking iff there exists a cycle, reachable from some initial state, that contains some accepting state. In this case, the intersection is not empty (the path from an initial state that traverses this cycle indefinitely, with labels from $\Sigma = 2^{\mathcal{P}}$ on its edges, belongs to both L_P and $L_{\neg \varphi}$), and $P \not\models \varphi$.

Combining partial order reduction with on-the-fly model-checking allows reduction in space and time by both methods. Let L' be the language of the reduced state graph G'. As seen in Theorem 3.5, the constructed reduced state space G' satisfies (1) $L' \subseteq L_P$, i.e., G' generates only interleaving sequences of P, and (2) for each $\xi \in L_P$ there exists $\xi' \in L'$ such that $states(v, \xi)$, $states(v', \xi')$ for some v, $v' \in T^\omega$, and $v \preceq_D^A v'$. It follows from (2) by theorem 3.6 that L' contains for each sequence ξ of L_P a sequence ξ' that has the same truth value, i.e., $\xi \models \varphi$ iff $\xi' \models \varphi$. Now, $L_P \cap L_{\neg \varphi} \neq \phi$ iff there exists some $\xi \in L_P \cap L_{\neg \varphi}$ iff there exists some ξ', $\xi \preceq_D^A \xi'$ such that $\xi' \in L' \cap L_{\neg \varphi}$. This proves that it is sufficient to check the nonemptiness of the intersection of the languages of the automata G' (rather than G) and B.

In the intersection of the reduced state graph G', seen as an automaton, and the automaton B, for each combined state $\langle x, y \rangle$, a B transition from the second component y of the state is paired with every operation in a $subset\, \mathcal{E}(x, y)$ of the program operations enabled from the program-state component x. Conditions **C1** and **C3'** are not changed, ignoring the new component y. Condition **C2** is now changed, taking into consideration the Büchi component y:

C2' For a state $\langle x, y \rangle$, if $\mathcal{E}(x, y)$ is a proper subset of the enabled process i operations at state x, no transition $\alpha \in \mathcal{E}(x, y)$, applied to x produces a state $\langle \alpha(x), y \rangle$ of $\mathcal{A}$ that is still open.

The procedure $ample(x, y)$ that calculates such a subset is similar to $ample(x)$ in Figure 2. It uses the additional paramenter y in checking for the new condition **C2'**. The only change is the frame at line 6 which is replaced by the following (the underlined condition is not needed under fairness assumption **F**):

| if $\underline{visible(\alpha)}$) or (exists node s' with $val(s') = \langle \alpha(x), y \rangle$ and $open(s')$) |

It is tempting to reduce the correctness of the on-the-fly algorithm to the on-line version **A1** that was presented in the previous section by projecting each combined state into its first (i.e., program) component. However, this does not work: for each program-state there might be several combined states with different second component. However, one can still make a similar reduction, from the on-the-fly version into a non-deterministic variant **A2** of **A1**, described below. In *italics* we give the actual choice done when the algorithm is combined with the property automaton:

1. Choose a number n, which is the maximal number of nodes that can have the same program-state value. *This number is actually the number of states of the Büchi automaton B.*

2. When checking in **R2** whether an operation α in a subset $\mathcal{E}(x)$ of the enabled operations closes a cycle (line 6 in Figure 2), if indeed a cycle would be closed by α because of reaching some state x that is the value of some open node (this excludes the case that x is also the value of some closed node, which allows

eliminating the cycle), and there are less than n nodes with value x, decide non-deterministically if to open a new node with value x or to discard the subset $\mathcal{E}(x)$ and choose another subset. *Actually, closing a cycle is done exactly when the combined state of the program and the Büchi automaton already exists and is open (i.e., is in the search stack).*

3. The algorithm **A2** *must* create an additional node s' for a state $x' = \alpha(x)$ for $x = val(s)$ if a proper subset of the enabled operations that contains α was returned by **R2** and there exist only *open* nodes with value x' (i.e., s' is needed to eliminate closing a cycle). However, **A2** can non-deterministically choose to create a new node even in case that a closed node with value x' exists, or in case that $ample(x)$ returns the set of all the enabled operations from x. *Actually, a new node is created iff there exists no node with the value of the new combined state.* If it was decided not to create a new node for x', and there exist multiple nodes with the value x', choose non-deterministically to connect the edge labeled α from s to one of them. This non-deterministic choice replaces lines 8-12 in Figure 1. *Actually, the connection choice is made to accord also with the value of the Büchi component.*[3]

The above changes to **A1** maintain the arguments in the proof and hence in the correctness of property **P3'**. It is easy to repeat the proof of Theorem 3.5 with respect to the new version of the algorithm, namely **A2+R2**.

The language of an automaton $\mathcal{A}$ is nonempty iff there exists an accepting state, accessible from an initial state, which is reachable from itself. The algorithm in [2] seeks in DFS order the accepting states. It stores them in its search stack, and then for each one of them, in reversed order (i.e., treating the last accepting state that was found first), it looks for a cycle. The cycle search is also done using a DFS algorithm. The execution of the two DFS algorithms, DFS1 and DFS2, are interleaved. The adaptation of this algorithm to partial order model-checking is by using the procedure $ample(x, y)$ as described above. This algorithm **A3** appears in Figure 3.

DFS1 (lines 1–20) is used to generate the intersection of the reduced state space and the checked automaton. Its correctness is based on using a projection from combined states to their first component, i.e., the program-states: the obtained projected structure is the same as the one produced by the above non-deterministic algorithm **A2**. When calling $ample(x, y')$ at line 4, the value of y' is a successor of the current $\mathcal{B}$ state y. That is, we apply first a Büchi transition β to y, such that the label of the transition β has the same subset of $\mathcal{P}$ as the current program-state x. Then the program operations $\mathcal{E}(x, y')$ returned by $ample(x, y')$ are combined with β such that for each $\alpha \in \mathcal{E}(x, y')$, $\langle \alpha, \beta \rangle$ is the $\mathcal{A}$ transitions taken from $\langle x, y \rangle$ to generate the new combined state $z = \langle \alpha(x), \beta(y) \rangle$ (line 5).

DFS2 (lines 21–30) searches for cycles through accepting states in the subgraph that was already generated by DFS1. DFS2 is only executed from an accepting node s after backtracking to it. Thus, from properties of the DFS algorithm, either a cycle through s was generated by DFS1, or all the nodes of $\mathcal{A}$ reachable from s where already generated by DFS1. This guarantees that the ability of the combined algorithm to find cycles through s, as proved in [2], is preserved. Alternatively, one can trade gain in space for time by regenerating the successors of $\langle x, y \rangle$ in DFS2 using $ample(x, y)$ and thus avoiding to store the edges (lines 8 and 12). The algorithm returns $true$ iff $P \not\models \varphi$, i.e., a reachable cycle (which is a counter example to $P \models \varphi$) was found.

---

[3]The intersection of G' with $\mathcal{B}$ might cause that a successor x' of the current program state x in G' will not be generated as no transition of $\mathcal{B}$ agrees with the label of x, even when there are program operations enabled from x. This (beneficially) eliminates expanding a node for x' and its successors. It corresponds in **A2** to choosing to close existing cycles rather than generating new multiple nodes.

```
1    proc DFS1(s):boolean
2      ⟨x, y⟩ := val(s); /* x = program-state, y = Büchi component */
3      forall β ∈ Σ, y' ∈ S s.t. same_label(x, β) and y --β--> y' ∈ δ do
4        forall α ∈ ample(x, y')
5          z := ⟨α(x), y'⟩;
6          if new(z) then
7            create_node(s', z);
8            create_edge(s, s');
9            set open(s'), unchecked2(s'); /* s' not participated in DFS2 */
10           if DFS1(s') then return true fi
11          else s' := node(z); /* s' is an existing node with value z */
12           create_edge(s, s') fi;
13       od
14     od;
15     if accepting(s) then /* if backtracked to an accepting node */
16       seed:=s; /* seek for a cycle through s */
17       return DFS2(s) fi /* do secondary search */
18     unset open(s); /* remove s from search stack and backtrack */
19     return false
20   end DFS1;

21   proc DFS2(t):boolean
22     forall t' ∈ succ(t) do /* use successors of t generated by DFS1 */
23       if t'= seed then return true ; /* a cycle was found */
24       if unchecked2(t') then /* if t' did not participate in DFS2 */
25         unset unchecked2(t'); /* mark that t' participated in DFS2 */
26         if DFS2(t') then return true fi
27       fi
28     od;
29     return false
30   end DFS2;
```

Figure 3: Algorithm **A3**: On-the-fly cycle-detection

In [2], an algorithm for reachable cycle detection that can deal with certain fairness assumptions is presented. It is based on finding num_proc cycles, one for satisfying the fairness condition for each process. This algorithm can also be combined with partial order model-checking. In this case, conditions **C1**, **C2'** and **C3** restrict the selection of subsets of operations returned by calling $ample(x, y)$. Certain fairness assumptions, such as strong fairness, require a substantial change of the algorithm in [2], and will be presented in the full version.

This can be used to check fairness assumptions which are at least as strong as (i.e., imply) **F**. However, as explained in [16], one has to add additional dependencies, so that the fairness assumption will satisfy that if $\xi \equiv_D \xi'$, then $\xi \models \psi$ iff $\xi' \models \psi$. That is, two equivalent sequences will be either both fair or both unfair. To achieve this, condition **C3** is applied also to the fairness assumption ψ the same way as it is applied to the checked property φ.

5 Implementation and Experimental Results

An instance of this algorithm was implemented by Gerard Holzmann as an extension of the model-checker SPIN [8]. SPIN [6] is a protocol validator that checks protocols and specifications written in the language PROMELA. The main criterion for choosing the particular implementation details was to avoid adding significant overhead in time and memory for making the reduction over the classical full search. In this way, the reduction can start to pay-off quickly, as no compensation for the additional cost is needed. This motivates the following implementation details.

A dependency relation was defined between pairs of PROMELA operations. It includes dependencies between operations referring to the same global objects, e.g., global variables, synchronous queues (but excluding asynchronous queues, where reads and writes can be commuted), operations of the same process, and operations that refer to an object that appears in the checked assertion (making all visible operations dependent as required in Theorems 3.3 and 3.6). However, no explicit representation of the dependency relation is calculated or stored. Its definition is merely used to prove that the actual algorithm used to calculate a subset of successors (*check_succ* at line 3 in Figure 2) indeed enforces the condition **C1**.

During pre-model-checking compilation of the checked PROMELA protocol and specification, each program control-state l of each process-definition is analyzed and annotated by one of three types of labels. These labels correspond to whether at run time, when this process is at control-state l, the set of enabled operations of this process satisfies condition **C1**:

safe It is already known at compile time that this process' set of enabled operations can be chosen.

⟨*maybe, C*⟩ the set of this processes' enabled operations can be chosen only if the precomputed condition C (which is one out of a small number of conditions) holds during run time. For example, if the control-state l includes only receive operations, such a condition C can be that all of their receiving queues are non-empty.

not_safe The set of this process' enabled operations cannot be chosen.

The results of checking the liveness property $\Box\Diamond at(m)$, i.e., that a label m is visited infinitely often (which was implemented by introducing a global variable that changes its value just before and after label m) for various protocols is summarized in the following table. Memory is given in Megabytes and time in seconds. The first line in each pair of line shows the measurements for the full search, while the second line shows the measurements for the reduced search. All measurements were made on a 40MHz SGI 4D/480S R3k processor, with 128 Mbyte of Memory.

Alg.	States	Trans.	Mem.	Time	Comment
leader	335919	1858549	66.6	124.5	leader election
	829	1521	1.2	0.1	in ring
sorting	287736	1787327	46.4	102.6	distributed sorting
	1541	2911	1.3	0.2	
urp	6491	24241	2.2	1.8	AT&T universal
	3301	7001	1.8	0.6	receiver
snoopy	287230	1342296	35.6	101.8	cache coherence
	131873	292519	17.5	29.0	
pftp	1601373	6492515	254.5	695.9	file copy
	419076	905644	68.1	105	

Acknowledgement. The author is grateful for Gerard Holzmann, who implemented the partial order reduction on SPIN, and Ramesh Bharadwaj for helping in debugging the system. Both of them and R. P. Kurshan gave many helpful comments.

References

[1] J. R. Büchi, On a decision method in restricted second order arithmetic, in E. Nagel et al. (eds.), Proceeding of the International Congress on Logic, Methodology and Philosophy of Science, Stanford, CA, Stanford University Press, 1960, 1–11.

[2] C. Courcoubetis, M. Vardi, P. Wolper, M, Yanakakis, Memory-efficient algorithms for the verification of temporal properties, Formal methods in system design 1 (1992) 275–288.

[3] J. C. Fernandez, L. Mounier, C. Jard, T. Jeron, On-the-fly verification of finite transition systems, Formal Methods in System Design 1 (1992), Kluwer, 251–273.

[4] P. Godefroid, Using partial orders to improve automatic verification methods, Computer Aided Verification 1990, DIMACS, Vol 3, 1991, 321–339.

[5] P. Godefroid, P. Wolper, A Partial Approach to Model Checking, $6th$ LICS, 1991, Amsterdam, 406–415.

[6] G. J. Holzmann, Design and Validation of Computer Protocols, Prentice Hall Software Series, 1992.

[7] G. J. Holzmann, P. Godefroid, D. Pirottin, Coverage preserving reduction strategies for reachability analysis, Proc. IFIP, Symp. on Protocol Specification, Testing, and Verification, June 1992, Orlando, U.S.A., 349-364.

[8] G. J. Holzmann, D. Peled, STREM: A static reduction method, Manuscript, 1994, available from the authors.

[9] S. Katz, D. Peled, Verification of distributed programs using representative interleaving sequences, Distributed Computing 6 (1992), 107–120.

[10] S. Katz, D. Peled, Defining conditional independence using collapses, Theoretical Computer Science 101 (1992), 337-359, a preliminary version appeared in BCS–FACS Workshop on Semantics for Concurrency, Leicester, England, July 1990, Springer, 262–280.

[11] R. P. Kurshan, Reducibility in analysis of coordination, Lecture Notes in Communication and Information, Springer, 103, 19–39.

[12] M. Z. Kwiatkowska, Fairness for non–interleaving concurrency, Phd. Thesis, Faculty of Science, University of Leicester, 1989.

[13] L. Lamport, What good is temporal logic, IFIP Congress, North Holland, 1983, 657–668.

[14] O. Lichtenstein, A. Pnueli, Checking that finite-state concurrent programs satisfy their linear specification, $11th$ ACM POPL, 1984, 97–107.

[15] A. Mazurkiewicz, Trace semantics, Advances in Petri Nets 1986, LNCS 255, Springer, 1987, 279–324.

[16] D, Peled, All from one, one for all, on model-checking using representatives, $5th$ international conference on Computer Aided Verification, Greece, 1993, LNCS, Springer, 409–423.

[17] A. Valmari, Stubborn sets for reduced state space generation, $10th$ International Conference on Application and Theory of Petri Nets, Vol. 2, 1–22, Bonn, 1989.

[18] A. Valmari, A Stubborn attack on state explosion, in E.M. Clarke, R.P. Kurshan (eds.), CAV'90, DIMACS, Vol 3, 1991, 25–42.

[19] A. Valmari, On-The-Fly Verification of stubborn sets, $5th$ CAV, Greece, 1993, LNCS 697, Springer, 397–408.

[20] P. Wolper, M.Y. Vardi, A.P. Sistla, Reasoning about infinite computation paths, Proceedings of $24th$ IEEE symposium on foundation of computer science, Tuscan, 1983, 185–194.

Improving Language Containment Using Fairness Graphs

Ramin Hojati[1] (UC Berkeley)
Robert Mueller-Thuns (Cadence Design Systems)
Robert K. Brayton (UC Berkeley)

Abstract

Language containment is one important approach to formal design verification. When working at a higher, more abstract level, additional unwanted behavior may be introduced in the model that can be excluded for the verification step using so-called fairness constraints. The language containment computation using Binary Decision Diagrams (BDD's) typically involves performing reachability analysis, early failure detection, and then applying a set of operators until convergence is achieved ([HTKB92]). The running time of the latter part (called the main computation) is correlated with the number of fairness constraints. In this paper, we introduce techniques which improve the efficiency of the main computation by analyzing a graph induced by the fairness constraints. This graph can be built efficiently using BDD's. We have implemented our algorithms in the verification system HSIS, and experimental results demonstrate the effectiveness of these ideas.

1 Introduction

Design Verification is the process of answering the question "Is what I specified what I wanted?" This is accomplished by specifying the system at a suitable level of abstraction and then proving properties of the system. For example, in a large design one can abstract the design to functional blocks and verify that the communication between blocks is correct. An example of a property may be that no two functional units write to the global bus at the same time, a so-called *safety property*.

The most widely used scheme for modeling a system is to use a set of interacting non-deterministic finite state machines. Equivalently, the system can be represented by its product machine. Recently, Binary Decision Diagrams (BDD's) ([Bry86]) have been used for representing FSM's. Using this scheme, the transition relation of each machine is represented by a BDD. Let $T_j(x_j, i_j, y_j)$ be the transition relation of the j-th FSM, where x_j represents the present state variable of the machine, y_j the next state, and i_j the set of inputs and outputs of the machine. We generally assume that the system is closed, so all inputs to a machine are produced by some other machine. The product machine is then represented by $T(x, i, y) = \prod_j T_j(x_j, i_j, y_j)$, x and y are the set of present and next state variables, and i is the rest of the variables (referred to as the *i/o variables*).

Abstraction may result in unwanted behaviors. Modeling unbounded but finite delays is an example. Here we want to allow the system to stay at a state s for some finite but unknown amount of time. The behaviors where the system is in s forever should be excluded. An easy abstraction is to allow infinite delays, but then all properties of the system should only be checked on traces where the system gets out of s infi-

---

1. The first author is supported by the Semiconductor Research Corporation under grant 94-DC-008.

nitely often if it enters s infinitely often. A second example of unwanted behavior is the modeling of schedulers. Assume that a set of processes are executed in a system controlled by a *fair scheduler*; one which disallows starvation of processes. An example of a fair scheduler is a round robin scheduler. To avoid having to model the details of a particular scheduler, system constraints can be used to disallow looking at traces where one process is blocked forever. Such constraints on a system's traces are called *fairness constraints (FCs)*. In our environment, abstraction and fairness constraints are allowed. Hence, for us a hardware system consists of a model of the hardware plus a set of fairness constraints.

Having specified the system, the next step is the verification of properties. For verification, one can use Computation Tree Logic (CTL) ([CES86]) formulas, or ω-automata ([VW86], [Kur87]). In this paper, we are concerned only with ω-automata verification. To verify a property expressed using ω-automata, we use *language containment (LC)*, which is the process of verifying that the language of a system is contained in the language of a property automaton. Language containment is performed by checking language emptiness of the intersection of the complement of the language of the property automaton and the language of the system.

The most general language containment environment used in practice is the *edge-Streett/edge-Rabin (eSeR) environment*, where the system is specified using edge-Streett automata and the property using edge-Rabin automata. The techniques we present in this paper are designed for the L-environment ([Kur87]), which is a subset of eSeR. It is shown in [HSB94] that the L-environment and eSeR have the same expressive power, however, eSeR can be (exponentially) more compact. The extension of our techniques to eSeR remains an open question.

To check for language containment, the product of the system and the complement of the property is built. The language containment check reduces to checking whether the language of this automaton is empty, which in turn is equivalent to answering the question of whether there is any run (i.e. an infinite path) satisfying all fairness constraints. Such runs are called *fair runs*. To do so, first the set of reachable states is computed, and the active set is initialized to this set. A technique known as *early failure detection*, which find easy failures quickly, is applied next. If no errors are found, then a subset of reachable states which cannot contain any fair runs is deleted from the active set. The last step is to apply a set of operators which trim portions of the current active set which cannot possibly contain any fair runs. We refer to this part of the computation as the *main computation*. The time-consuming operators in this set are known as *fair-path operators*, which perform a reachability computation for each fairness constraint. Hence, the running-time of the main computation is correlated with the number of fairness constraints.

Our hope is to make language containment with fairness constraints essentially as efficient as language containment without fairness constraints. Since in verification the reachable set is computed, and since early failure detection is relatively cheap, we need to speed up the last part of the algorithm to achieve our goal. In some cases, this last part is a significant portion of the running-time of the algorithm.

In this paper, we propose to take advantage of a graph induced by the fairness constraints, known as the *fairness graph*. If this graph is acyclic, then there are no fair runs. Hence, the main computation can be bypassed. If it contains cycles, then in some situations we can combine set of fairness constraints, and hence reduce the number of fairness constraints in the main computation. The problem of minimizing the number of fairness constraints at this step reduces to the *partition into forest problem*, which is NP-complete (GJ79]). So we use a greedy heuristic for it. We have implemented our techniques in the formal verification HSIS using BDD's. Based on limited experience, we observed that our techniques can reduce the run-time of the main computation when it is significant.

The paper's flow is as follows. Section 2 defines the preliminaries. Section 3 defines the fairness graph and states its properties. Section 4 presents our BDD based method for building the fairness graph. Section 5 formulates the optimization problem associated with clustering the cycle sets, states its NP-completeness, and describes our heuristic. Section 6 presents some experimental results. Conclusions and future work are given in section 7.

2 Preliminaries

Definition A *Finite State Machine (FSM)* or *Automaton* A is a tuple (Σ, Q, T, I), where Σ is a finite alphabet, Q is a finite set of states, T is a transition relation on $Q \times \Sigma \times Q$, and I is a set of initial states. A *run* r of an ω-string x in A is an infinite sequence of states, such that the first state $r_0 \in I$, and for all i, $(r_i, x_i, r_{i+1}) \in T$. The set of states occurring infinitely often in a run r, called the *infinitary set*, is denoted by $inf(r)$. Since the set of states Q is finite, $inf(r)$ is always a non-empty finite set. The set of all ω-strings which have a run in A is called the *behavior* or *language* of A.

It is routine to represent FSM's and sets of states using BDD's. We refer the reader to one such reference [Tou90].

In practice, one models a system using a set of FSM's. The language of the system, or the behavior of the system, is the intersection of the languages of all component FSM's, which can be represented using the product machine. Because of abstraction, one may add some unwanted behavior, which can be excluded using fairness constraints. In the L-environment, one uses L-processes for this task.

Definition An *L-process* is a FSM with a set of fairness constraints, which are a set of edges, known as *recur edges*, and a set of subsets of states, known as *cycle sets*. A string x is accepted if there exists a run of x, such that no recur edges is traversed infinitely often, and $inf(r)$ is not contained in any cycle set.

Consider the following example. Assume a FSM can get into a state s, in which it stays for 100 clock ticks. One might model this by putting a self-loop on s (note that the FSM can get out of state s non-deterministically at any point). This will reduce the state space, since otherwise 100 different states for s are needed. However, the problem is now that the FSM can stay in s forever. This problem is known as modeling *indefinite but finite delay*. To model this, we can use a cycle set. By putting a cycle set around s, we exclude the behavior that the FSM stays in s forever.

Hence, a system is specified as a set of interacting FSM's, each having its own set of fairness constraints. A behavior is legal if it is a *fair behavior* (one satisfying the fairness constraints) in every machine. The product L-process is the product FSM, with the fairness constraints of each FSM lifted to the product machine. Lifting means the following. Assume a cycle set is given. Each cycle set is a set S of states of a FSM P_i. The cycle set is changed to the set S' of those states of the product machine whose projection onto P_i is S. When BDD's are used to represent sets of states, lifting a set of states from a component machine to the product machine, is just thinking of the BDD as being defined over the set of variables of all components. Hence, no computations need to be done. It turns out that this product L-process accepts the intersection of the languages of the component L-processes.

To describe a property, one can use an L-automaton.

Definition An **L-automaton** is defined syntactically the same as an L-process, i.e. it is a finite automaton whose acceptance conditions are described by a set of recur edges and cycle sets. However, the interpretation is complementary. A string x is accepted by an L-automaton if there exists a run r of x such that either some recur edge is traversed infinitely often, or $inf(r)$ is contained in some cycle set.

The L-environment can be summarized as follows.

1. Express the system using a set of L-processes, or equivalently by their product L-process.

2. Describe the property using a strongly deterministic and complete L-automaton. A *complete* automaton is one which has a transition for every symbol from every state.

3. If we think of the deterministic L-automaton as an L-process, the language is complemented. So, it remains to check that the intersection of the languages of the product L-process describing the system and the L-process obtained from the property L-automaton is empty. This intersection can be represented using another L-process.

We have therefore reduced our task to checking that the $L(P) = \varnothing$, for some L-process P with cycle sets $c_1, ..., c_n$ and recur edges $Rec(x)$. To do this check, we compute the set of reachable states, and restrict the transition relation of the L-process to the set of reachable states. We then remove the recur edges. Call this machine $\tilde{P}$. Now, $L(P) = \varnothing$ iff no cycle of $\tilde{P}$ is contained in any of the cycle sets. We call such cycles *fair cycles*.

To check for the existence of fair cycles, we first check whether there are any cycles involving states in Γ, where $\Gamma = \bigcap_j \bar{c}_j$. Note that any such cycle is fair. If there are no such cycles, we delete all states in Γ from the set of reachable states and $\tilde{P}$. Call this machine Q. To check for fair cycles in Q, we apply a set of operators (which we will define below) until convergence is achieved. The set converged to is not empty iff there exists a fair cycle.

In what follows, let G be a graph, V the set of its vertices, $A \subseteq V$ a subset of its vertices, and $T(x, y)$ the transition (edge) relation of G.

Definition Let F be a monotone increasing k-ary predicate transformer. Define the *least fixpoint of F given Q*, denoted by $\mu(X, Q).FX$ where Q is an k-ary over sets $D_1, ..., D_k$, by the set $F^i(Q)$ such that $F(F^i(Q)) = F^i(Q)$. Intuitively, X is the variable we are recurring on, and Q is its initial value. Similarly, define the *greatest fixpoint* of a monotone decreasing k-ary predicate transformer F given Q, denoted by $\nu(X, Q).FX$, by the set $F^i(Q)$ such that $F(F^i(Q)) = F^i(Q)$.

Definition Let $S_1(A, y) = \exists x(A(x) \wedge T(x, y))$ and $S_1(x, A) = \exists y(A(y) \wedge T(x, y))$. Thus, $S_1(A, y)$ are the successors of $A(x)$ and $S_1(x, A)$ are the predecessors of $A(x)$.

Definition Let $R_1(A, y) = A(y) \vee S_1(A, y)$, $R_1(x, A) = A(x) \vee S_1(x, A)$, $R^*(A, y) = \mu(X, A).(R_1(X, y))$, $R^*(x, A) = \mu(X, A).(R_1(x, X))$. Note that the first two are "one-step" operators, while the last two compute the least fixed-point containing A. One should read $R^*(A, y)$ as the set of points reachable from A. Similarly

$R^*(x, A)$ is the set of points that can reach A. $R_1(A, y)$ and $R_1(x, A)$ are the one-step versions of these, respectively.

Definition We define two "stable set" operators, $S^*(A, y) = v(X, A) \cdot (S_1(X, y))$ and $S^*(x, A) = v(X, A) \cdot (S_1(x, X))$. One can think of the first as calculating the **backward stable set** contained in A (i.e. the set of all vertices which are reached by some vertex involved in some cycle in A), and the latter as the **forward stable set** of A (i.e. the set of all vertices which can reach some vertex involved in some cycle in A).

Definition Define the **forward fair-path operator** by $F(x) = \prod_j R^*(x, \bar{c}_j \wedge A)$, where A is the current active set. Note that if a state can reach a fair cycle, it is not deleted by this operator. Similarly, define the **backward fair-path operator** by $B(x) = \prod_j R^*(\bar{c}_j \wedge V, y)$. Note, if a state is reached by a fair cycle, it is not deleted by this operator.

One method for checking language containment is to apply forward and backward stable set, and forward and backward fair-path operators. If we converge to a non-empty set, then $L(P) = \emptyset$. Otherwise, it is not. The time-consuming part of this computation is the application of the fair-path operators. Each fair-path operator performs a reachability computation for each cycle set. Hence, we expect its running-time to be correlated strongly with the number of cycle sets. In the next few sections, we use the properties of a graph induced by the cycle sets in Q to reduce the run-time of this portion of the computation.

3 The Fairness Graph and Its Properties

In what follows, we assume Q is the state graph of an automaton with cycle sets $c_1, ..., c_n$, such that every node is contained in at least one cycle set (which is the case after early failure detection).

Definition The *fairness graph* Q_f induced by the cycle sets is a graph on n nodes $\tilde{c}_1, ..., \tilde{c}_n$ which has an edge from vertex $\tilde{c}_i$ to $\tilde{c}_j$ iff there exists an edge (u, v) in Q such that $v \in c_i$, $w \in c_j$, and $w \notin c_i$.

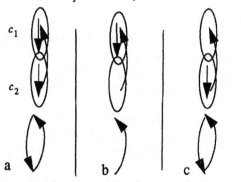

Figure 1

Examples In the above figure three examples of graphs and their fairness graphs are given. The original graphs are on top, and the fairness graphs are at bottom. Each oval around the vertices signifies a cycle set. Note that in example b, there is no edge from the first cycle set to the second one, since there is no edge from c_1 to c_2 whose end-point is not in c_1. Also, note that in example c, although there is no cycle in the original graph, there is a cycle in the fairness graph.

Definition Let a path in Q $x_1, ..., x_p$ be given. Let $c_1, ..., c_p$ be a sequence of cycle sets such that for all i, $x_i \in c_i$. We call this sequence a projection of a path. Let a *shadow of a path* be obtained from a projection of a path by deleting consecutive duplicate copies of a cycle set.

Intuitively, as we follow a path we go through a sequence of cycle sets. A shadow of a path is obtained by only counting those cycle sets which are different than the current one. Note that the shadow of a path is a sequence of states in its fairness graph. The following figure gives an example of a path and its shadows, which shows that the shadow of a path may not be unique, since the projection of a path may not be unique.

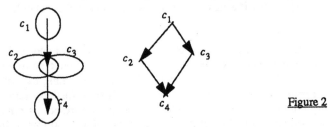

Figure 2

We have the following properties for Q_f, the fairness graph. The proofs of those lemmas which follow directly from the definition of the fairness graph are not given.

Lemma 1 There are no self-loops in Q_f.

Lemma 2 Let Γ be a cycle, a shadow of which visits $c_{i_1}, ..., c_{i_k}, c_{i_1}$ in order. Then, there is a corresponding cycle $\bar{c}_{i_1}, ..., \bar{c}_{i_k}, \bar{c}_{i_1}$ in Q_f.

Lemma 3 Every shadow of a fair cycle is a cycle in Q_f containing more than one state (recall that a fair cycle is a cycle not contained in any of the cycle sets).

Lemma 4 There may be a cycle $\bar{c}_{i_1}, ..., \bar{c}_{i_k}, \bar{c}_{i_1}$ in Q_f, but no corresponding cycle in Q that visits $c_{i_1}, ..., c_{i_k}, c_{i_1}$.

Proof In the above figure, c is an example.

Theorem 1 Let D denote the set of all vertices in Q_f which are not contained in any cycle. Let $Q' \subseteq Q$ be a new graph where all the vertices which are contained in some cycle set in D have been deleted. The cycle sets of Q' are the same cycle sets of Q except for those contained in D. Then, every fair cycle of Q is present in Q' (note that every fair cycle of Q' is clearly a cycle set in Q).

Proof It suffices to show that every state involved in some fair cycle in Q is present in Q'. Let $x \in Q$ be involved in some fair cycle. Every fair cycle of Q has a shadow which is a cycle in Q_f. Hence, all cycle sets containing x will not be deleted by the above procedure (QED).

Note that the above procedure gives us a method for deleting portions of the active set in which no fair behavior can exist. In some cases, as the following corollary shows, we may be able to decide the emptiness check without any further computations.

Corollary 1 If Q_f is acyclic, then there are no fair cycles in Q.

Corollary 1 If Q_f is acyclic, then there are no fair cycles in Q.

4 Building the Fairness Graph

Building the fairness graph is the most time-consuming part of the overhead of our algorithm. Note that if we were not using BDD's, building this graph would have been very time-consuming, since possibly all edges in the product L-process had to be visited. There are many ways to build this graph using BDD's, some performing better than others. In section 4.1, we first present a naive algorithm, and then present one which improves the performance of the first one. In section 4.2, we discuss ways of reflecting the structure of Q better in Q_f, by deleting states which cannot contain any fair cycles.

4.1 Algorithms

To build the fairness graph using BDD's, we introduce two new variables which take values from 1 to n. The first variable, denoted by x', corresponds to the present state of the fairness graph, whereas the second one, denoted by y', corresponds to the next state. We eventually build a transition relation $R(x', y')$, which represents the fairness graph. Let $A(x)$ represent the active set of states, which is originally the set of reachable states with the states in $\bigcap_i \bar{c}_i$ deleted. We assume the transition relation of Q, $T(x, y)$, is restricted to $A(x)$. The following lemma provides a straight-forward way to build the fairness graph.

Lemma 5 $\sum_i \sum_j \exists x \exists y (T(x, y) \wedge c_i(x) \wedge c_j(y) \wedge \overline{c_i(y)} \wedge (x' = i) \wedge (y' = j))$ computes Q_f. This computation does $2n^2$ quantifications, $5n^2$ Boolean AND's, and n^2 Boolean OR's.

Proof For every pair of cycle sets c_i and c_j the above computation asks whether there is an edge between them. If so, an edge (i, j) is added to the final relation. Hence, the relation represents Q_f. The number of operations follows by expanding the above sums, which has exactly n^2 terms (QED).

The improved version involves re-arranging the order of various operations.

Lemma 6 $\sum_j \exists y (\sum_i (\exists x (T(x, y) \wedge c_i(x))) \wedge \overline{c_i(y)} \wedge (x' = i)) \wedge c_j(y) \wedge (y' = j)$ computes Q_f. Moreover, it takes $2n$ quantifications, $5n$ Boolean AND's, and $2n$ Boolean OR's.

Proof It suffices to show that the above equation is algebraically equal to the one in the previous lemma. This follows by noticing that existential quantification and Boolean sums commute. To count the number of operations, note that $\sum_i (\exists x (T(x, y) \wedge c_i(x))) \wedge \overline{c_i(y)} \wedge (x' = i)$ is independent of j and has to computed only once. It requires n quantifications, $3n$ Boolean AND's, and n Boolean OR's. The outer computation takes n quantifications (once for each j), $2n$ Boolean AND's, and n Boolean OR's. The bounds follow (QED).

Algorithmically, we can think of our algorithm as consisting of two passes over the cycle sets, with the first pass computing $S(x', y) = \sum_i (\exists x T(x, y) \wedge c_i(x)) \wedge \overline{c_i(y)} \wedge (x' = i)$.

I. Let $S(x, y) = 0$.

For each cycle set c_i

   1. $U_i(y) = (\exists x\, (T(x,y) \wedge c_i(x))) \wedge \overline{c_i(y)}$.

   2. $S(x',y) = S(x',y) + (U_i(y) \wedge (x' = i))$.

The second pass takes $S(x',y)$ and returns the fairness graph $R(x',y')$. The algorithm is as follows.

II. Let $R(x',y') = 0$.

For each cycle set c_j

   1) $U_j(x') = \exists y\, (S(x',y) \wedge c_j(y))$

   2) $R(x',y') = R(x',y') + (U_j(x') \wedge (y' = j))$.

We remark that $U_i(y)$ is the set of next states from c_i not in c_i, and $U_j(x')$ is the set of cycle sets that have an edge, not contained entirely in c_j, to a state in c_j. We use these observations in the next section.

4.2 Improving the Structure of the Fairness Graph

In this section, we describe techniques which give a better model of Q in terms of Q_f. Consider example a in figure 3, where s_1, s_2, s_3, s_4 are the states and c_1, c_2, c_3, c_4 are the cycle sets. Assume in the first pass we process c_1, c_2, c_3, c_4 in that order. When processing c_4, we find that $U_4(y) = 0$, which means c_4 has no outgoing edges. We call such cycle sets **sink cycle sets**. Sink cycle sets cannot be involved in any fair cycles, and can be deleted from Q. Hence s_4 and c_4 are deleted. Now, s_2 has no outgoing edges, and can be deleted by the forward stable set operator (recall this operator deletes any states which cannot reach a cycle). Let **optimization 1** denote the deletion of sink cycle sets as well as the application of forward and backward stable set operators in the first pass. Optimization 1 in this example deletes s_2, s_4 and c_4 from Q.

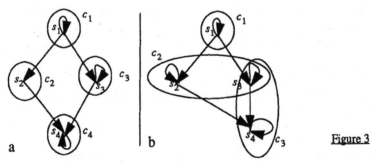

a   b   Figure 3

Now, consider example b in figure 3. Assume in the first pass c_1, c_2, c_3 are processed in order. Since c_3 is a sink cycle set, states s_3 and s_4 are deleted. The stable set operators cannot delete any more states in this case. When we processed c_2, we added the pair $(1, s_3)$ to $S(x',y)$, which means c_1 has an edge to s_3. But since s_3 is deleted from the graph, this is not the case any more. The question is whether we will have an edge $(1, 2)$ in $R(x',y')$. We call edges in $R(x',y')$ which do not exist in the final fairness graph, **fake edges**. Assume we modified our algorithm in step 1 of pass 2 to $U_j(x') = \exists y\, (S(x',y) \wedge c_j(y) \wedge A(y))$, where $A(y)$ is the current active set of states in Q. Note that in the first pass, we always restrict $T(x, y)$ to $A(x)$. We also assume that the stable set operators are originally applied to Q, before the building of fairness graph is started. We have the following important theorem.

Theorem 2 If optimization 1 is applied during the first pass, then $R(x', y')$ contains no fake edges.

Proof Let (u, v) be an edge from c_i to c_j in the original Q (example a, figure 4). Then, (i, j) is fake edge in $R(x', y')$ iff at least one of u or v is deleted by optimization 1. We distinguish between three cases.

Case 1. v is deleted by optimization 1. Then $v \notin c_j$ in the second pass, and the pair (i, c_j) gets deleted from $S(x', y)$ in pass 2, step 1, and does not gives rise to the edge (i, j) in $R(x', y')$.

Case 2. Only u is deleted in the first pass, and this occurs because u belongs to some sink cycle set c_k. We have that $v \notin c_k$. But, we have an edge from c_k to c_j, i.e. c_k is not a sink cycle set. We have reached a contradiction.

Case 3. Only u is deleted in the first pass, and this occurs because of application of a stable set operator after some sink cycle set c_k is deleted (example b, figure below). Since v is not deleted, v can reach a fair cycle. Hence, u can reach a fair cycle. Thus, u is not deleted by forward stable set operator. So, u is deleted by backward stable set operator. Since, before deletion of c_k, u could be reached by some cycle, there is a cycle in c_k which can reach u. But, since $u \notin c_k$, we have that c_k has an out-going edge, i.e. it is not a sink cycle set. We have reached a contradiction again (QED).

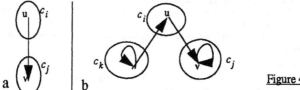

a   b   Figure 4

Now, consider the following example. Assume in the second pass, we process c_1, c_2, c_3 in that order. When processing c_2, we find that $U_2(x') = 0$, i.e. c_2 does not have any in-coming edges. We call such cycle sets, which cannot be involved in any fair cycles, *source cycle sets*. Let *optimization 2* be the deletion of source cycle sets and application of the stable set operators in the second pass.

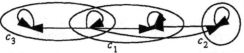

Figure 5

Lemma 7 Applying optimization 2 can result in fake edges.

Proof In figure 5, after c_2 is processed the edge $(1, 2)$ is added to $R(x', y')$. Then, in processing c_3, we find that $U_2(y') = 0$, so we can delete c_3. Thus $(1, 2)$ is a fake edge (QED).

Consider figure 6. After the fairness graph is built, we notice that c_3 is not involved in any cycle of the fairness graph. The states of such a cycle set cannot be involved in any fair cycle, and can be deleted from the state graph. Let *optimization 3* denote the deletion of such cycle sets and the application of the stable set operators, after the fairness graph is built.

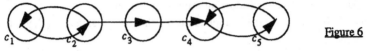

Figure 6

Lemma 8 Application of optimization 3 can result in fake edges.

Proof Figure 8 gives an example, where the state graph Q is on the left and the fairness graph Q_f on the right. Optimization 3 deletes c_2. By deleting the nodes of c_2 from Q, the edges $(1, 3)$ and $(3, 1)$ in Q_f become fake (QED).

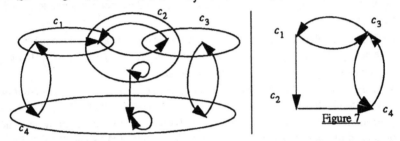

In summary, optimization 1 does not create fake edges, whereas optimizations 2 and 3 might. To deal with fake edges, we have implemented two algorithms. Because we have not run enough examples, we cannot which is preferred at this point. The first algorithm is as follows.

1. *Do the first pass with optimization 1.*
2. *Do the second pass with optimization 2. If any states are deleted, go back to step 1, and restart. Otherwise, go to step 3.*
3. *Perform optimization 3 on the fairness graph. If any states are deleted, go back to step 1, and restart the computation.*

Note that the first algorithm re-starts the computation as soon as any state is deleted by optimization 2 and 3. The second algorithm is given below.

1. *Do the first pass with optimization 1.*
2. *Do the second pass with optimization 2.*
3. *Perform optimization 3 on the fairness graph.*
4. *If any states are deleted in steps 2 or 3, go back to step1.*

This algorithm carries out the computation to the end, ignoring possible fake edges. At the end, if there is a possibility of fake edges (because of deletion of states by optimizations 2 or 3), the computation is re-started.

5 Clustering Cycle Sets

In this section, we formulate the problem of minimizing the number of fairness constraints. We show that the problem is NP-complete, and propose a heuristic solution.

5.1 The Clustering Problem

Definition An *acyclic cluster* is any sets of vertices in Q_f which do not contain a cycle.

Theorem 3 Let $S_1, ..., S_p$ be a partition of vertices of Q_f such that every S_i is an acyclic cluster. Let Q' have the same states as Q, but with p cycle sets $c'_1, ..., c'_p$, where c'_i is formed by taking the union of all cycle sets in S_i. Then, a cycle os fair in Q iff it is fair in Q'.

Proof Let Γ be a fair cycle in Q. The only way Γ is not a fair cycle in Q' is for Γ

to be entirely contained in some c'_i. Assume this is the case. By lemma 3, every shadow of Γ is a cycle in Q_f. Hence the shadow involving cycle sets only in S_i is a cycle involving only states in S_i. But this contradicts the assumption that each S_i is an acyclic cluster. Conversely, assume Γ is a fair cycle in Q'. The only way for Γ not to be a fair cycle in Q is for Γ to be contained entirely in some c_j. Let's say c_j is in some partition S_i. Since Γ is fair in Q', it is not entirely contained in S_i. Hence it cannot be contained in c_j (QED).

Note that the above theorem can be used to reduce the number of fairness constraints. We call this technique *clustering*. The associated optimization problem is to decompose a directed graph (Q_f) into a minimum set of acyclic clusters. This problem is NP-complete, since it is equivalent to the *Partition into Forest* problem described in [GJ79]. We now show how this problem can be reduced to the decomposition of each strongly connected component (SCC) of Q_f.

Lemma 9 Let the SCC's of Q_f be given. Assume each is decomposed separately into a set of acyclic clusters. The union of a set of acyclic clusters each from different SCC's is again an acyclic cluster.

Proof This follows from the definition of SCC's (QED).

Theorem 4 Assume each SCC of Q_f can be decomposed into a minimum of m_i acyclic clusters. Let $M = max\{m_i\}$. Then, M is the solution of the partition into forest problem.

Proof Clearly by lemma 9, a solution with M clusters can be constructed by clusters with at most one element from each SCC. Now assume M is not minium, and let a set of $M' < M$ acyclic clusters $S_1, ..., S_{M'}$ be given. We can built a clustering $W_1, ..., W_{M'}$ for each SCC W with at most M' clusters by taking $W_i = S_i \cap W$. This contradicts that m_i is a minimum solution for each SCC (QED).

The above theorem gives us a way of clustering vertices of Q_f by processing each SCC separately.

Example Consider the example in Figure 8, which shows a fairness graph with three SCC's: $\{1, 2, 3\}$, $\{4, 5, 6\}$, and $\{7, 8, 9, 10\}$. The first SCC can be decomposed into $\{1, (2, 3)\}$, the second into $\{4, (5, 6)\}$, the third into $\{(7, 8), (9, 10)\}$. Thus, $M = 2$. The final decomposition is $\{(1, 4, 7, 8), (2, 3, 5, 6, 9, 10)\}$, and we have reduced the cycle sets from 10 to 2.

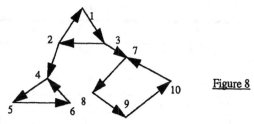

Figure 8

5.2 Heuristic Clustering

We use a simple recursive, greedy heuristic for the clustering of each SCC, which is described below.

1. If two vertices do not form a cycle, merge them into a hyper node.

2. If any merges were performed in step 1, call the routine recursively on the graph formed by taking each cluster as a node.

Note that in step 1, a vertex is merged with at most one other vertex. This allows many clusters to grow concurrently (as opposed of starting with one and continuously adding to it). In step 2, if any clustering was done in step 1, we form a new graph by taking the clusters found in step 1 as new vertices and putting an edge between them if the vertices in the cluster were connected in the original graph.

Remark We need to compare our heuristic to the heuristic of maximally growing a cluster until no more nodes can be added, and then continuing with the next cluster. We also propose to process the SCC's in decreasing size. Since, we would not need to decompose a small SCC into less than $M = max\{m_i\}$ seen so far, we can use this as an early bounding step to speed up processing.

6 Experiments

We have recently implemented our algorithms in HSIS, a formal verification system developed at UC Berkeley that supports language containment and model checking using BDD's. To test our algorithms, we used an abstracted description of a SUN Sparc memory model [Sparc-8], [DNP93]. Our system model is a memory system that is accessed by a number of processors, as in a typical shared memory multiprocessor system. The memory is modeled abstractly as a single port device that grants write access to the attached processors by randomly selecting one.

For our experiments we mapped an assembly code sequence (from [Sparc-8]) that implements spin locking to an automaton representing a processor. Another automaton models the memory and its random scheduling. To model general computation, both the automata for the processor and the memory have several states with self loops: the processor can stay in an idle state indefinitely long before attempting to acquire the lock; it can also wait indefinitely long returning to its computation loop, and stay in its critical section (after acquiring the lock) for an unspecified period of time. Each of these states becomes a one-element cycle set, since we want to exclude the behavior from the system where processors remain in one state forever. Similarly, we require that the memory does not always choose to service the same processor, i.e. the memory is "fair".

We verify the following liveness property: a processor will eventually be able to acquire the lock (*no starvation*). We verified the property for systems with 3 and 4 processors. The results are summarized in table 1. All times are in seconds and are measure on a DECsystem 5000/260 machine with 128 MB of memory.

Table 1:

num of processors	3	4
reachable states num	10685	143228
reachability	5.35	116.7
main comp. w/o f.g.	**5.47**	**61.45**
total time w/o f.g.	15.10	221.98
building the f.g	**1.45**	**14.81**
main comp. with f.g	**0.0**	**0.0**
total time with f.g.	11.11	175.1
improvement factor	3.77	4.15

Note that in this example, we were able to decide the language containment check by only analyzing the fairness graph. Hence, the main computation time with the fairness graph analysis is 0. The last row is obtained by taking the time of main computation without our technique, divided by the sum of building the fairness graph and the time for the main computation with our technique.

7 Conclusions and Future Work

Verification using language containment involves performing reachability, and then calling an algorithm whose running-time is proportional to the number of fairness constraints (called the main computation). If the number of fairness constraints is large, the main computation can be time-consuming. Our goal is to make verification using language containment basically as efficient as performing reachability. This work investigated ways to make use of the information inherent in the graph induced by the fairness constraints on a system, the *fairness graph*, to reduce the run-time of the main computation. We suggested techniques to construct and process the fairness graph. Our techniques are a useful complement to early failure detection, because they can help when a property passes. We used the example of an abstract memory model as a test-case to validate the techniques, and concluded that the methods hold promise. On this example, we observed a factor of 4 reduction in the run-time of the main computation. Because our implementation is preliminary, there may also be some scope to improve the efficiency of computing the fairness graph and the clustering.

References

[Bry86] R. E. Bryant, *"Graph Based Algorithms for Boolean Function Manipulation"*, IEEE Trans. on Computers, C-35(8):677-691, August 1986.

[CES86] E. M. Clarke, E. A. Emerson, and A. P. Sistla. *"Automatic Verification of Finite-State Concurrent Systems Using Temporal Logic Specifications"*, ACM Transactions on Programming Languages and Systems. 8(2), pp.244-263, 1986.

[DNP93] D. L. Dill, A. G. Nowatzyk, S. Park, *"Formal Specification and Verification of Abstract Memory Models"*, Conference on Hardware Description Languages, 1993.

[HTKB92] Ramin Hojati, Herve Touati, Robert P. Kurshan, Robert K. Brayton, *"Efficient ω-Regular Language Containment"*, Computer-Aided Verification, 1992.

[HSB94] Ramin Hojati, Vigyan Singhal, Robert Brayton, *"Edge-Streett/Edge-Rabin Automata Environment for Formal Verification Using Language Containment"*, UCB/ERL Tech. Report, M94/12, 1994.

[Kur87] R.P. Kurshan, *"Reducibility in Analysis of Coordination"*, In Discrete Event Systems: Models and Applications, volume 103 of LINCS, pages 19-39, 1987.

[GJ79] Michael R. Garey, David S. Johnson, *"Computers and Intractability, A Guide to the Theory of NP-Completeness"*, Freeman and Co., 1979

[Sparc-8] The SPARC Architecture Manual, Version 8, SPARC International Inc., Menlo Park, CA, 1992.

[Tou90] H. Touati, H. Savoj, B. Lin, R. K. Brayton, A. S. Vincentelli, *"Implicit State Enumeration of Finite State Machines Using BDD's"*, International Conference on Computer-Aided Design, 1990.

[VW86] M.Y. Vardi and P.L. Wolper, *"An Automata-Theoretic Approach to Program Verification"*, Logic in Computer Science, pages 332-334, 1986

A Parallel Algorithm for Relational Coarsest Partition Problems and Its Implementation *

Insup Lee and S. Rajasekaran

Department of Computer and Information Science

University of Pennsylvania

Philadelphia, PA 19104-6389

Abstract

Relational Coarsest Partition Problems (RCPPs) play a vital role in verifying concurrent systems. It is known that RCPPs are $\mathcal{P}$-complete and hence it may not be possible to design polylog time parallel algorithms for these problems.

In this paper, we present a parallel algorithm for RCPP, in which its associated label transition system is assumed to have m transitions and n states. This algorithm runs in $O(n^{1+\epsilon})$ time using $\frac{m}{n^\epsilon}$ EREW PRAM processors, for any fixed $\epsilon < 1$. This algorithm is analogous and optimal with respect to the sequential algorithm of Kanellakis and Smolka. The same algorithm runs in time $O(n \log n)$ using $\frac{m}{\log n} \log \log n$ CRCW PRAM processors. We also describe implementation and experimental results on performance of our algorithm.

1 Introduction

Relational Coarsest Partition Problems (RCPPs) play an important role in verifying concurrent systems in the form of equivalence checking. In their pioneering work, Kanellakis and Smolka [6] presented an efficient algorithm for the RCPP with multiple relations. Their algorithm had a run time of $O(mn)$, where m is the number of transitions and n is the number of states in the RCPP. This algorithm has been used in practice to verify systems with thousands of states. Our work is to extend the applicability of this algorithm with the use of parallelism.

In a recent work of Zhang and Smolka [9], an attempt has been made to parallelize the classical Kanellakis-Smolka algorithm. However, the main thrust of this work was from practical considerations. In particular, complexity analysis has not been provided

---

*This research was supported in part by NSF CCR93-11622 and DARPA/NSF CCR90-14621.

and was not the main concern of the paper. On the other hand, it has been shown that RCPP (even when there is only a single function) is $\mathcal{P}$-complete [1]. $\mathcal{P}$-complete problems are presumed to be problems that are hard to efficiently parallelize. It is widely believed that there may not exist polylog time parallel algorithms for any of the $\mathcal{P}$-complete problems that use only a polynomial number of processors.

Since RCPP has been proven to be $\mathcal{P}$-complete, we restrict our attention to designing polynomial time algorithms. In this paper we present a parallel algorithm for RCPP: This algorithm runs in $O(n^{1+\epsilon})$ time using $\frac{m}{n^{\epsilon}}$ EREW (Exclusive-Read Exclusive-Write) PRAM processors for any fixed $\epsilon < 1$. This algorithm is optimal with respect to the Kanellakis-Smolka algorithm. We say a parallel algorithm that runs in time T using P processors is optimal with respect to a sequential algorithm with a run time of S, if $PT = O(S)$, i.e., the *work done* by the parallel algorithm is asymptotically the same as that of the sequential algorithm. The same algorithm runs in time $O(n \log n)$ using $\frac{m}{\log n} \log \log n$ CRCW (Concurrent-Read Concurrent-Write) PRAM processors. The parallel algorithm described in this paper is for single-relation RCPP. It can, however, be easily extended for multiple-relation RCPP without changing the asymptotic run-time complexity or processor bound.

The rest of the paper is organized as follows. In Section 2, we state the problem and provide some useful facts about parallel computation. Section 3 gives details of our parallel algorithm and Section 4 describes an example that explains how our algorithm works. Section 5 presents analysis of our algorithm and Section 6 reports our implementation results. Section 7 concludes the paper.

2 Problem Statement

Definition 1 *A labeled transition system (LTS) M is $\langle Q, Q_0, T \rangle$, where Q is a set of states, $Q_0 \subseteq Q$ is a set of initial states, and $T \subseteq Q \times Q$ is a transition relation.*

Definition 2 *For any state $p \in Q$, let $T(p) = \{q \in Q | (p,q) \in T\}$. Also for any subset B of Q, let $T(B)$ stand for $\cup_{p \in B} T(p)$. Similarly define $T^{-1}(p)$ and $T^{-1}(B)$ for any $p \in Q$ and for any $B \subseteq Q$.*

The Relational Coarsest Partitioning Problem (RCPP) is defined as follows:

Input: An LTS $M = \langle Q, Q_0, T \rangle$ with a finite state set Q, an initial partition π_0 of Q and a relation T on $Q \times Q$.

Output: the coarsest (having the fewest blocks) partition $\pi = \{B_1, \cdots, B_l\}$ of Q such that

1. π is a refinement of π_0, and
2. for every p, q in block B_i, and for every block B_j in π,

$$T(p) \cap B_j \neq \emptyset \text{ iff } T(q) \cap B_j \neq \emptyset$$

That is, either $B_i \subseteq T^{-1}(B_j)$ or $B_i \cap T^{-1}(B_j) = \emptyset$.

2.1 Parallel Computation Models

A large number of parallel machine models have been proposed. Some of the widely accepted models are: 1) fixed connection machines, 2) shared memory models, 3) the boolean circuit model, and 4) the parallel comparison trees. Of these we'll focus on 1) and 2) only. The *time complexity* of a parallel machine is a function of its input size. Precisely, time complexity is a function $g(n)$ that is the maximum over all inputs of size n of the time elapsed when the first processor begins execution until the time the last processor stops execution.

A fixed connection network is a directed graph $G(V, E)$ whose nodes represent processors and whose edges represent communication links between processors. Usually we assume that the degree of each node is either a constant or a slowly increasing function of the number of nodes in the graph. Fixed connection networks are supposed to be the most practical models. The Connection Machine, Intel Hypercube, ILLIAC IV, Butterfly, etc. are examples of fixed connection machines.

In shared memory models (also known as PRAMs, i.e., Parallel Random Access Machines), processors work synchronously communicating with each other with the help of a common block of memory accessible by all. Each processor is a random access machine. Every step of the algorithm is an arithmetic operation, a comparison, or a memory access. Several conventions are possible to resolve read or write conflicts that might arise while accessing the shared memory. EREW (Exclusive Read Exclusive Write) PRAM is the shared memory model where no simultaneous read or write is allowed on any cell of the shared memory. CREW (Concurrent Read Exclusive Write) PRAM is a variation which permits concurrent read but not concurrent write. And finally, CRCW (Concurrent Read Concurrent Write) PRAM model allows both concurrent read and concurrent write. Write conflicts in the above models are taken care of with a priority scheme.

The parallel run time T of any algorithm for solving a given problem can not be less than $\frac{S}{P}$ where P is the number of processors employed and S is the run time of the best known sequential algorithm for solving the same problem. We say a parallel algorithm is *optimal* if it satisfies the equality: $PT = O(S)$. The product PT is referred to as *work done* by the parallel algorithm.

The model assumed in this paper is the PRAM. Though no PRAM machines exist, it is easy to describe algorithms on this model and usually algorithms developed for this model can be easily mapped on to more practical models.

2.2 Some Useful Facts

In this section, we state some well-known results which are used to analyze algorithms presented in this paper.

Lemma 1 *[3] If W is the total number of operations performed by all the processors using a parallel algorithm in time T, we can simulate this algorithm using P processors such that the new algorithm runs in time $\lfloor \frac{W}{P} \rfloor + T$.*

As a consequence of the above Lemma we can also get:

Lemma 2 *If a problem can be solved in time T using P processors, we can solve the same problem using P' processors (for any $P' \leq P$) in time $O\left(\frac{PT}{P'}\right)$.*

Given a sequence of numbers $k_1, k_2, \ldots, k_n$, the problem of *prefix sums computation* is to output the numbers $k_1, k_1 + k_2, \ldots, k_1 + k_2 + \ldots + k_n$. The following Lemma is a folklore [5]:

Lemma 3 *Prefix sums of a sequence of n numbers can be computed in $O(\log n)$ time using $\frac{n}{\log n}$ EREW PRAM processors.*

The following Lemma is due to Cole [4]

Lemma 4 *Sorting of n numbers can be done in $O(\log n)$ time using n EREW PRAM processors.*

The following Lemma concerns with the problem of sorting numbers from a small universe:

Lemma 5 *[2] n numbers in the range $[0, n^c]$ can be sorted in $O(\log n)$ time using $\frac{n}{\log n} \log \log n$ CRCW PRAM processors, as long as c is a constant.*

This problem can also be solved in $O(n^\epsilon)$ time for any fixed $\epsilon < 1$, using $\frac{n}{n^\epsilon}$ EREW PRAM processors.

3 A Parallel Algorithm Based on Kanellakis-Smolka Algorithm

The Kanellakis-Smolka algorithm runs sequentially in $O(nm)$ time, where n is the number of states and m is the number of transitions. The basic idea behind the Kanellakis-Smolka algorithm is to split a block in the current partition if not all states in the block can go to the same set of blocks.

Figure 1 outlines our parallel algorithm which is based on the Kanellakis-Smolka algorithm. The algorithm uses known parallel algorithms for sorting and prefix sums. We describe the data structures used in the algorithm and then the steps of the algorithm.

Data Structures. Let $T(p)$ stand for $\{q \in Q \mid (p, q) \in T\}$, i.e., $T(p)$ is the set of states to which there is a transition from p. We also define $T^{-1}(p)$ to be $\{q \in Q \mid (q, p) \in T\}$.

The current partition is represented as an array $PARTITION$. It is an array of size n with (block id, state) pairs. For example, a pair (i, q) represents that the state q currently belongs to the ith block. We maintain the array $PARTITION$ such that states belonging to the same block appear consecutively.

$\pi := \pi_0$; split := true

while split **do**

    split := false; let $\pi = \{B_1, B_2, \ldots, B_\ell\}$

    Unmark $B_1, B_2, \ldots, B_\ell$

    1. for $i := 1$ **to** n **in parallel do**

        $TEMP[i] := TSIZE[PARTITION[i].state]$

    2. Compute the prefix sums of $TEMP[1], TEMP[2], \ldots, TEMP[n]$

        Let the sums be $v_1, v_2, \ldots, v_n$

    3. for $i := 1$ **to** n **in parallel do**

        $s_i := PARTITION[i].state$

        Let $T[s_i]$ be $\{q_1, \ldots, q_k\}$

        for $j := 1$ **to** k **in parallel do**

    Let processor in-charge of transition (s_i, q_j) write $(B[s_i], V[s_i], B[q_j])$ in $L[v_{i-1} + j]$

    4. Sort the sequence L in lexicographic order.

    5. for $i := 1$ **to** m **in parallel do if** $L[i] = L[i+1]$ **then** $L[i] := 0$

    6. Compress the list L using a prefix computation

    7. for each block B_i $(1 \leq i \leq \ell)$ **in parallel do**

        for each j, $2 \leq j \leq n_i$ **in parallel do**

            if $[q_{i,j}] \neq [q_{i,1}]$ **then** mark B_i

    8. if there is at least one marked block **then**

        split := true; $\ell := \ell + 1$

        Pick one of the marked blocks (say B_i) arbitrarily

        for each p **in** B_i **do**

            if $[p] \neq [q_{i,1}]$ **then**

                $B[p] := \ell + 1$

                Change the corresponding entry in $PARTITION$ to $(p, \ell + 1)$

        /* $B_{\ell+1} := B_i - \{p \in B_i : [p] = [q_{i,1}]\}$ and $B_i := B_i - B_{\ell+1}$ */

        Using a prefix computation, modify $PARTITION$ such that all tuples

        corresponding to the same block are in successive positions.

        When the array $PARTITION$ is modified, positions of some

        states q's might change; inform the processors associated with

        the corresponding $T(q)$'s of this change.

Figure 1: Parallel Algorithm Based on Kanellakis-Smolka Algorithm

The array $TRANSITIONS$ is used to store the relation T of the LTS. In particular, the array is of size m and each entry contains the (from-state, to-state) pair. In the array $TRANSITIONS$, we store the transitions of $T(1)$, followed by the transitions of $T(2)$, and so on. $TSIZE$ is an array of size n such that $TSIZE[q]$ stands for $|T(q)|$ for each q in Q. Note that the arrays $TRANSITIONS$ and $TSIZE$ are never altered during the algorithm.

We also maintain an array B such that for each state p in Q, $B[p]$ is the id of a block to which p belongs in the current partition π. In addition, for each state $p \in Q$, we let $[p]$ stand for the set, $\{B[q] \mid q \in T(p)\}$. We emphasize here that no repetition of elements is permitted in $[p]$. For any state q in Q, we let $[T(q)]$ stand for the sequence $B[p_1], B[p_2], \ldots, B[p_t]$, where $T(q) = \{p_1, p_2, \ldots, p_t\}$. Notice that $[T(q)]$ can have multiple occurrences of the same element. Also, let $V[s]$ stand for the position of state s within its block. We let $n_i = |B_i|$ for any block B_i in the current partition and denote the jth element of block B_i by $q_{i,j}$.

As an example to illustrate our data structures, consider the following initial partition,

$$\pi_0 = \{\{a, b, c\}, \{d, e, f\}, \{g, h, i\}\}.$$

Let the transition relation T be defined as follows: $T(a) = \{d, f\}$, $T(b) = \{d\}$, $T(c) = \{e, f\}$, $T(d) = \{g, i\}$, $T(e) = \{a, b\}$, $T(f) = \{g\}$, $T(g) = \{a\}$, $T(h) = \{b, c, d\}$, $T(i) = \{a, b\}$. Table 1 shows the contents of $PARTITION$, $TRANSITIONS$, B, and $TSIZE$ at the beginning.

$PARTITION$	$(1, a)$	$(1, b)$	$(1, c)$	$(2, d)$	$(2, e)$	$(2, f)$	$\ldots$
$TRANSITIONS$	(a, d)	(a, f)	(b, d)	(c, e)	(c, f)	(d, g)	$\ldots$
B	1	1	1	2	2	2	$\ldots$
$TSIZE$	2	1	2	2	2	1	$\ldots$

Table 1: Contents of Data Structures: An Example

At the beginning, $PARTITION$ has tuples corresponding to the initial partition. The array $TRANSITIONS$ never gets modified in the algorithm. The array B is also initialized appropriately. For any state q, processors associated with $T(q)$ keep track of the position of state q in the array $PARTITION$.

The algorithm repeats as long as there is a possibility of splitting at least one of the blocks in the current partition. Steps 1-3 are to construct a sequence L of triples. Each state contributes a triple corresponding to each one of transitions going out of the state. If s_i is any state such that $T(s_i) = \{q_1, q_2, \ldots, q_k\}$, then the corresponding triples are $(B[s_i], V[s_i], B[q_j])$, for $j = 1, 2, \ldots, k$.

Steps 4-6 are to eliminate duplicates in L and compress the array L. At the end of Step 6, the array L contains $[p]$ for every state p in each block in the current partition. Furthermore, for each block $B = \{p_1, \ldots, p_k\}$, $[p_1], [p_2], \ldots, [p_k]$ appear consecutively in L.

Step 7 identifies blocks that can be split. Note that even if there is a single j such that $[q_{i,j}] \neq [q_{i,1}]$, we may end up splitting the block B_i and thus the block B_i is marked.

Step 8 picks one of the marked blocks arbitrarily and splits it. If the block B_i is chosen, then B_i is split into B_i and $B_{\ell+1}$, where $B_{\ell+1} = \{p \in B_i | [p] \neq [q_{i,1}]\}$ and B_i is updated to be $B_i - B_{\ell+1}$. After the splitting, we update $PARTITION$ such that states belonging to the same block appear consecutively. Note that we could have split in parallel all those blocks that are marked instead of just one such block as done in Step 8; even then, the worst case run-time of the algorithm would be the same.

4 An Illustrative Example

We now illustrate our algorithm with an example. The example considered is the same as above. The initial partition π_0 is given by $\{\{a, b, c\} \{d, e, f\} \{g, h, i\}\}$. The transition relation is defined as:

$$T(a) = \{d, f\}; \; T(b) = \{d\}; \; T(c) = \{e, f\}; \; T(d) = \{g, i\}; \; T(e) = \{a, b\};$$

$$T(f) = \{g\}; \; T(g) = \{a\}; \; T(h) = \{b, c, d\}; \; T(i) = \{a, b\}.$$

The initial contents of various data structures are shown in Table 1. We call each run of the *while* loop as a phase of the algorithm.

Phase I: At the end of Step 3 the list L looks like:

$$(1, 1, 2), (1, 1, 2), (1, 2, 2), (1, 3, 2), (1, 3, 2), (2, 1, 3), (2, 1, 3), (2, 2, 1), (2, 2, 1),$$

$$(2, 3, 3), (3, 1, 1), (3, 2, 1), (3, 2, 1), (3, 2, 2), (3, 3, 1), (3, 3, 1)$$

In Step 4, L is sorted in lexicographic order. The above L happens to be in sorted order already. Steps 5 and 6 compress L as follows.

$$(1, 1, 2), (1, 2, 2), (1, 3, 2), (2, 1, 3), (2, 2, 1), (2, 3, 3), (3, 1, 1), (3, 2, 1), (3, 2, 2), (3, 3, 1)$$

In Step 7, the algorithm realizes that: $[q_{2,2}] \neq [q_{2,1}]$; $[q_{3,2}] \neq [q_{3,1}]$. Therefore, the blocks B_2 and B_3 will be marked.

In Step 8, one of the marked blocks is picked arbitrarily. Let B_2 be the picked block. B_2 gets split into two blocks namely $\{d, f\}$ and $\{e\}$. $PARTITION$ gets modified to:

$$(1, a) \; (1, b) \; (1, c) \; (2, d) \; (2, f) \; (4, e) \; (3, g) \; (3, h) \; (3, i)$$

Phase II: The list L after Step 3 looks like:

$$(1, 1, 2), (1, 1, 2), (1, 2, 2), (1, 3, 4), (1, 3, 2), (2, 1, 3), (2, 1, 3), (2, 2, 3), (4, 1, 1),$$

$$(4,1,1), (3,1,1), (3,2,1), (3,2,1), (3,2,2), (3,3,1), (3,3,1)$$

After L gets sorted and compressed (in Steps 4 through 6), L becomes:

$$(1,1,2), (1,2,2), (1,3,2), (1,3,4), (2,1,3), (2,2,3),$$
$$(3,1,1), (3,2,1), (3,2,2), (3,3,1), (4,1,1)$$

In Step 7, blocks B_1 and B_3 get marked. In Step 8, one of the marked blocks (say B_1) gets chosen. As a result, $PARTITION$ gets modified as follows:

$$(1,a)\ (1,b)\ (5,c)\ (2,d)\ (2,f)\ (4,e)\ (3,g)\ (3,h)\ (3,i)$$

Phase III: The list L gets formed in Step 3:

$$(1,1,2), (1,1,2), (1,2,2), (5,1,4), (5,1,2), (2,1,3), (2,1,3), (2,2,3), (4,1,1),$$
$$(4,1,1), (3,1,1), (3,2,1), (3,2,5), (3,2,2), (3,3,1), (3,3,1)$$

In Steps 4 through 6, L gets modified as follows:

$$(1,1,2), (1,2,2), (2,1,3), (2,2,3), (3,1,1), (3,2,1), (3,2,2), (3,2,5), (3,3,1),$$
$$(4,1,1), (5,1,2)(5,1,4)$$

In Step 7, B_3 gets marked and hence is chosen in Step 8 for splitting. $PARTITION$ now becomes:

$$(1,a)\ (1,b)\ (5,c)\ (2,d)\ (2,f)\ (4,e)\ (3,g)\ (3,i)\ (6,h)$$

Phase IV: No block gets marked in this phase and hence the algorithm terminates to yield the final partition of: $\{\{a,b\}\ \{c\}\ \{d,f\}\ \{e\}\ \{g,i\}\ \{h\}\}$.

5 Analysis

We assume that there are $n + m$ processors, one for each state and one for each transition.

Step 1 takes $O(1)$ time using n processors. Steps 3,5,7 also take $O(1)$ time but need m processors. In Step 2, prefix computation can be done using $\frac{n}{\log n}$ EREW PRAM processors in $O(\log n)$ time (by Lemma 3). In Step 4, we need to sort m numbers in the range $[0, n^3]$, and hence, we apply Lemma 5 to infer that it can be done in $O(\log m) = O(\log n)$ time using $\frac{m}{\log n} \log\log n$ CRCW PRAM processors, or in n^ϵ time using $\frac{m}{n^\epsilon}$ EREW PRAM processors for any fixed $\epsilon < 1$. Step 6 takes $O(\log m) = O(\log n)$ time using $\frac{m}{\log n}$ EREW PRAM processors (by Lemma 3). In Step 8, prefix computation takes $O(\log n)$ time using $\frac{n}{\log n}$ EREW PRAM processors and the rest of the computation can be completed in $O(1)$ time using n processors.

Thus, each run of the while loop can be completed in either: 1) $O(\log n)$ time with a total work of $m \log\log n$ on the CRCW PRAM, or 2) $O(n^\epsilon)$ time with a total work of $O(m)$ on the EREW PRAM. Since the while loop can be executed at most n times, we get the following theorem (using Lemmas 1 and 2):

Theorem 1 *RCPP with m transitions and n states can be solved 1) in $O(n \log n)$ time using $\frac{m}{\log n} \log \log n$ CRCW PRAM processors, or 2) in $O(n^{1+\epsilon})$ time and $\frac{m}{n^\epsilon}$ EREW PRAM processors, for any fixed $\epsilon < 1$.*

The same algorithm can be modified easily to the case of multiple-relation RCPP to use quadruples instead of triples in the list L. The stated processor and time bounds still hold, where m is the total number of transitions in all the relations.

6 Implementation Details

We have implemented our parallel algorithm on two parallel machines, CM2 and CM5 of the Thinking Machines Corp. We employed the CM2 located in the CIS department of the University of Pennsylvania for program development. CM2 is a SIMD machine and has 4096 processing elements. Input and output are through a front end (which is a sun 3/60 work station). Each processing element is bit serial and can compute any boolean function that maps three bits into two bits. On the other hand, CM5 is a MIMD machine with 512 processing elements. Unlike the CM2 processors, CM5 processors are quite powerful; each processing element is comparable to a work station in computing power. We accessed the CM5 located in the CS department of University of Illinois at Urbana-Champaign through internet.

Both CM2 and CM5 provide a routing network for the processors to communicate. In CM2 the underlying routing network has a topology of a hypercube; whereas in CM5, the routing network takes the topology of a fat tree. There are special hardware to handle operations such as scan, broadcast, etc., in both of these machines.

In CM5, we can choose a subset of the processors to work with at any time. We have exploited this facility to study the scalability of our parallel program. Though CM2 supports virtual processors, it does not support selection of a subset. One could run programs written for CM2 on CM5 without much effort. We have coded our algorithm in C* (a parallel programming language supported by CM2 which is very similar to C). The same program runs on CM5. The main objective of this experiment was to study the behavior of the program when the number of processors used changes.

Input to the program was generated as follows: We fix the number of states (call it N) and the number of initial blocks in the Relational Coarsest Partition Problem. States in each block were chosen randomly (under a uniform distribution). Transitions were also picked randomly. Transition Probability, T_p, is a parameter that the user can choose. Each possible transition is picked with this probability.

Figure 2 shows the results of our experiment with $N = 10,000$ and $T_p = 0.05$. The number of initial blocks was 50. The program was run with various number of processors: 32, 64, 128, 256 and 512. For each processor configuration, the time indicated is the average of 5 independent runs.

Solid lines correspond to total execution time of the program, whereas the dotted lines correspond to the time spent on just sorting (in Step 4). 65 to 70 % of the total execution time is spent on sorting. Since the execution time of our program is always

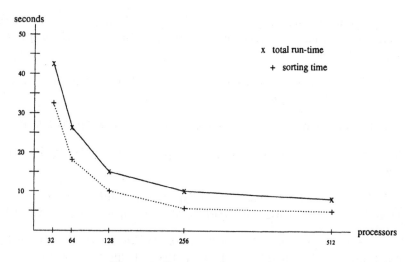

Figure 2: Execution Times on CM5

bounded below by how fast the parallel machine can sort, we are currently exploring ways of substituting sorting with some other operations.

7 Conclusions

We have presented a simple parallel algorithm for RCPP and its implementation. An interesting open problem is to design faster versions of this algorithm. The bottleneck in this algorithm is the use of sorting. Since RCPP is known to be $\mathcal{P}$-complete, a reasonable time to aim for will be $O(n^\epsilon)$, for any fixed $\epsilon < 1$. In [7], we present an efficient algorithm for RCPP which runs in time $O(n \log n)$ using $\frac{m}{n} \log n$ CREW PRAM processors. This algorithm is based on the sequential algorithm of Paige and Tarjan (whose run time is $O(m \log n)$) [8]. Due to lack of space, we are unable to provide details of this algorithm.

Acknowledgements

We are grateful to Inhye Kang for many stimulating discussions. We are also grateful to Angela Lai and D.R. Mani for their wonderful help in implementing our algorithm.

References

[1] C. Alvarez, J.L. Balcazar, J. Gabarro, and M. Santha. Parallel Complexity in the Design and Analysis of Concurrent Systems. In *PARLE '91. Parallel Architectures and Languages Europe, Vol 1*. Springer-Verlag LNCS 505, 1991.

[2] P.C.P. Bhatt, K. Diks, T. Hagerup, V.C. Prasad, T. Radzik, and S. Saxena. Improved Deterministic Parallel Integer Sorting. *Information and Computation*, pages 29–47, 1991.

[3] R.P. Brent. The Parallel Evaluation of General Arithmetic Expressions. *Journal of the ACM*, 21(2):201–208, 1974.

[4] R. Cole. Parallel Merge Sort. *SIAM Journal on Computing*, 17:770–785, 1988.

[5] J. Já Já. *Parallel Algorithms: Design and Analysis*. Addison-Wesley Publishers, 1992.

[6] P.C. Kanellakis and S.A. Smolka. CCS Expressions, Finite State Processes, and Three Problems of Equivalence. *Information and Computation*, 86:43–68, 1990.

[7] I. Lee and S. Rajasekaran. Parallel Algorithms for Relational Coarsest Partition Problems. Technical Report MS-CIS-93-71, Dept. of CIS, Univ. of Pennsylvania, July 1993.

[8] R. Paige and R.E. Tarjan. Three Partition Refinement Algorithms. *SIAM Journal on Computing*, 16(6):973–989, 1987.

[9] S. Zhang and S.A. Smolka. Towards efficient parallelization of equivalence checking algorithms. Unpublished Manuscript, 1993.

Another Look at LTL Model Checking [*]

E. Clarke[1], O. Grumberg[2], K. Hamaguchi[1]

1. Carnegie Mellon, Pittsburgh
2. The Technion, Haifa

Abstract. We show how LTL model checking can be reduced to CTL model checking with fairness constraints. Using this reduction, we also describe how to construct a *symbolic* LTL model checker that appears to be quite efficient in practice. In particular, we show how the SMV model checking system developed by McMillan [16] can be extended to permit LTL specifications. The results that we have obtained are quite surprising. For the examples we considered, the LTL model checker required at most twice as much time and space as the CTL model checker. Although additional examples still need to be tried, it appears that efficient LTL model checking is possible when the specifications are not excessively complicated.
Keywords: automatic verification, temporal logic, model checking, binary decision diagrams

1 Introduction

Over the past thirteen years there has been considerable research on efficient model checking algorithms for branching-time temporal logics like CTL (See [5] for a survey). Verification tools based on these algorithms have discovered non-trivial design errors in sequential circuits and protocols [10] and are now beginning to be used in industry. There has been relatively little research, however, on efficient model checking algorithms for linear-temporal logic (LTL), and practical verification tools are virtually non-existent. In fact, the question of whether it is possible to develop such tools has been argued for many years. Sistla and Clarke [17] showed in 1982 that the model checking problem for LTL was, in general, PSPACE complete. Later, Pnueli and Lichtenstein [14] gave an LTL model checking algorithm that was exponential in the size of the formula, but *linear* in the size of the model. Based on this result, they argued that the high complexity of LTL model checking might still be acceptable for short formulas. Vardi and Wolper [18] obtained a different algorithm based on ω-automata with roughly the same complexity. Unfortunately, the LTL algorithms appeared significantly more difficult to implement. Because of this, very few LTL model checkers were actually constructed. To the best of our knowledge, no experiments were made to determine how the CTL and LTL model checking algorithms actually compared in practice.

In this paper we show how LTL model checking can be reduced to CTL model checking with fairness constraints. We also describe how to construct a *symbolic* LTL model checker that appears to be quite efficient in practice. In particular, we show how the SMV model checking system developed by McMillan as part of his Ph.D. thesis [16] can be extended to permit LTL specifications. We have developed a translator T that takes an LTL formula f and constructs an SMV program $T(f)$ to build the tableau for f. The tableau construction that we use is similar to the one described in [4]. To check that f holds for some SMV program M, we combine the

[*] This research was sponsored in part by the Avionics Laboratory, Wright Research and Development Center, Aeronautical Systems Division (AFSC), U.S. Air Force, Wright-Patterson AFB, Ohio 45433-6543 under Contract F33615-90-C-1465, ARPA Order No. 7597 and in part by the National Science foundation under Grant No. CCR-9217549 and in part by the Semiconductor Research Corporation under Contract 92-DJ-294 and in part by the Wright Laboratory, Aeronautical Systems Center Air Force Materiel Command, USAF, and the Advanced Research Projects Agency (ARPA) under grant number F33615-93-1-1330. The third author was supported by a Kurata Research Grant and a Kyoto University Foundation Grant. The views and conclusions contained in this document are those of the authors and should not be interpreted as representing the official policies, either expressed or implied of the U.S. government.

text of $T = \mathcal{T}(\neg f)$ with the text of M to obtain a new SMV program $P = Prod(T, M)$. We add CTL fairness constraints to P in order to make sure that eventualities of the form $a \mathbf{U} b$ are actually fulfilled (i.e. to eliminate those paths along which $a \mathbf{U} b$ and a hold continuously, but b never holds). By checking an appropriate CTL formula on P we can find the set V_f of all of those states s such that f holds along every path that begins at s. The projection of V_f to the state variables of M gives the set of states where the formula f holds.

Note that our approach makes it unnecessary to modify SMV (or even understand how SMV is actually implemented). We have evaluated the approach on several standard SMV programs (including Martin's distributed mutual exclusion circuit [15] and the synchronous arbiter described in McMillan's thesis [16]). In order to make sure that the experiments were unbiased, we deliberately chose specifications which could be expressed in both CTL and LTL. The results that we obtained were quite surprising. For the examples we considered, the LTL model checker required at most twice as much time and space as the CTL model checker. Although additional examples still need to be tried, it appears that efficient LTL model checking is possible when the specifications are not excessively complicated. In the full paper we will describe how the same basic approach can be used to extend SMV for testing inclusion between various types of ω-automata.

2   Binary Decision Diagrams

Ordered binary decision diagrams (OBDDs) are a canonical form representation for boolean formulas [3]. They are often substantially more compact than traditional normal forms such as conjunctive normal form or disjunctive normal form, and they can be manipulated very efficiently. An OBDD is similar to a binary decision tree, but has the following properties.

- Its structure is a directed acyclic graph rather than a tree.
- A total order is placed on the occurrence of variables as the graph is traversed from root to leaf.
- No two subgraphs in the graph represents the same function.

Bryant showed that given a variable ordering, the OBDD representation for a boolean formula is unique.

We can implement various important logical operations using OBDDs. The function that restricts some argument x_i of the boolean function f to a constant value b, denoted by $f \mid_{x_i \leftarrow b}$, can be performed in time which is linear in the size of the original binary decision diagram [3]. The restriction algorithm allows us to compute the OBDD for the formula $\exists x f$ as $f \mid_{x \leftarrow 0} + f \mid_{x \leftarrow 1}$. All 16 two-argument logical operations can also be implemented efficiently on boolean functions that are represented as OBDDs. The complexity of these operations is linear in the size of the argument OBDDs [3]. Furthermore equivalence checking of two boolean functions can be done in constant time, by using a hash table properly[2].

OBDDs are extremely useful for obtaining concise representations of relations over finite domains [4, 16]. If R is n-ary relation over $\{0, 1\}$ then R can be represented by the OBDD for its *characteristic function*

$$f_R(x_1, \ldots, x_n) = 1 \text{ iff } R(x_1, \ldots, x_n).$$

Otherwise, let R be an n-ary relation over the finite domain D. Using an appropriate binary encoding of D, we can represent R by an OBDD.

3   Computation Tree Logics

We begin by describing the temporal logic CTL* [8, 9, 12], which can express both linear-time and branching-time properties. In this logic, a path quantifier, either $\mathbf{A}$ ("for all computation paths") or $\mathbf{E}$ ("for some computation paths") can prefix an assertion composed of arbitrary combinations of the usual linear-time operators $\mathbf{G}$ ("always"), $\mathbf{F}$ ("sometimes"), $\mathbf{X}$ ("nexttime"), and $\mathbf{U}$ ("until"). Both Linear Temporal Logic (LTL) and Computation Tree Logic (CTL) are included in CTL*.

There are two types of formulas in CTL*: *state formulas* (which are true in a specific state) and *path formulas* (which are true along a specific path). Let AP be the set of atomic proposition names. The syntax of state formulas is given by the following rules:

- If $p \in AP$, then p is a state formula.
- If f and g are state formulas, then $\neg f$ and $f \vee g$ are state formulas.
- If f is a path formula, then $\mathbf{E}(f)$ is a state formula.

Two additional rules are needed to specify the syntax of path formulas:

- If f is a state formula, then f is also a path formula.
- If f and g are path formulas, then $\neg f$, $f \vee g$, $\mathbf{X} f$, and $f \mathbf{U} g$ are path formulas.

CTL^* is the set of state formulas generated by the above rules.

We define the semantics of CTL^* with respect to a Kripke structure $M = \langle S, R, L \rangle$, where S is the set of states; $R \subseteq S \times S$ is the transition relation, which must be *total* (i.e., for all states $s \in S$ there exists a state $s' \in S$ such that $(s, s') \in R$); and $L : S \to \mathcal{P}(AP)$ is a function that labels each state with a set of atomic propositions true in that state. In this paper, we assume that all Kripke structures are *finite*.

A *path in M* is an infinite sequence of states, $\pi = s_0, s_1, \ldots$ such that for every $i \geq 0$, $(s_i, s_{i+1}) \in R$. We use π^i to denote the *suffix* of π starting at s_i. If f is a state formula, the notation $M, s \models f$ means that f holds at state s in the Kripke structure M. Similarly, if f is a path formula, $M, \pi \models f$ means that f holds along path π in Kripke structure M. When the Kripke structure M is clear from context, we will usually omit it. The relation $\models$ is defined inductively as follows (assuming that f_1 and f_2 are state formulas and g_1 and g_2 are path formulas):

1. $s \models p$ $\Leftrightarrow p \in L(s)$.
2. $s \models \neg f_1$ $\Leftrightarrow s \not\models f_1$.
3. $s \models f_1 \vee f_2$ $\Leftrightarrow s \models f_1$ or $s \models f_2$.
4. $s \models \mathbf{E}(g_1)$ $\Leftrightarrow$ there exists a path π starting with s such that $\pi \models g_1$.
5. $\pi \models f_1$ $\Leftrightarrow s$ is the first state of π and $s \models f_1$.
6. $\pi \models \neg g_1$ $\Leftrightarrow \pi \not\models g_1$.
7. $\pi \models g_1 \vee g_2$ $\Leftrightarrow \pi \models g_1$ or $\pi \models g_2$.
8. $\pi \models \mathbf{X} g_1$ $\Leftrightarrow \pi^1 \models g_1$.
9. $\pi \models g_1 \mathbf{U} g_2$ $\Leftrightarrow$ there exists a $k \geq 0$ such that $\pi^k \models g_2$ and for all $0 \leq j < k$, $\pi^j \models g_1$.

The following abbreviations are used in writing CTL^* formulas:

- $f \wedge g \equiv \neg(\neg f \vee \neg g)$ • $\mathbf{F} f \equiv true \mathbf{U} f$
- $\mathbf{A}(f) \equiv \neg \mathbf{E}(\neg f)$ • $\mathbf{G} f \equiv \neg \mathbf{F} \neg f$

CTL [1, 8] is a restricted subset of CTL^* that permits only branching-time operators—each of the linear-time operators $\mathbf{G}$, $\mathbf{F}$, $\mathbf{X}$, and $\mathbf{U}$ must be immediately preceded by a path quantifier. More precisely, CTL is the subset of CTL^* that is obtained if the following two rules are used to specify the syntax of path formulas.

- If f and g are state formulas, then $\mathbf{X} f$ and $f \mathbf{U} g$ are path formulas.
- If f is a path formula, then so is $\neg f$.

Linear temporal logic (LTL), on the other hand, will consist of formulas that have the form $\mathbf{A} f$ where f is a path formula in which the only state subformulas permitted are atomic propositions. More precisely, a path formula is either:

- an atomic proposition $p \in AP$.
- If f and g are path formulas, then $\neg f$, $f \vee g$, $\mathbf{X} f$, and $f \mathbf{U} g$ are path formulas.

There are eight basic CTL operators: **AX**, **EX**, **AG**, **EG**, **AF**, **EF**, **AU** and **EU**. Each of the eight operators can be expressed in terms of three operators **EX**, **EG**, and **EU**.

4 CTL Model Checking

CTL Model checking is the problem of finding the set of states in a state transition graph where a given CTL formula is true. One approach for solving this problem is a symbolic model checking using an OBDD to represent the transition relation of the graph. Assume that the transition relation is given as a boolean formula $R(\bar{v}, \bar{v}')$ in terms of current state variables $\bar{v} = (v_1, \ldots, v_n)$ and next state variables $\bar{v}' = (v_1', \ldots, v_n')$. The algorithm takes a CTL formula f, and the OBDD that represents $R(\bar{v}, \bar{v}')$. For each subformula g, the algorithm computes the states that satisfy g in a bottom-up manner. This step is performed by OBDD operations. The algorithm returns an OBDD that represents exactly those states of the system that satisfy the formula f.

Fairness constraints were introduced for checking the correctness of CTL formulas along fair computation paths. A *fairness constraint* can be an arbitrary set of states, usually described by a formula of the logic. A path is said to be *fair* with respect to a set of fairness constraints if each constraint holds *infinitely often* along the path. The path quantifiers in CTL formulas are then restricted to fair paths. The CTL model checking under given fairness constraints can also be performed using OBDD operations. As will be shown in the next section, LTL model checking can be reduced to CTL model checking under fairness constraints.

5   LTL Model Checking

In this section we consider the model checking problem for linear temporal logic. Let $\mathbf{A}\,f$ be a linear temporal logic formula. Thus, f is a *restricted path formula* in which the only state subformulas are atomic propositions. We wish to determine all of those states $s \in S$ such that $M, s \models \mathbf{A}\,f$. By definition $M, s \models \mathbf{A}\,f$ iff $M, s \models \neg\,\mathbf{E}\,\neg f$. Consequently, it is sufficient to be able to check the truth of formulas of the form $\mathbf{E}\,f$ where f is a restricted path formula. If the Kripke structure is represented explicitly as a state transition graph, this problem is known to be PSPACE-complete [17] in general.

Lichtenstein and Pnueli [14] developed an algorithm for the problem that was linear in the size of the model M and exponential in the length of the formula f. Although their algorithm was linear in the size of the model, it was still impractical for large examples because of the state explosion problem. As in the case of CTL model checking, representing the transition relation as an OBDD enables the procedure to be applied to much larger examples. The exponential complexity of their algorithm in terms of formula length is caused by a tableau construction which may require exponential space in the size of the formula.

Burch et. al developed a symbolic satisfiability algorithm for LTL [4], This algorithm is based on implicit tableau construction, which leads to an additional reduction in space and time. We also use this implicit technique in the following model checking algorithm. We begin with an informal description of the algorithm. Given a formula $\mathbf{E}\,f$ and a Kripke structure M, we construct a special Kripke structure T called the *tableau* for the path formula f. This structure includes *every* path that satisfies f. By composing T with M, we find the set of paths that appear in both T and M. A state in M will satisfy $\mathbf{E}\,f$ if and only if it is the start of a path in the composition that satisfies f. The CTL model checking procedure described in Section 4 is used to find these states.

We now describe the construction of the tableau T in detail. Let AP_f be the set of atomic propositions in f. The tableau associated with f is a structure $T = (S_T, R_T, L_T)$ with AP_f as its set of atomic propositions. Each state in the tableau is a set of *elementary* formulas obtained from f. The set of elementary subformulas of f is denoted by $el(f)$ and is defined recursively as follows:

- $el(p) = \{p\}$ if $p \in AP$.
- $el(\neg g) = el(g)$.
- $el(g \vee h) = el(g) \cup el(h)$.
- $el(\mathbf{X}\,g) = \{\mathbf{X}\,g\} \cup el(g)$.
- $el(g\ \mathbf{U}\ h) = \{\mathbf{X}(g\ \mathbf{U}\ h)\} \cup el(g) \cup el(h)$.

Thus, the set of states S_T of the tableau is $\mathcal{P}(el(f))$. The labeling function L_T is defined so that each state is labeled by the set of atomic propositions contained in the state.

In order to construct the transition relation R_T, we need an additional function sat that associates with each elementary subformula g of f a set of states in S_T. Intuitively, $sat(g)$ will be the set of states that satisfy g.

- $sat(g) = \{\sigma \mid g \in \sigma\}$ where $g \in el(f)$.
- $sat(\neg g) = \{\sigma \mid \sigma \notin sat(g)\}$.
- $sat(g \vee h) = sat(g) \cup sat(h)$.
- $sat(g\ \mathbf{U}\ h) = sat(h) \cup (sat(g) \cap sat(\mathbf{X}(g\ \mathbf{U}\ h)))$.

We want the transition relation to have the property that each elementary formula in a state is true in that state. Clearly, if $\mathbf{X}g$ is in some state σ, then all the successors of σ should satisfy

g. Furthermore, since we are dealing with LTL formulas, if $\mathbf{X}g$ is not in σ, then σ should satisfy $\neg\mathbf{X}g$. Hence, no successor of σ should satisfy g. The obvious definition for R_T is

$$R_T(\sigma, \sigma') = \bigwedge_{\mathbf{X}g \in el(f)} \sigma \in sat(\mathbf{X}\,g) \Leftrightarrow \sigma' \in sat(g).$$

Figure 1 gives the tableau for the formula $g = a\ \mathbf{U}\ b$. To reduce the number of edges, we connect two states σ and σ' with a bidirectional arrow if there is an edge from σ to σ' and also from σ' to σ. Each subset of $el(g)$ is a state of T. $sat(\mathbf{X}g) = \{1, 2, 3, 5\}$ since each of these states contains the formula $\mathbf{X}g$. $sat(g) = \{1, 2, 3, 4, 6\}$ since each of these states either contains b or contains a and $\mathbf{X}g$. There is a transition from each state in $sat(\mathbf{X}g)$ to each state in $sat(g)$ and from each state in the complement of $sat(\mathbf{X}g)$ to each state in the complement of $sat(g)$.

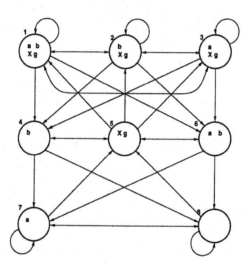

Fig. 1. Tableau for $a\ \mathbf{U}\ b$

Unfortunately, the definition of R_T does not guarantee that *eventuality* properties are fulfilled. We can see this behavior in Figure 1. Although state 3 belongs to $sat(g)$, the path that loops forever in state 3 does not satisfy the formula g since b never holds on that path. Consequently, an additional condition is necessary in order to identify those paths along which f holds. A path π that starts from a state $\sigma \in sat(f)$ will satisfy f if and only if

- For every subformula $g\ \mathbf{U}\ h$ of f and for every state σ on π, if $\sigma \in sat(g\ \mathbf{U}\ h)$ then either $\sigma \in sat(h)$ or there is a later state τ on π such that $\tau \in sat(h)$.

The definition of R_T can also cause T to have states which have no successors. We call such states *deadend* states. For example, a state $\{a, \mathbf{X}\,a, \mathbf{X}\,\neg a\}$ has no successors, because no state is in both $sat(a)$ and $sat(\neg a)$. Because of the semantics of LTL, all of the sequences that satisfy a path formula f must be infinite. Thus, if we remove the deadend states from the tableau of f, no path that satisfies f will be eliminated. Therefore, in the tableau of f, we can safely ignore finite sequences that terminate in deadend states.

In order to state the key property of the tableau construction, we must introduce some new notation. Let $\pi = s_0, s_1, \ldots$ be a path in a Kripke structure M, then $label(\pi) = L(s_0), L(s_1), \ldots$. Let $l = l_0, l_1, \ldots$ be a sequence of subsets of some set Σ and let $\Sigma' \subseteq \Sigma$. The *restriction* of l to Σ', denoted by $l\,|_{\Sigma'}$, is the sequence $l'_0, l'_1, \ldots$ where $l'_i = l_i \cap \Sigma'$ for every $i \geq 0$. The following theorem makes precise the intuitive claim that T includes every path which satisfies f.

Theorem 1. *Let T be the tableau for the path formula f. Then, for every Kripke structure M and every path π' of M, if $M, \pi' \models f$ then there is a path π in T that starts in a state in $sat(f)$, such that $label(\pi') \mid_{AP_f} = label(\pi)$.*

Next, we want to compute the product $P = (S, R, L)$ of the tableau $T = (S_T, R_T, L_T)$ and the Kripke structure $M = (S_M, R_M, L_M)$.

- $S = \{(\sigma, \sigma') \mid \sigma \in S_T, \sigma' \in S_M \text{ and } L_M(\sigma') \cap AP_f = L_T(\sigma)\}$.
- $R((\sigma, \sigma'), (\tau, \tau'))$ iff $R_T(\sigma, \tau)$ and $R_M(\sigma', \tau')$.
- $L((\sigma, \sigma')) = L_T(\sigma)$.

P may have deadend states, even if T contains no deadend states. However, it is not difficult to show that P contains exactly the infinite paths π'' for which there are infinite paths π in T and π' in M such that $label(\pi'') = label(\pi) = label(\pi') \mid_{AP_f}$. Thus, if we remove the deadend states from the product, no path that satisfies f will be eliminated. As a result, we can safely ignore finite sequences in P. We extend the function sat to be defined over the set of states of the product P by $(\sigma, \sigma') \in sat(g)$ if and only if $\sigma \in sat(g)$.

We next apply CTL model checking and find the set of all states V in P, $V \subseteq sat(f)$, that satisfy $\mathbf{EG}\, true$ with the fairness constraints

$$\{sat(\neg(g \mathbf{U} h) \vee h) \mid g \mathbf{U} h \text{ occurs in } f\}. \tag{1}$$

Each of the states in V is in $sat(f)$. Moreover, it is the start of an infinite path that satisfies all of the fairness constraints. These paths have the property that no subformula $g \mathbf{U} h$ holds almost always on the path while h remains false. The correctness of our construction is summarized by the following theorem.

Theorem 2. *$M, \sigma' \models \mathbf{E} f$ if and only if there is a state σ in T such that $(\sigma, \sigma') \in sat(f)$ and $P, (\sigma, \sigma') \models \mathbf{EG}\, True$ under fairness constraints $\{sat(\neg(g \mathbf{U} h) \vee h) \mid g \mathbf{U} h \text{ occurs in } f\}$.*

To illustrate this construction, we check the formula $g = a \mathbf{U} b$ on the Kripke structure M in Figure 2. The tableau T for this formula is given in Figure 1. If we compute the product P as described above, we obtain the Kripke structure shown in Figure 3. Although P contains deadend states $(4, 4')$ and $(8, 3')$, we can ignore those states. We use the CTL model checking algorithm to find the set V of states in $sat(g)$ that satisfy the formula $\mathbf{EG}\, true$ with the fairness constraint $sat(\neg(a \mathbf{U} b) \vee b)$. It is easy to see that the fairness constraint corresponds to the following set of states $\{(2, 4'), (5, 3'), (7, 1'), (6, 2'), (1, 2')\}$. Thus, every state in Figure 3 satisfies $\mathbf{EG}\, true$. However, only $(2, 4'), (3, 1'), (1, 2')$ and $(6, 2')$ are in $sat(g)$, so the states $1', 2'$, and $4'$ of M satisfy $\mathbf{E}\, g = \mathbf{E}[a \mathbf{U} b]$.

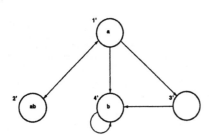

Fig. 2. Kripke Structure M

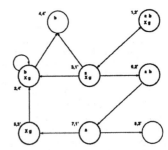

Fig. 3. The product P of the structure M and the tableau T

We now describe how the above procedure can be implemented using OBDDs. We assume that the transition relation for M is represented by an OBDD as in the previous section. In order to represent the transition relation for T in terms of OBDDs, we associate with each elementary formula g a state variable v_g. We describe the transition relation R_T as a boolean formula in terms of two copies $\bar{v}$ and $\bar{v}'$ of the state variables. The boolean formula is converted to an OBDD to obtain a concise representation of the tableau. When the composition P is constructed, it is convenient to separate out the state variables that appear in AP_f. The symbol $\bar{p}$ will be used to denote a boolean vector that assigns truth values to these state variables. Thus, each state in S_T will be represented by a pair $(\bar{p}, \bar{r})$, where $\bar{r}$ is a boolean vector that assigns values to the state variables that appear in the tableau but not in AP_f. A state in S_M will be denoted by a pair $(\bar{p}, \bar{q})$ where $\bar{q}$ is a boolean vector that assigns values to the state variables of M which are not mentioned in f. Thus, the transition relation R_P for the product of the two Kripke structures will be given by

$$R_P(\bar{p}, \bar{q}, \bar{r}, \bar{p}', \bar{q}', \bar{r}') = R_T(\bar{p}, \bar{r}, \bar{p}', \bar{r}') \wedge R_M(\bar{p}, \bar{q}, \bar{p}', \bar{q}').$$

We use the symbolic model checking algorithm that handles fairness constraints to find the set of states V that satisfy **EG** *true* with the fairness constraints given in (1). We must be careful because of the sequences in the product P that terminates in deadend states. Let $\{c_1, c_2, \ldots, c_l\}$ be fairness constaints. The symbolic model checking algorithm computes the set V as the greatest fixpoint of the following function $\tau : \mathcal{P}(S) \to \mathcal{P}(S)$:

$\tau(Z) = \{s \mid$ for any c_i, there exists a sequence of length one or greater from s to a state $s' \in sat(c_i) \cap Z\}$.

It is easy to see that, if V is computed in this manner, then every state $\sigma \in V$ will be the beginning of an infinite path that satisfies all of the fairness constraints $c_1, c_2, \ldots, c_l$. Suppose that a state σ is in V. Since $\sigma \in V = \tau(V)$, the definition of τ guarantees that there exists a sequence of length one or larger from σ to some state $\sigma_1 \in sat(c_1)$. Since $\sigma_1 \in V$, we can find a finite sequence from σ_1 to some $\sigma_2 \in sat(c_2)$. Thus, we can eventually find an infinite path π from σ which goes through states in $sat(c_i)$ infinitely often, for each c_i. The states in V are represented by boolean vectors of the form $(\bar{p}, \bar{q}, \bar{r})$. Thus, a state $(\bar{p}, \bar{q})$ in M satisfies $\mathbf{E} f$ if and only if there exists $\bar{r}$ such that $(\bar{p}, \bar{q}, \bar{r}) \in V$ and $(\bar{p}, \bar{r}) \in sat(f)$.

6 LTL Model Checking Using the SMV Model Checker

As stated in Section 5, LTL model checking can be reduced to CTL model checking under fairness constraints. If the tableau and the fairness constraints for a given LTL formula are represented implicitly as boolean formulas, we can perform symbolic LTL model checking using an existing symbolic model checker for CTL. We have developed a translator that enables the SMV model checker to handle LTL formulas. For a given LTL formula, the translator generates an SMV program for the corresponding tableau and fairness constraints. We can perform symbolic LTL model checking using the resulting SMV program. In this section, we describe how the translator works.

We begin with a brief description of the SMV model checker. SMV is a tool for checking that finite-state systems satisfy specifications given in CTL. It uses the OBDD-based symbolic model checking algorithm in Section 4. The language component of SMV is used to describe complex finite-state systems. Figure 4 shows an SMV program for the Kripke structure in Figure 2 and a specification $\mathbf{A}(a \mathbf{U} b)$. This example illustrates the basic features of SMV that are needed to explain the translation procedure. The syntax and semantics of the complete language are given in McMillan's thesis [16].

SMV users can decompose the description of a complex finite-state system into modules. Module definitions begin with the keyword MODULE. The module main is the top-level module. (The example in Figure 4 contains a single module; however, our translator can handle programs with multiple modules.) Variables are declared using the keyword VAR. In the example, a and b are boolean variables (line 3–4). The TRANS statements are used to define transitions of the model (lines 5–8). In the TRANS statements, next(g) is obtained from g by replacing each state variable v in g by the corresponding next state variable v'. For example, next(a & !b) means $a' \wedge \neg b'$ where a' are b' are the next state variables for a and b, respectively. Thus, each TRANS statement determines a propositional formula that relates the original state variables and the next state variables. The transition relation for an SMV program is obtained by taking the conjunction of these formulas. CTL formulas are declared as specifications using the keyword SPEC (line 9).

Next, we describe the translation algorithm. Suppose that we have an SMV program with an LTL formula $\mathbf{A} f$, instead of a CTL formula, as its specification. As stated in Section 5, it is

```
-- Kripke structure

MODULE
                        :
                        :
MODULE
                        :
                        :
MODULE        main
                        :
                        :
```

```
1   MODULE main   -- simple program

2   VAR

3   a: boolean;
4   b: boolean;

5   TRANS ( a & !b) -> next(!(a & !b))
6   TRANS ( a &  b) -> next(a & !b)
7   TRANS (!a &  b) -> next(!a & b)
8   TRANS (!a & !b) -> next(!a & b)

9   SPEC A[a U b]
```

Fig. 4. Simple SMV program

```
-- LTL formula
SPEC          A f
```

Fig. 5. An SMV program

sufficient to handle a formula $E \neg f$. The translator replaces $A f$ with an SMV description of the tableau and the fairness constraints for $\neg f$. The translation of the SMV program in Figure 5 is shown in Figure 6. The translation follows the general procedure outlined in Section 5:

1. Associate a state variable with each elementary formula of $\neg f$.
2. Represent the transition relation of the tableau for $\neg f$ as a boolean formula in terms of the state variables.
3. Represent fairness constraints as boolean formulas in terms of the state variables.
4. Generate a CTL specification.

In the first step, the formula f is negated and expanded to a formula in which the only operators are $\vee, \neg, X, U$. The parse tree of $\neg f$ is traversed to find its elementary formulas. If a node associated with formula $X g$ (or $g U h$) is visited, then the corresponding elementary formula $X g$ (or $X(g U h)$) is stored in the list el_list. The translator declares a new variable $EL_{X g}$ for each formula $X g$ in the list el_list. Since atomic propositions are already declared in the original SMV program, they are not declared again.

In order to generate descriptions for the transition relation and the fairness constraints, we have to construct the characteristic function S_h of $sat(h)$ for each subformula or elementary formula h in $\neg f$. The translator builds these functions using a DEFINE statement[2]. The translator traverses the parse tree of $\neg f$, and generates the appropriate SMV statements at each node.

$$S_h := p; \qquad\qquad \text{if } p \text{ is an atomic proposition.}$$
$$S_h := EL_h; \qquad\qquad \text{if } h \text{ is elementary formula } X g \text{ in } el\_list.$$
$$S_h := !S_g; \qquad\qquad \text{if } h = \neg g.$$
$$S_h := S_{g_1} \mid S_{g_2}; \qquad \text{if } h = g_1 \vee g_2.$$
$$S_h := S_{g_2} \mid \left(S_{g_1} \& S_{X(g_1 U g_2)} \right); \text{ if } h = g_1 U g_2.$$

The transition relation can be described in terms of the characteristic functions as follows:

$$\bigwedge_{X g \in el(f)} S_{X g}(\bar{v}) \Leftrightarrow S_g(\bar{v}')$$

---

[2] This statement associates a symbol with an SMV expression. When the symbol appears in the program, it is replaced with the expression.

The expression $S_g(\bar{v}')$ is represented in SMV by `next(S_g)`. The translator constructs a formula $S_{\mathbf{X}_g}$ = `next` (S_g) for each $\mathbf{X} g$ in *el_list*. These formulas are combined in a TRANS statement to give the transition relation for the tableau.

```
TRANS
  ( S_{X_{g1}} = next (S_{g1}) ) &
  ( S_{X_{g2}} = next (S_{g2}) ) &
            ⋮
  ( S_{X_{gN}} = next (S_{gN}) )
```

Likewise, the translator traverses the parse tree and generates an SMV FAIRNESS constraint for each node associated with a formula of form $g \mathbf{U} h$:

FAIRNESS !$S_{g\mathbf{U}h}$ | S_h

Finally, the translator generates an SMV SPEC statement. From Theorem 2, it is clear that the formula $\mathbf{E} \neg f$ can be checked using the the specification $S_{\neg f} \wedge \mathbf{EG}\, True$. Thus, in order to check the LTL formula $\mathbf{A} f = \neg \mathbf{E} \neg f$, the translator constructs an SMV SPEC statement for $\neg(S_{\neg f} \wedge \mathbf{EG}\, True)$.

We illustrate the translation procedure by applying it to the simple example in Figure 4. The result of this procedure is shown in Figure 7. The statements in lines 1 through 8 come from the original SMV program, while the statements in lines 9 through 19 are generated by the tableau construction for $a \mathbf{U} b$. The translation procedure first determines that a, b and $\mathbf{X}(a \mathbf{U} b)$ are elementary formulas and causes the state variable EL_X_a_U_b to be declared for $\mathbf{X}(a \mathbf{U} b)$ in line 10. Next, the DEFINE statement in lines 12 through 16 is constructed for the characteristic functions of $sat(a)$, $sat(b)$, $sat(\mathbf{X}(a \mathbf{U} b))$, $sat(a \mathbf{U} b)$ and $sat(\neg a \mathbf{U} b)$. The Trans statement in line 17 causes the transition relation for the tableau to be constructed, and line 18 contains the fairness constraint for $a \mathbf{U} b$. Finally, the specification to be checked is given by the 'SPEC' statement in line 19.

7 Experimental Results

This section describes the experimental results that we obtained for symbolic LTL model checking. In order to compare the performance of LTL model checking with CTL model checking, we used two sequential circuit designs whose specifications can be described in both LTL and CTL.

The first example is a distributed mutual exclusion(DME) circuit designed by Alain Martin[15]. The DME circuit is a speed-independent token ring, which consists of identical arbiter cells. A user of the DME circuit obtains exclusive access to the resource via request and acknowledge signals. We assume aribitrary delay for all gates in the circuit. Each gate is modeled as a finite-state machine that non-deterministically decides either to recompute its output or remain unchanged. We verify the correctness of the following two specifications:

1. *(Safety)* No two users are acknowledged simultaneously.
2. *(Liveness)* All requests are eventually acknowledged.

The safety specification is given by the formula

$$\mathbf{AG} \bigwedge_{1 \leq i < j \leq n} \neg(ack_i \wedge ack_j),$$

where ack_i means that user i is acknowledged. This formula is both an LTL formula and a CTL formula. In the experiments for this specification, infinite delays are allowed at each gate. In other words, the output value of each gate can remain unchanged forever.

Next, we verify that requests are eventually acknowledged. We only check this specification with respect to a single user (user 1). In this case the LTL specification has the form:

$$\mathbf{AG}(req_1 \rightarrow \mathbf{F}\, ack_1)$$

This formula is equivalent to the CTL formula:

$$\mathbf{AG}(req_1 \rightarrow \mathbf{AF}\, ack_1)$$

```
-- Kripke structure

MODULE
              ⋮

MODULE        main
              ⋮
```

```
-- Tableau for f
VAR       -- new variables
          EL_X_{g_1} : boolean;
          EL_X_{g_2} : boolean;
              ⋮
          EL_X_{g_N} : boolean;

DEFINE    -- characteristic function
          S_{h_1} := ⋯;
          S_{h_2} := ⋯;
              ⋮
          S_{h_M} := ⋯;

TRANS     -- transition relation
          ( S_X_{g_1} = next (S_{g_1}) ) &
          ( S_X_{g_2} = next (S_{g_2}) ) &
              ⋮
          ( S_X_{g_N} = next (S_{g_N}) )

-- fairness constraints
FAIRNESS  !S_{g'_1}U_{h'_1} | S_{h'_1}
FAIRNESS  !S_{g'_2}U_{h'_2} | S_{h'_2}
              ⋮
FAIRNESS  !S_{g'_3}U_{h'_3} | S_{h'_3}

-- new specification
SPEC   !(S_{¬f} & EG true)
```

Fig. 6. Translator output for SMV program

```
1  MODULE main  -- simple program

2  VAR

3  a: boolean;
4  b: boolean;

5  TRANS ( a & !b) -> next(!(a & !b))
6  TRANS ( a &  b) -> next(a & !b)
7  TRANS (!a &  b) -> next(!a & b)
8  TRANS (!a & !b) -> next(!a & b)

9  VAR

10 EL_X_a_U_b : boolean;

11 DEFINE

12 S_a       := a;
13 S_b       := b;
14 S_X_a_U_b  := EL_X_a_U_b;
15 S_a_U_b    := S_b | (S_a & S_X_a_U_b);
16 S_NOT_a_U_b := !S_a_U_b;

17 TRANS     S_X_a_U_b = next(S_a_U_b)

18 FAIRNESS    !S_a_U_b | b

19 SPEC       !(S_NOT_a_U_b & EG true)
```

Fig. 7. Translator output for simple SMV program

If infinite delays are allowed at each gate, these formulas are not true. In order to overcome this problem we use a fairness constraint which ensures that the output of the gate is reevaluated infinitely often.

SMV provides several options to perform model checking. We verified the circuit using the following approach.

- A single OBDD is constructed for the transition relation of the circuit.
- The reachable states of the circuit are determined, and evaluation of the CTL operators is restricted to these states.
- At each step in the forward search, the transition relation is restricted to the set of reachable

states. The *Restrict* function of Coudert, Madre and Berthet [11] is used for this purpose.

Table 1 summarizes the experimental results for the safety specification, and Table 2 summarizes the results for the liveness specification. The columns show the number of the cells (#cell), the maximum number of OBDD nodes used at any given time (#nodes), the run time on SPARC station 10 (time), the size of the transition relation in OBDD nodes (trans.) and the number of the reachable states (#reachable states). In the experiment for the safety specification, we observe that the number of reachable states for LTL model checking is twice as large as for CTL model checking. The increase in allocated OBDD nodes and run time is less than 10%. In the experiments for the liveness specification, the number of the reachable states is four times larger for LTL model checking, while the increase in space and time is 1.5–3 times larger.

#cell	#nodes		#time(sec)		trans.		#reachable states	
	CTL	LTL	CTL	LTL	CTL	LTL	CTL	LTL
3	11326	11362	17.9	20.5	2778	2781	6579	13158
4	13458	15357	47.5	49.4	4757	4760	75172	150344
5	22321	22348	100.5	104.4	6760	6763	802425	1.60485e+06
6	25869	27318	182.3	193.6	8763	8766	8.2166e+06	1.64332e+07
7	28413	33310	326.4	329.3	10766	10769	8.1784e+07	1.63568e+08
8	44322	44369	509.2	526.3	12769	12772	7.97393e+08	1.59479e+09
9	49702	49755	794.0	794.8	14772	14775	7.65302e+09	1.53060e+10
10	55082	55141	1125.2	1362.7	16775	16778	7.30144e+10	1.46029e+11

Table 1. Safety specification for the DME circuit

#cell	#nodes		#time(sec)		trans.		#reachable states	
	CTL	LTL	CTL	LTL	CTL	LTL	CTL	LTL
3	12721	33940	426.1	1260.5	2778	3004	6579	26316
4	26541	72029	2553.2	6096.7	4757	4983	75172	300688
5	47346	120299	9623.1	21950.1	6760	6986	802425	3.2097e+06
6	92080	183043	36995.3	66502.5	8763	8989	8.2166e+06	3.28664e+07
7	163867	263380	97807.1	191990.0	10766	10992	8.1784e+07	3.27136e+08

Table 2. Liveness specification for the DME circuit

The second example is a synchronous bus arbiter which is described in McMillan's thesis [16]. This circuit is composed of a *daisy chain* of identical arbiter cells. The requester with the highest priority receives an acknowledgement from the arbiter under normal operation, while a round-robin scheme is applied when the bus traffic becomes very heavy. Each cell is modeled by a deterministic machine, so the whole arbiter circuit is also a deterministic machine. The specifications in this case are essentially the same as in the case of the DME circuit discussed previously:

1. (*Safety*) No two users are acknowledged simultaneously.
2. (*Liveness*) All requests are eventually acknowledged.

In fact, exactly the same LTL and CTL specifications can be used.

In the experiments using SMV, we used the options to construct single transition relations, and to compute reachable states before model checking. Table 3 shows the experimental results for the safety specification and Table 4 shows the results for the liveness specification. For the safety specification we observe that the number of reachable states for LTL model checking checking is twice as large as for CTL model checking. The number of the allocated OBDD nodes and run time both increase by a factor of 1.5. In the second experiment, the number of the reachable states is four times larger for LTL model checking. The amount of space and time that is required is 1.5–2 times larger.

#cell	#nodes		#time(sec)		trans.		#reachable states	
	CTL	LTL	CTL	LTL	CTL	LTL	CTL	LTL
3	384	734	0.08	0.1	80	122	384	768
4	654	1279	0.1	0.1	112	218	2048	4096
5	987	1913	0.11	0.15	144	318	10240	20480
6	1383	2628	0.13	0.18	176	418	49152	98304
7	1842	3424	0.16	0.21	208	518	229376	458752
8	2364	4301	0.16	0.26	240	618	1.04858e+06	2.09715e+06
9	2949	5259	0.16	0.33	272	718	4.71859e+06	9.43718e+06
10	3597	6298	0.21	0.33	304	818	2.09715e+07	4.19430e+07
11	4308	7418	0.21	0.41	336	918	9.22747e+07	1.84549e+08
12	5082	8619	0.31	0.45	368	1018	4.02653e+08	8.05306e+08

Table 3. Safety specification for the synchronous arbiter

#cell	#nodes		#time(sec)		trans.		#reachable states	
	CTL	LTL	CTL	LTL	CTL	LTL	CTL	LTL
3	996	2159	0.10	0.26	80	134	384	1536
4	1531	3137	0.20	0.36	112	196	2048	8192
5	2155	4254	0.38	0.43	144	258	10240	40960
6	2867	5483	0.43	0.48	176	320	49152	196608
7	3667	6820	0.48	0.61	208	382	229376	917504
8	4555	8266	0.53	0.81	240	444	1.04858e+06	4.1943e+06
9	5531	9821	0.71	1.01	272	506	4.71859e+06	1.88744e+07
10	6595	10000	0.83	1.23	304	568	2.09715e+07	8.38861e+07
11	7747	10001	1.00	1.46	336	630	9.22747e+07	3.69099e+08
12	8987	10052	1.16	1.71	368	692	4.02653e+08	1.61061e+09

Table 4. Liveness specification for the synchronous arbiter

8 Directions for Future Research

Certainly the most important thing that remains to be done is to try additional examples. Based on the two examples that we have considered in detail so far, it appears that efficient LTL model checking is possible when the formula that is being checked is not excessively complicated. This does not mean that LTL will take the place of CTL in model checking applications. Many other problems, like testing inclusion and equivalence between various types omega-automata [7], can also be reduced to CTL model checking. LTL, on the other hand, does not appear to have this flexibility. Moreover, in many of the applications of model checking to verification, it is important to be able to assert the existence of a path that satisfies some property. For example, *absence of deadlock* might be expressed by the CTL formula **AG EF** *start* (Regardless of what state the program enters, there exists a computation leading back to the *start* state). Neither this formula nor its negation can be expressed in LTL [6], so LTL model checking techniques cannot be used to decide whether the formula is true or not. Ideally, it should be possible to reason about linear-time and branching-time properties in the same logic (say, CTL*). We believe this goal can potentially be realized by extending the techniques discusssed in this paper. Emerson and Lei [13] have shown how to reduce CTL* model checking to LTL model checking. If the transformation outlined in this paper can be extended to incorporate their reduction, then it should be possible to develop a model checker that can handle both types of properties.

References

1. M. Ben-Ari, Z. Manna, and A. Pnueli. The temporal logic of branching time. *Acta Informatica*, 20:207–226, 1983.
2. K. S. Brace, R. L. Rudell, and R. E. Bryant. Efficient implementation of a BDD package. In *Proceedings of the 27th ACM/IEEE Design Automation Conference.* IEEE Computer Society Press, June 1990.
3. R. E. Bryant. Graph-based algorithms for boolean function manipulation. *IEEE Transactions on Computers*, C-35(8), 1986.
4. J. R. Burch, E. M. Clarke, K. L. McMillan, D. L. Dill, and L. J. Hwang. Symbolic model checking: 10^{20} states and beyond. *Information and Computation*, 98(2):142–170, June 1992.
5. E. Clarke, O. Grumberg, and D. Long. Verification tools for finite-state concurrent systems systems. In *A Decade of Concurrency*, Noordwijkerhout, The Netherlands, June 1993. To appear in Springer Lecture Notes In Computer Science.
6. E. M. Clarke and I. A. Draghicescu. Expressibility results for linear time and branching time logics. In *Linear Time, Branching Time, and Partial Order in Logics and Models for Concurrency*, volume 354, pages 428–437. Springer-Verlag: Lecture Notes in Computer Science, 1988.
7. E. M. Clarke, I. A. Draghicescu, and R. P. Kurshan. A unified approach for showing language containment and equivalence between various types of ω-automata. *Information Processing Letters*, 46:301–308, 1993.
8. E. M. Clarke and E. A. Emerson. Synthesis of synchronization skeletons for branching time temporal logic. In *Logic of Programs: Workshop, Yorktown Heights, NY, May 1981*, volume 131 of *Lecture Notes in Computer Science*. Springer-Verlag, 1981.
9. E. M. Clarke, E. A. Emerson, and A. P. Sistla. Automatic verification of finite-state concurrent systems using temporal logic specifications. *ACM Transactions on Programming Languages and Systems*, 8(2):244–263, 1986.
10. E. M. Clarke, O. Grumberg, H. Hiraishi, S. Jha, D. E. Long, K. L. McMillan, and L. A. Ness. Verification of the Futurebus+ cache coherence protocol. In L. Claesen, editor, *Proceedings of the Eleventh International Symposium on Computer Hardware Description Languages and their Applications.* North-Holland, April 1993.
11. O. Coudert, J. C. Madre, and C. Berthet. Verifying temporal properties of sequential machines without building their state diagrams. In R. P. Kurshan and E. M. Clarke, editors, *Proceedings of the 1990 Workshop on Computer-Aided Verification*, June 1990.
12. E. A. Emerson and J. Y. Halpern. "Sometimes" and "Not Never" revisited: On branching time versus linear time. *Journal of the ACM*, 33:151–178, 1986.
13. E.A. Emerson and Chin Laung Lei. Modalities for model checking: Branching time strikes back. *Twelfth Symposium on Principles of Programming Languages*, New Orleans, La., January 1985.
14. O. Lichtenstein and A. Pnueli. Checking that finite state concurrent programs satisfy their linear specification. In *Proceedings of the Twelfth Annual ACM Symposium on Principles of Programming Languages*, January 1985.
15. A. J. Martin. The design of a self-timed circuit for distributed mutual exclusion. In H. Fuchs, editor, *Proceedings of the 1985 Chapel Hill Conference on Very Large Scale Integration*, 1985.
16. K. L. McMillan. *Symbolic Model Checking: An Approach to the State Explosion Problem*. PhD thesis, Carnegie Mellon University, 1992.
17. A. P. Sistla and E.M. Clarke. Complexity of propositional temporal logics. *Journal of the ACM*, 32(3):733–749, July 1986.
18. M. Y. Vardi and P. Wolper. An automata-theoretic approach to automatic program verification. In *Proceedings of the First Annual Symposium on Logic in Computer Science*. IEEE Computer Society Press, June 1986.

The Mobility Workbench[*]
— A Tool for the π-Calculus —

Björn Victor[†]   Faron Moller[‡]

Abstract

In this paper we describe the first prototype version of the Mobility Workbench (MWB), an automated tool for manipulating and analyzing mobile concurrent systems (those with evolving connectivity structures) described in the π-calculus. The main feature of this version of the MWB is checking open bisimulation equivalences. We illustrate the MWB with an example automated analysis of a handover protocol for a mobile telephone system.

Dedicated to Ellen on the occasion of her birth.

1   Introduction

Process algebra is the general study of distributed concurrent systems in an algebraic framework. There have been many successful models formulated within this framework, one representative example being Milner's CCS [10]. Each approach has added to more than a decade of fruitful discoveries on the mathematical foundations of concurrent processes, so that now it is the case that these theories can be applied in practice, perhaps using automated tools of which there are many; for a useful survey see [9]. Certainly, as systems become more complex, it becomes necessary to invoke the use of automated tools to aid in their analyses. These tools however exploit the fact that properties of *finite-state* systems are decidable, and hence they cannot be used except for this simple class of systems.

A shortcoming of these process algebras, which was perhaps necessary for the development of such a complex field, is that they enforce restrictions on the nature of the systems which they attempt to model. One such restriction is the inability to model evolving communication structures. However this particular shortcoming is now being tackled within the CCS framework with the π-calculus [13], an extension of CCS which allows for the modelling of *mobility* within systems, the ability for systems to dynamically alter their communication structures. The foremost problem with such an extension is the greatly increased complexity of the analysis of systems. One may say that our understanding of finite-state systems is rather complete now; however, with extra constructs within the algebra it becomes nontrivial to even define, let alone to then decide, semantic equivalences between systems.

---

[*]Research supported by ESPRIT BRA Grants 6454: CONFER and 7166: CONCUR2.

[†]Dept of Computer Systems, Uppsala University, Box 325, S-751 05 Uppsala, Sweden, and Swedish Institute for Computer Science.

[‡]Dept of Computer Science, University of Edinburgh, The King's Buildings, Edinburgh.

In this paper we describe the MWB (Mobility Workbench), a tool for manipulating and analyzing mobile concurrent systems described in the π-calculus. In the current version, the basic functionality is to decide the *open bisimulation equivalences* of Sangiorgi [16], for agents in the monadic π-calculus with the original positive match operator. This is decidable for π-calculus agents with *finite control* — analogous to CCS finite-state agents — which do not admit parallel composition within recursively defined agents. There are various other analysis routines implemented, including commands for finding deadlocks and for interactively simulating agents.

The outline of the paper is as follows. We start in Section 2 with a brief presentation of the π-calculus, its syntax and semantic definition, as well as the definition of equality of agents. We also briefly describe aspects of the implementation of equality checking. We then present the MWB in Section 3, demonstrating its main utilities on simple examples. In Section 4 we describe an extended realistic case study, the automated verification of a part of a mobile telephone protocol. Finally in Section 5 we describe future development plans for the MWB.

2 Mobile Processes: The π-calculus

In this section we give a brief presentation of the syntax and semantics of the π-calculus, as well as a description of the open bisimulation equivalences and efficient characterisations for these which will be used in the implementation. For fuller treatments of these topics we refer to [13, 16].

There are two entities in the π-calculus: *names* (ranged over by x,y,z,w,v,u), and *processes* (ranged over by P,Q,R). The syntax of the π-calculus is given by the following BNF equation

$$P ::= 0 \mid A(x_1, \ldots, x_k) \mid \alpha.P \mid [x = y]P \mid P_1 + P_2 \mid P_1|P_2 \mid (\nu x)P$$

where A ranges over some set of variables with associated nonnegative arities k, and α represents an input $x(y)$, a free output $\overline{x}y$, a bound output $\overline{x}(y)$, or a silent event τ. Briefly, 0 represents an inactive process; each process variable $A(x_1, \ldots, x_k)$ has a corresponding definitional body P; matching $[x = y]P$ is read as "*if $x = y$ then P*"; sum $P_1 + P_2$ offers the choice of P_1 or P_2; composition $P_1|P_2$ places the two processes P_1 and P_2 side-by-side in parallel execution; restriction $(\nu x)P$ hides the name x from the environment of P; and action prefixing $\alpha.P$ performs the relevant input, output or silent transition, thus evolving into P. (Bound output $\overline{x}(y).P$ is in fact simply shorthand for the expression $(\nu y)\overline{x}y.P$.)

The definitions of *free* and *bound* names are standard ($x(y).P$ and $(\nu y)P$ bind y), and we shall write $fn(P)$ and $fn(\alpha)$ for the free names of P and α; $bn(P)$ and $bn(\alpha)$ for the bound names of P and α; and $n(P)$ and $n(\alpha)$ for the names of P and α. The definitions for substitution and alpha conversion are equally standard,

$$\text{pre}: \frac{}{\alpha.P \xrightarrow{\alpha} P} \qquad\qquad \text{match}: \frac{P \xrightarrow{\alpha} P'}{[x = x]P \xrightarrow{\alpha} P'}$$

$$\text{Ide}: \frac{P\{\tilde{y}/\tilde{x}\} \xrightarrow{\alpha} P'}{A(\tilde{y}) \xrightarrow{\alpha} P'} \left(A(\tilde{x}) \stackrel{\text{def}}{=} P\right) \qquad \text{sum}: \frac{P \xrightarrow{\alpha} P'}{P + Q \xrightarrow{\alpha} P'}$$

$$\text{par}: \frac{P \xrightarrow{\alpha} P'}{P|Q \xrightarrow{\alpha} P'|Q} \left(bn(\alpha) \cap fn(Q) = \emptyset\right) \quad \text{com}: \frac{P \xrightarrow{\overline{x}y} P', \; Q \xrightarrow{x(z)} Q'}{P|Q \xrightarrow{\tau} P'|Q'\{y/z\}}$$

$$\text{open}: \frac{P \xrightarrow{\overline{x}y} P'}{(\nu y)P \xrightarrow{\overline{x}(y)} P'} \left(x \neq y\right) \qquad \text{close}: \frac{P \xrightarrow{\overline{x}(y)} P', \; Q \xrightarrow{x(y)} Q'}{P|Q \xrightarrow{\tau} (\nu y)\left(P'|Q'\right)}$$

$$\text{res}: \frac{P \xrightarrow{\alpha} P'}{(\nu x)P \xrightarrow{\alpha} (\nu x)P'} \left(x \notin n(\alpha)\right)$$

Figure 1: Semantic derivation rules.

with renaming possible to avoid name capture. We shall always identify processes or transitions which differ only on bound names.

In Figure 1 we present the operational rules for the π-calculus. We have omitted symmetric rules for sum, par, com and close. From this definition of the single-step transition system we can define the weak transition system which abstracts from silent transitions in the usual fashion: $\Longrightarrow$ represents $(\xrightarrow{\tau})^*$ and $\stackrel{\hat{\alpha}}{\Longrightarrow}$ represents $\Longrightarrow$ when $\alpha = \tau$ and $\Longrightarrow \xrightarrow{\alpha} \Longrightarrow$ when $\alpha \neq \tau$.

2.1 Open Bisimulation

There are several ways in which one can define bisimilarity of π-calculus terms, varying on the attitude taken towards name instantiation. Notable among these are the *early* and *late* bisimulations of [13]. We choose to concentrate on the elegant notion of *open bisimilarity* of [16] — which is finer than the previous equivalences — for several reasons. Firstly, the strong version is a congruence. In particular, it is preserved by input prefix, unlike either of early or late equivalence. Its naturality is attested to by a simple axiomatisation. Most importantly, it has an efficient characterisation (described below) which we exploit in the implementation.

Definition 2.1 *A binary relation $\mathcal{R}$ between process terms is a* strong open bisimulation *if whenever $(P, Q) \in \mathcal{R}$ then for each substitution σ from names to names,*

- *if $P\sigma \xrightarrow{\alpha} P'$ then $Q\sigma \xrightarrow{\alpha} Q'$ for some Q' with $(P', Q') \in \mathcal{R}$;*

- *if $Q\sigma \xrightarrow{\alpha} Q'$ then $P\sigma \xrightarrow{\alpha} P'$ for some P' with $(P', Q') \in \mathcal{R}$.*

P and Q are strongly open bisimilar, written $P \sim Q$, if $(P, Q) \in \mathcal{R}$ for some strong open bisimulation $\mathcal{R}$.

If we replace $\xrightarrow{\alpha}$ by $\xRightarrow{\widehat{\alpha}}$ in the consequents of the two clauses in the above definition, then P and Q are (weak) open bisimilar, written $P \approx Q$.

This definition is actually only adequate for the calculus without restriction. The inclusion of restriction requires a treatment of *distinctions* [13], symmetric and irreflexive binary relations over names stipulating when names are not allowed to be equated by instantiation. We can however define open bisimilarity with respect to distinctions; these will be referred to by $\sim_D$ and $\approx_D$ for the strong and weak relations, respectively. For example, $[x = y]z.\mathbf{0} \sim_{\{(x,y)\}} \mathbf{0}$ though these two terms are not equal by the original definition of $\sim$. We do not go into these definitions in detail here, leaving the reader instead to consult [13, 16], but we note that the open bisimilarity relations correspond to the cases when the distinctions D are empty.

2.2 An Efficient Characterisation of Open Bisimulation

The definition of open bisimilarity (as with those for early and late equivalence) involves universal quantifications over substitutions, which make a direct implementation infeasible. However, there is an alternate characterisation for open bisimilarity based on the transition system presented in Figure 2 which is similar to the approach of [8]. The transitions here are of the form $P \overset{M,\alpha}{\rightsquigarrow} P'$, where M intuitively represents the least condition (set of name identities) under which action α can occur. Thus for example we have $[x = y]\alpha.P \overset{x=y,\alpha}{\rightsquigarrow} P$. These composite transitions (M, α) are ranged over by μ, and we write $n(M, \alpha)$ for $n(M) \cup n(\alpha)$ and similarly for $bn(M, \alpha)$. We shall always assume that the condition $x = x$ is ignored in the rules match, com and close, so that for example $[x = x]P \overset{\mu}{\rightsquigarrow} P'$ whenever $P \overset{\mu}{\rightsquigarrow} P'$.

We also define a weak transition system: $\overset{M,\alpha}{\rightsquigarrow}$ represents $\left(\overset{N_1,\tau}{\rightsquigarrow} \cdots \overset{N_n,\tau}{\rightsquigarrow}\right) \overset{L,\alpha}{\rightsquigarrow} \left(\overset{K_1,\tau}{\rightsquigarrow} \cdots \overset{K_m,\tau}{\rightsquigarrow}\right)$ for $n, m \geq 0$, where $M = L \wedge \bigwedge_i N_i \wedge \bigwedge_i K_i$ if $\alpha \neq \tau$, and $\left(\overset{N_1,\tau}{\rightsquigarrow} \cdots \overset{N_n,\tau}{\rightsquigarrow}\right)$ for $n \geq 0$, where $M = \bigwedge_i N_i$ if $\alpha = \tau$, where in either case no name bound in α occurs in the accumulated condition M.

In the following definition, we denote by σ_M the substitution on names induced by the equivalence classes associated with the equivalence relation corresponding to M; we select one representative of each class and map the other members of the class to it. We also use $\Rightarrow$ to denote logical implication, and $\equiv$ to denote equality modulo alpha conversion.

$$
\text{pre} : \frac{}{\alpha.P \overset{\text{true},\alpha}{\rightsquigarrow} P} \qquad\qquad \text{match} : \frac{P \overset{M,\alpha}{\rightsquigarrow} P'}{[x=y]P \overset{M \wedge x=y,\alpha}{\rightsquigarrow} P'} \left(x, y \notin bn(\alpha) \right)
$$

$$
\text{Ide} : \frac{P\{\tilde{y}/\tilde{x}\} \overset{\mu}{\rightsquigarrow} P'}{A(\tilde{y}) \overset{\mu}{\rightsquigarrow} P'} \left(A(\tilde{x}) \overset{\text{def}}{=} P \right) \qquad\qquad \text{sum} : \frac{P \overset{\mu}{\rightsquigarrow} P'}{P+Q \overset{\mu}{\rightsquigarrow} P'}
$$

$$
\text{par} : \frac{P \overset{\mu}{\rightsquigarrow} P'}{P|Q \overset{\mu}{\rightsquigarrow} P'|Q} \left(bn(\mu) \cap fn(Q) = \emptyset \right) \qquad \text{com} : \frac{P \overset{M,\overline{x}y}{\rightsquigarrow} P', \; Q \overset{N,w(z)}{\rightsquigarrow} Q'}{P|Q \overset{M \wedge N \wedge x=w,\tau}{\rightsquigarrow} P'|Q'\{y/z\}}
$$

$$
\text{open} : \frac{P \overset{M,\overline{x}y}{\rightsquigarrow} P'}{(\nu y)P \overset{M,\overline{x}(y)}{\rightsquigarrow} P'} \left(y \notin n(M) \cup \{x\} \right) \quad \text{close} : \frac{P \overset{M,\overline{x}(y)}{\rightsquigarrow} P', \; Q \overset{N,w(y)}{\rightsquigarrow} Q'}{P|Q \overset{M \wedge N \wedge x=w,\tau}{\rightsquigarrow} (\nu y)\left(P'|Q' \right)}
$$

$$
\text{res} : \frac{P \overset{\mu}{\rightsquigarrow} P'}{(\nu x)P \overset{\mu}{\rightsquigarrow} (\nu x)P'} \left(x \notin n(\mu) \right)
$$

Figure 2: Alternate transition system.

Definition 2.2 *A binary relation* $\mathcal{R}$ *between process terms is a strong* $\asymp$-*bisimulation if whenever* $(P,Q) \in \mathcal{R}$ *then*

- *if* $P \overset{M,\alpha}{\rightsquigarrow} P'$ *then* $Q \overset{N,\beta}{\rightsquigarrow} Q'$ *for some* N, β *and* Q' *with* $M \Rightarrow N$, $\alpha\sigma_M \equiv \beta\sigma_M$ *and* $(P'\sigma_M, Q'\sigma_M) \in \mathcal{R}$;

- *if* $Q \overset{M,\alpha}{\rightsquigarrow} Q'$ *then* $P \overset{N,\beta}{\rightsquigarrow} P'$ *for some* N, β *and* P' *with* $M \Rightarrow N$, $\alpha\sigma_M \equiv \beta\sigma_M$ *and* $(P'\sigma_M, Q'\sigma_M) \in \mathcal{R}$.

P and Q are strongly $\asymp$-bisimilar, written $P \asymp Q$, if $(P,Q) \in \mathcal{R}$ for some strong $\asymp$-bisimulation $\mathcal{R}$.

If we replace $\overset{N,\beta}{\rightsquigarrow}$ by $\overset{N,\beta}{\Rrightarrow}$ in the above definition, then P and Q are (weak) $\asymp$-bisimilar, written $P \succeq\!\!\!\asymp Q$.

Again this definition is valid only for the subcalculus without restriction, but again we can define $\asymp$-bisimilarity with respect to distinctions, and again we leave the details to [16].

The following theorems from [16] and [18] respectively, are what we are particularly interested in.

Theorem 2.3 (Sangiorgi) $\sim_D$ *coincides with* $\asymp_D$

Theorem 2.4 (Victor) $\approx_D$ *coincides with* $\succeq\!\!\!\asymp_D$.

2.3 Algorithmic Aspects

The usual technique for deciding bisimilarity of (finite-state) systems is to construct the state space of the two systems in question and then perform a partition refinement algorithm to try to distinguish the two start states. However, this technique is inapplicable in the case of the π-calculus due to the problems of name instantiation.

For example, the two terms

$$P(x) \stackrel{\text{def}}{=} x(y).\overline{y}y.P(y) \qquad Q(x) \stackrel{\text{def}}{=} x(y).\Big(\overline{y}y.Q(y) \ + \ [x=y]\overline{y}y.Q(y)\Big)$$

are clearly open bisimilar. In contrast with the late and early equivalences, with $\asymp$-equivalence it is enough to instantiate the bound name of an input with a single fresh name (a name x is *fresh* with respect to a transition $P \stackrel{\mu}{\leadsto} Q$ if $x \notin fn(Q) - bn(\mu)$). But then $P(x)$ has a minimal state space consisting of only two states, namely itself and the intermediate state $\overline{x}x.P(x)$ attained by instantiating the y by x, thus performing the input action $x(x)$. Regardless of how the state space of $Q(x)$ is generated, it cannot be equated to the above state space of $P(x)$, as its first transition cannot instantiate y to x, due to the appearance of x in the ensuing process. To match the two, the state space of $P(x)$ must be extended to match that of $Q(x)$.

Thus the implementation of the bisimulation algorithm is by necessity an *"on-the-fly"* algorithm [4]: the state spaces of the two systems in question are generated together during the construction of the candidate bisimulation relation which equates them.

Beyond this, the algorithm implemented in the MWB follows Definition 2.2 (both its strong and weak versions) closely: given two agents P and Q and a relation $\mathcal{R}$, check if (P, Q) is already in the relation. If so, return the relation unchanged. Otherwise, for each transition of P, find a (strong or weak) transition of Q such that the conditions match appropriately and the actions are equivalent under the substitution σ induced by the first, larger, condition. Make the transitions instantiate the same bound name (alpha-converting the derivatives), and assuming that P and Q are equivalent (by adding (P, Q) to the relation), apply the substitution σ to the derivatives and recurse over them using the extended relation. This either fails, which causes the current recursion to try the next transition of Q (or to fail if no such transition exists), or returns a relation relating the two derivatives, which is used in subsequent recursive calls. Finally, when all transitions of P have been matched by Q, match each transition of Q with a (strong or weak) transition of P in the same way, returning the resulting relation. (This is the approach in the absence of distinctions; in the more general case, distinctions are handled in a suitable fashion.)

3   The Mobility Workbench

The basic functionality of the MWB is to decide (strong and weak) open bisimilarity. Amongst other things, it can also be used to find deadlocks and for interactively simulating agents.

In Figure 3 we have a sample session which demonstrates some simple usage. Note that we render ν as $\hat{}$ and $\overline{x}y$ as `'x<y>` when typing in ASCII format. First we define an agent `Buf1` implementing a one-place buffer, then another, `Buf2`, implementing a two-place buffer by composing two instances of `Buf1`, and finally three agents, `Buf20`, `Buf21` and `Buf22`, together implementing a two-place buffer without parallel composition.

We proceed with this example by comparing the two implementations for weak equality. The MWB responds by saying that they are equivalent and that it found a bisimulation relation with 18 tuples, and asks us if we want to inspect it. We respond positively and the MWB prints out the relation as a list of pairs of agents with associated distinction sets.

We then simulate the behaviour of the agent `Buf2(i,o)`. The MWB presents the possible transitions, along with their least necessary conditions (if not trivial), and prompts the user to select one of them. After having a single choice on the first two steps, we then get a choice of three transitions; the first which is possible only if the names `i` and `o` are the same; the second which uses a new name `~v0` since all other known names are free and thus can't be reused; and the third, which simply outputs the value we read.

Next, we change the definition of `Buf22` to introduce a possible deadlock and again check for weak equivalence between `Buf2(i,o)` and `Buf20(i,o)`. This time we find that they are not equivalent, and proceed by looking for deadlocks in `Buf20(i,o)`; as the MWB finds deadlocked agents, it tells us the agent and the transition trace that leads to the deadlocked agent.

Finally we try equating `Buf2(i,o)` and `Buf20(i,o)` under the proviso that `i` is different from all other free names of the two agents (namely `o`). Under this distinction, the deadlocks don't appear, and the MWB reports that they are again equivalent.

4   An Extended Example: Mobile Telephones

As a case study, we have specified and verified the core of the handover protocol intended to be used in the GSM Public Land Mobile Network (PLMN) proposed by the European Telecommunication Standards Institute (ETSI). The formal specification of the protocol, and its service specification, are due to Orava and Parrow [14], who also verified the protocol algebraically. Fredlund and Orava [5] later verified the protocol automatically by specifying the protocol in LOTOS [17], which was translated to labelled transition systems using the Cæsar tool [6], which were in turn minimized using the Aldébaran tool [3], and finally compared

The Mobility Workbench
(Preliminary version 0.86, built Tue Oct 12 15:27:55 MET 1993)


```
MWB> agent Buf1(i,o) = i(x).'o<x>.Buf1(i,o)
MWB> agent Buf2(i,o) = (^m)(Buf1(i,m) | Buf1(m,o))
MWB> agent Buf20(i,o) = i(x).Buf21(i,o,x)
MWB> agent Buf21(i,o,x) = i(y).Buf22(i,o,x,y) + 'o<x>.Buf20(i,o)
MWB> agent Buf22(i,o,x,y) = 'o<x>.Buf21(i,o,y)

MWB> weq Buf2(i,o) Buf20(i,o)
The two agents are related.
Relation size = 18.  Do you want to see it?  (y or n) y
R = < Buf2(i,o), Buf20(i,o) >  {}
      < (^y)(Buf1(i,y) | Buf1(y,o)), Buf20(i,o) >  {}
      < (^y)('y<m>.Buf1(i,y) | 'o<x>.Buf1(y,o)), Buf22(i,o,x,m) >  {}
      . . .


MWB> step Buf2(i,o)
   0:  -- i(x) --> (^m)('m<x>.Buf1(i,m) | Buf1(m,o))
Step> 0
   0:  -- t --> (^m)(Buf1(i,m) | 'o<x>.Buf1(m,o))
Step> 0
   0:  -- [i=o],t --> (^m)('m<x>.Buf1(i,m) | Buf1(m,o))
   1:  -- i(~v0) --> (^m)('m<~v0>.Buf1(i,m) | 'o<x>.Buf1(m,o))
   2:  -- 'o<x> --> (^m)(Buf1(i,m) | Buf1(m,o))
Step> 1
   0:  -- 'o<x> --> (^m)('m<~v0>.Buf1(i,m) | Buf1(m,o))
Step> quit

MWB> agent Buf22(i,o,x,y) = 'o<x>.Buf21(i,o,y) + [i=o]t.0

MWB> weq Buf2(i,o) Buf20(i,o)
The two agents are NOT related.

MWB> deadlocks Buf20(i,o)
Deadlock found in 0, reachable by 3 transitions:
   -- i(x) -- i(y) -- [i=o],t -->
Deadlock found in 0, reachable by 5 transitions:
   -- i(x) -- i(y) -- 'o<x> -- i(~v0) -- [i=o],t -->
Deadlock found in 0, reachable by 7 transitions:
   -- i(x) -- i(y) -- 'o<x> -- i(~v0) -- 'o<y> -- i(y) -- [i=o],t -->

MWB> weqd (i) Buf2(i,o) Buf20(i,o)
The two agents are related.
Relation size = 8.  Do you want to see it?  (y or n) y
R = < Buf2(i,o), Buf20(i,o) >  {i#o}
      < (^y)(Buf1(i,y) | Buf1(y,o)), Buf20(i,o) >  {i#o}
      < (^y)('y<m>.Buf1(i,y) | 'o<x>.Buf1(y,o)), Buf22(i,o,x,m) >  {i#o}
      . . .
```


Figure 3: A simple sample session with the MWB.

using Aldébaran. The following informal presentation of the protocol is based on the presentation in [5].

The PLMN is a cellular system which can be seen as consisting of Mobile Stations (MSs), Base Stations (BSs), and Mobile Switching Centres (MSCs). The MS, mounted in e.g. a car, provides service to an end user. The BS manages the interface between the MS and a stationary network, controlling all radio communication within a geographical area (a cell). All communication with the MS in a cell is routed through the BS responsible for the cell. The MSC manages a set of BSs, and communicates with them and with other MSCs using a stationary network.

When a MS moves across a cell boundary, the handover procedure changes the communication partner of the MS from the BS of the old cell to the BS of the new cell, ensuring that the MS is constantly in contact with the MSC. The MSC initiates the handover by transmitting a *handover command* message to the MS via the old BS. The handover command message contains parameters enabling the MS to locate the new BS. When transmitting this message the MSC suspends transmission of all messages except for messages related to the handover procedure. When the MS receives the handover command message, it disconnects the old radio links and initiates the new radio links. To establish these connections the MS sends *handover access* messages to the new BS, in order to synchronize with the new BS. When the connections are successfully established, the BS sends a *handover complete* message to the MSC via the new BS. When this message has been received, the network resumes normal operations and releases the old radio links, which are now free and can be allocated to another MS.

In Figure 4 we present a π-calculus specification of the protocol. This is drawn from [14], but in this presentation we omit the failure handling aspects of the protocol. In Figure 5 we present the MWB code for this specification, which differs from Figure 4 in that the argument lists of agent identifier definitions must contain all free names which appear in the agent.

Correctness of this specification would come from showing that it matched some (ideally simple) service specification which would clearly define the desired behaviour of the system. In Figures 6 and 7, we present the service specification of the handover protocol and its rendering into MWB code, respectively.

When checking the protocol specification (System) against the more abstract service specification (Spec), we must express the fact that the parameters in, out, ho_acc, ho_com, data, ho_cmd and ch_rel are constants, i.e. they are distinct from all other free names. This is done by using the weqd (weak open bisimulation with distinctions) command of the MWB, in the following way:

```
weqd (i,o,acc,com,data,cmd,rel)
     Spec(i,o)   System(i,o,acc,com,data,cmd,rel)
```

$$CC(f_a, f_p, l) \stackrel{\text{def}}{=} \quad \text{in}(v).\overline{f}_a\text{data}.\overline{f}_a v.CC(f_a, f_p, l)$$
$$+ l(m_{new}).\overline{f}_a\text{ho\_cmd}.\overline{f}_a m_{new}.f_p(c).$$
$$[c = \text{ho\_com}]\overline{f}_a\text{ch\_rel}.f_a(m_{old}).\overline{l}m_{old}.CC(f_p, f_a, l)$$

$$HC(l, m) \stackrel{\text{def}}{=} \quad \overline{l}m.l(m).HC(l, m)$$

$$MSC(f_a, f_p, m) \stackrel{\text{def}}{=} \quad (\nu l)\Big(HC(l, m) \mid CC(f_a, f_p, l)\Big)$$

$$BS_a(f, m) \stackrel{\text{def}}{=} \quad f(c).\Big(\; [c = \text{data}]f(v).\overline{m}\text{data}.\overline{m}v.BS_a(f, m)$$
$$+ [c = \text{ho\_cmd}]f(v).\overline{m}\text{ho\_cmd}.\overline{m}v.$$
$$f(c).[c = \text{ch\_rel}]\overline{f}m.BS_p(f, m)\Big)$$

$$BS_p(f, m) \stackrel{\text{def}}{=} \quad m(c).[c = \text{ho\_acc}]\overline{f}\text{ho\_com}.BS_a(f, m)$$

$$MS(m) \stackrel{\text{def}}{=} \quad m(c).\Big(\; [c = \text{data}]m(v).\overline{\text{out}}v.MS(m)$$
$$+ [c = \text{ho\_cmd}]m(m_{new}).\overline{m}_{new}\text{ho\_acc}.MS(m_{new})\Big)$$

$$P(f_a, f_p) \stackrel{\text{def}}{=} \quad (\nu m)\Big(MSC(f_a, f_p, m) \mid BS_p(f_p, m)\Big)$$

$$Q(f_a) \stackrel{\text{def}}{=} \quad (\nu m)\Big(BS_a(f_a, m) \mid MS(m)\Big)$$

$$System \stackrel{\text{def}}{=} \quad (\nu f_a)(\nu f_p)\Big(P(f_a, f_p) \mid Q(f_a)\Big)$$

Figure 4: A formal specification of the handover procedure. After [14].

The bisimulation found by the MWB has 823 tuples. Running on a Sun SPARCstation 2 with 32Mb memory, it took approximately 61 CPU hours (user mode). The ML heap size was at most 17430 kb (well below the physical memory size of the machine). This appears to be rather extreme, and indeed the current prototype version of the MWB was not written with efficiency in focus. However, the new version of MWB being developed (see Section 5) finds a bisimulation with 249 tuples (leaving out alpha-equivalents) in just over 6 minutes CPU time on a SPARCstation 10, using 135 Mb heap space.

5 Future Development

On the theoretical side, the weak equivalence $\approx_D$ should be further investigated, e.g. regarding its axiomatization and its relationship to the late and early weak equivalences.

The prototype version of the MWB described here leaves room for many improvements. One deficiency is that it only handles the monadic π-calculus, where only one name can be sent or received atomically, while the polyadic π-calculus [11] generalises communication to allow zero or more names. The polyadic π-calculus

```
MWB> agent CC(fa,fp,l,in,data,ho_cmd,ho_com,ch_rel) =
        in(v).'fa<data>.'fa<v>.CC(fa,fp,l,in,data,ho_cmd,ho_com,ch_rel)
      + l(mnew).'fa<ho_cmd>.'fa<mnew>.fp(c).  [c=ho_com]'fa<ch_rel>.
                    fa(mold).'l<mold>.CC(fp,fa,l,in,data,ho_cmd,ho_com,ch_rel)

MWB> agent HC(l,m) = 'l<m>.l(m).HC(l,m)

MWB> agent MSC(fa,fp,m,in,data,ho_cmd,ho_com,ch_rel) =
        (^l)(HC(l,m) | CC(fa,fp,l,in,data,ho_cmd,ho_com,ch_rel))

MWB> agent BSa(f,m,ho_acc,ho_com,data,ho_cmd,ch_rel) =
        f(c).([c=data]f(v).'m<data>.'m<v>.BSa(f,m,ho_acc,ho_com,data,ho_cmd,ch_rel)
          + [c=ho_cmd]f(v).'m<ho_cmd>.'m<v>.f(c).
                    [c=ch_rel]'f<m>.BSp(f,m,ho_acc,ho_com,data,ho_cmd,ch_rel))

MWB> agent BSp(f,m,ho_acc,ho_com,data,ho_cmd,ch_rel) =
        m(c).[c=ho_acc]'f<ho_com>.BSa(f,m,ho_acc,ho_com,data,ho_cmd,ch_rel)

MWB> agent MS(m,data,ho_cmd,out,ho_acc) =
        m(c).( [c=data]m(v).'out<v>.MS(m,data,ho_cmd,out,ho_acc)
          +[c=ho_cmd]m(mnew).'mnew<ho_acc>.MS(mnew,data,ho_cmd,out,ho_acc))

MWB> agent P(fa,fp,in,ho_acc,ho_com,data,ho_cmd,ch_rel) =
        (^m)( MSC(fa,fp,m,in,data,ho_cmd,ho_com,ch_rel)
          | BSp(fp,m,ho_acc,ho_com,data,ho_cmd,ch_rel))

MWB> agent Q(fa,out,ho_acc,ho_com,data,ho_cmd,ch_rel) =
        (^m)( BSa(fa,m,ho_acc,ho_com,data,ho_cmd,ch_rel)
          | MS(m,data,ho_cmd,out,ho_acc))

MWB> agent System(in,out,ho_acc,ho_com,data,ho_cmd,ch_rel) =
        (^fa)(^fp)( P(fa,fp,in,ho_acc,ho_com,data,ho_cmd,ch_rel)
              | Q(fa,out,ho_acc,ho_com,data,ho_cmd,ch_rel))
```

Figure 5: The MWB code for the protocol specification.

also allows a notion of sorts, roughly analogous to the notion of types in functional programming.

Another shortcoming of the current version of the MWB is of course its inefficiency. We are currently developing a new version of the MWB with efficiency as one of its goals; e.g. using de Bruijn indices [1] to represent names (making alpha conversion unnecessary), and using hash tables to record the possible transitions of an agent instead of computing them each time they are needed. This version will handle the polyadic π-calculus, and will include the sort inference algorithm developed by Gay [7]. The model checking algorithm due to Dam [2] is also being implemented. As hinted at in the previous section, this new version is providing great gains in efficiency. In particular, with the new version we are able to verify the full handover protocol as presented in [14] in less than 11 CPU minutes.

On the longer term, we expect the tool to evolve with the needs of users: adding more "utility" commands, e.g., for minimizing agents, finding distinguishing

$$Spec \quad \overset{\text{def}}{=} \quad in(v).S_1(v) \ + \ \tau.Spec$$

$$S_1(v) \quad \overset{\text{def}}{=} \quad in(v).S_2(v_1, v) \ + \ \overline{out}v_1.Spec \ + \ \tau.\overline{out}v_1.Spec$$

$$S_2(v_1, v_2) \quad \overset{\text{def}}{=} \quad in(v).S_3(v_1, v_2, v) \ + \ \overline{out}v_1.S_1(v_2) \ + \ \tau.\overline{out}v_1.\overline{out}v_2.Spec$$

$$S_3(v_1, v_2, v_3) \quad \overset{\text{def}}{=} \quad \overline{out}v_1.S_2(v_2, v_3)$$

Figure 6: A formal specification of the service specification.

```
MWB> agent Spec(in,out) = in(v).S1(v,in,out) + t.Spec(in,out)
MWB> agent S1(v1,in,out) =
          in(v).S2(v1,v,in,out) + 'out<v1>.Spec(in,out) + t.'out<v1>.Spec(in,out)
MWB> agent S2(v1,v2,in,out) =
          in(v).S3(v1,v2,v,in,out) + 'out<v1>.S1(v2,in,out) +
t.'out<v1>.'out<v2>.Spec(in,out)
MWB> agent S3(v1,v2,v3,in,out) = 'out<v1>.S2(v2,v3,in,out)

MWB> weqd (i,o,acc,com,data,cmd,rel) Spec(i,o) System(i,o,acc,com,data,cmd,rel)
The two agents are related.
Relation size = 823.  Do you want to see it?  (y or n) n
```

Figure 7: The MWB code for the service specification.

formulae, etc. We would also like to see graphical interfaces based on Sangiorgi's tree representation [16], Parrow's Interaction Diagrams [15], and Milner's π-nets [12].

We would also like to face the intractability of the problem of deciding equivalence for more general π-calculus agents (without finite control). The inequality problem is semidecidable — and indeed due to the "on-the-fly" implementation of our algorithm we can provide such results — but the equality problem requires a tool running in some sort of semi-automatic mode, asking the human user for assistance at different points, e.g. for choosing strategy and tactics to solve a given problem.

Acknowledgement

We would like to thank Davide Sangiorgi for initiating the theoretical developments explored in this paper, and for providing us with numerous comments and corrections. Joachim Parrow and Lars-åke Fredlund also contributed useful comments on early drafts of this paper, and Lars-åke provided model code from which the first author learned a great deal of Standard ML.

References

[1] N.G. de Bruijn. Lambda Calculus Notation with Nameless Dummies, a Tool for Automatic Formula Manipulation, with Application to the Church-Rosser Theorem. *Indagationes Mathematicae*, 34:381–392. North-Holland, 1972.

[2] M. Dam. Model Checking Mobile Processes. In Proceedings of CONCUR'93. *Lecture Notes in Computer Science* 715. E. Best (ed). pp22–36. Springer-Verlag, 1993.

[3] J-C. Fernandez. Aldébaran: A tool for verification of communicating processes. Technical Report RTC 14, IMAG, Grenoble, 1989.

[4] J-C. Fernandez and L. Mounier. "On-the-fly" verification of behavioural equivalences and preorders. In Proceedings of CAV'91. 1991.

[5] L.-å. Fredlund and F. Orava. Modelling Dynamic Communication Structures in LOTOS. In Proceedings of FORTE'91, K.R. Parker and G.A. Rose (eds). pp185–200. North-Holland, 1992.

[6] H. Garavel and J. Sifakis. Compilation and verification of LOTOS specifications. In Proceedings of Protocol Specification, Testing, and Verification X, 1990.

[7] S.J. Gay. A Sort Inference Algorithm for the Polyadic π-Calculus. In Proceedings of 20th ACM Symp. on Principles of Programming Languages, ACM Press, 1993.

[8] M. Hennessy and H. Lin. Symbolic bisimulations. Research Report TR1/92. University of Sussex, 1992.

[9] E. Madelaine. Verification tools from the CONCUR project. *Bulletin of the European Association of Theoretical Computer Science* 47, pp110–126, June 1992.

[10] R. Milner. **Communication and Concurrency**. Prentice-Hall, 1989.

[11] R. Milner. The polyadic π-calculus: a tutorial. Research Report ECS-LFCS-91-180. University of Edinburgh, October 1991.

[12] R. Milner. Action Structures for the π-Calculus. Research Report ECS-LFCS-93-264. University of Edinburgh, May 1993.

[13] R. Milner, J. Parrow and D. Walker. A calculus of mobile processes (Parts I and II). *Journal of Information and Computation*, 100:1–77, September 1992.

[14] F. Orava and J. Parrow. An algebraic verification of a mobile network. *Formal Aspects of Computing*, 4:497–543, 1992.

[15] J. Parrow. Interaction Diagrams. Swedish Institute of Computer Science Research Report R93:06, 1993. (To appear in Proceedings of REX'93, Springer-Verlag.)

[16] D. Sangiorgi. A theory of bisimulation for the π-calculus. In Proceedings of CONCUR'93. *Lecture Notes in Computer Science* 715. E. Best (ed). pp127–142. Springer-Verlag, 1993.

[17] P.H.J. van Eijk, C.A. Vissers and M. Diaz (eds). **The Formal Description Technique LOTOS**. North-Holland, 1989.

[18] B. Victor. Forthcoming licentiate thesis, Uppsala University, 1994.

Compositional semantics of ESTEREL and verification by compositional reductions. *

R. de Simone and A. Ressouche

INRIA Sophia-Antipolis
B.P.93
06902 Sophia-Antipolis Cdx
FRANCE

Abstract. We present a compositional semantics of the ESTEREL synchronous reactive language, in the process algebraic style of *Structured Operational Semantics*. We then study its interplay with various reductional transformations on the underlying automata model, focusing on compositionality and congruence properties. These properties allow early nested reductions to take place at intermediate stages during the construction of a (reduced) model, a key point in cutting down the combinatorial explosion which plagues verification of parallel programs.

We consider the following transformations: bisimulation minimisation (state quotient), hiding of signals made invisible (abstraction), trimming of behaviours disallowed from external context (filtering). We illustrate part of the approach on a simple hardware bus arbiter specification. The verification method was implemented in STRL-MAUTO, a version of the AUTO tool customized to the "synchronous reactive" structure of actions.

1   Introduction

Verification of parallel systems is often plagued by the famous combinatorial blow-up arising in the representation of the global state space model. Many approaches have been proposed to cut down this complexity. One of the most fruitful is *compositional reduction*, by which subsystems minimisation is carried as early as possible along the process structure. This is made possible when the parallel systems are expressed in a process algebraic syntax and endowed with a compositional form of *Structured Operational Semantics*. Such an approach has been followed for the "classical" process algebras CCS, MEIJE, LOTOS amongst others, leading to tool support.

We want here to adapt these techniques -known as *verification by reductions*- to the specific case of *synchronous reactive systems*, and more specifically to the ESTEREL language (in its abstract, process-algebraic form). The main distinction in terms of *transition system* interpretation in between such formalisms and usual process algebras lays in the structure of action labels, which are now triples of: *necessarily present* signals; *necessarily absent* signals; emitted signals (reaction). An event needs to obey certain coherency properties, but must not necessarily be complete, in the sense that some signals may remain undetermined in it, indifferently present or absent. This allows economic representation of several possible input flows at once. We call *reactive*

---

* This work was partly supported by the ESPRIT BRA project Concur2

automata the finite transition systems bearing such type of labels. They correspond esentially to a rephrasing of *multiple input/output sequential machines* to an algebraic formulation of behaviour labels.

We provide the behavioural semantic interpretation of ESTEREL into reactive automata in a *SOS* style that matches our compositional purposes. This semantics differs in presentation from the original [1]. Here the input flow is synthesized upward from components to parallel products, rather than *inherited* by subprocesses from a global one. A keypoint in this compositional approach resides in the proper definition of *well-caused* processes, answering the well-known problem of signals being emitted as a result of their own presence/absence. To this end we add to each transition of reactive systems a *causal relation*, to remain acyclic, thereby completing the full definition of our reactive automata.

Next we define a number of reduction transformations on reactive automata, including *signal hiding* (abstraction), *symbolic bisimulation* (equivalence quotient), and *context filtering* (extended restriction), and study their compositional properties w.r.t. our operational semantics. We take *compositionality* in a broader sense, wondering *how much* reduction can actually be distributed along components prior to composition by a language operator (typically the `parallel` construct).

Symbolic bisimulation of reactive automata is more refined than usual plain bisimulation, and as a new notion seems very interesting in its own rights: consider the simple case of a state p, with $p \xrightarrow{i?.o!} p_1$ and $p \xrightarrow{i\#.o!} p_2$ while p_1 and p_2 are bisimulating each other (here $i\#$ means: i *is tested as absent*). This state p is somehow bisimilar to the one with a single transition labeled simply $o!$ leading to a state equivalent to p_1 and p_2, as any presence value for i has the same effect.

Finally we illustrate part of the approach on a simple hardware *bus arbiter* modeled in our language.

2 SOS behavioural semantics of ESTEREL

We now provide a compositional structured operational semantics for a large subset of (pure) ESTEREL. Extension to the full language is straightforward. We start by defining *events*, to become then the behavioural labels of *reactive automata*'s transitions. A structural interpretation of the process algebraic operators into reactive automata transformers is then given.

2.1 Reactive & broadcast structure of actions

In ESTEREL, as other synchronous reactive formalisms, behaviours are composed of events: an *event* E consists of a set of occurring input signals together with a set of output signals emitted back *in reaction*, synchronously with the inputs. Input signals not occurring in an event are supposed to be absent. We shall depart from this usual representation in that our events will explicitly specify *absent* inputs in addition to present ones, and leave unspecified signals when their presence value is immaterial for the transition to be taken. It should be clear that this approach allows to *factor* many possible input behaviours in a single transition description.

Signals will be supposed in the sequel to belong to some *finite* interface set $\mathcal{A}$. We use a predefined specific signal `_exit` to indicate (sequential) termination, which is thus dealt with internally much as other signal. Still, it can only be emitted, never received.

Definition 2.1 *An event E is a triple $(I, J, O) \in 2^A \times 2^A \times 2^A$ such that I and J are disjoint, and $(O \setminus \{\_exit\}) \subset I$. An event consists of: received signals in I (which were tested as present in the course of the transition); forbidden signal in J (tested as absent); emitted signals in O. We let $\mathcal{E}$ be the set of events.*

An event is called saturated *if $I \cup J = A$.*

An input event is simply a couple $(I, J) \in 2^A \times 2^A$ with $I \cap J = \emptyset$. We let $\mathcal{IE}$ be the set of input events.

Notation 2.1 *We typically write $E = I?.J\#.O!$ instead of (I, J, O) to reinforce type discrimination, and further, e.g. $i_1?.i_2?.j\#$ instead of $\{i_1, i_2\}?.\{j\}\#.\emptyset!$ (the product notation indicates simultaneity). We note $E \subset E'$ when $I \subset I', J \subset J', O \subset O'$, and $E \setminus \{S\}$ for $(I \setminus \{S\})?.(J \setminus \{S\})\#.(O \setminus \{S\})!$.*

We chose to include O in I. As an alternative one could use $I' = I \setminus O$ instead. This latter representation is more compact, but certain logical analogies we shall need in the sequel fit better the first choice.

ESTEREL (introduced below) is an imperative language containing an explicit parallel operator. This operator is *synchronous*, in the sense that both subprocesses proceeds at the same pace. Actions are broadcast, so that a given signal must be perceived consistently (as present or absent) by all parallel processes. At the event level, this introduces a specific definition of simultaneous product.

Definition 2.2 *Two events $E_1 = I_1?.J_1\#.O_1!$ and $E_2 = I_2?.J_2\#.O_2!$ are compatible, noted $E_1 \uparrow E_2$ iff:*

$$(J_1 \cap I_2) = \emptyset = (J_2 \cap I_1)$$

The simultaneity product of two compatible events E_1 and E_2, noted $E_1.E_2$ is the event $E = (I_1 \cup I_2)?.(J_1 \cup J_2)\#.(O_1 \setminus \{\_exit\} \cup O_2 \setminus \{\_exit\} \cup (\{\_exit\} \cap O_1 \cap O_2))$

The treatment of $\_exit$ in products insures *distributed termination*: a parallel process terminates only when both sides do. Note in particular that $(E \subset E') \Rightarrow (E \uparrow E')$ and $E_1, E_2 \subset E_1.E_2$.

With a logical *and* interpretation of product, the input part of an event (the I and J sets) builds exactly a characteristic propositional formula for saturated input events, as a conjunction of literals and negated literals. The simultaneity product respects this interpretation. On the other hand the analogy is much less clear on the output part of the event: a signal not in O is *not* irrelevant (it is certainly not emitted yet), but not incompatible in simultaneous product with this emission on the other part. We shall pursue this logical interpretation further, but *on signal receptions* only.

Definition 2.3 *A logical event expression is a couple (f, O) where f is a propositional formula (based on input signal names as atomic propositions), and O is a finite set of emitted signals. We note L_A the class of logical event expressions on A.*

A logical event expression represents a finite set of reactive events (as a characteristic function). They can be retrieved back by simple prenex disjunctive normal form expansion of f, under simultaneous product interpretation of each summand.

Definition 2.4 *By extension, an event $E = I?.J\#.O$ is called compatible with a saturated imput event $E_0 = I_0?.J_0\#$ if $I_0 \cap J = \emptyset = (J_0 \cap I) \setminus O$*

In short, compatibility here only allows E to suppose **more** present signals than provided from outside, and only when emitted by the process itself, which builds partly its own environment.

2.2 Reactive automata

Definition 2.5 *A reactive automaton is a structure* $(Q, A, T, init)$ *where* Q *is a (finite) set of states, init* $\in Q$ *is the initial state,* A *is a (finite) signal interface (or sort), and* $T \subset (Q \times Ev_A \times Q)$ *is a transition relation labeled by reactive events on* A.

A logical reactive automaton is the similar structure, only with the transition labels in L_A *instead of* Ev_A.

Reactive automata are tightly linked to *sequential machines* in hardware theory, or various models of reactive formalisms [3]. Sequential machines are automata equipped with two functions, *Out* and *Next*, mapping respectively a couple $(State, IE)$, $IE \in \mathcal{IE}$ to a (synchronous) output event and to a next state. Our reactive automata, while modeling in essence the same objects, insist on algebraic structure of events in a way that will prove useful to define the semantics. A reactive automaton defines executions, from saturated input events describing the environment's offer as follows: a compatible input event occurring in an outgoing transition is selected, which sets an output event and a resulting state, these of the transition. A specific form of determinism is called for here.

Definition 2.6 *A reactive automaton is* input-deterministic *iff two distinct transitions leaving the same state cannot be compatible. Formally:*

$$\forall p \in Q, \forall t = (p, E, p_1), t' = (p, E', p_2) \in T, \neg(E \uparrow E')$$

Definition 2.7 *A reactive automaton is* input-complete *iff it can react to any incoming input event, up to the fact that the process can itself partly build its own environment, by raising signals. Formally:*

$$\forall E_0 \ saturated \ \in \mathcal{E}, \forall p \in Q, \exists E, (p \xrightarrow{E} \ ), E \uparrow E_0$$

One should note that input-determinism does **not** imply the uniqueness of such E in the definition of input-completeness (different solutions may not be mutually compatible). Uniquenes will be gained from additional causality requirements later.

2.3 Syntax of ESTEREL as a Process Algebra

Esterel is a programming language designed for structured modeling of reactive systems (or sequential machines), as described in the previous section. Reactive systems are open systems, with successive event reactions taking place in (discrete) instants of time. The main novel features of Esterel are: explicit signal handling (raising/testing); explicit parallelism for modular decomposition; explicit atomicity defining successive (logical discrete) instants; internal cooperation by local signaling; priority handling by watchdog constraints (again on the signals themselves). Therefore Esterel turns the concept of *signals*, central to the model, into its main structuring paradigm for programming. Apart from these novel features, Esterel contains more classical procedural constructs, sequential composition and loops. General recursion is disallowed to guarantee the finite state model property. We left out the **trap/exit** and value-passing mechanisms of the full language.

The syntax of the Esterel algebra is the following:

$$P = \textbf{stop} \mid \textbf{nothing} \mid \textbf{emit S} \mid P\|P \mid P;P \mid \textbf{present S then } P \textbf{ else } P \textbf{ end} \mid (P)$$
$$\mid \textbf{do } P \textbf{ watching S} \mid \textbf{signal S in } P \textbf{ end} \mid \textbf{loop } P \textbf{ end} \mid P[\textbf{S/S}]$$

where S figures a syntactic class of mere signal names, to be instantiated in $\mathcal{A}$..

Formal semantics is provided below. We give now some informal concerns that led to its design. First of all, *sequential composition* is **not** an atomicity operator; a module can proceed in sequence inside the same instant. In Esterel jargon it is said that *semicolon is infinitely fast.* This assumption corresponds in hardware circuits to the fact that, at this logical-gate level of description, the output(s) of a gate is synchronous with its input(s), and can be "instantly" wired to other gates. Given this, each module is possibly endowed with quite complex behaviours in a single atomic step, raising and testing several signals in causal fashion. An informal reaction description may start like: *provided signal I occurs, then test signal J, and if J is here then emit O in answer, else if K* .... Of course signals may also be tested independently, if in parallel. As a limitation to this interpretation, it must discard ill-caused modules, in which a signal emission would result from its own presence value. This corresponds to *races* in circuits.

The **stop** construct implements atomicity. A module, or better said all its parallel branches, will maximally progress in sequence until next **stop** points, providing a common *end-of-reaction.* The **do/watching** construct implements priority preemption: the body is executed at successive instants *until* the guarding signal occurs, killing the body. For technical reasons this preemption is not active at the very same instant when control reaches this (sub)term.

2.4 Ill-caused processes

As already mentioned, internal signal chatters may cause ill-caused situations. We describe below three generic pathological examples, which must be eliminated as causally incorrect. This calls for the introduction of an additional *dependence* relation in between signals, which must be kept acyclic as a strict partial order.

signal S in (present S then emit S else emit a) This process has two external behaviours: $\emptyset$ and $a!$, depending on non-causal choices about S below the signal declaration. It is therefore not *input-deterministic*;

signal S in (present S then nothing else emit S) This process has no external behaviour, as S is emitted iff it is not here! It is therefore not *input-complete*;

signal S in signal T in (present S else emit T‖ present T else emit S) One cannot find here any notion of smaller or greater fixpoint solution, due to possible reaction to signal absence.

Events and dependencies could be represented in a single, partially commutative structure. We prefer our interpretation using separate *events/dependence relations* objects since the dependence flow relation is not based on signals names, but occurrences of them (that is, the dependence may vary from transitions to transitions). Also, sufficient criteria for uniform causal correctness are known at static semantic level, which solve in practice the causality problem (by approximation) at a different level. We shall not enter details here.

In the sequel we let R be a binary relation on signals, and R^+ its non-reflexive transitive closure. Only a signal emitted as a consequence of its own testing will introduce reflexive couples in R^+, breaking the partial strict order property.

2.5 SOS behavioural semantics of ESTEREL as a Process Algebra

We let S, T, ... be variables ranging over signal names in $\mathcal{A}$.. We let $P, Q, \ldots$ be variables ranging over programs (syntaxic states), and E, E_1, E_2 range over events. The sets I,

J and O always refer implicitly to an event E, possibly with matching subscripts. The deduction rules will "prove" behaviours of the shape $P \xrightarrow{E,R} Q$ in a natural deduction style.

We now proceed with the semantic rules for each language construct.

$$\text{stop} \xrightarrow{\emptyset?.\emptyset\#.\emptyset! \ , \ \emptyset} \text{nothing} \tag{1}$$

$$\text{nothing} \xrightarrow{\emptyset?.\emptyset\#.\{\_exit\}! \ , \ \emptyset} \text{nothing} \tag{2}$$

$$\text{emit } S \xrightarrow{\{S\}?.\emptyset\#.\{S, \_exit\}! \ , \ \emptyset} \text{nothing} \tag{3}$$

$$\frac{P \xrightarrow{E \ , \ R} P', \quad \_exit \notin O}{P;Q \xrightarrow{E \ , \ R} P';Q} \tag{4}$$

$$\frac{P \xrightarrow{E_1 \ , \ R_1} P', Q \xrightarrow{E_2 \ , \ R_2} Q', \_exit \in O_1, \ E_1 \uparrow E_2, \ O_2 \cap I_1 = \emptyset}{P;Q \xrightarrow{(E_1 \setminus \{\_exit\}).E_2 \ , \ (R_1 \cup R_2 \cup ((I_1 \cup J_1) \times O_2)^+)} Q'} \tag{5}$$

$$\frac{P \xrightarrow{E_1 \ , \ R_1} P', \quad Q \xrightarrow{E_2 \ , \ R_2} Q', \quad E_1 \uparrow E_2}{P\|Q \xrightarrow{E_1.E_2 \ , \ (R_1 \cup R_2)^+} P'\|Q'} \tag{6}$$

$$\frac{P \xrightarrow{E \ , \ R} P', \quad S \notin J\#}{\text{present } S \text{ then } P \text{ else } Q \text{ end} \xrightarrow{(I \cup \{S\})?.J\#.O! \ , \ (R \cup (\{S\} \times O))^+} P'} \tag{7}$$

$$\frac{Q \xrightarrow{E \ , \ R} Q', \quad S \notin I?}{\text{present } S \text{ then } P \text{ else } Q \text{ end} \xrightarrow{I?.(J \cup \{S\})\#.O! \ , \ (R \cup (\{S\} \times O))^+} Q'} \tag{8}$$

$$\frac{P \xrightarrow{E \ , \ R} P', \quad (S \in I? \Rightarrow S \in O!), \quad \neg R^+(S,S)}{\text{signal } S \text{ in } P \xrightarrow{E \setminus \{S\} \ , \ R \setminus (\{S\} \times \mathcal{A} \cup \mathcal{A} \times \{S\})} \text{signal } S \text{ in } P'} \tag{9}$$

$$\frac{P \xrightarrow{E \ , \ R} P', \quad \_exit \notin O}{\text{loop } P \text{ end} \xrightarrow{E \ , \ R} P'; \text{loop } P \text{ end}} \tag{10}$$

$$\frac{P \xrightarrow{E \ , \ R} P'}{\text{do } P \text{ watching } S \xrightarrow{E \ , \ R} \text{present } S \text{ then nothing else do } P' \text{ watching } S} \tag{11}$$

The form of rule 9 is sufficient because all operators maintain *coherency* of events, implying already that $(S \in J\#) \Rightarrow \neg(S \in O!)$. Rules 10 and 11 only perform unfolding, in a tail-recursion way that respects rationality, and thus finite state representation. The loop construct assumes its body not to terminate in its initial reaction, to avoid unguardedness. The watching construct does not become active at the instant the instruction gets control.

The previous semantics associates a (finite) *reactive automaton* $\tilde{P}$ with every process term P.

Definition 2.8 *An* ESTEREL *process is* well-caused *if for all deduction of a global transition in the reactive automaton, the causal relation R^+ remains acyclic at all nodes of the deduction tree.*

This definition in essence asks for R to remain a strict partial order for all subterms of a given precess, except for those which are never used in deductions (dead code).

Proposition 2.1 *Let P be a well-caused* ESTEREL *process and $\tilde{P}$ its associated reactive automaton. Then $\forall p$ state of $\tilde{P}$, $\forall E_0$ saturated input event, $\exists E$ unique, $p \xrightarrow{E}$ and $E \uparrow E_0$. Then of course $\tilde{P}$ is in particular input-complete and input-deterministic.*

Proof (sketch) By structural case analysis. The proof uses the following lemma

Lemma 2.1 *Let $(E, R), (E', R')$ label transitions outgoing from the same state in $\tilde{P}$. Suppose $\exists s \in \mathcal{A}$, $E = s\#.(E \setminus \{s\})$, $E' = s?.s!.(E' \setminus \{s\})$ and $(E \setminus \{s\}) \uparrow (E' \setminus \{s\})$. Then either s is a minimal point in both R and R', or R' has a cycle on s*

The lemma allows to settle the problem of input-determinism for the **signal** declaration operator. Other cases of interest are the **present** test operator, where it is proved that to refute input-completeness in the global automaton, one must run into a cycle in the dependence because of the phenomenon encountered in the previous counterexamples.

3 Compositionality of reductions

3.1 General framework

We now introduce 3 distinct types of reductions/transformations on reactive automata, and study the corresponding compositionality properties of the language constructs. We take here the word *compositionality* in a broad sense: our reductions will in general include parameters, and these will adapt from global processes to subterms. In fact the study of *compositionality* will largely amount to the characterisation of parameter transformations which allow "good" subautomata reductions prior to composition. These transformations may also rely on static semantics information gathered from the subterms themselves (typically, their external signal interface). A formal definition of optimality of reductions in general is out of scope. Instead we will argue in each case.

While compositionality will indicate where reductions *may* take place to simplify subsystems before combining them, there is a general trade-off between the expected gain and the time spent in the reduction phase. In general the parallel constructs are the most critical syntaxic locations where prior simplification is helpful, while other operators will only pass transformed parameters down the syntax tree. We will not study further such strategies.

We now introduce our three classes of reductions:

behaviour abstraction identifying similar (sets of sequences of) behaviours, for instance by *hiding* signals. In this paper we shall only deal with this latter special case. The parameters are thus *sets of signals* (to retain visible). Different choices of visible signals will provide different *partial views* of a system.

state minimisation according to behavioural equivalences, or bisimulations [6]. This reduction does not involve parameters.

behaviour filtering constraining (sequences of) behaviours according to some environment, itself described as a finite state structure. Here parameters are quite complex, consisting of automata with logical event acceptors as labels.

3.2 Hiding and Compositionality

Let V be a set of *visible signals*, and H_V the transformation function which erases signals not in V from transition events. Obviously H_V may decrease the number of transitions by merging these which only differ on invisible signals. It can also loose determinism, and keeps the number of states unchanged.

The basic compositionality problem with this transformation is that a signal declared hidden could in general be tested several times in the deduction tree of a given transition. Identical choices are imperative then, so the signal should be reintroduced as visible (but only in the necessary scope).

A signal is called free when occurring outside the scope of a local declaration **signal S in** .We note $\mathcal{I}_P$ (respectively $\mathcal{O}_P$) the sets of *free* signals occurring in a **present S** or **watching S** subterm of P(respectively in an **emit S**).

We let $Com(P,Q) \equiv_{def} ((\mathcal{I}_P \cup \mathcal{O}_P) \cap \mathcal{I}_Q) \cup ((\mathcal{I}_Q \cup \mathcal{O}_Q) \cap \mathcal{I}_P)$. This auxiliary function provides the list of signals whose presence value is shared by P and Q.

Proposition 3.1

$$H_V(\textbf{stop}) = \textbf{stop}$$
$$H_V(\textbf{nothing}) = \textbf{nothing}$$
$$H_V(\textbf{emit S}) = \textbf{nothing} \;\; if\; S \notin V$$
$$\textbf{emit S} \;\; otherwise$$
$$H_V(\textbf{loop } P) = \textbf{loop } H_V(P)$$
$$H_V(\textbf{signal S in } P) = H_V\,(\textbf{signal S in } H_V(P))$$
$$H_V(P\|Q) = H_V\left(H_{V \cup Com(P,Q)}(P)\|H_{V \cup Com(P,Q)}(Q)\right)$$
$$H_V(P;Q) = H_V\left(H_{V \cup Com(P,Q)}(P);H_{V \cup Com(P,Q)}(Q)\right)$$
$$H_V(\textbf{do } P \textbf{ watching S}) = \textbf{do } H_V(P) \textbf{ watching S } if\; S \notin (\mathcal{I}_P \cup \mathcal{O}_P)$$
$$H_V\left(\textbf{do } H_{V \cup \{S\}}(P) \textbf{ watching S}\right) \;\; otherwise$$
$$H_V(\textbf{present S then } P\textbf{else } Q) = \textbf{present S then } H_V(P)\textbf{else } H_V(Q)$$
$$if\; S \notin (\mathcal{I}_P \cup \mathcal{O}_P \cup \mathcal{I}_Q \cup \mathcal{O}_Q)$$
$$H_V\left(\textbf{present S then } H_{V \cup \{S\}}(P)\textbf{else } H_{V \cup \{S\}}(Q)\right)$$
$$otherwise.$$

According to these identities, one may choose to descend some hidings down the syntax tree. It may come as a surprise that **signal S** construct does not add S to **V**, but this shall in general be dealt with by nested parallel constructs. A related problem lies in the "best binary" parallel division of n subterms in parallel, which we shall ignore in this paper.

3.3 Symbolic Bisimulation and Compositionality

Plain strong bisimulation on automata is simply the coarsest equivalence in between states that respects event abilities. It is now a well-established notion, see e.g. [6]. Plain strong bisimulation is trivially a congruence w.r.t. all ESTEREL operators, as their semantics is given in a purely behavioural format. Also it permutes with the *hiding* reduction defined above, and is compatible with non-determinism.

In *reactive* automata, an event may represent *several* possible reactions due to unsaturated input signals. This makes room for a new *symbolic* (strong) bisimulation, when different combination of events would actually build up plain bisimulation on saturated reactions.

As an example, consider the very simple reactive automaton
$$P = (\{s_0, s_1\}, s_0, \{i, o\}, T = \{(s_0, i?.o!, s_1), (s_0, i\#.o!, s_1)\}).$$
Its output and resulting state are the same no matter which transition is taken, no matter which presence value the signal i may take. We want to identify P with the simpler
$$Q = (\{s_0, s_1\}, s_0, \{i, o\}, T = \{(s_0, o!, s_1)\}).$$
About the dependence relation here: any potential dependence (i, o) should be discarded. As Q is equally determisitic or complete as P, this is harmless, and can only safely allow more processes to be causally "correct".

This example can be generalized to all cases where transitions have labeling events with identical outputs and *symbolic bisimilar* resulting states (not just identical). Then the input events can be grouped in *sums of products*, and more generally into propositional formulas (justifying the introduction of *logical reactive automata*). The equality problem for propositional boolean functions have been of course extensively studied, either recently through *Binary Decision Diagrams*'s canonical forms, or in the past with *Prime Implicants* theory.

Notation 3.1 *Let B be a (symmetric) binary relation on states. We note IE_{B,O,p_0} the function $S \to \mathcal{IE}$ defined by*

$$IE_{B,O,p_0}(p) = \{I?.J\# \ / \ \exists p', B(p_0, p') \text{ and } p \xrightarrow{\ I?.J\#.O!\ } p'\}$$

Definition 3.1 *Let $A = (S, s_{init}, Ev_\mathcal{A}, T)$ be a reactive automaton. A (symmetric) binary relation $B \subset S \times S$ is a symbolic bisimulation iff:*

$$\forall (p, q) \in B, \forall O \subset \mathcal{A}_{out}, \forall r \in S, \sum_{I_k?.J_k\# \in IE_{B,O,r}(p)} I_k?.J_k\# = \sum_{I_k?.J_k\# \in IE_{B,O,r}(q)} I_k?.J_k\#$$

Proposition 3.2 *(Strong) Symbolic Bisimulation is a congruence with respect to all operators of ESTEREL on well-caused processes.*

Logical reactive automata have canonical form w.r.t. symbolic bisimulation up to equivalence of boolean formula, and the corresponding minimisation is compositional w.r.t. all operators of ESTEREL (on well-caused processes).

As already mentioned, BDDs provide a framework where equivalence of boolean formulae becomes identity. The situation is more complex for simple reactive automata, since there is no unique minimal "sum of products" form for a boolean formula.

3.4 Context Filtering and Compositionality

Often in practice one is only interested to the behaviour of a given reactive system according to a supposed *context*, representing all possible reaction offers from a supposed environment. Such features are partially provided in ESTEREL or LUSTRE [2] at programming level.

Also, when putting two reactive processes in parallel, it can be highly beneficial to first create each only in a context compatible with the other, to limit the size of subcomponents. This involves a creative step, where one guesses context properties of sibling processes. Nevertheless our approach will formally validate these assumptions in a second phase.

Distributing global contexts *Context-dependent* bisimulations [4] introduced a notion of relativity of behaviours in a process algebraic setting. Similarly, so-called *Don'tCare* sets provided relativity of behaviours, this time for hardware circuit descriptions. Inspiring from these two sources, we introduce *context filtering* as a reduction of behaviours.

Definition 3.2 *A reactive context C is an automaton $(S, s_{init}, \mathcal{A}, T)$ where the transition labels are (satisfiable) propositional formulae based on the basic predicates $S?, S \in \mathcal{A}_{in}$ and $S!, S \in \mathcal{A}_{out}$*

Thus, the only difference with logical reactive automata is the symmetric treatment of output and input signals. In particular a context may impose that a signal *cannot* be emitted in a given reaction. We do not require a reactive context in general to be input-complete or input-deterministic.

We now proceed to define the F_C filtering transformation.

Definition 3.3 *Let C be a context and P a logical reactive automaton. The filtering of P by C (noted $F_C(P)$) is the automaton defined by the SOS rule:*

$$\frac{P \xrightarrow{(f, O!)} P' \quad C \xrightarrow{g} C' \quad (f \wedge g/O) \text{ satisfiable}}{F_C(P) \xrightarrow{(f \wedge g/O, O!)} F_{C'}(P')}$$

where g/O is the propositional formula in which all basic predicates $S!$ have been replaced by: **true** *if $S \in O$,* **false** *otherwise.*

Notice that while filtering may restrict transitions and reachable states, it constructs an automaton based on the *cartesian product* of the process with the context. The definition could be carried to simple reactive automata by expanding the resulting labels into sums of products, then sums into several transitions.

Our compositionality concern now is to distribute C along the structure of the process. Due to the complexity of the parameters here (operational contexts), we shall only show how a context C can be distributed from a parallel process to a context $Left(C)$ for its left subcomponent (*Right* case is symmetric). The transformation uses static information on the right components (its *output* sort, noted $\mathcal{O}_{Right}$).

We call a basic predicate *in positive position* in a propositional formula if it appears inside an *even* number of negations.

Definition 3.4

$$Left(C) = (S, s_{init}, \mathcal{A}, T')$$

where T' is deduced from T by replacing in each labeling formula all basic predicates in $\mathcal{O}_{Right}$ in positive position by **true**.

In essence, our definition consists in weakening the demand for outputs, to those which *may not* be left to the other side to emit.

We briefly sketch the transformation for other operators. Distributing contexts alongside sequential products is hard somehow: the head process should live with the same context, while the tail processes should respect *any* postfix part of it (making *any* context state potentially initial). Positive outputs are dealt with in the same fashion as for parallel construct. The **present** test operator selects parts of the context according to compatibiliy of the initial transitions. The **watching** operator "cuts" the context after non-initial transitions assuming presence of the signal.

Local contexts As previously mentioned, providing "local" contexts for subcomponents in parallel can be very efficient in cutting down complexity (an illustration for the Bus arbiter example described below is provided in figure 4). But validity of these "creative" local assumptions must then be checked

A general way to check this validity, proposed by Larsen, is to keep track of - only- these transitions which cross from the "allowed" part to the "inaccessible" part of the filtered reactive automaton. These transitions are recorded *without* their target states (and further behaviours). The validity check consists in verifying that no such transitions is then used as part of a filtered global behaviour.

Formally, we complete the specification of F_C with another rule.

Definition 3.5 *Let C be a context and note Ref_C (for "refused by C") be the formula*

$\neg \bigvee g_i$, *over all g_i such that $C \xrightarrow{g_i}$. Let* **Error** *be a new distinguished "syntactic" state. Then*

$$\frac{P \xrightarrow{(f, O!)} P' \quad (f \wedge Ref_C/O) \text{ satisfiable}}{F_C(P) \xrightarrow{(f \wedge Ref_C/O, O!)} \textbf{Error}}$$

Checking for presence of constant **Error** in the global system will then indicate inappropriate local context assumptions.

In the *deterministic, causally correct* reactive case an even simpler solution exists: if a behaviour which could effectively be part of a global behaviour is improperly filtered at a subsystem level, then no other global behaviour can make up for the loss, so that the global reactive automaton becomes input-incomplete. This provides a simple test, to be performed only once, for the validity of local filter applications. Of course the determinacy requirements implies that this improved method cannot be combined with *hiding* for instance.

3.5 A Bus Arbiter example

"Gate-Level" description The purpose of this synchronous circuit is to select a single output **AckOut**$_i$ amongst possibly several **RequestIn**$_j$ inputs received from n users. It chooses basically the one of lowest index, but ensures fairness through use of a token ring: the owner of the ring gets priority over the regular previous selection system, which it must therefore disallow (or "override"). The circuit "Gate-Level" description is provided in figures 1 (for the basic component) and 2 for the ring, here of 4 components ($n = 4$). The Override(In/Out) wire design instantly propagates down the cells and, in the end, aborts the GrantIn signals through an auxiliary SignalInverter module. Another auxiliary Init module provides the seminal Token.

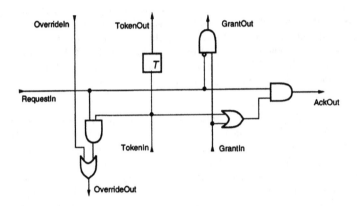

Fig. 1. The basic element

The programs in figure 3 provides an ESTEREL encoding of the description. We shall not carry the equivalence proof here, which would require recalling the formal definition of sequential machine interpretation of circuits. We added derived constructs to ESTEREL for sake of concision:

halt ≡ loop stop end,

do P watching immediate S ≡ present S then nothing else do P watching S,

await [immediate] S ≡ do halt watching [immediate] S,

every S do P ≡ await S; loop (do (P;halt) watching S).

Also, the keyword run introduces non-recursive module instantiation, possibly involving alphabetic renaming.

4   Conclusions

We presented a compositional semantics of ESTEREL and studied its combination with several transformation and reductions operations on labeled transition systems, allowing the same verificational approach on reactive systems as we knew worked well on "classical" CCS-like process algebras. We obtained this through a structure of behaviours taking signal absence into account. This work could be compared with compositional semantics of ARGOS [5], a formalism close to STATECHARTS. Our approach allows non-determinism in some constructions.

The structural syntax of ESTEREL allows to state (and solve) the problem of compositional reduction, and thus to deal with state explosion in a way orthogonal to symbolic model-checking techniques. These symbolic structures (*BDDs*) could still serve to

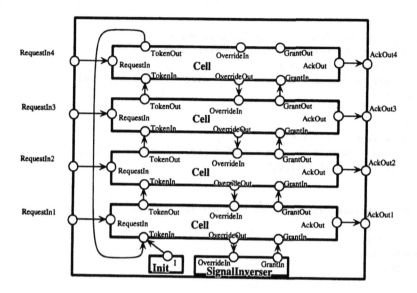

Fig. 2. The Token ring network

implement event labels in processes and contexts, and support symbolic bisimulation checking.

Finally we should stress that during our early experiments the local filterings proved perhaps the most efficient technique for cutting down complexity, while not automatically sound. There we started viewing our approach as a useful *user-guided proof-assistant*, with human creativity providing extra assumptions (on local signals) to speed up construction and cut down space requirements, while the system still performs quite fancy semantic automatic reductions (such as *bisimulation minimisation* or *signals hiding abstraction*).

References

1. G. Berry and G. Gonthier. The Esterel synchronous programming language: design, semantics, implementation. *Science of Computer Programming*, 1992.
2. N. Halbwachs, F. Lagnier, and C. Ratel. An experience in proving regular networks of processes by modular model checking. *Acta Informatica*, 1992.
3. D. Harel. Statecharts: A visual approach to complex systems. *SCP*, 1987.
4. K. Larsen. *Context-dependent Bisimulation between Processes*. PhD thesis, Edinburgh University, 1986.
5. F. Maraninchi. Operational and Compositional Semantics of Synchronous Automaton Generation. In *CONCUR'92*, volume LNCS 630, 1992.
6. R. Milner. *Communication and Concurrency*. Prentice Hall, 1989.

```
module Cell:
input RequestIn, GrantIn, TokenIn, OverrideIn;
output GrantOut, TokenOut, AckOut, OverrideOut;
 every immediate GrantIn do
   present RequestIn then emit AckOut else emit GrantOut end
 end
||
 loop
   await immediate TokenIn;
   present RequestIn then emit AckOut; emit OverrideOut end;
   await tick; emit TokenOut
 end
||
 every immediate OverrideIn do emit OverrideOut end
.
module Init:
output Token;
emit Token
.
module SignalInverser:
input Override;
output Grant;
loop
  present Override else emit Grant end;
  await tick
end.
module Arbiter4:
input RequestIn1, RequestIn2, RequestIn3, RequestIn4;
output AckOut1, AckOut2, AckOut3, AckOut4;
signal G1, G2, G3, G4, O1, O2, O3, O4, T1, T2, T3, T4, Gnull, Onull in
      run Init[signal T1/Token]
  ||  run SignalInverser[signal O1/Override, G1/Grant]
  ||  run Cell[signal RequestIn1/RequestIn, AckOut1/AckOut,
                      G1/GrantIn, G2/GrantOut, T1/TokenIn, T2/TokenOut,
                      O1/OverrideOut, O2/OverrideIn]
  ||  run Cell[signal RequestIn2/RequestIn, AckOut2/AckOut,
                      G2/GrantIn, G3/GrantOut, T2/TokenIn, T3/TokenOut,
                      O2/OverrideOut, O3/OverrideIn]
  ||  run Cell[signal RequestIn3/RequestIn, AckOut3/AckOut,
                      G3/GrantIn, G4/GrantOut, T3/TokenIn, T4/TokenOut,
                      O3/OverrideOut, O4/OverrideIn]
  ||  run Cell[signal RequestIn4/RequestIn, AckOut4/AckOut,
                      G4/GrantIn, Gnull/GrantOut, T4/TokenIn, T1/TokenOut,
                      O4/OverrideOut, Onull/OverrideIn]
end.
```

Fig. 3. Elementary modules and 4-cells network.

Model Checking Using
Adaptive State and Data Abstraction

(extended abstract)

Dennis Dams[1], Rob Gerth[1†],
Gert Döhmen[2‡], Ronald Herrmann[2], Peter Kelb[2],
Hergen Pargmann[3§]

[1] Eindhoven University of Technology, Dept. of Math. and Computing Science, P.O. Box 513, 5600 MB Eindhoven, The Netherlands. Email: {wsindd,robg}@win.tue.nl
[2] OFFIS, Westerstraße 10–12, 26111 Oldenburg, Germany. Email: Peter.Kelb@arbi.informatik.uni-oldenburg.de
[3] University of Oldenburg, 26121 Oldenburg, Germany

Abstract. We present a partitioning algorithm for checking ACTL specifications that distinguishes between states only if this is necessary to ascertain the specification. This algorithm is then generalized to also abstract from the variable values in the states. Here, too, the values between which the algorithm distinguishes are determined by what is needed to decide whether or not the specification holds. The resulting algorithm is being implemented in an ROBDD based model checker for VHDL/S.

Keywords: model checking, ACTL, abstract interpretation, state partitioning, binary decision diagrams (BDDs)

1 Introduction

The major stumbling block for successful application of model checking to complex systems is the size of the state graphs of these systems. This is the so-called *state explosion* problem. Although an impressive step forward has been made by the introduction of ROBDD based techniques [BCM+90], this step has not moved us beyond the block; rather, it has pushed it further ahead. In order to deal with the state explosion problem in a fundamental way, we need versatile abstraction methods, that allow the abstraction from any details that are not relevant to the property being checked.

Examples of such methods are the reduction of models by collapsing states which are bisimulation equivalent [BFH+92] and the partial order approaches that allow parts of the state graph caused by different interleavings of independent, parallel actions to be ignored [Val91, GW91, Pel93]. These methods are safe for large classes of specifications. The first one applies to any property that cannot distinguish between bisimilar states — which is in fact the case for most common specification languages. The latter methods basically apply to linear temporal logic specifications; but also see [GKPP94].

However, if the specific set of properties to be verfied is known beforehand, many more details will become irrelevant and much better reductions can in principle be

---

† Currently working in ESPRIT project P6021 "Building Correct Reactive Systems (REACT)".
‡ Currently working in ESPRIT project P6128 "Formal Methods in Hardware Design (FORMAT)".
§ Currently working in Projekt "Informationssysteme".

effected. The problem then is to find out which are the relevant details needed to check some given set of properties. In the light of the quest for automated methods, we need efficient algorithms to perform this task.

In [DGG93], these problems are dealt with by an approach which is based on iterative refinement of the model under construction by constructing ever finer partitions of the concrete state space; as such, it generalizes the state partitioning method of [BFH$^+$92]. Starting with an ACTL specification[4] φ to be verified, the full model C and an abstract model that contains no details, the model is successively refined until it contains enough information to either prove or disprove the formula. Each of the models that is generated in such a sequence of iterative refinements has the property that it *preserves* φ; i.e., whenever φ holds in the abstract model A, it also holds in the full model C, called the *concrete model* henceforth. When φ does *not* hold however, there are two possible reasons. One is that the abstract model does not yet contain enough detail, although φ does hold in the concrete; the other is that φ does not hold in the concrete model. Further refinement of the model will then bear out which one of these cases is true: either there will be a point in the refinement process where φ becomes true in the abstract model, or at some point no further refinement is possible—the model has become *stable*—while φ is still false. It is shown that in the latter case, φ is false in the concrete model as well; i.e., stable models *strongly preserve* satisfaction of specifications.

An obvious factor determining the success of an abstraction methodology is the possibility it provides to construct abstractions in a *direct* fashion, i.e., without intermediate construction of the complete detailed model. Otherwise, the target of avoiding the state explosion would clearly be missed. Yet, all the above mentioned abstraction methods require access to the concrete transition relation, which still may be prohibitively expensive for the state partitioning based methods.

This paper starts by formulating a state partitioning algorithm for ACTL within an abstract interpretation framework. The basic algorithm is then generalized so that the concrete transition relation can be abstracted as well; but in such a way that the algorithm will *automatically* adapt the abstraction until it can decide truth or falsehood of the specification. Both the state splittings as well as the abstractions of the transition relation are governed by what is needed to establish the validity of the specification.

We obtain a two level approach. On the first level, an abstraction R_d of the concrete transition relation R_c is choosen, yielding abstract transition systems $\mathcal{D}$, in such a way that satisfaction of the specification φ is preserved. Then, on the second level, the partitioning algorithm constructs models A that, in their turn, are abstractions of $\mathcal{D}$. This algorithm terminates either if the specification φ is satisfied in the model A just constructed, or if A becomes stable. In the latter case φ may still not hold, from which we conclude that φ is invalid in $\mathcal{D}$ as well. However, $\mathcal{D}$ itself is an abstraction of the concrete model C and we cannot immediately conclude that φ is false in C. Thus we face the problem of separating 'true' counter examples to the satisfaction of φ from artifacts caused by the first level abstraction. In the paper we show that it is possible to determine whether a counter example is genuine or not. Assuming we do not have a true counter example, we must change the first-level abstraction $\mathcal{D}$ so as to include more detail; i.e., the choosen abstraction $\mathcal{D}$ must be refined. Then the second-level partitioning algorithm

---

[4] ACTL is the universal fragment of CTL.

again will iteratively construct models that now are abstractions of the new $\mathcal{D}$. As we obviously want to retain what we have computed—$\mathcal{D}$ and the stable $\mathcal{A}$—two other problems are raised: how can the previously stable model $\mathcal{A}$ be used as the starting point for the partitioning algorithm and how can a first-level abstraction $\mathcal{D}$ be adapted rather than recomputed. We provide answers by defining a family of abstractions—depth-k abstractions—and by concretely showing how models can be adapted and re-used.

The algorithm is being implemented in a tool for the verification of VHDL/S code. VHDL/S [HSD+93] is a language developed in the ESPRIT project 6128 FORMAT. The goal of FORMAT is to provide an environment for the efficient development of correct VHDL designs, where correctness is pursued along two different lines: by synthesis and verification. VHDL/S integrates four different and self-contained linguistic paradigms: VHDL, state based specifications [Har87], symbolic timing diagrams [SD93], and temporal logic. The former two are operational, the latter two are declarative in nature. State based specifications are translated into ROBDDs [HK94], while specifications written in VHDL are first translated into Petri Nets, which have actions (e.g., assignments) associated with their transitions. In a second step, these nets are translated into ROBDDs so that symbolic model checking can be applied. The iterative refinement algorithm is fit into this second translation phase, as this is the point at which the state explosion occurs.

The implementation of the state partitioning algorithm is also ROBDD-based, with the obvious advantage that many previously implemented modules of the system can be reused. A key point here is that the first-level abstraction allows us to approximate the concrete model. We exploit this by limiting the size of ROBDDs, so that concrete states are approximated by sets of states. Furthermore, by interpreting the actions that are associated with transitions in nets over such abstract states, an approximation of the transition relation is obtained.

The next section gives some background material. In Sect. 3 we develop the ACTL partitioning algorithm and extend it to use data abstraction in Sect. 4. A sketch of an ROBDD implementation embedded in a VHDL/S model checker would have occupied the penultimate section, were it not for the page limitation imposed by Springer. Finally, in Sect. 5 we draw some conclusions and point to future work.

2 Preliminaries

ACTL and ECTL We assume some countable set of local propositional symbols Prop = $\{p, q, \ldots\}$. We define ACTL (universal Computation Tree Logic) in its positive normal form in which negations only apply to propositions. The set of *well-formed formulae* (written wff) is defined as follows

— for $p \in$ Prop, p and $\neg p$ are wff,
— if φ and ψ are wff, then so are $\varphi \vee \psi$ and $\varphi \wedge \psi$,
— if φ and ψ are wff, then so are $AX\varphi$, $AU(\varphi, \psi)$ and $AV(\varphi, \psi)$.

ECTL (existential Computation Tree Logic) is defined as $\{\neg\varphi \mid \varphi \in ACTL\}$.

The AV-modality is needed because the use of negation is constrained. Otherwise, we would have had $AV(\varphi, \psi) \equiv A\neg U(\neg\varphi, \neg\psi)$ as can be gleaned from the satisfaction definition below. Write AW to denote either AU or AV. Let Atoms(φ) be the set of subformulae of φ that are either propositions or of the form $AX\psi$ or $AW(\psi, \psi')$.

Transition systems and satisfaction ACTL is intended to express properties about computations which are generated by transition systems $T = (V, I, R)$ where V is some set of states, $I \subseteq V$ is the set of initial states and R is the transition relation which is always assumed to be *total* to circumvent some technicalities. A *path* σ in T is an infinite sequence $\sigma = s_0 s_1 \cdots$ of states such that $s_i \, R \, s_{i+1}$ for every i. Write σ_n for state s_n in σ. An *s-path* is a path that starts at state s. Define the *precondition function*, pre_R, associated with R by $pre_R(D) = \{c \mid \exists d \in D \ c \, R \, d\}$. Write *pre* if the transition relation is clear.

As usual, we need a *valuation function* $\mathbf{V} \colon V \to 2^{\text{Prop}}$ to define which propositions are true in which states. A (Kripke) model M is a pair $(T, \mathbf{V})$ of a transition system and a valuation function. $M, s \models \varphi$ denotes that the formula φ is *true* at the state s in the transition system T with valuation $\mathbf{V}$. Its inductive definition follows:

- $M, s \models p$ iff $p \in \mathbf{V}(s)$, for $p \in \text{Prop}$,
- $M, s \models \neg p$ iff $p \notin \mathbf{V}(s)$, for $p \in \text{Prop}$,
- $M, s \models \varphi \vee \psi$ iff $M, s \models \varphi$ or $M, s \models \psi$,
- $M, s \models \varphi \wedge \psi$ iff $M, s \models \varphi$ and $M, s \models \psi$,
- $M, s \models \text{AX}\varphi$ iff $M, \sigma_1 \models \varphi$ for every s-path σ,
- $M, s \models \text{AU}(\varphi, \psi)$ iff for every s-path σ there is a $k \geq 0$ such that $M, \sigma_k \models \psi$ and $M, \sigma_i \models \varphi$ for every $i < k$,
- $M, s \models \text{AV}(\varphi, \psi)$ iff there is no s-path σ such that for some $k \geq 0$ $M, \sigma_k \models \neg\psi$ and $M, \sigma_i \models \neg\varphi$ for every $i < k$.

For $S \subseteq V$, define $M, S \models \varphi$ by $\forall s \in S \ M, s \models \varphi$. $M \models \varphi$ denotes $M, I \models \varphi$. When clear from the context, we omit M.

For $\text{AU}(\varphi, \psi)$ and $\text{AV}(\varphi, \psi)$-formulae, we define *approximants* as follows:

$$\text{AU}_0(\varphi, \psi) = \text{false} \quad \text{AU}_{i+1}(\varphi, \psi) = \psi \vee (\varphi \wedge \text{AXAU}_i(\varphi, \psi))$$
$$\text{AV}_0(\varphi, \psi) = \text{true} \quad \text{AV}_{i+1}(\varphi, \psi) = \psi \wedge (\varphi \vee \text{AXAV}_i(\varphi, \psi))$$

If the transition system T has N states then for every state s

$$M, s \models \left(\text{AU}(\varphi, \psi) \equiv \bigvee_{i < N} \text{AU}_i(\varphi, \psi)\right) \wedge \left(\text{AV}(\varphi, \psi) \equiv \bigwedge_{i < N} \text{AV}_i(\varphi, \psi)\right). \tag{1}$$

So, we also have $M, s \models \text{AW}_N(\varphi, \psi) \equiv \text{AW}_{N+i}(\varphi, \psi)$ for any $i > 0$. In other words, on finite transition systems the truth of AW-formulae is determined by a finite set of approximants.

Abstract Interpretation Many of the results and constructions below are most easily expressed using the language of *Abstract Interpretation* [CC77]; a general framework to define static analyses of programs. The basic tenet is that the operations of a programming language which operate on concrete values can be mimicked by corresponding abstract operations defined over abstract values that describe sets of concrete values.

The starting point is choosing a set of *abstract states*, V_a. Each abstract state $a \in V_a$ describes a set of concrete states. Conversely, every set $C \subseteq V_c$ of concrete states has a 'best', or most precise description. This is formalized via a *concretization function* $\gamma \colon V_a \to 2^{V_c}$ and an *abstraction function* $\alpha \colon 2^{V_c} \to V_a$. For each a, $\gamma(a)$ is the set of all concrete states described by a; for each $C \subseteq V_c$, $\alpha(C)$ is the most precise description in the sense that $C \subseteq \gamma(\alpha(C))$ and $C \subseteq \gamma(a)$ implies $\gamma(\alpha(C)) \subseteq \gamma(a)$ for any $a \in V_a$. Thus, $\alpha(C)$ is the least description of C w.r.t. the *approximation ordering* $\preceq$ on V_c defined by $a \preceq b$ iff $\gamma(a) \subseteq \gamma(b)$. A given γ uniquely determines an appropriate α (if it exists) by setting $\alpha(C)$ to be the least a such that $\gamma(a) \supseteq C$. The α thus defined is written $\gamma^\flat$. We mention that, similarly, α determines a unique appropriate γ as well.

These requirements are often captured by saying that (α, γ) is a *Galois insertion* from $(2^{V_c}, \subseteq)$ to $(V_a, \preceq)$: (i) α and γ are total and monotonic, (ii) for every $C \in 2^{V_c}$ we have $(\gamma \circ \alpha)(C) \supseteq C$, and (iii) for every $a \in V_a$ we have $(\alpha \circ \gamma)(a) = a$.

Given such an abstract interpretation of the data, functions $f : V_c \to V_c$ can be described by *safe* abstract interpretations $f_a : V_a \to V_a$ that satisfy $f_a(a) \succeq \alpha(f(\gamma(a)))$.[5] In particular, there is a *precise* abstract interpretation of f defined by $\bar{f}_a = \alpha \circ f \circ \gamma$ and f_a is safe just in case $f_a \succeq \bar{f}_a$ (pointwise). Safeness means that given a description of the parameter, f_a yields a description of the result value.

A static analysis can then be viewed as an abstract execution of the program in which data and operations are abstractly interpreted, yielding a description of any concrete execution.

Binary decision diagrams Reduced Ordered BBDs [Bry86, Bry92] are a way to economically represent boolean functions in a canonical way. Although for most boolean functions the size of their ROBDD representation is exponentially large, in most practical cases the ROBDDs are sufficiently small. This, together with the fact that boolean operations, equivalence and tautology checking can be done very efficiently on ROBDDs, is the reason why ROBDDs are so popular. ROBDDs only supply a canonical representation relative to an arbitrary but fixed ordering on the boolean (input) variables and this ordering greatly influences the size of the ROBDDs.

The use of ROBDDs in model checking is based on coding transition relations as boolean functions. Given a transition system (V, I, R), take vectors $\mathbf{x}, \mathbf{x}'$ of boolean variables long enough to code for all states in V (e.g., take $|\mathbf{x}|, |\mathbf{x}'| \geq {}^2\log(|V|)$). Then, define a boolean function $\ulcorner R \urcorner(\mathbf{x}, \mathbf{x}')$[6] by

$$\ulcorner R \urcorner(\mathbf{x}, \mathbf{x}') = 1 \iff \exists x, x' \in V \ \ x \, R \, x' \ \& \ \beta(x) = \mathbf{x} \ \& \ \beta(x') = \mathbf{x}' \, ,$$

where β is the coding function that maps states to bit strings.

Symbolic model checking is based on such ROBDD representations [BCM+90]. The basic operation that needs to be done is computing preconditions, $pre(C)$, which translates into computing *relational products* $\exists \mathbf{x}' (\ulcorner R \urcorner(\mathbf{x}, \mathbf{x}') \wedge \ulcorner C \urcorner(\mathbf{x}'))$. Obviously, to represent a set as an ROBDD we use its characteristic predicate.

3 ACTL Partitioning

The aim is to develop an algorithm that allows verifying an ACTL-specification without generating the complete state-graph of the system to be verified. Specifically, we want to verify the specification using an *abstraction* of the state-graph. The type of abstraction

---

[5] $f(C) = \{f(c) \mid c \in C\}$.

[6] We usually do not make a distinction between the boolean function $\ulcorner R \urcorner$ and its ROBDD representation.

that we have in mind is characterized by the following statement based on results from [DGG94].

An *abstraction* of $C = (V_c, I_c, R_c)$ is a transition system $A = (V_a, I_a, R_a)$ for which there is a *concretization function* $\gamma: V_a \to 2^{V_c}$ such that R_a satisfies

$$\forall a, b \in V_a \ (\ \exists c \in \gamma(a) \ \exists d \in \gamma(b) \ \ c \ R_c \ d \ \Rightarrow \ a \ R_a \ b \ )$$

and I_a satisfies $\forall c \in I_c \exists a \in I_a \ c \in \gamma(a)$. Satisfaction of a proposition p in an abstract state a is defined by

$$a \models p \ \text{ iff } \ \gamma(a) \models p \ . \tag{2}$$

Satisfaction of other formulae is then defined as usual.

For such abstract systems, the logic ACTL is *weakly preserved*:

$$\forall \varphi \in ACTL, \ a \in V_a \ \ A, a \models \varphi \ \Rightarrow \ C, \gamma(a) \models \varphi \tag{3}$$

Write γ_C to explicitly indicate that the transition system C is being abstracted.

If γ is part of a Galois Insertion (α, γ) (as will always be the case in this paper), then we may rephrase the above statement by saying that in our context the proper notion of precise abstraction of a (transition) relation R_c w.r.t. (α, γ) is defined by

$$R_a = \{(a, b) \mid \exists c, d \in V_c \ \ c \in \gamma(a) \ \& \ c \ R_c \ d \ \& \ \alpha(d) = b\} \ .$$

Again, see [DGG94] for more details.

As every transition system trivially is an abstraction of itself (up to the difference between states c and singletons $\{c\}$, which is irrelevant in this context), there is no formal distinction between 'concrete' and 'abstract' transition systems. The existence of the concretization function implies that we can view the states in an abstraction as predicates over the concrete states; which we shall often do (and, hence, take $V_a = 2^{V_c}$). By this interpretation, the concretization function is fixed as the standard interpretation of predicates over V_c.

3.1 The Basic Partitioning Algorithm

Since abstractions only weakly preserve ACTL (i.e., in (3) the implication in the other direction does not hold in general), there is a potential problem if a specification φ happens not to be satisfied in the concrete system, because in general we cannot draw that conclusion given some abstraction. The reason can be gleaned from (2)[7]: we may well have that $A, a \not\models \varphi$ while there is a concrete state $c \in \gamma(a)$ for which $C, c \models \varphi$. Such a formula is said to be *not determined* in a.

Is it possible to decide whether $C \not\models \varphi$ by analysing an abstraction A? The inductive nature of the satisfaction definition suggests that this should be possible in case all subformulae of φ are determined in (all states in) A. A closer look reveals the following [DGG93]:

---

[7] We stress that this definition of satisfaction is forced by the aim to have weak preservation.

Define the *companion* of φ, Comp(φ) as

$$\text{Atoms}(\varphi) \setminus \{\text{AW}(\psi, \psi') \mid \psi, \psi' \in \text{ACTL}\}$$
$$\cup \{\text{AW}_i(\psi, \psi') \mid i > 0, \ \text{AW}(\psi, \psi') \in \text{Atoms}(\varphi)\} \ .$$

If every formula in Comp(φ) is determined in $\mathcal{A}$, then φ is strongly preserved in $\mathcal{A}$:

$$\mathcal{A}, a \models \varphi \Longleftrightarrow \mathcal{C}, \gamma(\alpha) \models \varphi \ . \tag{4}$$

Such companion sets can be stratified: on the lowest level are the propositions; on each higher level one finds formulae of the form $\text{AX}\psi$ where ψ is some boolean combination of formulae of lower levels.

This suggests a strategy to construct an abstract model for some given φ in which φ is strongly preserved: start with some abstraction and 'add' more detail by partitioning, or splitting, states in which some formula is not determined, into parts in which it is; one part in which the formula holds and the other part in which it does not. The structure of Comp(φ) ensures that one can always partition states relative to formulae of the form $\text{AX}\psi$ with ψ already determined (except for the first step during which the splitting is relative to propositions). In [DGG93] we developed algorithms along this line for constructing minimal abstractions in which every $\varphi \in \text{ACTL}$ is determined and also for single formulae. Figure 1 gives a generalized version of the single-formulae *partitioning* algorithm.

The pseudo code uses a number of primitive operations, that must satisfy certain requirements.

$\mathcal{A} := $ "Initial abstraction"
for $p \in \text{Atoms}(\varphi) \cap \text{Prop}$ **do** $\mathcal{A} := \text{split}(\mathcal{A}, p, \text{pre})$ **od**
$F := \text{Comp}(\varphi) \setminus \text{Prop}$
repeat
    pick $\psi \in \min_{\mathcal{A}}(F); \quad F := F \setminus \{\psi\}$
    $s := \text{splitter}(\text{pre}, \psi); \quad \mathcal{A} := \text{split}(\mathcal{A}, s, \text{pre})$
until $\text{stable}_\varphi(\mathcal{A})$

Fig. 1. Partitioning algorithm for $\varphi \in \text{ACTL}$

Ordering on F The algorithm splits states with respect to the minimal elements of F. So, the requirement on the ordering is that if $\text{AX}\psi \in \min_{\mathcal{A}}(F)$ then ψ should be determined in $\mathcal{A}$.

An abstraction is partitioned w.r.t. a formula ψ in two steps.

Splitter This function determines the states in which ψ holds. Consequently, it satisfies $a = \text{splitter}(\text{pre}, \psi)$ iff $\gamma(a) = \{c \mid \mathcal{C}, c \models \psi\}$. As $\psi = \text{AX}\psi'$, we can compute $\text{splitter}(\text{pre}, \psi)$ as the characteristic predicate of $\{c \mid \forall d \ c \ R_c \ d \rightarrow d \in \gamma(||\psi'||)\}$.

Here, we have confused abstract states and predicates; furthermore, $||\psi'||$ denotes the characteristic predicate of ψ', which has been computed in a previous iteration. Define the *precondition function* pre by $\text{pre}(b) = a$ iff $\gamma(a) = \{c \mid \exists d \ c \ R_c \ d \wedge d \in \gamma(b)\}$, or, more abstractly, as $\gamma^b \circ pre_c \circ \gamma$.[8] Thus, there is essentially no distinction between pre and pre_c. Later on this will change. Now we can define $\text{splitter}(\text{pre}, \psi) = \neg\text{pre}(\neg\psi')$.

Splitting Next, we split all abstract states and compute the abstract transition relation. We have $\text{split}(\mathcal{A}, s, \text{pre}) = (V'_a, I'_a, R'_a)$ where $V'_a = \{a \wedge s, \ a \wedge \neg s \mid a \in V_a\}$, $I'_a = \{a' \mid \gamma(a') \cap I \neq \emptyset, \ a' \in V'_a\}$ and $R'_a = \{(a', b') \mid a' \wedge \text{pre}(b') \neq \text{false}, \ a', b' \in V'_a\}$. If we expand definitions, we find that $a' \wedge \text{pre}(b') \neq \text{false}$ rewrites to $\exists c \in \gamma(a') \ \exists d \in \gamma(b') \ c \ R_c \ d$ as should be the case. Here, too, we have confused abstract states and predicates.

Termination We may stop partitioning states either when φ becomes valid in some abstraction or when $\text{Comp}(\varphi)$ becomes determined. Hence, we may take

$$\text{stable}_\varphi(\mathcal{A}) = (\mathcal{A} \models \varphi) \vee (\text{Comp}(\varphi) \text{ is determined in } \mathcal{A}) \ .$$

Even if $\mathcal{C} \not\models \varphi$, the partitioning algorithm will terminate for finite-state systems: although it is true that any atom $\text{AW}(\psi, \psi')$ will contribute an infinite set of approximants to the companion, by (1) there will only be finitely many among them which will cause states to split; i.e., at most the first N approximants, where N is the number of concrete states.

$\text{Comp}(\varphi)$ is determined if no companion formula causes an additional split. Satisfaction of φ can be checked with any ACTL (or CTL) model checker.

Optimizations Even on this abstract level there are some optimizations to the basic algorithm possible. We briefly mention some. It is possible to *update* the abstract transition relation after a splitting instead of recomputing it and if the transition relation is deterministic then the computation of splitters simplifies.

More importantly, as there is a notion of initial state in which specifications should hold, it pays off to do a simultaneous reachability analysis while splitting. The reason is that abstractions preserve non-reachability; i.e., if some abstract state a becomes unreachable in an abstraction then every concrete state in $\gamma(a)$ will be unreachable in the concrete model. This is expected to greatly reduce the size of the models. See [DGG93] for details.

4 Data Abstraction

The above method abstracts from the concrete states which induces an abstraction of the transition relation. However, to compute pre, (parts of) the concrete transition relation is needed and this can be quite expensive in terms of both space and time. On the other hand, we may perform the partitioning algorithm w.r.t. an underlying transition that is already an abstraction of $\mathcal{C}$, thus making computing pre easier. Transitivity of abstractions guarantees weak preservation. We obtain two levels of abstraction: the concrete transition system $\mathcal{C}$ is first 'data abstracted' into $\mathcal{D} = (V_d, R_d)$; the partitioning

---

[8] Formally speaking, we have thus defined pre to be the precise abstract interpretation of pre_c (i.e., of pre_{R_c}) w.r.t the Galois insertion (γ^b, γ).

algorithm then computes an abstraction of $\mathcal{D}$ (w.r.t. a pre that is determined by R_d). An abstract transition between two abstract states A and A' is illustrated in the picture on this page.

The stable abstraction of $\mathcal{D}$ computed by the partitioning algorithm for some specification φ is strongly preserving w.r.t. $\mathcal{D}$ (i.e., (4) holds when $\mathcal{D}$ is substituted for $\mathcal{C}$). Unfortunately, this does not imply that there is strong preservation w.r.t. $\mathcal{C}$. More precisely, we have strong preservation w.r.t. a larger transition relation on the concrete states:

Let $\mathcal{A}$ be a stable abstraction of $\mathcal{D}$ as computed by the algorithm and let $\mathcal{D}$ be an abstraction of $\mathcal{C}$ with concretization function γ^d. Define $R_{\mathsf{pre}} = \{(c, d) \mid c, d \in V_c,\ c \in \gamma^d(\mathsf{pre}((\gamma^d)^b(d)))\}$. Note that $R_c \subseteq R_{\mathsf{pre}}$. We have
(i) (V_c, R_{pre}) is an abstraction of $\mathcal{C}$, and
(ii) $\mathcal{A} \models \varphi$ iff $(V_c, R_{\mathsf{pre}}) \models \varphi$.

An iterative process is suggested: choose some initial data abstraction $\mathcal{D}$ and compute a stable abstraction $\mathcal{A}$. If it does not satisfy the specification, choose a new data abstraction $\mathcal{D}'$ that is more detailed in the sense that $R_{\mathsf{pre}'} \subseteq R_{\mathsf{pre}}$ (obviously, we still should have safeness: $R_{\mathsf{pre}'} \supseteq R_c$) and start the algorithm again, but now with the (previously) stable model $\mathcal{A}$ as initial abstraction.

If $\mathcal{C} \not\models \varphi$ then it seems that in the end we still need to choose pre to be precise in order to draw that conclusion. However, it turns out that stable abstractions satisfy one more property.

Let $\mathcal{D}$ be a data abstraction of $\mathcal{C}$ and let $\mathcal{A}$ be a stable abstraction of $\mathcal{D}$ as obtained by partitioning. An edge $(A, A') \in R_a$ in $\mathcal{A}$ is called *exact* if it satisfies

$$\forall a \in \gamma_{\mathcal{D}}(A)\ \ \forall c \in \gamma^d(a)\ \ \exists a' \in \gamma_{\mathcal{D}}(A')\ \exists d \in \gamma^d(a')\ \ c\ R_c\ d\ .$$

A path $\mathbf{A} = A_0 A_1 \cdots$ is *exact* if every edge (A_i, A_{i+1}) on $\mathbf{A}$ is. Then $C \not\models \varphi$ provided there is an exact path $\mathbf{A}$ in $\mathcal{A}$ which is a counterexample for φ and such that the propositions appearing in φ are determined in the states on $\mathbf{A}$.

The exactness condition is illustrated in the picture to the right. Note that for each a and c a corresponding a' and d must be found.

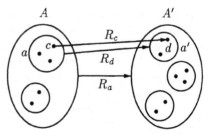

As the algorithm performs its first model check only after the abstract model has been split w.r.t. the propositions, the determinacy constraint is automatically satisfied. Hence, if we can detect whether or not a transition is exact when computing R'_a during a split($\mathcal{A}, s$, pre) operation, then we may still be able to conclude that $\mathcal{C} \not\models \varphi$ without splitting w.r.t. the precise pre.

This leads to the algorithm in Fig. 2. In the exit condition we have written $\mathcal{A}^-$ for the abstraction in which only the marked (exact) edges have been retained. As $\neg\varphi$ is an ECTL-formula, $\mathcal{A}^- \not\models \varphi$ can only hold if there is a path $\mathbf{A}$ in $\mathcal{A}^-$ that witnesses $\neg\varphi$ and, hence, is a counterexample to $\mathcal{A} \models \varphi$. By construction of $\mathcal{A}^-$, $\mathbf{A}$ only contains marked edges and hence is exact.

> Choose initial data abstraction and a corresponding safe **pre**
> **do**
>> Execute the partitioning algorithm (while marking exact edges)
>> **exit if** $(\mathcal{A} \models \varphi) \vee (\mathcal{A}^- \not\models \varphi)$
>> Choose a new data abstraction and corresponding safe **pre**
>> Unmark the edges of $\mathcal{A}$
> **od**

Fig. 2. Partitioning algorithm for $\varphi \in$ ACTL with data abstraction

---

Remaining questions are how to mark edges and how to choose pre's. Especially to the latter question there is obviously no univocal answer. We discuss one possibility.

4.1 Depth-k Abstraction

The precondition function pre_c of $\mathcal{C}$ is determined by its transition relation: $pre_c(D) = \{c \mid \exists d \in D \ c \ R_c \ d\}$. So, the choice of abstraction is determined by the aim to lessen the complexity of calculating R_c, and the requirements that it supports the detection of exact edges and that it should be possible to change pre by 'adapting' the current one rather than by recomputing it afresh.

Now, (concrete) transition systems arise as the interpretation of programs. Programs can be abstractly viewed as defining *state-transformers* $t(\mathbf{x}, \mathbf{x}')$ where $\mathbf{x}$ is a vector of program variables and t specifies the relation between the new values $\mathbf{x}'$ of these variables and the old ones that a program step enforces. A program's state is thus a tuple of values in some domain Val (an n-tuple if $\mathbf{x}$ contains n variables). The transition relation associated with it is then $R_t = \{(\mathbf{v}, \mathbf{v}') \mid \mathbf{v}, \mathbf{v}' \in Val^n, \ t(\mathbf{v}, \mathbf{v}')\}$. For ease of exposition we take $Val \subseteq \mathbf{N}$. Obviously, if the program is finite state, i.e., if R_t is finite, then Val is included in some initial segment of $\mathbf{N}$.

Abstraction The abstractions that we propose to use are based on restricting the precision with which variable values are recorded. The idea behind this is that the ROBDDs used to represent sets of states will have a limited depth. To support edge marking we need to know whether a variable's abstract value is precise or not. This leads to abstract value domains $Val_a^k = \{r, (r, \mathsf{O}) \mid r < 2^k, r \in Val\} \cup \{\mathsf{T}\}$ for $k > 0$ where O indicates an 'overflow' in the sense that the value is too large to be represented with k bits; $Val_a^0 = \{\mathsf{T}\}$. Abstract values of the form r represent concrete values smaller than 2^k precisely; a value (r, O) means that there is overflow but that the least k bits are correct and have value r; and T indicates absence of any knowledge. The concretization and abstraction function (i.e., the γ^d and $(\gamma^d)^\flat$ from before) for $k > 0$ are

$$\gamma_C^k(a) = \begin{cases} \{a\}, & \text{if } a < 2^k \\ \left\{ n \ \middle| \ \begin{array}{l} n \geq 2^k, \\ n \bmod 2^k = r \end{array} \right\}, & \text{if } a = (r, \mathsf{O}) \\ \mathbf{N}, & \text{if } a = \mathsf{T} \end{cases}, \quad \alpha_C^k(N) = \begin{cases} r, & \text{if } N = \{r\} \ \& \ r < 2^k \\ (r, \mathsf{O}), & \text{if } N \bmod 2^k = \{r\} \\ & \quad \& \ \neg(N < 2^k) \\ \mathsf{T}, & \text{otherwise} \end{cases}$$

In fact, α_C^k as just defined equals $(\gamma_C^k)^\flat$.

Partitioning algorithm Our implementation of the algorithm in Fig. 2 will use these depth-k abstractions: each variable value is represented in Val_a^k for some k. It uses the precise abstract interpretation t^k of t on Val_a^k: $t^k(a,b)$ iff $\exists c,d \in V_c$ $c \in \gamma^k(a)$ & $t(c,d)$ & $\alpha^k(d) = b$. To be precise, given the concrete system $C = (V_c, I_c, R_c)$ (with $V_c \subseteq \mathbf{N}^n$), define $\mathcal{A}^k = (V_a^k, I_a^k, R_a^k)$ by $V_a^k = (Val_a^k)^n$, $I_a^k = \{\alpha^k(c) \mid c \in I_c\}$ and $R_a^k = \{(\mathbf{a}, \mathbf{a}') \mid t^k(\mathbf{a}, \mathbf{a}')\}$. The algorithm constructs models stable w.r.t. $\mathcal{A}^k$ for ever larger values of k; whence, $\mathrm{pre}^k = \gamma_{\mathcal{A}^k}^b \circ \mathrm{pre}_{R_a^k} \circ \gamma_{\mathcal{A}^k}$ is the precondition function used when splitting w.r.t. $\mathcal{A}^k$.

Correctness It works, because not only are the $\mathcal{A}^k$ abstractions of C, but they also form a hierarchy in the sense that $\mathcal{A}^k$ is an abstraction of $\mathcal{A}^l$ if $k \leq l$. This follows from Val_a^k being an abstract interpretation of Val_a^l for $k \leq l$ via the concretization function $\gamma_l^k: Val_a^k \rightarrow 2^{Val_a^l}$ defined as

$$
\gamma_l^k(a) = \begin{cases}
\{a\}, & \text{if } a < 2^k \\
\{n \mid 2^k \leq n < 2^l,\ n \bmod 2^k = r\} \cup \\
\quad \{(n, \mathsf{O}) \mid 2^k \leq n < 2^l,\ n \bmod 2^k = r\}, & \text{if } a = (r, \mathsf{O}) \\
\mathsf{T}, & a = \mathsf{T}
\end{cases}
$$

We also have $\gamma^k = \gamma^l \circ \gamma_l^k$. Hence, an abstraction of $\mathcal{A}^k$ can be transformed into a transition system over V_a^l by replacing every V_a^k state a by the V_a^l states in $\gamma_l^k(a)$. Thus we obtain the initalization for the next iteration of the partitioning algorithm.

As for termination, consider $\mathcal{A}^K$ where K is chosen such that concrete values are at most 2^K. (If C is finite state then such a K obviously exists.) We have $I_a^K = I_c$, i.e., the abstractions are precise, and R_a^K restricted to the precise abstract values coincides with R_c. Hence, the parts of C and $\mathcal{A}^K$ that are reachable from the initial nodes are isomorphic. Because furthermore the stable abstraction computed by the partitioning algorithm w.r.t. pre^K is strongly preserving w.r.t. $\mathcal{A}^K$, we have $\mathcal{A} \models \varphi$ iff $C \models \varphi$.

This seems to prove termination of the algorithm but there is a catch: If $C \not\models \varphi$, then the algorithm terminates not if $\mathcal{A} \models \varphi$ but if $\mathcal{A}^- \not\models \varphi$ (where $\mathcal{A}$ is the stable abstraction of $\mathcal{A}^K$). In other words, the algorithm terminates if marking satisfies the following property:

> If $\mathcal{A}$ is a stable abstraction of $\mathcal{A}^K$ as computed by the partitioning algorithm, and if the concrete values are at most 2^K, then every edge in the reachable part of $\mathcal{A}$ is marked. (5)

Marking states The stable model that the algorithm attempts to construct is in its turn an abstraction of $\mathcal{A}^k$. Hence, the abstract states A will in fact be subsets of (formally: predicates over) $(Val_a^k)^n$. It is edges between these subsets that are possibly going to be marked. Given two abstract states A and B, there is an edge between them if $A \cap \mathrm{pre}^k(B) \neq \mathrm{false}$. From this we can only conclude that $\exists a \in \gamma_{\mathcal{A}^k}(A)\ \exists b \in \gamma_{\mathcal{A}^k}(B)\ a\ R_a^k\ b$, while $a\ R_a^k\ b$ gives that $\exists c \in \gamma^k(a)\ \exists d \in \gamma^k(b)\ c\ R_c\ d$.

If $A \subseteq \mathrm{pre}^k(B)$ then we know that

$$
\forall a \in \gamma_{\mathcal{A}^k}(A)\ \exists c \in \gamma^k(a)\ \exists b \in \gamma_{\mathcal{A}^k}(B)\ \exists d \in \gamma^k(b)\ c\ R_c\ d\ .
$$

Comparing this with the definition of exactness of an edge, the relevant question is: 'when can we replace $\exists c \in \gamma^k(a)$ by $\forall c \in \gamma^k(a)$ in this formula?' One obvious answer is when $|\gamma^k(a)| = 1$, i.e., when the concrete values of the variable are represented precisely. Another case in which we can replace it, is when we have a state transformer such as $x' = 0$. If A 'is' the predicate $x = (1, O) \wedge y = 2$ and B equals $x = 0 \wedge y = 2$, then the edge from A to B is exact although A contains abstract overflow values. Call a variable x a *don't care* for B just in case $\mathsf{pre}^k(B)$ is independent of x (interpreted as a predicate). Then, the criterion for marking an edge (A, B) becomes

$A \subseteq \mathsf{pre}^k(B)$ and for any variable x, either x is represented precisely in every $a \in \gamma_{A^k}(A)$ or x is a don't care for $\mathsf{pre}^k(B)$.

It is straightforward to show that under this marking condition every reachable edge will be marked if the depth-k abstraction gives enough precision as expressed in Property (5).

Note that marking as defined here is a 'safe' approximation of exactness. I.e., a marked edge is guaranteed to be exact but not necessarily vice versa. Clearly, marking every exact edge can only be done if the concrete transition relation is known.

Computing abstract relations As for ease of computation, obviously, the fewer bits we use to represent the values, the more efficient computing the abstraction becomes. Also, note that if $t^k(\mathbf{a}, \mathbf{a}')$ holds for 'precise' abstract values, i.e., for values *not* of the form (r, O) or $\top$, then $t^l(\mathbf{a}, \mathbf{a}')$ holds as well for any $l \geq k$. Moreover, for operations like addition or multiplication even if the result of the operation must be represented as (r, O) in Val_a^k, the lower bits r stay the same if the operation is interpreted in a more precise Val_a^l. Hence, it is possible to 'extend' pre^k to pre^l rather than to recompute it.

5 Conclusions and Future Work

We have presented an ACTL state partitioning algorithm that, for a given formula φ, computes the 'coarsest' abstraction that allows the truth of φ to be determined. This algorithm has been extended with a second orthogonal abstraction scheme that allows the automated choice of data abstraction relative to which the states are partitioned.

The algorithm described above is being implemented as part of a VHDL/S verifier developed in ESPRIT project 6128 FORMAT. After the total realization of the algorithm, practical experience has to show how far the existing limits of automatic verification have been pushed ahead. We should also investigate how already computed information can be reused more extensively. We are thinking not only of incrementally extending the ROBDDs that describe the effect of the VHDL/S code fragments when the algorithm changes the depth-k abstraction, but also of integrating model checking the specification φ with the partitioning process: each partitioning step causes an additional subformula of φ to be determined; the model checker needs to do the same. Finally, although the depth-k abstraction seems to be a good candidate for step-wise refinement of abstraction schemes, there may be good alternatives as well. The work done on the field of abstract interpretation of programs for static analysis can help for future applications in the area of automatic verification.

Acknowledgments The first two authors wish to thank OFFIS and the University of Oldenburg for their hospitality. We also thank Werner Damm for the intensive 'bull session' he organized and participated in, during which the ideas on which this paper is based were developed.

References

[BCM+90] J. R. Burch, E. M. Clarke, K. L. Mcmillan, D. L. Dill, and J. Hwang. Symbolic model checking: 10^{20} states and beyond. In *Proceedings of the Fifth Anual IEEE Symposium on Logic in Computer Science (LICS)*, 1990.

[BFH+92] A. Bouajjani, J.-C. Fernandez, N. Halbwachs, P. Raymond, and C. Ratel. Minimal state graph generation. *Science of Computer Programmming*, 18(3):247–271, 1992.

[Bry86] R. E. Bryant. Graph-based algorithms for boolean function manipulation. *Transactions on Computers*, C-35:677–691, 1986.

[Bry92] R. E. Bryant. Symbolic boolean manipulation with ordered binary-decision diagrams. *ACM Computing Surveys*, 24(3):293–318, 1992.

[CC77] P. Cousot and R. Cousot. Abstract interpretation: a unified lattice model for static analysis of programs by constructing or approximation of fixed points. In *Proceedings of the Fourth Annual ACM Symposium on Principles of Programming Languages (POPL)*, pages 238–252. ACM, 1977.

[DGG93] D. Dams, R. Gerth, and O. Grumberg. Generation of reduced models for checking fragments of CTL. In C. Courcoubetis, editor, *Proceedings of the Fifth Conference on Computer-Aided Verification*, volume 697 of *Lecture Notes in Computer Science*. Springer-Verlag, 1993.

[DGG94] D. Dams, O. Grumberg, and R. Gerth. Abstract interpretation of reactive systems: abstractions preserving ∀CTL*, ∃CTL* and CTL*. In *Proceedings of PROCOMET*, IFIP. North-Holland, 1994. To appear.

[GKPP94] R. Gerth, R. Kuiper, D. Peled, and W. Penczek. A partial order approach to branching time logic model checking, 1994. Submitted.

[GW91] P. Godefroid and P. Wolper. A partial approach to model checking. In *Proceedings of the Sixth Anual IEEE Symposium on Logic in Computer Science (LICS)*, 1991.

[Har87] D. Harel. Statecharts: A visual formalism for complex systems. *Science of Computer Programming*, 8, 1987.

[HK94] J. Helbig and P. Kelb. An OBDD representation of statecharts, 1994. To appear in EDAC94.

[HSD+93] J. Helbig, R. Schlör, W. Damm, G. Döhmen, and P. Kelb. VHDL/S—integrating statecharts, timing diagrams and VHDL. *Microprocessing and Microprogramming*, 38:571–580, 1993.

[Pel93] D. Peled. All from one, one for all, on model-checking using representatives. In *Proceedings of the Fifth International Conference on Computer-Aided Verification*, Lecture Notes in Computer Science, pages 409–423. Springer-Verlag, 1993.

[SD93] R. Schlör and W. Damm. Specification and verification of system-level hardware designs using timing diagrams. In *EDAC93*, 1993.

[Val91] A. Valmari. A stubborn attack on state explosion. In *Proceedings of the Second Conference on Computer-Aided Verification*, Lecture Notes in Computer Science. Springer-Verlag, 1991.

Automatic Verification of Timed Circuits

Tomas G. Rokicki* and Chris J. Myers**

Stanford University
Stanford, CA 94305

Abstract. This paper presents a new formalism and a new algorithm for verifying timed circuits. The formalism, called *orbital nets*, allows hierarchical verification based on a behavioral semantics of timed trace theory. We present improvements to a geometric timing algorithm that take advantage of concurrency by using partial orders to reduce the time and space requirements of verification. This algorithm has been fully automated and incorporated into a design system for timed circuits, and experimental results demonstrate that this verification algorithm is practical for realistic examples.

1 Introduction

Timing considerations are critical in circuit design. In the design of *timed circuits*, timing information is taken into account resulting in simpler circuits than their *speed-independent* counterparts in which gate delays are assumed to be unbounded [1]. Timing information must also be considered for designs with a mixed synchronous/asynchronous environment. Finally, even in speed-independent circuit design, timing must be considered when verifying the *isochronic fork* and *atomic gate* assumptions [2] once the circuit is laid out and delay parameters can be estimated.

Timed circuit verification is difficult both because circuit elements and circuit communication must be accurately modeled, and because timing considerations introduce another exponential factor of complexity. To address these problems, we start with a formalism of synchronized finite-state agents, modeled by labeled safe Petri nets, with structural constructions for *receptiveness* and *failures*. To this formalism, we add timing such that structural constructions on nets correspond to behavioral operations from timed trace theory. Finally, we introduce a new verification algorithm that allows efficient exploration of the entire timed state space.

Untimed trace theory for circuit verification originated with Rem, Snepscheut, and Udding [3] and was extended by Dill [4]; we provide structural constructions and syntactic shorthands for labeled safe Petri nets that correspond to the behavioral semantics operations. Burch [5] extended trace theory semantics to timed circuits; we extend this work with an operational formalism that allows timing in the specification, and thus hierarchical timed verification.

* Supported by a John and Fannie Hertz Fellowship and an ARCS fellowship. Currently at Hewlett-Packard Laboratories, Palo Alto, CA 94306.
** Supported by an NSF Fellowship, ONR Grant no. N00014-89-J-3036, and research grants from the Center for Integrated Systems, Stanford University and the Semiconductor Research Corporation, Contract no. 93-DJ-205.

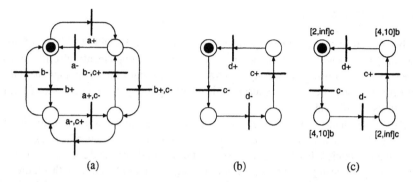

Fig. 1. (a) A fast NAND gate with inputs a and b and output c, (b) a delay buffer with input c and output d, and (c) with timing requirements annotated b for behavior and c for constraint.

Dill [6], Lewis [7], and Berthomieu and Diaz [8] originated geometric state space exploration, and it has become an active area of research [9, 10, 11]. We improve these techniques for systems with concurrency using partial orders and apply them to circuit verification. Recent work by Yoneda et. al. [12] also presented efficient verification through partial order considerations. Our work differs in that our formalism includes notions of specification, circuit composition, and receptiveness which enable us to perform efficient verification on nontrivial timed circuit examples; to our knowledge, neither timed automata nor Time Petri nets have been used in this fashion. Finally, our verification procedure computes a preorder conformance relation, allowing hierarchical verification.

2 Orbital Nets

In this section, we describe *orbital nets* and show how timed trace theory can be used as a behavioral semantics for verification. For brevity, we assume the reader is familiar with Petri nets [13] and with trace theory [4].

Orbital nets are based on labeled safe Petri nets extended with automatic net constructions and syntactic shorthands for composition and receptiveness [15]. The net constructions allow us to retain relatively straightforward operational semantics, while the syntactic shorthands allow us to compose the nets without an exponential blow-up in net size. Orbital nets also allow us to easily mix behavior and environmental requirements even at the gate model level.

For a large class of speed-independent and delay-insensitive designs, any hazard is potentially fatal [14], so simple delay models that are easy to integrate into gate models suffice. With the more complex delay models required for modeling real-time circuit delay, such integration is no longer easy or straightforward. Labeling each transition in an orbital net with a (possibly empty) set of actions remedies this difficulty by separating the modeling of the function of a gate from its delay behavior. As an example, the net corresponding to an infinitely fast NAND gate is given in figure 1(a).

2.1 Behavioral Semantics

For behavioral semantics, we adopt trace theory as defined by Dill [4]. Using a framework of Petri nets with synchronization and receptiveness, implementing verification of trace structure conformance is straightforward. Determining whether an implementation conforms to a specification is reduced to determining if any of a specific set of failure transitions can be enabled.

Dill's trace theory is based on sequences of actions, but our nets allow transitions to be labeled with sets of actions. A trace theory based on sequences of sets of actions yields a conformance relation that distinguishes, for instance, interleaved and concurrent actions. In addition, composing a net that uses interleaving on a pair of actions with another net that has those same actions labeling one transition may lead to an unintended deadlock. We do not attempt to resolve the complexities that arise in use of such a trace theory. Instead, we define conservative structural conditions on the use of labels consisting of sets of actions that allow us to use Dill's trace theory. For instance, we cannot perform a trace theory analysis of the fast NAND gate given above, but when we compose that model with the simple buffer given in figure 1(b) and hide the internal wire, the resulting net is conformation equivalent to the model Dill presents in [4].

2.2 Timing Requirements

Timing is associated with an orbital net place as a timing requirement consisting of a lower bound, an upper bound, and a type (min, max, type). The lower bound is a non-negative integer and the upper bound is an integer greater than or equal to the lower bound, or ∞. Since real values can be expressed as rationals within any required accuracy, restricting the bounds of timing requirements to be integers does not decrease the expressiveness of orbital nets. Since there are only a finite number of timing parameters, if any are rational, we can multiply all of them by the least common denominator.

There are two types of timing requirements: *constraint* and *behavior*. If any transitions in the postset of a place are labeled with an input action, then the timing requirement is of type constraint, otherwise it is of type behavior. Informally, the distinction between behavior and constraint timing requirements follows precisely the difference between input and output actions. The net for the untimed buffer shown in figure 1(b) only describes the successful traces, which are all the prefixes of $(c-, d-, c+, d+)^*$. The receptiveness construction adds failure traces caused by an input event that deviates from this pattern. The net generates only *successful* output sequences, but it accepts *all* input sequences, dividing them into success and failure traces. Consider now the buffer pictured in figure 1(c), in which our buffer model has been extended with timing requirements. The behavioral place labeled [4, 10] indicates that an output will occur between 4 and 10 time units after the preceding input occurs; no traces violating this requirement will be generated by the net. The constraint places labeled [2, ∞] indicate that, while the net will accept any timing between an output event and the succeeding input event, only those that have a delay of 2 time units or more are successful traces.

The delay model just described is an extremely simple delay model that suffices for many types of circuits. More complex delay models can and have been constructed, modeling more accurately the behavior of a gate under hazard conditions; for these, separation of gate models into combinational function and delay behavior is essential.

Each transition can have at most one behavior place in its preset. The intent is that each behavior place represents a single nondeterministic choice of delay that cannot be affected by external state or other behavior places. Since behavior places precede outputs and each wire can be an output for only a single model, circuit composition satisfies this naturally if each component model does. In general, this restriction does not limit expressiveness; other timing semantics can be simulated with appropriate net constructions.

2.3 Timed Operational Semantics

Each token that resides in a constraint or behavior place has a time-valued *age* parameter which advances with time and describes how long the token has resided in the place. The function *max-advance* on a marking is defined as the minimum value of $(max - age)$ for all marked behavior places, or ∞ if there are no marked behavior places. This upper limit on time advancement maintains the ages of all behavior place tokens below the maximum allowed by their range. Time is advanced by increasing the ages of all tokens in timed places by precisely the same amount.

A transition is *untimed-enabled* if all places in its preset have tokens. A transition is *timed-enabled* if it is untimed-enabled and if any input place with a behavior timing requirement has a token with $age \geq min$. Transition firings are assumed to be instantaneous; any number of transitions can fire without time advancing. Before firing a transition, the constraint places in the entire net are checked, and if any contains a token with $age > max$, this firing is marked as a failure. The tokens in the places in the preset of the fired transition are removed. The ages of the tokens removed from constraint places are checked, and if $age < min$, this firing is marked as a failure. Tokens are then put into the places in the postset of the fired transition, and all tokens put into timed places are assigned an age of zero. After the firing of a transition, every marked behavior place must have a transition in its postset that is untimed-enabled in the new state; if this condition is not satisfied, the firing is a failure. This requirement ensures that every token in a behavior place is consumable in all states in which its timing conditions are met, and thus the age at which the token is consumed cannot be controlled by external state.

With these semantics, untimed constructions for receptiveness and synchronization apply unchanged to the timed case. In addition, the trace theory operation of *mirroring* is also preserved, allowing hierarchical verification.

To discuss the verification procedure, we define a *timed firing sequence* to be a sequence of pairs of transition firings and time values. For simplicity, we shall assume the time value represents a non-negative duration since the last transition firing. Executing a timed firing sequence α on an orbital net results in the timed state $fire(\alpha)$. The set of legal firing sequences is defined recursively as follows. The empty sequence is a legal firing sequence. For a legal firing sequence α and for every value of τ such that $\tau \leq max\text{-}advance(fire(\alpha))$, then $\alpha, (\phi, \tau)$ is a legal firing sequence, where ϕ represents an 'empty' firing. In addition, if transition t is timed-enabled, then $\alpha, (t, 0)$ is also a legal firing sequence.

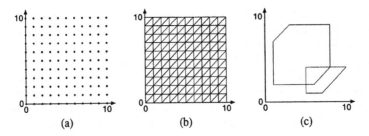

Fig. 2. (a) Discrete, (b) unit cube, and (c) geometric timed state space representations.

We define an *untimed state* to be an orbital net marking ignoring the ages of the tokens. The function *untime* returns an *untimed firing sequence* from a timed firing sequence by stripping the timing and removing any ϕ firings. The reachable state space is the range of the function *fire* over the legal firing sequences.

3 Discrete and Unit-Cube Time Verification

The basic idea behind finite-state verification is that, if the reachable state space is finite or has a finite representation, only a finite subset of the firing sequences needs to be considered to compute the set of reachable states. In our timing semantics, the ages of tokens can be real values, so the state space is infinite. In order to perform finite-state verification, we must either restrict the set of values that these ages can attain, or group the timed states into a finite number of equivalence classes. Discrete time verification uses the first approach, while Alur's unit-cube technique uses the second. (Infinite upper bounds are easily handled; we shall omit the details here [15].)

The first approach is justified by the proof that considering only integer event times gives a full characterization of the continuous time behavior of an orbital net [15]. This proof is similar to one given by Henzinger, et. al. in [16] for timed transition systems. With this result and the operational semantics given above, finite-state verification techniques can now be employed. Indeed, this discrete time approach was taken by Burch for verifying timed circuits [5].

Let us assume the number of distinct untimed states in an orbital net is $|S|$. If the maximum value of any timing requirement is k, and there are at most n timed tokens in any state (this value is trivially bounded by the size of our safe net), the size of the state space represented by discrete points, as in figure 2(a), is $|S|(k+1)^n$.

If the equivalence between discrete and continuous time does not hold for a particular formalism, it is still possible to perform finite-state real-time verification, using Alur's unit-cube technique [17]. He considers equivalence classes of timed states with the same integral clock values and a particular linear ordering of the fractional values of the clocks. For the two-dimensional case, the equivalence classes are pictured in figure 2(b); every point, line segment, and interior triangle is an equivalence class. The worst-case size of the state space for his method is asymptotically

$$|S|\frac{n!}{\ln 2}\left(\frac{k}{\ln 2}\right)^n 4^{1/k},$$

which is worse than the discrete time method by more than $n!$ [15].

Both of these techniques, however, are of little more than theoretical interest, because the size of the state space increases exponentially with the maximum value of the timing requirements. In general, during verification, every possible integer firing time must be considered for every transition from each state. For a circuit with timing values accurate to two significant digits, with up to six independent concurrent pending events, the state space is easily in excess of 10^{12} states—well beyond the capabilities of most finite-state verification techniques. Our experimental results indicate that the number of discrete states can be astronomical.

4 Geometric Timing Verification

In this section, we discuss a known time verification technique, *geometric timing*, that usually performs well in practice, even though the worst-case performance is much worse than either the discrete or the unit cube approaches [6, 7, 8, 9, 10, 11].

4.1 Geometric Regions

Rather than consider at each step a single discrete timed state, or a minimum equivalence class of timed states, the geometric timing method considers a large number of timed states in parallel. Specifically, convex geometric regions of timed states represented by upper and lower bounds on specific clock values and on the differences between pairs of specific clock values are used as the representation of the timed state space. Two sample regions are given in figure 2(c). The set of such constraints is usually represented by a matrix a, where the constraints on clocks $\{c_1 \ldots c_n\}$ are of the form $c_i - c_j \leq a_{ji}$. A fictitious clock c_0 that is always exactly zero is introduced so that upper and lower limits on a particular clock can be represented in the same form [6].

For any convex region that can be represented by such a matrix, there are many matrices that represent the same convex region. The process of *canonicalization* [6] can be performed to yield a matrix such that every constraint is maximally tight; if the constraint system represented by the matrix has any solutions, then there is precisely one canonicalized matrix representing that region. This canonicalization can be performed with Floyd's algorithm, which runs in time $O(n^3)$. In general, since only incremental changes are made to the matrix during verification, specializations of Floyd's algorithm that run in time $O(n^2)$ suffice [15].

4.2 Geometric Regions as Aggregates of Discrete Timed States

Since integer-valued timed sequences accurately model the behavior of orbital nets, it is sufficient to show that verification with geometric regions considers precisely the same set of states that discrete verification does. This is accomplished by giving the correlation between each operation in discrete time verification and in geometric timing verification. We do not discuss the aspects of verification that do not consider time, since they are the same in both cases. For simplicity, we consider each geometric region as a collection of discrete timed states that are handled in parallel. While such a perspective is not necessary, it may be conducive to an understanding of the method.

The geometric region technique operates over an untimed firing sequence α, calculating directly the full set of timed states reachable from all timed firing sequences β that satisfy $untime(\beta) = \alpha$. Thus, rather than separately considering every possible occurrence time for a particular transition in α during verification, in one step the geometric region method considers all possible occurrence times. We describe how it works for a single transition occurrence, assuming it works for the predecessor sequence; the trivial base case and structural induction on sequences completes the proof for all sequences.

With discrete time verification, determining whether a particular transition is timed-enabled entails comparing the token ages with known constants. With geometric regions, we determine the subset of the timed states in the region for which the particular transition is enabled. This can be performed by introducing the enabling conditions on the transition as additional constraints on the region and recanonicalizing. For orbital nets, these conditions describe a convex region in the appropriate form, and it is easy to show that the intersection of two such convex regions is a convex region of the same form. Canonicalization by definition does not reduce the set of timed states represented.

After selecting an enabled transition, firing that transition involves removing some set of timed tokens and introducing new timed tokens. In the discrete case, removing timed tokens involves discarding their ages. With geometric regions, removing timed tokens involves projection of the system of constraints to eliminate a particular set of variables. Introducing new timed tokens in the discrete case sets the ages of these tokens to zero; in the geometric region case, we introduce a new set of variables equal to zero.

After firing a transition, we allow time to advance; this corresponds to ϕ firings. In the discrete case, advancing time involves adding some number t to all token ages. In the geometric case, advancing time involves extruding the geometric region in the $c_1 = c_2 = \cdots = c_n$ direction, subject to *max-advance*, which itself is a convex region.

The other operations, such as checking for constraint place violations, involve simple inspections of the geometric region; since each inequality is maximally constrained, there exists some solution for which that inequality is an equation.

4.3 Performance of Geometric Timing Verification

Verification based on geometric regions can be very efficient. In particular, if a timed system does not exhibit much concurrency, our examples show that there is often very close to one geometric region for every untimed state. The circuit examples in the first part of table 1 illustrate this; for these examples, standard geometric timing runs very rapidly, even when the timing parameters are very large.

On the other hand, some examples require an extreme number of geometric regions. The adverse example adv4x40 shown in figure 3, using standard geometric timing techniques, generates an incredible 219,977,777 distinct geometric regions. This is more than either the number of discrete timed states or unit-cube equivalence classes.

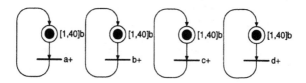

Fig. 3. The adverse example adv4x40 with $n = 4$ and $k = 40$.

5 Partial Order Timing Verification

We improve this method for systems with concurrency using *partial order timing*. The major source of blow-up in the adverse example is the way the standard geometric timing algorithm calculates exactly the set of timed states reachable from a sequence of transition firings; the transition firings are linearly ordered, even if they are concurrent in the system being evaluated. That is, if two concurrent transitions start clocks, the constraints between the ages of the two clocks will reflect the linear order that the transitions were fired in the original sequence. In general, if there are n concurrent transitions that reset clocks visible in the resulting timed state, there are $n!$ different sequences that need to be considered, each of which leads to a distinct geometric region. For this reason, it is important to distinguish the causal ordering of transitions from the non-causal ordering caused by the selection of a particular firing sequence.

5.1 Concurrency, Causality, and Processes

A *process* is an acyclic, choice-free net created from a Petri net and a firing sequence. The process reflects the causality and concurrency inherent in that firing sequence. Initially, it contains a single transition with places in its postset corresponding to each token in the initial marking. Transitions are added in the same order as they occur in the firing sequence. For each transition in the firing sequence, a correspondingly labeled transition is added to the process. A set of arcs into the transition are connected from the most recently added places in the process corresponding to places in the preset of the transition in the original Petri net. Finally, a new set of places corresponding to the places in the postset of the transition in the original net are added, and these places are connected to the new transition. The function *process* takes a sequence and returns the corresponding process. The resulting process for the firing of the sequence $[a+, b+]$ in our adverse example is shown in figure 4.

Every place and every transition in the created process, except the first, correspond to some place and some transition in the original net. Every place and every transition in the original net correspond to zero or more places and transitions in the process. A process explicitly represents the concurrency in a particular firing sequence. That is, a particular process corresponds to many different firing sequences that differ only in the interleavings of concurrent transitions; every such firing sequence fires the same set of transitions and leads to the same final untimed state. Thus, the process shown in figure 4 corresponds both to the trace $[a+, b+]$ and to the trace $[b+, a+]$.

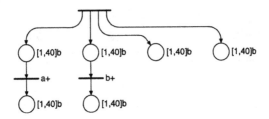

Fig. 4. One process from the adverse example.

5.2 Verification with Partial Order Timing

In general, for a process with simple timing constraints between pairs of transitions, we may calculate the minimum and maximum separation between all pairs of transitions in time $O(n^3)$ using Floyd's algorithm. If none of the places are timed, then the process is a partial order on the events. That is, an untimed place p introduces only the constraint $fire\text{-}time(\bullet p) \leq fire\text{-}time(p\bullet)$. Because the constraint places do not affect the firing of transitions, such places introduce the same kind of causality relation. On the other hand, for each behavior place p in the resulting process with a timing requirement of $(min, max, behavior)$, two constraints are introduced. The first reflects the minimum separation, $fire\text{-}time(\bullet p) - fire\text{-}time(p\bullet) \leq -min$. The second reflects the maximum separation, $fire\text{-}time(p\bullet) - fire\text{-}time(\bullet p) \leq max$. All constraints introduced in this fashion for a given process must be satisfied.

Performing this operation on the process in figure 4 determines that $a+$ can follow the initial transition by between 1 and 40 time units, and $b+$ can also follow the initial transition by between 1 and 40 time units. The time separation between $a+$ and $b+$ must be between -39 and 39 time units, inclusive.

From this information, the full set of geometric regions reachable with this process is calculated. Specifically, for two clocks c_i and c_j created by the firing of transitions t_i and t_j, the constraint on $c_j - c_i$, or entry a_{ij} in our constraint matrix, is the maximum separation from t_j to t_i. The minimum and maximum ages of the clocks are derived from the previous process in the same manner as for the geometric method.

The partial order technique operates over an untimed firing sequence α, calculating directly the full set of timed states reachable from any timed firing sequence β such that $process(untime(\beta)) = process(\alpha)$. Thus, rather than separately considering every interleaving of concurrent transitions, in one step the partial order method considers all possible interleavings. For untimed verification, different interleavings result in the same state. For timed verification, different interleavings usually result in different sets of timed states, with different future behavior, leading to a combinatorial explosion of timed regions for each untimed state. Representing, as a single constraint matrix, the union of all timed states reachable from all possible interleavings, therefore, dramatically reduces the size of the state space representation. In fact, the partial order method typically reduces the number of timed regions for each untimed state to a value close to one.

Verification proceeds just as it does for the previous methods based on sequences, except that, for each sequence, the algorithm constructs the corresponding process. With depth-first search, this is done incrementally. The algorithm also incrementally calculates a constraint matrix that stores the firing time relationship among the transitions. The algorithm then calculates the geometric region corresponding to this process by adding the upper and lower bounds on the clocks; after canonicalizing this matrix, it has produced the full set of reachable states for that process.

Our process calculation can never introduce states that are not reachable by a sequence, since for any timed state in the process, some ordering of the transitions in the process is a sequence that generates the same timed state. In addition, because the algorithm considers the firing of every transition enabled in every reachable geometric region, it visits all successors and thus the entire timed state graph. Therefore, the same set of reachable states is obtained using process-based region calculations as is obtained using sequence-based region calculations.

5.3 Efficiency Considerations

The number of transitions in the process is equal to the length of the firing sequence plus one, and it increases with the depth of our search. Calculating the minimum separations between the occurrence times in our process, even with our incremental $O(n^2)$ approach, becomes prohibitively expensive as the firing sequence lengthens. In addition, the algorithm needs a constraint matrix for each step; this would require a tremendous amount of storage during depth-first search.

To keep n bounded as the depth of our search increases, the algorithm determines what prefix, if any, of our process can safely be ignored. The algorithm can eliminate any transitions that no longer affect future calculations. In general, the algorithm can eliminate a variable from any set of equations or inequalities whenever it has produced the full set of equations or inequalities that use that variable. Since all constraints introduced through the firing of a transition are associated with places connecting the new transition to the old, once a transition in our process no longer has any places in its postset which do not have a transition in their postset, it is eliminated from our constraint matrix. Thus, our n is—at most—the number of tokens in the original net at any given time, plus one for the current transition.

This technique has a more expensive transition firing computation, so for the simplest examples with little concurrency, it is slower; but, it dramatically reduces the number of interleavings considered and also the number of geometric regions for each untimed state when there are concurrent transitions. Because the number of geometric regions is typically small, a further optimization is possible. Rather than backtracking only when an identical geometric region is found, verification can backtrack whenever a new geometric region is a subset of a previously seen geometric region. Comparing two geometric regions for inclusion can be performed in $O(n^2)$ time.

6 Experimental Results

The verification procedure described in the previous section has been automated in the tool Orbits. This tool has been incorporated into the design system for timed circuits ATACS described in [1]. Experimental results are given in table 1. The left four columns

Table 1. Experimental results. Time values are given in seconds. An entry of time indicates that the verification did not complete within two hours, and an entry of space indicates that the verification ran out of space before completing.

Examples	Startup time	Net nodes	Untimed states	Discrete states	Geometric timing regions	time	Partial order regions	time
MMUopt	0.31	293	22	734	28	0.02	22	0.04
MMUunopt	0.24	212	33	3245	43	0.04	33	0.06
dram	1.16	1335	96	5697	120	0.36	96	0.39
pipe	0.16	126	15	11657	18	0.01	15	0.02
scsictrl	0.18	248	16	200	21	0.03	16	0.03
scsi1	12.67	15477	170	1247	294	3.49	170	3.69
scsi1BRK	14.35	15674	197	1481	351	9.47	205	10.46
scsi2	10.31	11496	155	1029	186	6.69	159	6.81
scsi2BRK	10.14	11372	159	1052	193	6.64	163	6.70
scsi3BRK	12.67	14866	492	10319	1123	27.11	653	28.42
tsbm	2.84	4115	292	46212	965	2.65	443	2.58
tsbmBRK2	0.78	730	392	1.33e7	1789	3.60	550	1.83
tsbmBRK4	0.68	782	312	1.14e7	1047	2.71	476	1.95
adv3x40	0.05	6	1	68921	1.52e5	164.99	1	0.01
adv4x40	0.03	8	1	2.83e6	space		1	0.01
adv50x40	0.27	100	1	4.36e80	space		1	60.21
phil3	0.19	149	144	27806	758	0.77	188	0.36
phil4	0.22	197	1152	9.82e5	time		1541	6.98
phil5	0.25	245	9840	3.47e7	time		14039	159.40
seitz	0.41	355	344	2.92e13	3234	5.48	416	1.22
seitz2	0.55	624	4572	5.48e19	space		5820	29.79
kyy5	2.46	1484	5266	>1e20	space		6083	56.74
kyy15	1.97	1484	18357	>1e20	space		20250	321.47

indicate values that are the same for geometric and partial order timing. The startup time is the time required to parse the input and construct the appropriate orbital net. The number of net nodes is the sum of the places and transitions in the resulting orbital net. The third column gives the number of untimed states. The fourth column gives the number of discrete states, after all timing parameters are divided by their greatest common divisor. The next four columns give the number of geometric regions and the run time in seconds for verification using standard geometric timing and partial order timing, respectively.

The first half of table 1 consists of circuits created by ATACS taken from examples in [1] and [18]. These examples do not exhibit much concurrency, and the number of geometric regions created by the geometric region method is close to the number of untimed states. These circuits are composed of typically less than a dozen complex gates, but some of these gates have as many as ten inputs; it is these large gates that cause large orbital nets.

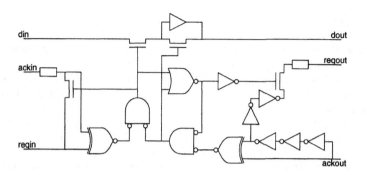

Fig. 5. The Seitz queue element, a small timed circuit.

The second half of the table consists of other circuits and systems that exhibit a high degree of concurrency. For example, seitz1 is pictured in Figure 5; seitz2 is two connected copies of this circuit. The kyy examples [19] have thirty-seven gates and timing parameters given to three significant digits. Where the examples ran out of time or space using the geometric method, often the verification was far from done. For the seitz2 example, after one hour of CPU time, only 1,404 of the 4,572 untimed states have been seen, yet 473,202 distinct geometric regions have been encountered. One particular untimed state has 13,275 distinct geometric regions at this point. Partial order timing for this example finds the entire state space as 5,820 geometric regions in one half minute of CPU time. Examples were run on an HP9000/735 with 144 megabytes of memory using CScheme 7.3.

7 Conclusion

We have introduced orbital nets which extend safe Petri nets to provide an efficient formalism for modeling timed circuit behavior. Discrete methods provide the best known worst-case complexity for timed verification, but generally fail in practice due to exponential blowup on the timing parameters. Geometric methods work well in practice, but fail for highly concurrent systems. Thus, we improve upon geometric methods using partial order timing. Our examples show that partial order timing can handle much larger examples than the standard geometric methods.

Acknowledgments

We would like to thank David Dill, Jerry Burch, Daniel Weise, Peter Beerel, and Teresa Meng of Stanford University and Ludmila Cherkasova and Vadim Kotov of Hewlett-Packard Laboratories for their discussions and comments during the development of Orbits.

References

1. Chris J. Myers and Teresa H.-Y. Meng. Synthesis of timed asynchronous circuits. *IEEE Transactions on VLSI Systems*, 1(2):106–119, June 1993.
2. Alain J. Martin. Programming in VLSI: From communicating processes to delay-insensitive VLSI circuits. In C.A.R. Hoare, editor, *UT Year of Programming Institute on Concurrent Programming*. Addison-Wesley, 1990.
3. Martin Rem, Jan L. A. van de Snepscheut, and Jan Tijmen Udding. Trace theory and the definition of hierarchical components. In R. Bryant, editor, *Third Caltech Conference on VLSI*, pages 225–239. Computer Science Press, Inc., 1983.
4. David L. Dill. Trace theory for automatic hierarchial verification of speed-independent circuits. ACM Distinguished Dissertations, 1989.
5. Jerry R. Burch. *Trace Algebra for Automatic Verification of Real-Time Concurrent Systems*. PhD thesis, Carnegie Mellon University, 1992.
6. David L. Dill. Timing assumptions and verification of finite-state concurrent systems. In *Proceedings of the Workshop on Automatic Verification Methods for Finite-State Systems*, June 1989.
7. Harry R. Lewis. Finite-state analysis of asynchronous circuits with bounded temporal uncertainty. Technical report, Harvard University, July 1989.
8. Bernard Berthomieu and Michel Diaz. Modeling and verification of time dependent systems using time petri nets. *IEEE Transactions on Software Engineering*, 17(3), March 1991.
9. R. Alur, C. Courcoubetis, D. Dill, N. Halbwachs, and H. Wong-Toi. An implementation of three algorithms for timing verification based on automata emptiness. In *Proceedings of the Real-Time Systems Symposium*, pages 157–166. IEEE Computer Society Press, 1992.
10. Thomas Henzinger, Xavier Nicollin, Joseph Sifakis, and Sergio Yovine. Symbolic model-checking for real-time systems. In *Proceedings of the 7th Symposium Logics in Computers Science*. IEEE Computer Society Press, 1992.
11. Nicolas Halbwachs. Delay analysis in synchronous programs. In Costas Courcoubetis, editor, *Computer Aided Verification*, pages 333–346. Springer-Verlag, 1993.
12. Tomohiro Yoneda, Atsufumi Shibayama, Bernd-Hologer Schlingloff, and Edmund M. Clarke. Efficient verification of parallel real-time systems. In Costas Courcoubetis, editor, *Computer Aided Verification*, pages 321–332. Springer-Verlag, 1993.
13. James L. Peterson. *Petri Net Theory and the Modeling of Systems*. Prentice Hall, 1981.
14. Peter A. Beerel and Teresa H.-Y. Meng. Semi-modularity and testability of speed-independent circuits. *INTEGRATION, the VLSI journal*, 13(3):301–322, September 1992.
15. Tomas G. Rokicki. *Representing and Modeling Circuits*. PhD thesis, Stanford University, 1993.
16. Thomas A. Henzinger, Zohar Manna, and Amir Pnueli. What good are digital clocks? In *ICALP 92: Automata, Languages, and Programming*, pages 545–547. Springer-Verlag, 1992.
17. Rajeev Alur. *Techniques for Automatic Verification of Real-Time Systems*. PhD thesis, Stanford University, August 1991.
18. Chris J. Myers and Teresa H.-Y. Meng. Automatic hazard-free decomposition of high-fanin gates in asynchronous circuit synthesis. To be published.
19. Ken Yun. Private communication, 1993.

Springer-Verlag
and the Environment

We at Springer-Verlag firmly believe that an international science publisher has a special obligation to the environment, and our corporate policies consistently reflect this conviction.

We also expect our business partners – paper mills, printers, packaging manufacturers, etc. – to commit themselves to using environmentally friendly materials and production processes.

The paper in this book is made from low- or no-chlorine pulp and is acid free, in conformance with international standards for paper permanency.

Lecture Notes in Computer Science

For information about Vols. 1–739
please contact your bookseller or Springer-Verlag

Vol. 777: K. von Luck, H. Marburger (Eds.), Management and Processing of Complex Data Structures. Proceedings, 1994. VII, 220 pages. 1994.

Vol. 778: M. Bonuccelli, P. Crescenzi, R. Petreschi (Eds.), Algorithms and Complexity. Proceedings, 1994. VIII, 222 pages. 1994.

Vol. 779: M. Jarke, J. Bubenko, K. Jeffery (Eds.), Advances in Database Technology — EDBT '94. Proceedings, 1994. XII, 406 pages. 1994.

Vol. 780: J. J. Joyce, C.-J. H. Seger (Eds.), Higher Order Logic Theorem Proving and Its Applications. Proceedings, 1993. X, 518 pages. 1994.

Vol. 781: G. Cohen, S. Litsyn, A. Lobstein, G. Zémor (Eds.), Algebraic Coding. Proceedings, 1993. XII, 326 pages. 1994.

Vol. 782: J. Gutknecht (Ed.), Programming Languages and System Architectures. Proceedings, 1994. X, 344 pages. 1994.

Vol. 783: C. G. Günther (Ed.), Mobile Communications. Proceedings, 1994. XVI, 564 pages. 1994.

Vol. 784: F. Bergadano, L. De Raedt (Eds.), Machine Learning: ECML-94. Proceedings, 1994. XI, 439 pages. 1994. (Subseries LNAI).

Vol. 785: H. Ehrig, F. Orejas (Eds.), Recent Trends in Data Type Specification. Proceedings, 1992. VIII, 350 pages. 1994.

Vol. 786: P. A. Fritzson (Ed.), Compiler Construction. Proceedings, 1994. XI, 451 pages. 1994.

Vol. 787: S. Tison (Ed.), Trees in Algebra and Programming – CAAP '94. Proceedings, 1994. X, 351 pages. 1994.

Vol. 788: D. Sannella (Ed.), Programming Languages and Systems – ESOP '94. Proceedings, 1994. VIII, 516 pages. 1994.

Vol. 789: M. Hagiya, J. C. Mitchell (Eds.), Theoretical Aspects of Computer Software. Proceedings, 1994. XI, 887 pages. 1994.

Vol. 790: J. van Leeuwen (Ed.), Graph-Theoretic Concepts in Computer Science. Proceedings, 1993. IX, 431 pages. 1994.

Vol. 791: R. Guerraoui, O. Nierstrasz, M. Riveill (Eds.), Object-Based Distributed Programming. Proceedings, 1993. VII, 262 pages. 1994.

Vol. 792: N. D. Jones, M. Hagiya, M. Sato (Eds.), Logic, Language and Computation. XII, 269 pages. 1994.

Vol. 793: T. A. Gulliver, N. P. Secord (Eds.), Information Theory and Applications. Proceedings, 1993. XI, 394 pages. 1994.

Vol. 794: G. Haring, G. Kotsis (Eds.), Computer Performance Evaluation. Proceedings, 1994. X, 464 pages. 1994.

Vol. 795: W. A. Hunt, Jr., FM8501: A Verified Microprocessor. XIII, 333 pages. 1994.

Vol. 796: W. Gentzsch, U. Harms (Eds.), High-Performance Computing and Networking. Proceedings, 1994, Vol. I. XXI, 453 pages. 1994.

Vol. 797: W. Gentzsch, U. Harms (Eds.), High-Performance Computing and Networking. Proceedings, 1994, Vol. II. XXII, 519 pages. 1994.

Vol. 798: R. Dyckhoff (Ed.), Extensions of Logic Programming. Proceedings, 1993. VIII, 362 pages. 1994.

Vol. 799: M. P. Singh, Multiagent Systems. XXIII, 168 pages. 1994. (Subseries LNAI).

Vol. 800: J.-O. Eklundh (Ed.), Computer Vision – ECCV '94. Proceedings 1994, Vol. I. XVIII, 603 pages. 1994.

Vol. 801: J.-O. Eklundh (Ed.), Computer Vision – ECCV '94. Proceedings 1994, Vol. II. XV, 485 pages. 1994.

Vol. 802: S. Brookes, M. Main, A. Melton, M. Mislove, D. Schmidt (Eds.), Mathematical Foundations of Programming Semantics. Proceedings, 1993. IX, 647 pages. 1994.

Vol. 803: J. W. de Bakker, W.-P. de Roever, G. Rozenberg (Eds.), A Decade of Concurrency. Proceedings, 1993. VII, 683 pages. 1994.

Vol. 804: D. Hernández, Qualitative Representation of Spatial Knowledge. IX, 202 pages. 1994. (Subseries LNAI).

Vol. 805: M. Cosnard, A. Ferreira, J. Peters (Eds.), Parallel and Distributed Computing. Proceedings, 1994. X, 280 pages. 1994.

Vol. 806: H. Barendregt, T. Nipkow (Eds.), Types for Proofs and Programs. VIII, 383 pages. 1994.

Vol. 807: M. Crochemore, D. Gusfield (Eds.), Combinatorial Pattern Matching. Proceedings, 1994. VIII, 326 pages. 1994.

Vol. 808: M. Masuch, L. Pólos (Eds.), Knowledge Representation and Reasoning Under Uncertainty. VII, 237 pages. 1994. (Subseries LNAI).

Vol. 809: R. Anderson (Ed.), Fast Software Encryption. Proceedings, 1993. IX, 223 pages. 1994.

Vol. 810: G. Lakemeyer, B. Nebel (Eds.), Foundations of Knowledge Representation and Reasoning. VIII, 355 pages. 1994. (Subseries LNAI).

Vol. 811: G. Wijers, S. Brinkkemper, T. Wasserman (Eds.), Advanced Information Systems Engineering. Proceedings, 1994. XI, 420 pages. 1994.

Vol. 812: J. Karhumäki, H. Maurer, G. Rozenberg (Eds.), Results and Trends in Theoretical Computer Science. Proceedings, 1994. X, 445 pages. 1994.

Vol. 813: A. Nerode, Yu. N. Matiyasevich (Eds.), Logical Foundations of Computer Science. Proceedings, 1994. IX, 392 pages. 1994.

Vol. 814: A. Bundy (Ed.), Automated Deduction—CADE-12. Proceedings, 1994. XVI, 848 pages. 1994. (Subseries LNAI).

Vol. 815: R. Valette (Ed.), Application and Theory of Petri Nets 1994. Proceedings, 1994. IX, 587 pages. 1994.

Vol. 817: C. Halatsis, D. Maritsas, G. Philokyprou, S. Theodoridis (Eds.), PARLE '94. Parallel Architectures and Languages Europe. Proceedings, 1994. XV, 837 pages. 1994.

Vol. 818: D. L. Dill (Ed.), Computer Aided Verification. Proceedings, 1994. IX, 480 pages. 1994.

Vol. 819: W. Litwin, T. Risch (Eds.), Applications of Databases. Proceedings, 1994. XII, 471 pages. 1994.

Vol. 820: S. Abiteboul, E. Shamir (Eds.), Automata, Languages and Programming. Proceedings, 1994. XIII, 644pages. 1994.

Vol. 821: M. Tokoro, R. Pareschi (Eds.), Object-Oriented Programming. Proceedings, 1994. XI, 535 pages. 1994.